Graduate Texts in Mathematics 161

Springer
New York
Berlin
Heidelberg
Barcelona
Budapest
Hong Kong
London
Milan
Paris
Santa Clara
Singapore
Tokyo

Graduate Texts in Mathematics

1 TAKEUTI/ZARING. Introduction to Axiomatic Set Theory. 2nd ed.
2 OXTOBY. Measure and Category. 2nd ed.
3 SCHAEFER. Topological Vector Spaces.
4 HILTON/STAMMBACH. A Course in Homological Algebra.
5 MAC LANE. Categories for the Working Mathematician.
6 HUGHES/PIPER. Projective Planes.
7 SERRE. A Course in Arithmetic.
8 TAKEUTI/ZARING. Axiomatic Set Theory.
9 HUMPHREYS. Introduction to Lie Algebras and Representation Theory.
10 COHEN. A Course in Simple Homotopy Theory.
11 CONWAY. Functions of One Complex Variable I. 2nd ed.
12 BEALS. Advanced Mathematical Analysis.
13 ANDERSON/FULLER. Rings and Categories of Modules. 2nd ed.
14 GOLUBITSKY/GUILLEMIN. Stable Mappings and Their Singularities.
15 BERBERIAN. Lectures in Functional Analysis and Operator Theory.
16 WINTER. The Structure of Fields.
17 ROSENBLATT. Random Processes. 2nd ed.
18 HALMOS. Measure Theory.
19 HALMOS. A Hilbert Space Problem Book. 2nd ed.
20 HUSEMOLLER. Fibre Bundles. 3rd ed.
21 HUMPHREYS. Linear Algebraic Groups.
22 BARNES/MACK. An Algebraic Introduction to Mathematical Logic.
23 GREUB. Linear Algebra. 4th ed.
24 HOLMES. Geometric Functional Analysis and Its Applications.
25 HEWITT/STROMBERG. Real and Abstract Analysis.
26 MANES. Algebraic Theories.
27 KELLEY. General Topology.
28 ZARISKI/SAMUEL. Commutative Algebra. Vol.I.
29 ZARISKI/SAMUEL. Commutative Algebra. Vol.II.
30 JACOBSON. Lectures in Abstract Algebra I. Basic Concepts.
31 JACOBSON. Lectures in Abstract Algebra II. Linear Algebra.
32 JACOBSON. Lectures in Abstract Algebra III. Theory of Fields and Galois Theory.
33 HIRSCH. Differential Topology.
34 SPITZER. Principles of Random Walk. 2nd ed.
35 WERMER. Banach Algebras and Several Complex Variables. 2nd ed.
36 KELLEY/NAMIOKA et al. Linear Topological Spaces.
37 MONK. Mathematical Logic.
38 GRAUERT/FRITZSCHE. Several Complex Variables.
39 ARVESON. An Invitation to C^*-Algebras.
40 KEMENY/SNELL/KNAPP. Denumerable Markov Chains. 2nd ed.
41 APOSTOL. Modular Functions and Dirichlet Series in Number Theory. 2nd ed.
42 SERRE. Linear Representations of Finite Groups.
43 GILLMAN/JERISON. Rings of Continuous Functions.
44 KENDIG. Elementary Algebraic Geometry.
45 LOÈVE. Probability Theory I. 4th ed.
46 LOÈVE. Probability Theory II. 4th ed.
47 MOISE. Geometric Topology in Dimensions 2 and 3.
48 SACHS/WU. General Relativity for Mathematicians.
49 GRUENBERG/WEIR. Linear Geometry. 2nd ed.
50 EDWARDS. Fermat's Last Theorem.
51 KLINGENBERG. A Course in Differential Geometry.
52 HARTSHORNE. Algebraic Geometry.
53 MANIN. A Course in Mathematical Logic.
54 GRAVER/WATKINS. Combinatorics with Emphasis on the Theory of Graphs.
55 BROWN/PEARCY. Introduction to Operator Theory I: Elements of Functional Analysis.
56 MASSEY. Algebraic Topology: An Introduction.
57 CROWELL/FOX. Introduction to Knot Theory.
58 KOBLITZ. p-adic Numbers, p-adic Analysis, and Zeta-Functions. 2nd ed.
59 LANG. Cyclotomic Fields.
60 ARNOLD. Mathematical Methods in Classical Mechanics. 2nd ed.

continued after index

Peter Borwein Tamás Erdélyi

Polynomials and Polynomial Inequalities

Springer

Peter Borwein
Department of Mathematics
and Statistics
Simon Fraser University
Burnaby, B.C., V5A 1S6
Canada

Tamás Erdélyi
Department of Mathematics
Texas A&M University
College Station, TX 77843
USA

Mathematics Subject Classification (1991): 26Cxx, 26D05, 12Dxx, 30C10, 33C45

Library of Congress Cataloging-in-Publication Data
Borwein, Peter B.
Polynomials and polynomial inequalities/Peter Borwein, Tamás Erdélyi.
p. cm. – (Graduate texts in mathematics; 161)
Includes bibliographical references and index.
ISBN 0-387-94509-1 (alk. paper)
1. Polynomials. 2. Inequalities (Mathematics) I. Erdélyi, Tamás. II. Title. III. Series.
QA241.B775 1995
515′.252–dc20 95-8374

Printed on acid-free paper.

Production managed by Francine McNeill; manufacturing supervised by Joe Quatela.
Photocomposed copy prepared using Springer-Verlag's macro.
Printed and bound by R.R. Donnelley and Sons, Harrisonburg, VA.
Printed in the United States of America.

9 8 7 6 5 4 3 2 1

ISBN 0-387-94509-1 Springer-Verlag New York Berlin Heidelberg

For Theresa, Sophie, Alexandra, Jennifer, and Erika

Preface

Polynomials pervade mathematics, and much that is beautiful in mathematics is related to polynomials. Virtually every branch of mathematics, from algebraic number theory and algebraic geometry to applied analysis, Fourier analysis, and computer science, has its corpus of theory arising from the study of polynomials. Historically, questions relating to polynomials, for example, the solution of polynomial equations, gave rise to some of the most important problems of the day. The subject is now much too large to attempt an encyclopedic coverage.

The body of material we choose to explore concerns primarily polynomials as they arise in analysis, and the techniques of the book are primarily analytic. While the connecting thread is the polynomial, this is an analysis book. The polynomials and rational functions we are concerned with are almost exclusively of a single variable.

We assume at most a senior undergraduate familiarity with real and complex analysis (indeed in most places much less is required). However, the material is often tersely presented, with much mathematics explored in the exercises, some of which are quite hard, many of which are supplied with copious hints, some with complete proofs. Well over half the material in the book is presented in the exercises. The reader is encouraged to at least browse through these. We have been much influenced by Pólya and Szegő's classic "Problems and Theorems in Analysis" in our approach to the exercises. (Though unlike Pólya and Szegő we chose to incorporate the hints with the exercises.)

The book is mostly self-contained. The text, without the exercises, provides an introduction to the material, but much of the richness is reserved for the exercises. We have attempted to highlight the parts of the theory and the techniques we find most attractive. So, for example, Müntz's lovely characterization of when the span of a set of monomials is dense is explored in some detail. This result epitomizes the best of the subject: an attractive and nontrivial result with several attractive and nontrivial proofs.

There are excellent books on orthogonal polynomials, Chebyshev polynomials, Chebyshev systems, and the geometry of polynomials, to name but a few of the topics we cover, and it is not our intent to rewrite any of these. Of necessity and taste, some of this material is presented, and we have attempted to provide some access to these bodies of mathematics. Much of the material in the later chapters is recent and cannot be found in book form elsewhere.

Students who wish to study from this book are encouraged to sample widely from the exercises. This is definitely "hands on" material. There is too much material for a single semester graduate course, though such a course may be based on Sections 1.1 through 5.1, plus a selection from later sections and appendices. Most of the material after Section 5.1 may be read independently.

Not all objects labeled with "E" are exercises. Some are examples. Sometimes no question is asked because none is intended. Occasionally exercises include a statement like, "for a proof see ... "; this is usually an indication that the reader is not expected to provide a proof.

Some of the exercises are long because they present a body of material. Examples of this include E.11 of Section 2.1 on the transfinite diameter of a set and E.11 of Section 2.3 on the solvability of the moment problem. Some of the exercises are quite technical. Some of the technical exercises, like E.4 of Section 2.4, are included, in detail, because they present results that are hard to access elsewhere.

Acknowledgments

We would like to thank Dick Askey, Weiyu Chen, Carl de Boor, Karl Dilcher, Jens Happe, András Kroó, Doron Lubinsky, Gua-Hua Min, Paul Nevai, Allan Pinkus, József Szabados, Vilmos Totik, and Richard Varga. Their thoughtful and helpful comments made this a better book. We would also like to thank Judith Borwein, Maria Fe Elder, and Chiara Veronesi for their expert assistance with the preparation of the manuscript.

Contents

1

Introduction and Basic Properties

Overview

The most basic and important theorem concerning polynomials is the Fundamental Theorem of Algebra. This theorem, which tells us that every polynomial factors completely over the complex numbers, is the starting point for this book. Some of the intricate relationships between the location of the zeros of a polynomial and its coefficients are explored in Section 2. The equally intricate relationships between the zeros of a polynomial and the zeros of its derivative or integral are the subject of Section 1.3. This chapter serves as a general introduction to the body of theory known as the geometry of polynomials. Highlights of this chapter include the Fundamental Theorem of Algebra, the Eneström-Kakeya theorem, Lucas' theorem, and Walsh's two-circle theorem.

1.1 Polynomials and Rational Functions

The focus for this book is the polynomial of a single variable. This is an extended notion of the polynomial, as we will see later, but the most important examples are the algebraic and trigonometric polynomials, which we now define. The complex $(n+1)$-dimensional vector space of algebraic polynomials of degree at most n with complex coefficients is denoted by $\mathcal{P}_n^c$.

If $\mathbb{C}$ denotes the set of complex numbers, then

$$\mathcal{P}_n^c := \left\{ p : p(z) = \sum_{k=0}^{n} a_k z^k, \quad a_k \in \mathbb{C} \right\}. \tag{1.1.1}$$

When we restrict our attention to polynomials with real coefficients we will use the notation

$$\mathcal{P}_n := \left\{ p : p(z) = \sum_{k=0}^{n} a_k z^k, \quad a_k \in \mathbb{R} \right\}, \tag{1.1.2}$$

where $\mathbb{R}$ is the set of real numbers. Rational functions of type (m, n) with complex coefficients are then defined by

$$\mathcal{R}_{m,n}^c := \left\{ \frac{p}{q} : p \in \mathcal{P}_m^c, q \in \mathcal{P}_n^c \right\}, \tag{1.1.3}$$

while their real cousins are denoted by

$$\mathcal{R}_{m,n} := \left\{ \frac{p}{q} : p \in \mathcal{P}_m, q \in \mathcal{P}_n \right\}. \tag{1.1.4}$$

The distinction between the real and complex cases is particularly important for rational functions (see E.4).

The set of trigonometric polynomials $\mathcal{T}_n^c$ is defined by

$$\mathcal{T}_n^c := \left\{ t : t(\theta) := \sum_{k=-n}^{n} a_k e^{ik\theta}, \quad a_k \in \mathbb{C} \right\}. \tag{1.1.5}$$

A real trigonometric polynomial of degree at most n is an element of $\mathcal{T}_n^c$ taking only real values on the real line. We denote by $\mathcal{T}_n$ the set of all real trigonometric polynomials of degree at most n. Other characterizations of $\mathcal{T}_n$ are given in E.9. Note that if $z := e^{i\theta}$, then an arbitrary element of $\mathcal{T}_n^c$ is of the form

$$z^{-n} \sum_{k=0}^{2n} b_k z^k, \qquad b_k \in \mathbb{C} \tag{1.1.6}$$

and so many properties of trigonometric polynomials reduce to the study of algebraic polynomials of twice the degree on the unit circle in $\mathbb{C}$.

The most basic theorem of this book, and arguably the most basic nonelementary theorem of mathematics, is the Fundamental Theorem of Algebra. It says that a polynomial of exact degree n (that is, an element of $\mathcal{P}_n^c \backslash \mathcal{P}_{n-1}^c$) has exactly n complex zeros counted according to their multiplicities.

Theorem 1.1.1 (Fundamental Theorem of Algebra). *If*

$$p(z) := \sum_{i=0}^{n} a_i z^i, \qquad a_i \in \mathbb{C}, \quad a_n \neq 0,$$

then there exist $\alpha_1, \alpha_2, \ldots, \alpha_n \in \mathbb{C}$ *such that*

$$p(z) = a_n \prod_{i=1}^{n} (z - \alpha_i).$$

Here the multiplicity of the zero at α_i is the number of times it is repeated. So, for example,

$$(z-1)^3(z+i)^2$$

is a polynomial of degree 5 with a zero of multiplicity 3 at 1 and with a zero of multiplicity 2 at $-i$. The polynomial

$$p(z) := \sum_{i=0}^{n} a_i z^i, \qquad a_i \in \mathbb{C}, \quad a_n \neq 0$$

is called *monic* if its *leading coefficient* a_n equals 1. There are many proofs of the Fundamental Theorem of Algebra based on elementary properties of complex functions (see Theorem 1.2.1 and E.4 of Section 1.2). We will explore this theorem more substantially in the next section of this chapter.

Comments, Exercises, and Examples.

The importance of the solution of polynomial equations in the history of mathematics is hard to overestimate. The Greeks of the classical period understood quadratic equations (at least when both roots were positive) but could not solve cubics. The explicit solutions of the cubic and quartic equations in the sixteenth century were due to Niccolo Tartaglia (ca 1500–1557), Ludovico Ferrari (1522–1565), and Scipione del Ferro (ca 1465–1526) and were popularized by the publication in 1545 of the "Ars Magna" of Girolamo Cardano (1501–1576). The exact priorities are not entirely clear, but del Ferro probably has the strongest claim on the solution of the cubic. These discoveries gave western mathematics an enormous boost in part because they represented one of the first really major improvements on Greek mathematics. The impossibility of finding the zeros of a polynomial of degree at least 5, in general, by a formula containing additions, subtractions, multiplications, divisions, and radicals would await Niels Henrik Abel (1802–1829) and his 1824 publication of "On the Algebraic Resolution of Equations." Indeed, so much algebra, including Galois theory, analysis, and particularly complex analysis, is born out of these ideas that it is hard to imagine how the flow of mathematics might have proceeded without these issues being raised. For further history, see Boyer [68].

E.1 Explicit Solutions.

a] Quadratic Equations. Verify that the quadratic polynomial x^2+bx+c has zeros at

$$\frac{-b-\sqrt{b^2-4c}}{2}\,, \qquad \frac{-b+\sqrt{b^2-4c}}{2}\,.$$

b] Cubic Equations. Verify that the cubic polynomial x^3+bx+c has zeros at

$$\alpha+\beta\,, \quad -\left(\frac{\alpha+\beta}{2}\right)+i\sqrt{3}\left(\frac{\alpha-\beta}{2}\right), \quad -\left(\frac{\alpha+\beta}{2}\right)-i\sqrt{3}\left(\frac{\alpha-\beta}{2}\right),$$

where

$$\alpha=\sqrt[3]{\frac{-c}{2}+\sqrt{\frac{c^2}{4}+\frac{b^3}{27}}}$$

and

$$\beta=\sqrt[3]{\frac{-c}{2}-\sqrt{\frac{c^2}{4}+\frac{b^3}{27}}}\,.$$

c] Show that an arbitrary cubic polynomial, x^3+ax^2+bx+c, can be transformed into a cubic polynomial as in part b] by a transformation $x\mapsto ex+f$.

d] Observe that if the polynomial x^3+bx+c has three distinct real zeros, then α and β are necessarily nonreal and hence $4b^3+27c^2$ is negative. So, in this simplest of cases one is forced to deal with complex numbers (which was a serious technical problem in the sixteenth century).

e] Quartic Equations. The quartic polynomial $x^4+ax^3+bx^2+cx+d$ has zeros at

$$-\frac{a}{4}+\frac{R}{2}\pm\frac{\alpha}{2}\,, \qquad -\frac{a}{4}+\frac{R}{2}\pm\frac{\beta}{2}\,,$$

where

$$R=\sqrt{\frac{a^2}{4}-b+y}\,,$$

y is any root of the resolvent cubic

$$y^3-by^2+(ac+4d)y-a^2d+4bd-c^2\,,$$

and

$$\alpha,\beta=\sqrt{\frac{3a^2}{4}-R^2-2b\pm\frac{4ab-8c-a^3}{4R}}\,, \qquad R\neq 0\,,$$

while

$$\alpha,\beta=\sqrt{\frac{3a^2}{4}-2b\pm 2\sqrt{y^2-4d}}\,, \qquad R=0\,.$$

These unwieldy equations are quite useful in conjunction with any symbolic manipulation package.

E.2 Newton's Identities. Write

$$(x-\alpha_1)(x-\alpha_2)\cdots(x-\alpha_n) = x^n - c_1x^{n-1} + c_2x^{n-2} - \cdots + (-1)^n c_n\,.$$

The coefficients c_k are, by definition, the *elementary symmetric functions* in the variables $\alpha_1,\ldots,\alpha_n$.

a] For positive integers k, let

$$s_k := \alpha_1^k + \alpha_2^k + \cdots + \alpha_n^k\,.$$

Prove that

$$s_k = (-1)^{k+1}kc_k + (-1)^k \sum_{j=1}^{k-1}(-1)^j c_{k-j}s_j\,, \qquad k \le n$$

and

$$s_k = (-1)^{k+1} \sum_{j=k-n}^{k-1}(-1)^j c_{k-j}s_j\,, \qquad k > n\,.$$

Here, and in what follows, an empty sum is understood to be 0.

A *polynomial of n variables* is a function that is a polynomial in each of its variables. A *symmetric polynomial of n variables* is a polynomial of n variables that is invariant under any permutation of the variables.

b] Show by induction that any symmetric polynomial in n variables (with integer coefficients) may be written uniquely as a polynomial (with integer coefficients) in the elementary symmetric functions $f_1, f_2, \ldots, f_n$.

Hint: For a symmetric polynomial f in n variables, let

$$\sigma(f) := (\nu_1, \nu_2, \ldots, \nu_n)\,, \qquad \nu_1 \ge \nu_2 \ge \cdots \ge \nu_n \ge 0$$

if

$$f(x_1, x_2, \ldots, x_n) = \sum_{\alpha_1=0}^{\nu_1}\sum_{\alpha_2=0}^{\nu_2}\cdots\sum_{\alpha_n=0}^{\nu_n} c_{\alpha_1,\alpha_2,\ldots,\alpha_n} x_1^{\alpha_1}x_2^{\alpha_2}\cdots x_n^{\alpha_n}$$

and $c_{\nu_1,\nu_2,\ldots,\nu_n} \ne 0$. If

$$\sigma(f) = (\nu_1, \nu_2, \ldots, \nu_n) \quad \text{and} \quad \sigma(g) = (\widetilde{\nu}_1, \widetilde{\nu}_2, \ldots, \widetilde{\nu}_n)\,,$$

then let $\sigma(f) < \sigma(g)$ if $\nu_j \le \widetilde{\nu}_j$ for each j with a strict inequality for at least one index. This gives a (partial) well ordering of symmetric polynomials in n variables, that is, every set of symmetric polynomials in n variables has a minimal element. Now use induction on $\sigma(f)$. □

c] Show that

$$\left(\frac{1+\sqrt{5}}{2}\right)^k \to 0 \pmod 1 .$$

(By convergence to zero (mod 1) we mean that the quantity approaches integral values.)

Hint: Consider the integers

$$s_k := \alpha_1^k + \alpha_2^k ,$$

where $\alpha_1 := \frac{1}{2}(1+\sqrt{5})$ and $\alpha_2 := \frac{1}{2}(1-\sqrt{5})$. □

d] Find another algebraic integer α with the property that

$$\alpha^k \to 0 \pmod 1 .$$

Such numbers are called *Salem numbers* (see Salem [63]). It is an open problem whether any nonalgebraic numbers $\alpha > 1$ satisfy $\alpha^k \to 0 \pmod 1$.

E.3 Norms on $\mathcal{P}_n$. $\mathcal{P}_n$ is a vector space of dimension $n+1$ over $\mathbb{R}$. Hence $\mathcal{P}_n$ equipped with any norm is isomorphic to the Euclidean vector space $\mathbb{R}^{n+1}$, and these norms are equivalent to each other. Similarly, $\mathcal{P}_n^c$ is a vector space of dimension $n+1$ over $\mathbb{C}$. Hence $\mathcal{P}_n^c$ equipped with any norm is isomorphic to the Euclidean vector space $\mathbb{C}^{n+1}$, so these norms are also equivalent to each other. Let

$$p_n(x) := \sum_{k=0}^n a_k x^k, \qquad a_k \in \mathbb{R} .$$

Some common norms on $\mathcal{P}_n$ and $\mathcal{P}_n^c$ are

$$\|p\|_A := \sup_{x\in A} |p(x)| \qquad \textit{supremum norm}$$

$$:= \|p\|_{L_\infty(A)} \qquad L_\infty \textit{ norm}$$

$$\|p\|_{L_p(A)} := \left(\int_A |p(t)|^p \, dt\right)^{1/p} \qquad L_p \textit{ norm, } p \geq 1$$

$$\|p\|_{l_\infty} := \max_k \{|a_k|\} \qquad l_\infty \textit{ norm}$$

$$\|p\|_{l_p} := \left(\sum_{k=0}^n |a_k|^p\right)^{1/p} \qquad l_p \textit{ norm, } p \geq 1 .$$

In the first case A must contain $n+1$ distinct points. In the second case A must have positive measure.

a] Conclude that there exist constants C_1, C_2, and C_3 depending only on n so that

$$\|p'\|_{[-1,1]} \le C_1 \|p\|_{[-1,1]}\,,$$

$$\sum_{i=0}^{n} |a_i| \le C_2 \|p\|_{[-1,1]}\,,$$

$$\|p\|_{[-1,1]} \le C_3 \|p\|_{L_2[-1,1]}$$

for every $p \in \mathcal{P}_n^c$, and, in particular, for every $p \in \mathcal{P}_n$.

These inequalities will be revisited in detail in later chapters, where precise estimates are given in terms of n.

b] Show that there exist extremal polynomials for each of the above inequalities. That is, for example,

$$\sup_{0 \ne p \in \mathcal{P}_n} \frac{\|p'\|_{[-1,1]}}{\|p\|_{[-1,1]}}$$

is achieved.

E.4 On $\mathcal{R}_{n,m}$.

a] $\mathcal{R}_{n,m}$ is not a vector space because it is not closed under addition.

b] Partial Fraction Decomposition. Let $r_{n,m} \in \mathcal{R}_{n,m}^c$ be of the form

$$\frac{p(x)}{\prod_{k=1}^{m'} (x - \alpha_k)^{m_k}}\,, \qquad p \in \mathcal{P}_n^c\,, \quad \alpha_k \text{ distinct}\,, \quad p(\alpha_k) \ne 0\,.$$

Then there is a unique representation of the form

$$r_{n,m}(x) = q(x) + \sum_{k=1}^{m'} \sum_{j=1}^{m_k} \frac{a_{k,j}}{(x - \alpha_k)^j}\,, \qquad q \in \mathcal{P}_{n-m}^c\,, \quad a_{k,j} \in \mathbb{C}$$

(if $m > n$, then $\mathcal{P}_{n-m}^c$ is meant to be $\{0\}$).

Hint: Consider the type and dimension of expressions of the above form. □

c] Show that if

$$r_{n,m} \in \mathcal{R}_{n,m}^c\,,$$

then

$$\mathrm{Re}(r_{n,m}(\cdot)) \in \mathcal{R}_{n+m,2m}\,.$$

This is an important observation because in some problems a rational function in $\mathcal{R}_{n,n}^c$ can behave more like an element of $\mathcal{R}_{2n,2n}$ than $\mathcal{R}_{n,n}$.

E.5 Horner's Rule.

a] We have

$$\sum_{i=0}^{n} a_i x^i = (\cdots((a_n x + a_{n-1})x + a_{n-2})x + \cdots + a_1)x + a_0 .$$

So every polynomial of degree n can be evaluated by using at most n additions and n multiplications. (The converse is clearly not true; consider x^{2^n}.)

b] Show that every rational function of type $(n-1, n)$ can be put in a form so that it can be evaluated by using n divisions and n additions.

E.6 Lagrange Interpolation. Let z_i and y_i be arbitrary complex numbers except that the z_i must be distinct ($z_i \neq z_j$, for $i \neq j$). Let

$$l_k(z) := \frac{\prod_{i=0, i\neq k}^{n} (z - z_i)}{\prod_{i=0, i\neq k}^{n} (z_k - z_i)}, \qquad k = 0, 1, \ldots, n .$$

a] Show that there exists a unique $p \in \mathcal{P}_n^c$ that takes $n+1$ specified values at $n+1$ specified points, that is,

$$p(z_i) = y_i , \qquad i = 0, 1, \ldots, n .$$

This $p \in \mathcal{P}_n^c$ is of the form

$$p(z) = \sum_{k=0}^{n} y_k l_k(z)$$

and is called the *Lagrange interpolation polynomial.*

If all the z_i and y_i are real, then this unique interpolation polynomial is in $\mathcal{P}_n$.

b] Let

$$\omega(z) := \prod_{i=0}^{n} (z - z_i) .$$

Show that l_k is of the form

$$l_k(z) = \frac{\omega(z)}{(z - z_k)\omega'(z_k)}$$

and

$$p(z) = \sum_{k=0}^{n} \frac{y_k \omega(z)}{(z - z_k)\omega'(z_k)} .$$

c] An Error Estimate. Assume that the points $z_i \in [a,b]$, $i = 0,1,\dots,n$, are distinct and $f \in C^{n+1}[a,b]$ (that is, f is an $n+1$ times continuously differentiable real-valued function on $[a,b]$). Let $p \in \mathcal{P}_n$ be the Lagrange interpolation polynomial satisfying

$$p(z_i) = f(z_i)\,, \qquad i = 0,1,\dots,n\,.$$

Show that for every $x \in [a,b]$ there is a point $\xi \in (a,b)$ so that

$$f(x) - p(x) = \frac{1}{(n+1)!}\, f^{(n+1)}(\xi)\,\omega(x)\,.$$

Hence

$$\|f - p\|_{[a,b]} \le \frac{1}{(n+1)!}\, \|f^{(n+1)}\|_{[a,b]}\, \|\omega\|_{[a,b]}\,.$$

Hint: Choose λ so that $\varphi := f - p - \lambda w$ vanishes at x, that is,

$$\lambda := (f(x) - p(x))/\omega(x)\,.$$

Then repeated applications of Rolle's theorem yield that

$$\varphi^{(n+1)} = f^{(n+1)} - \lambda(n+1)!$$

has a zero ξ in (a,b). □

E.7 Hermite Interpolation.

a] Let $z_i \in \mathbb{C}$, $i = 1,2,\dots k$, be distinct. Let m_i, $i = 1,2,\dots,k$, be positive integers with $n+1 := \sum_{i=1}^k m_i$, and let

$$y_{i,j} \in \mathbb{C}\,, \quad i = 1,2,\dots,k\,, \quad j = 0,1,\dots,m_i - 1$$

be fixed. Show that there is a unique $p \in \mathcal{P}_n^c$, called the *Hermite interpolation polynomial*, so that

$$p^{(j)}(z_i) = y_{i,j}\,, \quad i = 1,2,\dots,k\,, \quad j = 0,1,\dots,m_i - 1\,.$$

If all the z_i and $y_{i,j}$ are real, then this unique interpolation polynomial is in $\mathcal{P}_{n-1}$.

Hint: Use induction on n. □

b] Assume that the points $z_i \in [a,b]$ are distinct and $f \in C^n[a,b]$. Let $p \in \mathcal{P}_{n-1}$ be the Hermite interpolation polynomial satisfying

$$p(z_i) = f^{(j)}(z_i)\,, \quad i = 1,2,\dots,k\,, \quad j = 0,1,\dots,m_i - 1\,.$$

Show that for every $x \in [a,b]$ there is a point $\xi \in (a,b)$ so that

$$f(x) - p(x) = \frac{1}{n!} f^{(n)}(\xi)\,\omega(x)$$

with

$$\omega(x) := \prod_{i=1}^{k} (x - x_i)^{m_i - 1}\,.$$

Hence

$$\|f - p\|_{[a,b]} \le \frac{1}{n!} \|f^{(n)}\|_{[a,b]} \|\omega\|_{[a,b]}\,.$$

Hint: Follow the hint given for E.6 c]. □

Polynomial interpolation and related topics are studied thoroughly in Davis [75]; Lorentz, Jetter, and Riemenschneider [83]; and Szabados and Vértesi [92].

E.8 On the Zeros of a $p \in \mathcal{P}_n$. Show that if $p \in \mathcal{P}_n$, then the nonreal zeros of p form conjugate pairs (that is, if z is a zero of p, then so is $\overline{z}$).

E.9 Factorization of Trigonometric Polynomials.

a] Show that $t \in \mathcal{T}_n$ (or $t \in \mathcal{T}_n^c$) if and only if t is of the form

$$t(z) = a_0 + \sum_{k=1}^{n} (a_k \cos kz + b_k \sin kz)\,, \qquad a_k, b_k \in \mathbb{R} \text{ (or } \mathbb{C})\,.$$

b] Show that if $t \in \mathcal{T}_n \backslash \mathcal{T}_{n-1}$, then there are numbers $z_1, z_2, \dots, z_{2n}$ and $0 \ne c \in \mathbb{C}$ such that

$$t(z) = c \prod_{j=1}^{2n} \sin \frac{z - z_j}{2}\,.$$

Show also that the nonreal zeros z_j of t form conjugate pairs.

E.10 Newton Interpolation and Integer-Valued Polynomials. Let $\Delta^k f(x)$ be defined inductively by

$$\Delta^0 f(x) := f(x)\,, \quad \Delta f(x) = \Delta^1 f(x) := f(x+1) - f(x)$$

and

$$\Delta^{k+1} f(x) := \Delta(\Delta^k f(x))\,, \qquad k = 1, 2, \dots\,.$$

Let

$$\binom{x}{k} := \frac{x(x-1)\cdots(x-k+1)}{k!}\,.$$

a] Show that $\binom{x}{k}$ is a polynomial of degree k that takes integer values at all integers.

b] Let f be an m times differentiable function on $[a, a+m]$. Show that there is a $\xi \in (a, a+m)$ such that

$$\Delta^m f(a) = f^{(m)}(\xi) .$$

c] Show that if $p \in \mathcal{P}_n^c$, then

$$p(x) = \sum_{k=0}^{n} \Delta^k p(0) \binom{x}{k} .$$

d] Suppose $p \in \mathcal{P}_n^c$ is integer-valued at all integers. Show that

$$p(x) = \sum_{k=0}^{n} a_k \binom{x}{k}$$

for some integers $a_0, a_1, \dots, a_n$. Note that this characterizes such polynomials.

e] Show that if $p \in \mathcal{P}_n^c$ takes integer values at $n+1$ consecutive integers, then p takes integer values at every integer.

f] Suppose $c \in \mathbb{R}$ and n^c is an integer for every $n \in \mathbb{N}$. Use part b] to show that c is a nonnegative integer.

1.2 The Fundamental Theorem of Algebra

The following theorem is a quantitative version of the Fundamental Theorem of Algebra due to Cauchy [1829]. We offer a proof that does not assume the Fundamental Theorem of Algebra, but does require some elementary complex analysis.

Theorem 1.2.1. *The polynomial*

$$p(z) := a_n z^n + a_{n-1} z^{n-1} + \cdots + a_0 \in \mathcal{P}_n^c , \qquad a_n \neq 0$$

has exactly n zeros. These all lie in the open disk of radius r centered at the origin, where

$$r := 1 + \max_{0 \le k \le n-1} \frac{|a_k|}{|a_n|} .$$

Proof. We may suppose that $a_0 \neq 0$, or we may first divide by z^k for some k. Now observe that

$$g(x) := |a_0| + |a_1|x + \cdots + |a_{n-1}|x^{n-1} - |a_n|x^n$$

satisfies $g(0) > 0$ and $\lim_{x\to\infty} g(x) = -\infty$. So by the intermediate value theorem, g has a zero in $(0, \infty)$ (which is, on considering $(g(x)/x^n)'$, in fact unique). Let s be this zero. Then for $|z| > s$,

$$|p(z) - a_n z^n| \le |a_0| + |a_1 z| + \cdots + |a_{n-1}z^{n-1}| < |a_n z^n|. \tag{1.2.1}$$

This, by Rouché's theorem (see E.1), shows that $p(z)$ and $a_n z^n$ have exactly the same number of zeros, namely, n, in any disk of radius greater than s.

It remains to observe that if $x \ge r$, then $g(x) < 0$ so $s < r$. Indeed,

$$\begin{aligned} g(x) &\le |a_n|x^n \left(-1 + \left(\max_{k=0,\dots,n-1} \frac{|a_k|}{|a_n|}\right) \sum_{k=0}^{n-1} x^{k-n}\right) \\ &< |a_n|x^n \left(-1 + \left(\max_{k=0,\dots,n-1} \frac{|a_k|}{|a_n|}\right) \frac{1}{x-1}\right) \\ &\le 0 \end{aligned}$$

for

$$x \ge 1 + \max_{k=0,\dots,n-1} \frac{|a_k|}{|a_n|}.$$

□

The exact relationship between the coefficients of a polynomial and the location of its zeros is very complicated. Of course, the more information we have about the coefficients, the better the results we can hope for. The following pretty theorem emphasizes this:

Theorem 1.2.2 (Eneström-Kakeya). *If*

$$p(z) := a_n z^n + a_{n-1}z^{n-1} + \cdots + a_0$$

with

$$a_0 \ge a_1 \ge \cdots \ge a_n > 0,$$

then all the zeros of p lie outside the open unit disk.

Proof. Consider

$$(1-z)p(z) = a_0 + (a_1 - a_0)z + \cdots + (a_n - a_{n-1})z^n - a_n z^{n+1}.$$

Then

$$|(1-z)p(z)| \ge a_0 - [(a_0 - a_1)|z| + \cdots + (a_{n-1} - a_n)|z|^n + a_n|z|^{n+1}].$$

Since $a_k - a_{k+1} \ge 0$, the right-hand expression above decreases as $|z|$ increases. Thus, for $|z| < 1$,

$$|(1-z)p(z)| > a_0 - [(a_0 - a_1) + \cdots + (a_{n-1} - a_n) + a_n] = 0,$$

and the result follows. □

Corollary 1.2.3. *Suppose*

$$p(z) := a_n z^n + a_{n-1} z^{n-1} + \cdots + a_0$$

with $a_k > 0$ *for each* k. *Then all the zeros of* p *lie in the annulus*

$$r_1 := \min_{k=0,\dots,n-1} \frac{a_k}{a_{k+1}} \le |z| \le \max_{k=0,\dots,n-1} \frac{a_k}{a_{k+1}} =: r_2 .$$

Proof. Apply Theorem 1.2.2 to $p(r_1 z)$ and $z^n p(r_2/z)$. □

This is a theme with many variations, some of which are explored in the exercises.

Theorem 1.2.4. *Suppose* $p > 1$, $q > 1$, *and* $p^{-1} + q^{-1} = 1$. *Then the polynomial* $h \in \mathcal{P}_n^c$ *of the form*

$$h(z) = a_n z^n + a_{n-1} z^{n-1} + \cdots + a_0 , \qquad a_n \neq 0$$

has all its zeros in the disk $\{z \in \mathbb{C} : |z| \le r\}$, *where*

$$r := \left\{ 1 + \left(\sum_{j=0}^{n-1} \frac{|a_j|^p}{|a_n|^p} \right)^{q/p} \right\}^{1/q} .$$

Proof. See E.6. □

Comments, Exercises, and Examples.

The Fundamental Theorem of Algebra appears to have been given its name by Gauss, although the result was familiar long before; it resisted rigorous proof by d'Alembert (1740), Euler (1749), and Lagrange (1772). It was more commonly formulated as a real theorem, namely: *every real polynomial factors completely into real linear or quadratic factors.* (This is an essential result for the integration of rational functions.) Girard has a claim to priority of formulation. In his "Invention Nouvelle en L'Algèbra" of 1629 he wrote "every equation of degree n has as many solutions as the exponent of the highest term." Gauss gave the first satisfactory proof in 1799 in his doctoral dissertation, and he gave three more proofs during his lifetime. His first proof, while titled "A new proof that every rational integral function of one variable can be resolved into real factors of the first or second degree," was in fact the first more-or-less satisfactory proof. Gauss' first proof is a geometric argument that the real and imaginary parts of a polynomial, u and v, have the property that the curves $u = 0$ and $v = 0$ intersect, and by modern standards has some topological problems. His third proof of 1816 amounts to showing that

$$\int_{|z|=r} \frac{p'(z)}{p(z)}\,dz$$

must vanish if p has no roots, which leads to a contradiction and is a genuinely analytic proof (see Boyer [68], Burton [85], and Gauss [1866]).

An almost purely algebraic proof using Galois theory, but based on ideas of Legendre, may be found in Stewart [73].

The "geometry of polynomials" is extensively studied in Marden [66] and Walsh [50], where most of the results of the section and much more may be accessed. See also Barbeau [89] and Pólya and Szegő [76].

Theorem 1.2.2 is due to Kakeya [12]. It is a special case of Corollary 1.2.3, due to Eneström [1893]. The Eneström-Kakeya theorem and related matters are studied thoroughly in Anderson, Saff, and Varga [79] and [81] and in Varga and Wu Wen-da [85], and a number of interesting properties are explored. For example, it is shown in the first of the above papers that the zeros of all p satisfying the assumption of Corollary 1.2.3 are dense in the annulus $\{z \in \mathbb{C} : r_1 \le |z| \le r_2\}$.

E.1 Basic Theorems in Complex Analysis. We collect a few of the basic theorems of complex analysis that we need. (Proofs may be found in any complex variables text such as Ahlfors [53] or Ash [71].)

a] Cauchy's Integral Formula. Let $D_r := \{z \in \mathbb{C} : |z| < r\}$. Suppose f is analytic on D_r and continuous on the closure $\overline{D}_r$ of D_r. Let ∂D_r denote the boundary of D_r. Then

$$0 = \int_{\partial D_r} f(t)\,dt\,,$$

$$f(z) = \frac{1}{2\pi i}\int_{\partial D_r} \frac{f(t)}{t-z}\,dt\,, \qquad z \in D_r\,,$$

and

$$f^{(n)}(z) = \frac{n!}{2\pi i}\int_{\partial D_r} \frac{f(t)}{(t-z)^{n+1}}\,dt\,, \qquad z \in D_r\,.$$

Unless otherwise specified, integration on a simple closed curve is taken anticlockwise. (We may replace ∂D_r and D_r by any simple closed curve and its interior, respectively, though for most of our applications circles suffice.)

b] Rouché's Theorem. *Suppose f and g are analytic inside and on a simple closed path γ (for most purposes we may use γ a circle). If*

$$|f(z) - g(z)| < |f(z)|$$

for every $z \in \gamma$, then f and g have the same number of zeros inside γ (counting multiplicities).

A function analytic on $\mathbb{C}$ is called *entire.*

c] **Liouville's Theorem.** *A bounded entire function is constant.*

d] **Maximum Principle.** *An analytic function on an open set $U \subset \mathbb{C}$ assumes its maximum modulus on the boundary. Moreover, if f is analytic and takes at least two distinct values on an open connected set $U \subset \mathbb{C}$, then*

$$|f(z)| < \sup_{z \in U} |f(z)|, \qquad z \in U.$$

e] **Unicity Theorem.** *Suppose f and g are analytic on an open connected set U. Suppose f and g agree on S, where S is an infinite compact subset of U, then f and g agree everywhere on U.*

E.2 Division.

a] Suppose p is a polynomial of degree n and $p(\alpha) = 0$. Then there exists a polynomial q of degree $n-1$ such that

$$p(x) = (x - \alpha)q(x).$$

Hint: Consider the usual division algorithm for polynomials. □

b] A polynomial of degree n has at most n roots.

This is the easier part of the Fundamental Theorem of Algebra. The remaining content is that every nonconstant polynomial has at least one complex root.

The next exercise develops the basic complex analysis tools mostly for polynomials on circles. The point of this exercise is to note that the proofs in this case are particularly straightforward.

E.3 Polynomial Complex Analysis.

a] Deduce Cauchy's integral formula for polynomials on circles.

Hint: Integrate z^n on ∂D_r. □

b] If $p(z) = a_n \prod_{i=1}^{n}(z-\alpha_i)$, then the number of indices i for which $|\alpha_i| < r$ is

$$\frac{1}{2\pi i}\int_{\partial D_r} \frac{p'(z)}{p(z)}\, dz,$$

provided no α_i lies on ∂D_r.

Hint: We have

$$\frac{p'(z)}{p(z)} = \sum_{i=1}^{n} \frac{1}{z - \alpha_i}$$

and

$$\frac{1}{2\pi i}\int_{\partial D_r} \frac{dz}{z - \alpha_i} = \begin{cases} 1 & \text{if } |\alpha_i| < r \\ 0 & \text{if } |\alpha_i| > r. \end{cases}$$

□

c] Deduce Rouché's theorem from part b] for polynomials f and g given by their factorizations, and for circles γ.

Hint: Let $h := 1 + (g - f)/f$. So $fh = g$ and

$$\frac{1}{2\pi i}\int_{\partial D_r} \frac{g'(z)}{g(z)}\,dz = \frac{1}{2\pi i}\int_{\partial D_r} \frac{f'(z)}{f(z)}\,dz + \frac{1}{2\pi i}\int_{\partial D_r} \frac{h'(z)}{h(z)}\,dz\,.$$

Show that the last integral is zero by expanding h^{-1} and applying b]. □

d] Deduce E.1 c] and E.1 d] from E.1 a].

e] Observe that the unicity theorem can be sharpened for polynomials as follows. If $p, q \in \mathcal{P}_n^c$ and $p(z) = q(z)$ for $n+1$ distinct values of $z \in \mathbb{C}$, then p and q are identical, that is, $p(z) = q(z)$ for every $z \in \mathbb{C}$. Equivalently, a polynomial $p \in \mathcal{P}_n^c$ is either identically 0 or has at most n zeros. (This is trivial from the Fundamental Theorem of Algebra, but as in E.2, it does not require it.)

E.4 The Fundamental Theorem of Algebra. *Every nonconstant polynomial has at least one complex zero.*

Prove this directly from Liouville's theorem.

E.5 Pellet's Theorem. *Suppose* $a_p \neq 0$, $|a_{p+1}| + \cdots + |a_n| > 0$, *and*

$$g(x) := |a_0| + |a_1|x + \cdots + |a_{p-1}|x^{p-1} - |a_p|x^p + |a_{p+1}|x^{p+1} + \cdots + |a_n|x^n$$

has exactly two positive zeros $s_1 < s_2$. *Then*

$$f(z) := a_n z^n + a_{n-1}z^{n-1} + \cdots + a_0 \in \mathcal{P}_n^c$$

has exactly p *zeros in the disk* $\{z \in \mathbb{C} : |z| \leq s_1\}$ *and no zeros in the annulus* $\{z \in \mathbb{C} : s_1 < |z| < s_2\}$.

Proof. Let $s_1 < t < s_2$. Then $g(t) < 0$, that is,

$$\sum_{\substack{j=0\\ j\neq p}}^{n} |a_j|t^j < |a_p|t^p\,.$$

Now apply Rouché's theorem to the functions

$$F(z) := a_p z^p \qquad \text{and} \qquad G(z) := \sum_{j=0}^{n} a_j z^j\,.$$

□

E.6 Proof of Theorem 1.2.4.

a] Hölder's inequality (see E.7 of Section 2.2) asserts that

$$\sum_{k=1}^{n} |a_k b_k| \leq \left(\sum_{k=1}^{n} |a_k|^p \right)^{1/p} \left(\sum_{k=1}^{n} |b_k|^q \right)^{1/q},$$

where $p^{-1} + q^{-1} = 1$ and $p \geq 1$. So if

$$p(z) := a_n z^n + a_{n-1} z^{n-1} + \cdots + a_0 \in \mathcal{P}_n^c,$$

then

$$\sum_{k=1}^{n-1} |a_k||z|^k \leq \left(\sum_{k=0}^{n-1} |a_k|^p \right)^{1/p} \left(\sum_{k=0}^{n-1} |z|^{kq} \right)^{1/q}.$$

b] Thus, for $|z| > 1$,

$$\begin{aligned}
|p(z)| &\geq |a_n||z|^n - \sum_{k=0}^{n-1} |a_k||z|^k \\
&\geq |a_n||z|^n \left\{ 1 - \left(\sum_{k=0}^{n-1} \left| \frac{a_k}{a_n} \right|^p \right)^{1/p} \left(\sum_{k=0}^{n-1} \frac{|z|^{kq}}{|z|^{nq}} \right)^{1/q} \right\} \\
&\geq |a_n||z|^n \left\{ 1 - \left(\sum_{k=0}^{n-1} \left| \frac{a_k}{a_n} \right|^p \right)^{1/p} \frac{1}{(|z|^q - 1)^{1/q}} \right\}.
\end{aligned}$$

c] When is the last expression positive?

E.7 The Number of Positive Zeros of a Polynomial. Suppose

$$p(z) := \sum_{j=0}^{n} a_j z^j$$

has m positive real roots. Then

$$m^2 \leq 2n \log \left(\frac{|a_0| + |a_1| + \cdots + |a_n|}{\sqrt{|a_0 a_n|}} \right).$$

This result is due to Schur though the proof more or less follows Erdős and Turán [50]. It requires using Müntz's theorem from Chapter 4.

a] Suppose

$$p(z) = a_n \prod_{k=1}^{n} (z - r_k e^{i\theta_k})$$

and

$$q(z) := \prod_{k=1}^{n} (z - e^{i\theta_k})\,.$$

Note that for $|z| = 1$,

$$\frac{|z - re^{i\theta}|^2}{|r|} \geq |z - e^{i\theta}|\,.$$

Use this to deduce that

$$|q(z)|^2 \leq \frac{|p(z)|^2}{|a_0 a_n|} \leq \left(\frac{|a_0| + |a_1| + \cdots + |a_n|}{\sqrt{|a_0 a_n|}} \right)^2$$

whenever $|z| = 1$.

b] Since p has m positive real roots q has m roots at 1. Use the change of variables $x := z + z^{-1}$ applied to $z^n q(z^{-1}) q(z)$ to show that

$$\begin{aligned}
&\|q\|^2_{\{|z|=1\}} \\
&\quad \geq \min_{\{b_k\}} \|(z-1)^m (z^{n-m} + b_{n-m-1} z^{n-m-1} + \cdots + b_1 z + b_0)\|^2_{\{|z|=1\}} \\
&\quad \geq \min_{\{c_k\}} \|x^m (x^{n-m} + c_{n-m-1} x^{n-m-1} + \cdots + c_1 x + c_0)\|_{[0,4]} \\
&\quad = 4^n \min_{\{d_k\}} \|x^m (x^{n-m} + d_{n-m-1} x^{n-m-1} + \cdots + d_1 x + d_0)\|_{[0,1]} \\
&\quad \geq \frac{4^n}{\sqrt{2n+1}\binom{2n}{n+m}}\,,
\end{aligned}$$

where the last inequality follows by E.2 c] of Section 4.2.

c] Show that

$$\log\left(\frac{4^n}{\sqrt{2n+1}\binom{2n}{n+m}} \right) \geq m^2/n$$

and finish the proof of the main result.

1.3 Zeros of the Derivative

The most basic and important theorem linking the zeros of the derivative of a polynomial to the zeros of the polynomial is variously attributed to Gauss, Lucas, Grace, and others, but is usually called Lucas' theorem [1874].

Theorem 1.3.1 (Lucas' Theorem). *Let $p \in \mathcal{P}_n^c$. All the zeros of p' are contained in the closed convex hull of the set of zeros of p.*

The proof of this theorem follows immediately from the following lemma by considering the intersection of the halfplanes containing the convex hull of the zeros of p.

Lemma 1.3.2. *Let $p \in \mathcal{P}_n^c$. If p has all its zeros in a closed halfplane, then p_n' also has all its zeros in the same closed halfplane.*

Proof. On consideration of the effect of the transformation $z \mapsto \alpha z + \beta$, by which any closed halfplane may be mapped to $\overline{H}_l := \{z : \operatorname{Re}(z) \le 0\}$, it suffices to prove the lemma under the assumption that p has all its zeros in $\overline{H}_l$. If p has all its zeros in $\overline{H}_l$, then

$$\frac{p'(z)}{p(z)} = \sum_{k=1}^{n} \frac{1}{z - \alpha_k}, \qquad \alpha_k \in \overline{H}_l .$$

But if $z \in H_r := \{z \in \mathbb{C} : \operatorname{Re}(z) > 0\}$, then

$$\frac{1}{z - \alpha_k} \in H_r \quad \text{for each} \quad \alpha_k \in \overline{H}_l ,$$

and it follows that

$$\sum_{k=1}^{n} \frac{1}{z - \alpha_k} \in H_r .$$

In particular,

$$\sum_{k=1}^{n} \frac{1}{z - \alpha_k} \ne 0 ,$$

which finishes the proof. □

There is a sharpening of Lucas' theorem for real polynomials formulated by Jensen. We need to introduce the notion of *Jensen circles* for a polynomial $p \in \mathcal{P}_n$. For $p \in \mathcal{P}_n$ the nonreal roots of p come in conjugate pairs. For each such pair, $\alpha + i\beta$, $\alpha - i\beta$, form the circle centered at α with radius $|\beta|$. So this circle centered on the x-axis at α has $\alpha + i\beta$ and $\alpha - i\beta$ on the opposite ends of its perpendicular diameter. The collection of all such circles are called the Jensen circles for p.

Theorem 1.3.3 (Jensen's Theorem). *Let $p \in \mathcal{P}_n$. Each nonreal zero of p' lies in or on some Jensen circle for p.*

The proof, which is similar to the proof of Lucas' theorem, is left for the reader as E.3.

We state the following pretty generalization of Lucas' theorem due to Walsh [21]. The proof is left as E.4. Proofs can also be found in Marden [66] and Pólya and Szegő [76].

Theorem 1.3.4 (Walsh's Two-Circle Theorem). *Suppose $p \in \mathcal{P}_n^c$ has all its n zeros in the disk D_1 with center c_1 and radius r_1. Suppose $q \in \mathcal{P}_m^c$ has all its m zeros in the disk D_2 with center c_2 and radius r_2. Then*

a] *All the zeros of $(pq)'$ lie in $D_1 \cup D_2 \cup D_3$, where D_3 is the disk with center c_3 and radius r_3 given by*

$$c_3 := \frac{nc_2 + mc_1}{n+m}, \qquad r_3 := \frac{nr_2 + mr_1}{n+m}.$$

b] *Suppose $(n \neq m)$. Then all the zeros of $(p/q)'$ lie in $D_1 \cup D_2 \cup D_3$, where D_3 is the disk with center c_3 and radius r_3 given by*

$$c_3 := \frac{nc_2 - mc_1}{n-m}, \qquad r_3 := \frac{nr_2 + mr_1}{|n-m|}.$$

Comments, Exercises, and Examples.

Lucas proved his theorem in 1874, although it is an easy and obvious consequence of an earlier result of Gauss. Jensen's theorem is formulated in Jensen [13] and proved in Walsh [20]. Much more concerning the geometry of zeros of the derivative can be found in Marden [66].

E.1 A Remark on Lucas' Theorem. Show that $p' \in \mathcal{P}_n^c$ has a zero α on the boundary of the convex hull of the zeros of p if and only if α is a multiple zero of p.

E.2 Laguerre's Theorem. *Suppose $p \in \mathcal{P}_n^c$ has all its zeros in a disk D. Let $\zeta \in \mathbb{C}$. Let w be any zero of*

$$q(z) := np(z) + (\zeta - z)p'(z)$$

(q is called the polar derivative of p with respect to ζ).

a] If $\zeta \notin D$, then w lies in D.

Hint: Consider $r(z) := p(z)(z-\zeta)^{-n}$, where p has all its zeros in D and $\zeta \notin D$. Then

$$\frac{r'(z)}{r(z)} = \frac{p'(z)}{p(z)} + \frac{n}{\zeta - z}$$

and if $q(w) = 0$ with $w \notin D$, then $r'(w) = 0$. Now observe that r is of the form

$$r(z) = s\left(\frac{1}{z-\zeta}\right), \qquad s \in \mathcal{P}_n^c,$$

where $s'((w-\zeta)^{-1}) = 0$. Note that $\zeta \notin D$ implies that

$$\widetilde{D} := \{(z-\zeta)^{-1} : z \in D\}$$

is a disk. Then s has all its zeros in $\widetilde{D}$ and so does s' by Lucas' theorem. However, $w \notin D$ implies $(w-\zeta)^{-1} \notin \widetilde{D}$, so $s'((w-\zeta)^{-1}) \neq 0$, a contradiction. □

b] If $p(w) \neq 0$, then any circle through w and ζ either passes through all the zeros of p_n or separates them.

E.3 Proof of Jensen's Theorem. Prove Theorem 1.3.3.

Hint: Suppose $p \in \mathcal{P}_n \setminus \mathcal{P}_{n-1}$ and denote the zeros of p by $z_1, z_2, \ldots, z_n$. Then

$$\frac{p'(z)}{p(z)} = \sum_{k=1}^{n} \frac{1}{z - z_k}\,.$$

If $z_k = \alpha_k + i\beta_k$ with $\alpha_k, \beta_k \in \mathbb{R}$, and $z = x + iy$ with $x, y \in \mathbb{R}$, then

$$\begin{aligned}&\operatorname{Im}\left(\frac{1}{x+iy-\alpha_k-i\beta_k} + \frac{1}{x+iy-\alpha_k+i\beta_k}\right)\\ &= \frac{-2y((x-\alpha_k)^2+y^2-\beta_k^2)}{((x-\alpha_k)^2+(y-\beta_k)^2)\cdot((x-\alpha_k)^2+(y+\beta_k)^2)}\,,\end{aligned}$$

and so outside all the Jensen circles and off the x-axis,

$$\operatorname{sign}\left(\operatorname{Im}\left(\frac{p'(z)}{p_n(z)}\right)\right) = -\operatorname{sign}(y) \neq 0\,.$$

□

E.4 Proof of Walsh's Theorem. Prove Theorem 1.3.4.

a] Prove Theorem 1.3.4 a].

Hint: Let z_0 be a zero of $p'q + q'p$ outside D_1 and D_2. Let

$$\zeta_1 := z_0 - \frac{np(z_0)}{p'(z_0)} \quad \text{and} \quad \zeta_2 := z_0 - \frac{mq(z_0)}{q'(z_0)}$$

($p'(z_0) \neq 0$ and $q'(z_0) \neq 0$ by Lucas' theorem). Observe that $\zeta_1 \in D_1$ and $\zeta_2 \in D_2$ by E.2, and

$$z_0 = \frac{n\zeta_2 + m\zeta_1}{n+m}\,.$$

□

b] Prove Theorem 1.3.4 b].

Hint: Proceed as in the hint to part a], starting from a zero z_0 of $p'q - q'p$ outside D_1 and D_2. □

c] If in Theorem 1.3.4 a] D_1, D_2, and D_3 are disjoint, then D_1 contains $n-1$ zeros, D_2 contains $m-1$ zeros, and D_3 contains 1 zero of $(pq)'$.

Hint: By a continuity argument we may reduce the general case to the case where $p(z) = (z-c_1)^n$ and $q(z) = (z-c_2)^m$. □

d] If in Theorem 1.3.4 b] $n = m$ and D_1 and D_2 are disjoint, then $D_1 \cup D_2$ contains all the zeros of $(p/q)'$.

E.5 Real Zeros and Poles.

a] If all the zeros of $p \in \mathcal{P}_n$ are real, then all the zeros of p_n' are also real.

b] Suppose all the zeros of both $p \in \mathcal{P}_n$ and $q \in \mathcal{P}_m$ are real, and all the zeros of p_n are smaller than any of the zeros of q_n. Show that all the zeros of $(p/q)'$ are real.

Hint: Consider the graph of

$$\frac{(p/q)'}{(p/q)} = \frac{p'}{p} - \frac{q'}{q}\,.$$

□

Define $W(p)$, the *Wronskian* of p, by

$$\begin{aligned} W(p)(z) &= p(z)p''(z) - (p'(z))^2 \\ &= \begin{vmatrix} p(z) & p'(z) \\ p'(z) & p''(z) \end{vmatrix} \\ &= p^2(z)\left(\frac{p'(z)}{p(z)}\right)'. \end{aligned}$$

c] Prove that if $p \in \mathcal{P}_n$ has only distinct real zeros, then $W(p)$ has no real zeros.

In Craven, Csordas, and Smith [87] it is conjectured that, for $p \in \mathcal{P}_n$, the number of real zeros of $W(p)/p^2$ does not exceed the number of nonreal zeros of p (a question they attribute to Gauss).

d] Let $p \in \mathcal{P}_n$. Show that any real zero of $W(p)$ lies in or on a Jensen circle of p.

Proof. See Dilcher [91]. □

e] Show that Lucas' theorem does not hold for rational functions.

Hint: Consider $r(x) = x/(\alpha^2 - x^2)$. □

The next exercise is a weak form of Descartes' rule of signs.

E.6 Positive Zeros of Müntz Polynomials. Suppose $\delta_0 < \delta_1 < \cdots < \delta_n$ and

$$f(x) := a_0x^{\delta_0} + a_1x^{\delta_1} + \cdots + a_nx^{\delta_n}, \quad a_k \in \mathbb{R}\,.$$

Show that either $f = 0$ or f has at most n zeros in $(0, \infty)$.

Hint: Proceed by induction on n. □

E.7 Apolar Polynomials and Szegő's Theorem. Two polynomials

$$f(x) := \sum_{k=0}^{n} a_k \binom{n}{k} x^k\,, \qquad a_n \neq 0$$

and

$$g(x) := \sum_{k=0}^{n} b_k \binom{n}{k} x^k\,, \qquad b_n \neq 0$$

are called *apolar* if

$$\sum_{k=0}^{n} (-1)^k a_k b_{n-k} \binom{n}{k} = 0\,.$$

a] A Theorem of Grace [02]. *Suppose that f and g are apolar polynomials. If f has all its zeros in a (closed or open) disk D, then g has at least one zero in D.*

Hint: Let $\alpha_1, \alpha_2, \dots, \alpha_n$ and $\beta_1, \beta_2, \dots, \beta_n$ denote the zeros of f and g, respectively. Suppose that the zeros of g are all outside D. Let

$$f_1(x) := nf(x) + (\beta_1 - x)f'(x)$$

and for $k = 2, 3, \dots, n$, let

$$f_k(x) := (n - k + 1)f_{k-1}(x) + (\beta_k - x)f_k'(x)\,.$$

Then, by E.2, each f_k has all its zeros in D. Now compute

$$f_{n-1}(\beta_n) = \frac{n!}{b_n}\left(\binom{n}{0} a_0 b_n - \binom{n}{1} a_1 b_{n-1} + \cdots + (-1)^n \binom{n}{n} a_n b_0\right) = 0\,,$$

where the vanishing follows by apolarity. This is a contradiction. □

b] If f and g are apolar, then the closed convex hull of the zeros of f intersects the closed convex hull of the zeros of g.

c] A Theorem of Szegő [22]. *Suppose*

$$f(x) := \sum_{k=0}^{n} a_k \binom{n}{k} x^k\,, \qquad a_n \neq 0\,,$$

$$g(x) := \sum_{k=0}^{n} b_k \binom{n}{k} x^k\,, \qquad b_n \neq 0\,,$$

and

$$h(x) := \sum_{k=0}^{n} a_k b_k \binom{n}{k} x^k\,.$$

Suppose f has all its zeros in a closed disk $\overline{D}$, and g has zeros $\beta_1, \dots, \beta_n$. Then all the zeros of h are of the form $\beta_i \gamma_i$ with $\gamma_i \in \overline{D}$.

Hint: Suppose δ is a zero of h. Then

$$\sum_{k=0}^{n} a_k b_k \binom{n}{k} \delta^k = 0 .$$

So the polynomial

$$r(x) := \sum_{k=0}^{n} (-1)^k \binom{n}{k} b_k \delta^k x^{n-k}$$

is apolar to f, and thus has a zero α in D. But then $\alpha = -\delta/\beta_i$ for some i since $r(x) = x^n g(-\delta/x)$. □

E.8 Zeros of the Integral. Suppose $p \in \mathcal{P}_n \setminus \mathcal{P}_{n-1}$ has all its zeros in $\overline{D}_1 := \{z \in \mathbb{C} : |z| \le 1\}$.

a] Show that the polynomial q defined by $q(x) := \int_0^x p(t)\,dt$ has all its zeros in $\overline{D}_2 := \{z \in \mathbb{C} : |z| \le 2\}$.

Hint: Apply E.7 c]. Take

$$f(x) := p(x), \quad g(x) := \sum_{k=0}^{n} \binom{n}{k} \frac{x^k}{k+1} .$$

Then

$$h(x) = \frac{1}{x} \int_0^x p(t)\,dt .$$

Note that $g(x) = (n+1)^{-1} x^{-1} ((1+x)^{n+1} - 1)$ has all its zeros in D_2. □

b] Show that

$$q(x) := \int_0^x \int_0^{t_{m-1}} \int_0^{t_{m-2}} \cdots \int_0^{t_1} p(t)\,dt\,dt_1 \cdots dt_{m-2}\,dt_{m-1}$$

has all its zeros in $\overline{D}_{r_{m,n}} := \{z \in \mathbb{C} : |z| \le r_{m,n}\}$, where $r_{m,n} \le m+1$ is the zero of

$$\sum_{k=0}^{n} \binom{m+n}{m+k} x^k$$

with the largest modulus. Note that q is the mth integral of p normalized so that the constants of integration are all zero.

Proof. See Borwein, Chen, and Dilcher [95]. □

E.9 Grace's Complex Version of Rolle's Theorem. *Suppose α and β are zeros of $p \in \mathcal{P}_n \setminus \mathcal{P}_{n-1}$. Then p' has at least one zero in the disk*

$$\overline{D}(c,r) := \{z \in \mathbb{C} : |z-c| \leq r\},$$

where

$$c := \frac{\alpha+\beta}{2} \qquad \textit{and} \qquad r := \frac{|\alpha-\beta|}{2}\cot\frac{\pi}{n}.$$

Hint: Assume, without loss of generality, that $\alpha = -1$ and $\beta = 1$. Let

$$p'(x) = \sum_{k=0}^{n-1} a_k x^k, \quad \text{that is}, \quad p(x) = c + \sum_{k=0}^{n-1} a_k \frac{x^{k+1}}{k+1}.$$

Apply E.7 a]. Note that

$$0 = \frac{p(1)-p(-1)}{2} = a_0 + \frac{a_2}{3} + \frac{a_4}{5} + \cdots.$$

So

$$f(z) := (z-1)^n - (z+1)^n$$

and p' are apolar. □

E.10 Corollaries of Szegő's Theorem. *Suppose*

$$f(z) := \binom{n}{0}a_0 + \binom{n}{1}a_1 z + \cdots + \binom{n}{n}a_n z^n,$$

$$g(z) := \binom{n}{0}b_0 + \binom{n}{1}b_1 z + \cdots + \binom{n}{n}b_n z^n,$$

and

$$h(z) := \binom{n}{0}a_0 b_0 + \binom{n}{1}a_1 b_1 z + \cdots + \binom{n}{n}a_n b_n z^n$$

with $a_n b_n \neq 0$.

a] *If f has all its zeros in a convex set S containing 0 and g has all its zeros in $[-1,0]$, then h has all its roots in S.*

b] *If f and g have all their zeros in $[-1,0]$, then so does h.*

E.11 Another Corollary of Szegő's Theorem. *If $\sum_{k=0}^{n} a_k z^k$ has all its zeros in $\overline{D}_1 := \{z \in \mathbb{C} : |z| \leq 1\}$, then so does $\sum_{k=0}^{n} \frac{a_k z^k}{\binom{n}{k}}$. In particular, $\sum_{k=0}^{n} \frac{z^k}{\binom{n}{k}}$ has all its zeros in $\overline{D}_1$.*

The results of the next exercise were first proved by M. Riesz (see, for example, Mignotte [92]) and were rediscovered by Walker [93].

E.12 Consecutive Zeros of p' for $p \in \mathcal{P}_n$ with Real Zeros. For a polynomial

$$p(x) := \prod_{i=1}^{n} (x - \alpha_i)\,, \quad \alpha_1 < \alpha_2 < \cdots < \alpha_n\,, \quad n \geq 2$$

with only real zeros, let

$$\Delta(p) := \min_{1 \leq i \leq n-1} (\alpha_{i+1} - \alpha_i)\,.$$

By Rolle's theorem

$$p'(x) = n \prod_{i=1}^{n-1} (x - \beta_i)\,, \quad \alpha_1 < \beta_1 < \alpha_2 < \beta_2 < \cdots < \beta_{n-1} < \alpha_n\,.$$

a] Suppose $n \geq 3$. Prove that $\Delta(p) < \Delta(p')$.

Outline. It is required to show that $\beta_j - \beta_{j-1} > \Delta(p)$ for each $j \geq 2$. Let $2 \leq j \leq n$ be fixed. Since

$$\frac{f'(\beta_j)}{f(\beta_j)} = \frac{f'(\beta_{j-1})}{f(\beta_{j-1})} = 0$$

we have

$$\sum_{i=1}^{n} \frac{1}{(\beta_{j-1} - \alpha_i)(\beta_j - \alpha_i)} = 0\,.$$

Now let $u_j := \alpha_j - \beta_{j-1}$, $v_j := \beta_j - \alpha_j$. Also for each i, let $d_i := \alpha_j - \alpha_{j-i}$, $e_i := \alpha_{j+i} - \alpha_j$. Then the above can be rewritten as

$$\sum_{i=1}^{j-1} \frac{1}{(d_i - u_j)(d_i + v_j)} + \frac{1}{(-u_j v_j)} + \sum_{i=1}^{n-j} \frac{1}{(e_i + u_j)(e_i - v_j)} = 0\,.$$

Define

$$F(u, v) := \sum_{i=1}^{j-1} \frac{uv}{(d_i - u)(d_i + v)} + \sum_{i=1}^{n-j} \frac{uv}{(e_i + u)(e_i - v)}\,.$$

Note that F is increasing in each variable $(0 \leq u < d_1,\ 0 \leq v < e_1)$ and observe that

$$F(u_j, v_j) = 1\,.$$

To prove the result, it suffices to show that if u and v are nonnegative numbers satisfying $u + v = \Delta(p)$, then $F(u, v) < 1$.

Now show that

$$\begin{aligned}
F(u,v) \le & uv\left(\sum_{i=1}^{j-1}\frac{1}{(d_i-u)(d_{i+1}-u)}+\sum_{i=1}^{n-j}\frac{1}{(e_{i+1}-v)(e_i-v)}\right)\\
\le & \frac{uv}{\Delta(p)}\left(\sum_{i=1}^{j-1}\left(\frac{1}{d_i-u}-\frac{1}{d_{i+1}-u}\right)+\sum_{i=1}^{n-j}\left(\frac{1}{e_i-v}-\frac{1}{e_{i+1}-v}\right)\right)\\
< & \frac{uv}{\Delta(p)}\left(\frac{1}{d_1-u}+\frac{1}{e_1-v}\right)\le\frac{uv}{\Delta(p)}\left(\frac{1}{v}+\frac{1}{u}\right)\le 1
\end{aligned}$$

whenever u and v are nonnegative numbers satisfying $u+v=\Delta(p)$. □

b] Suppose $n \ge 3$ and $\gamma \in \mathbb{R}$. Show that $\Delta(p'-\gamma p)$ has only real zeros and $\Delta(p'-\gamma p) > \Delta(p)$.

c] What happens when p has only real zeros but they are not necessarily distinct?

E.13 Fejér's Theorem on the Zeros of Müntz Polynomials. The following pretty results of Fejér may also be found in Pólya and Szegő [76]:

Suppose that $(\lambda_k)_{k=0}^{\infty}$ is an increasing sequence of nonnegative integers with $\lambda_0 = 0$.

a] Let

$$p(z) := \sum_{k=0}^{n} a_k z^{\lambda_k}, \qquad a_k \in \mathbb{C}, \quad a_0 a_1 \neq 0.$$

Then p has at least one zero $z_0 \in \mathbb{C}$ so that

$$|z_0| \le \left(\frac{\lambda_2\lambda_3\cdots\lambda_n}{(\lambda_2-\lambda_1)(\lambda_3-\lambda_1)\cdots(\lambda_n-\lambda_1)}\right)^{1/\lambda_1}\left|\frac{a_0}{a_1}\right|^{1/\lambda_1}.$$

Outline. We say that $z_1 \in \mathbb{C}$ is not less than $z_2 \in \mathbb{C}$ if $|z_2| \le |z_1|$. Studying $q(z) := z^{\lambda_n}p(z^{-1})$, we need to show that the largest zero of

$$q(z) = a_0 x^{\lambda_n} + \sum_{k=1}^{n} a_k x^{\lambda_n-\lambda_k}$$

is not less than

$$\left(\frac{(\lambda_2-\lambda_1)(\lambda_3-\lambda_1)\cdots(\lambda_n-\lambda_1)}{\lambda_2\lambda_3\cdots\lambda_n}\right)^{1/\lambda_1}\left|\frac{a_1}{a_0}\right|^{1/\lambda_1}.$$

We prove this statement by induction on n. The statement is obviously true for $n=1$. Now assume that the statement is true for $n-1$. It follows from Lucas' theorem that if q is a polynomial with complex coefficients, then the largest zero of q' is not greater than the largest zero of q.

By the above corollary of Lucas' theorem, it is sufficient to prove that the largest zero of

$$z^{\lambda_{n-1}-\lambda_n+1}q'(z) = \lambda_n a_0 z^{\lambda_{n-1}} + \sum_{k=1}^{n-1}(\lambda_n - \lambda_k)a_k z^{\lambda_{n-1}-\lambda_k}$$

is not less than

$$\left(\frac{(\lambda_2-\lambda_1)(\lambda_3-\lambda_1)\cdots(\lambda_n-\lambda_1)}{\lambda_2\lambda_3\cdots\lambda_n}\right)^{1/\lambda_1}\left|\frac{a_1}{a_0}\right|^{1/\lambda_1}.$$

However, this is true by the inductive hypothesis. □

b] Suppose

$$f(z) = \sum_{k=0}^{\infty} a_k z^{\lambda_k}\,, \qquad a_k \in \mathbb{C}$$

is an entire function so that $\sum_{k=1}^{\infty} 1/\lambda_k < \infty$, that is, the entire function f satisfies the *Fejér gap condition*. Show that there is a $z_0 \in \mathbb{C}$ so that $f(z_0) = 0$.

Hint: Use part a]. □

2

Some Special Polynomials

Overview

Chebyshev polynomials are introduced and their central role in problems in the uniform norm on $[-1,1]$ is explored. Sequences of orthogonal functions are then examined in some generality, although our primary interest is in orthogonal polynomials (and rational functions). The third section of this chapter is concerned with orthogonal polynomials; it introduces the most classical of these. These polynomials satisfy many extremal properties, similar to those of the Chebyshev polynomials, but with respect to (weighted) L_2 norms. The final section of the chapter deals with polynomials with positive coefficients in various bases.

2.1 Chebyshev Polynomials

The ubiquitous Chebyshev polynomials lie at the heart of many analytic problems, particularly problems in $C[a,b]$, the space of real-valued continuous functions equipped with the uniform (supremum) norm, $\|\cdot\|_{[a,b]}$. Throughout this book, for any real- or complex-valued function f defined on $[a,b]$,

$$\|f\|_{[a,b]} := \sup_{x\in[a,b]} |f(x)|\,.$$

The Chebyshev polynomials are defined by

$$
\begin{aligned}
T_n(x) &:= \cos(n \arccos x)\,, && x \in [-1,1]\,, \\
\text{(2.1.1)} \qquad &= \frac{1}{2}\left((x+\sqrt{x^2-1})^n + (x-\sqrt{x^2-1})^n\right), && x \in \mathbb{C}\,, \\
&= \frac{n}{2}\sum_{k=0}^{\lfloor n/2 \rfloor} (-1)^k \frac{(n-k-1)!}{k!(n-2k)!}(2x)^{n-2k}\,, && x \in \mathbb{C}\,.
\end{aligned}
$$

These elementary equivalences are left for the reader (see E.1). The nth Chebyshev polynomial has the following equioscillation property on $[-1,1]$. There exist $n+1$ points $\zeta_i \in [-1,1]$ with $-1 = \zeta_n < \zeta_{n-1} < \cdots < \zeta_0 = 1$ so that

$$
\text{(2.1.2)} \qquad T_n(\zeta_j) = (-1)^{n-j}\|T_n\|_{[-1,1]} = (-1)^{n-j}\,, \qquad j = 0,1,\ldots,n\,.
$$

In other words $T_n \in \mathcal{P}_n$ takes the values $\pm\|T_n\|_{[-1,1]}$ with alternating sign the maximum possible number of times on $[-1,1]$. (These extreme points are just the points $\cos(k\pi/n)$, $k = 0,1,\ldots,n$.) The Chebyshev polynomial T_n satisfies the following extremal property:

Theorem 2.1.1. *We have*

$$
\min_{p \in \mathcal{P}_{n-1}^c} \|x^n - p(x)\|_{[-1,1]} = \|2^{1-n}T_n\|_{[-1,1]} = 2^{1-n}\,,
$$

where the minimum is uniquely attained by $p(x) = x^n - 2^{1-n}T_n(x)$.

Proof. Observe that, while the minimum is taken over $\mathcal{P}_{n-1}^c$, we need only consider $p \in \mathcal{P}_{n-1}$, since taking the real part of a $p \in \mathcal{P}_{n-1}^c$ can only improve the estimate. From the above formulas for T_n we have

$$
2^{1-n}T_n(x) = x^n + s(x)\,, \qquad s \in \mathcal{P}_{n-1}\,.
$$

Now suppose there exists $q \in \mathcal{P}_{n-1}$ with

$$
\text{(2.1.3)} \qquad \|x^n - q(x)\|_{[-1,1]} < 2^{1-n}\,.
$$

Then

$$
2^{1-n}T_n(x) - (x^n - q(x)) = s(x) + q(x) \in \mathcal{P}_{n-1}
$$

changes sign between any two consecutive extrema of T_n, hence it has at least n zeros in $(-1,1)$, and thus it must vanish identically. This contradicts (2.1.3), and we are done up to proving uniqueness (this is left as E.2). □

Comments, Exercises, and Examples.

The Chebyshev polynomials T_n are named after the versatile Russian mathematician, P. L. Chebyshev (1821–1894). The T comes from the spelling Tchebychef (or some such variant; there are many in the literature). A wealth of information on these polynomials may be found in Rivlin [90]. Throughout later sections of this book the Chebyshev polynomials will keep recurring. The initial exercises explore elementary properties of the Chebyshev polynomials.

Erdős [39] proved that for $t \in \mathcal{T}_n$ with $\|t\|_{\mathbb{R}} \le 1$, the length of the graph of t on $[0, 2\pi]$ is the longest if and only if t is of the form $t(\theta) = \cos(n\theta + \alpha)$ with some $\alpha \in \mathbb{R}$ (see E.6). He conjectured that for any $p \in \mathcal{P}_n$ with $\|p\|_{[-1,1]} \le 1$, the maximum arc length is attained by the nth Chebyshev polynomial T_n. This is proved in Bojanov [82b]. Kristiansen [79] also claims a proof. In E.9 the reducibility of T_n is considered, and in E.11 the basic properties of the transfinite diameter are established.

E.1 Basic Properties.

a] Establish the equivalence of the three representations of T_n given in equation (2.1.1).

Hint: $\cos n\theta = \frac{1}{2}[(\cos\theta + i\sin\theta)^n + ((\cos\theta - i\sin\theta)^n]$. To get the third representation, use E.3 b]. □

b] The zeros of T_n are precisely the points

$$x_k = \cos\frac{(2k-1)\pi}{2n}, \qquad k = 1, 2, \dots, n.$$

c] The extrema of $T_n(x)$ in $[-1,1]$ are precisely the points

$$\zeta_k = \cos\frac{k\pi}{n}, \qquad k = 0, 1, \dots, n.$$

d] Observe that the zeros of T_n and T_{n+1} interlace, as do the extrema.

E.2 Uniqueness of the Minimum in Theorem 2.1.1. Prove the uniqueness of the minimum in Theorem 2.1.1.

Hint: Assume that $q \in \mathcal{P}_{n-1}^c$ and

$$\|x^n - q(x)\|_{[-1,1]} \le 2^{1-n}.$$

Then

$$h(x) := 2^{1-n}T_n(x) - \mathrm{Re}(x^n - q(x))$$

defines a polynomial from $\mathcal{P}_{n-1}$ on $\mathbb{R}$ having at least n zeros (counted according to their multiplicities). Thus

$$2^{1-n}T_n(x) = \mathrm{Re}(x^n - q(x)), \qquad x \in \mathbb{R},$$

which, together with the previous inequality, implies that $q(x)$ is real whenever $T_n(x) = \pm 1$. Now E.6 of Section 1.1 (Lagrange interpolation) yields that q has real coefficients. Hence

$$2^{1-n}T_n(x) = x^n - q(x)\,, \qquad x \in \mathbb{R}\,.$$

□

E.3 Further Properties of T_n.

a] **Composition.** Show that $T_{nm}(x) = T_n(T_m(x))$.

b] **Three-Term Recursion.** Show that

$$T_n(x) = 2xT_{n-1}(x) - T_{n-2}(x)\,, \qquad n = 2, 3, \ldots\,.$$

c] Verify that

$$\begin{aligned}
T_0(x) &= 1 \\
T_1(x) &= x \\
T_2(x) &= 2x^2 - 1 \\
T_3(x) &= 4x^3 - 3x \\
T_4(x) &= 8x^4 - 8x^2 + 1 \\
T_5(x) &= 16x^5 - 20x^3 + 5x\,.
\end{aligned}$$

Note that T_n is even for n even and odd for n odd.

d] **Another Formula for T_n.** Show that $T_n(x) = \cosh(n \cosh^{-1}(x))$ for every $x \in \mathbb{R} \setminus [-1, 1]$.

e] **Differential Equation.** Show that

$$(1 - x^2)T_n''(x) - xT_n'(x) + n^2 T_n(x) = 0\,.$$

f] **An Identity.** Show that

$$T_n(x) = \frac{T_{n+1}'(x)}{2n+2} - \frac{T_{n-1}'(x)}{2n-2}\,.$$

g] **Orthogonality.** Show that

$$\frac{2}{\pi}\int_{-1}^{1} \frac{T_n(x)T_m(x)dx}{\sqrt{1-x^2}} = \delta_{n,m} := \begin{cases} 0, & n \neq m \\ 1, & n = m > 0\,. \end{cases}$$

h] **Generating Function.** Show that

$$\frac{1 - yx}{1 - 2yx + y^2} = \sum_{n=0}^{\infty} T_n(x)y^n\,, \qquad x \in [-1, 1]\,, \quad |y| < 1\,.$$

Hint: Set $x = \cos\theta$ and sum. □

i] Another Representation of T_n. Show that

$$T_n(x) = \sum_{k=0}^{\lfloor n/2 \rfloor} \binom{n}{2k} x^{n-2k}(x^2-1)^k .$$

j] Another Identity. Show that

$$T_n\left(\tfrac{1}{2}(x+x^{-1})\right) = \tfrac{1}{2}(x^n + x^{-n}) .$$

E.4 Approximation to x^k on $[0,1]$.

a] Let $T_n^*(x) = T_n(2x-1)$ be the nth Chebyshev polynomial shifted to the interval $[0,1]$. Suppose

$$T_n^*(x) = \sum_{k=0}^{n} b_k x^k .$$

Show that for each $k = 0, 1, \ldots, n$,

$$\min_{c_j \in \mathbb{R}} \|x^k - \sum_{\substack{j=0 \\ j \neq k}}^{n} c_j x^j\|_{[0,1]} = \|b_k^{-1} T_n^*\|_{[0,1]} .$$

Hint: Proceed as in the proof of Theorem 2.1.1 and use E.6 of Section 1.3. □

b] Why does this not hold for T_n on $[-1,1]$?

E.5 A Composition Characterization. Suppose $(p_n)_{n=1}^{\infty}$ is a sequence of polynomials of degree n and for all positive integers n and m

$$p_n \circ p_m = p_{n \cdot m} .$$

Then there exists a linear transformation $w(x) = \alpha x + \beta$ so that

$$w \circ p_n \circ w^{-1} = x^n , \qquad n = 1, 2, \ldots$$

or

$$w \circ p_n \circ w^{-1} = T_n , \qquad n = 1, 2, \ldots .$$

This result is due to Block and Thielman [51]. The proof outlined in this exercise follows Rivlin [90].

a] Let

$$q(x) := a_0 + a_1 x + a_2 x^2\,, \quad a_2 \neq 0\,, \quad \text{and} \quad v(x) := \frac{x}{a_2} - \frac{a_1}{2a_2}\,.$$

Then

$$v^{-1}(q(v(x))) = x^2 + c \quad \text{with} \quad c := a_0 a_2 + (a_1/2) - (a_1^2/4)\,.$$

b] Let $q(x) = a_0 + a_1 x + a_2 x^2$, $a_2 \neq 0$. Then there is at most one polynomial p_n of degree exactly n so that

$$p_n(q(x)) = q(p_n(x))\,.$$

Hint: By a] we may assume $q(x) = x^2 + c$. Now suppose $r,\ s \in \mathcal{P}_n^c \backslash \mathcal{P}_{n-1}^c$,

$$r(x^2 + c) = r^2(x) + c$$

and

$$s(x^2 + c) = s^2(x) + c\,.$$

Then $u := r - s \in \mathcal{P}_{n-1}^c$ satisfies

$$u(x^2 + c) = u(x)(r(x) + s(x))$$

from which we deduce, by comparing degrees on both sides, that $n = 0$. (Note that the above conditions imply r and s monic.) □

c] Finish the proof of the initial statement of this exercise.

This is a special case of a more general theorem of Ritt [23] that classifies all rational functions r and s that commute in the sense that $r \circ s = s \circ r$.

d] Another Composition Characterization. Suppose $p \in \mathcal{P}_n$ has the property that the closure of the set

$$I_p := \{z \in \mathbb{C} : p^{[k]}(z) = 0 \text{ for some } k = 1, 2, \ldots\}$$

is the interval $[-1, 1]$, where $p^{[k]}$ is the kth iterate of p, that is,

$$p^{[1]} := p \quad \text{and} \quad p^{[k]} := p \circ p^{[k-1]} \quad \text{for} \quad k = 2, 3, \ldots\,.$$

Then $p(x) = \pm T_n(x)$.

e] Let

$$r_n(x) = \tan(n \tan^{-1}(x))\,.$$

Show that r_n is a rational function in $\mathcal{R}_{n,n}$, and observe that

$$r_n \circ r_m = r_{n \cdot m}\,.$$

E.6 Trigonometric Polynomials of Longest Arc Length. Theorem 5.1.3 (Bernstein-Szegő inequality) asserts that

$$t'(\theta)^2 + n^2 t^2(\theta) \le n^2 \|t\|_{\mathbb{R}}^2$$

for every $t \in \mathcal{T}_n$ and $\theta \in \mathbb{R}$. Use this to prove the following result of Erdős. For $t \in \mathcal{T}_n$ with $\|t\|_{\mathbb{R}} \le 1$, the length of the graph of t on $[0, 2\pi]$ is the longest if and only if it is of the form $t(\theta) = \cos(n\theta + \alpha)$ for some $\alpha \in \mathbb{R}$.

Hint: Suppose $t \in \mathcal{T}_n$ with $\|t\|_{\mathbb{R}} = 1$. Let $s(\theta) := \cos n\theta$. If

$$-1 < t(\theta_1) = s(\theta_2) < 1$$

holds, then by the Bernstein-Szegő inequality (see also E.5 of Section 5.1)

$$|t'(\theta_1)| \le n(1 - t^2(\theta_1))^{1/2} = n(1 - s^2(\theta_2))^{1/2} = |s'(\theta_2)|,$$

and if equality holds for one pair of θ_1, θ_2, then it holds for all pairs, and $t(\theta) \equiv \cos(n\theta + \alpha)$ for some $\alpha \in \mathbb{R}$. Suppose $t_n(\theta) \not\equiv \cos(n\theta + \alpha)$. Let τ and σ be monotone arcs of the graphs of $y = t(\theta)$ and $y = s(\theta)$, respectively, with endpoints of each having the same ordinates y_1 and y_2. Let $|\tau|$ and $|\sigma|$ be the length of τ and σ, respectively, and let $|\tau_x|$ and $|\sigma_x|$ be the length of the projection of τ and σ, respectively, on the x-axis. Show that

$$|\tau| < |\sigma| + (|\tau_x| - |\sigma_x|)$$

by approximating τ and σ by a polygonal line corresponding to a subdivision of the interval with endpoints, y_1 and y_2 on the y-axis. □

E.7 Monic Polynomials with Minimal Norm on an Interval.

a] The unique monic polynomial $p \in \mathcal{P}_n^c$ minimizing $\|p\|_{[a,b]}$ is given by

$$p(x) = 2\left(\frac{b-a}{4}\right)^n T_n\left(\frac{2x - a - b}{b - a}\right).$$

b] Let $0 < a < b$. Find all monic polynomials $p \in \mathcal{P}_n^c$ minimizing

$$\|p\|_{[-b,-a]\cup[a,b]}.$$

(For two intervals of different lengths this is a much harder problem. The problem was originally due to Zolotarev and is solved in terms of elliptic functions. See Todd [88], Fischer [92], and Peherstorfer [87].)

E.8 Lower Bound for the Norm of Polynomials on the Unit Disk. Let D be the open unit disk of $\mathbb{C}$. Show that

$$\|a_0 + a_1 z + \cdots + a_n z^n\|_D \geq \max_{0 \leq k \leq n} |a_k|$$

for arbitrary complex numbers $a_0, a_1, \ldots, a_n$. Thus z^n plays the role of the nth Chebyshev polynomial on the unit disk.

Hint: If $p(z) := a_0 + a_1 z + \cdots + a_n z^n$, then

$$a_m = \frac{1}{2\pi i} \int_{\partial D} \frac{p(z)}{z^{m+1}}\, dz\,.$$

□

The next exercise supposes some familiarity with the rudiments of reducibility over $\mathbb{Q}$ and basic properties, such as irreducibility of cyclotomic polynomials over $\mathbb{Q}$ (see Clark [71]). Details of the following observation of Schur's are in Rivlin [90].

E.9 On the Reducibility of T_n over $\mathbb{Q}$. Let $n \in \mathbb{N}$ be fixed.

a] The zeros of $T_n(x/2)$ are all of the form

$$x_j := e^{(2j-1)i\pi/(2n)} + e^{-(2j-1)i\pi/(2n)}\,, \qquad j = 1, 2, \ldots, n\,.$$

b] If $n \geq 3$ and ζ is a primitive nth root of unity, then $\zeta + \zeta^{-1}$ is of degree $\varphi(n)/2$. (Here φ is the Euler φ function.)

c] Thus if T_n is irreducible over $\mathbb{Q}$, then n must be a power of 2.

d] For a positive integer h, let

$$F_h(x) := \prod_{\substack{j=1 \\ \gcd(2j-1,2n)=h}}^{n} (x - x_j)\,.$$

(Here $\gcd(m,n)$ denotes the greatest common divisor of m and n.) Show that if h is odd, then F_h is irreducible over $\mathbb{Q}$.

e] The Factorization of T_n.

$$2T_n(x/2) = \prod_{h|n\,,\, h\,\mathrm{odd}} F_h(x)\,.$$

So if n is odd, T_n has $\varphi(n)$ factors, while if n is even, then T_n has $\varphi(m)$ factors, where m is the largest odd divisor of n.

f] Let $n \geq 3$ be odd. Then $T_n(x)/x$ is irreducible over $\mathbb{Q}$ if and only if n is prime.

E.10 Chebyshev Polynomials of the Second Kind. Let the *Chebyshev polynomials of the second kind* be defined by

$$U_{n-1}(x) := \frac{1}{n} T_n'(x) = \frac{\sin n\theta}{\sin \theta}, \qquad x = \cos\theta .$$

a] $U_n(x) = 2T_n(x) + U_{n-2}(x)$.

b] $T_n(x) = U_n(x) - xU_{n-1}(x)$.

c] $U_n(x) = \sum_{k=0}^{n} x^k T_{n-k}(x)$.

d] $U_n(x) = \dfrac{\left(x+\sqrt{x^2-1}\right)^{n+1} - \left(x-\sqrt{x^2-1}\right)^{n+1}}{2\sqrt{x^2-1}}$.

e] **Orthogonality.** Show that

$$\frac{2}{\pi}\int_{-1}^{1} U_n(x)U_m(x)\sqrt{1-x^2}\,dx = \delta_{n,m} := \begin{cases} 0, & n \neq m \\ 1, & n = m > 0 . \end{cases}$$

f] **Three-Term Recursion.** Show that

$$U_0(x) = 1, \quad U_1(x) = 2x,$$

$$U_n(x) = 2xU_{n-1}(x) - U_{n-2}(x), \quad n = 2, 3, \ldots .$$

(Note that this is the same recursion as for T_n.)

g] **The Coefficients of U_n.** Show that

$$U_n(x) = \sum_{k=0}^{\lfloor n/2 \rfloor} (-1)^k \binom{n-k}{k} (2x)^{n-2k} .$$

h] **Another Form of U_n.** Show that

$$U_n(x) = \sum_{k=0}^{\lfloor n/2 \rfloor} (-1)^k \binom{n+1}{2k+1} x^{n-2k}(x^2-1)^k .$$

The concepts of transfinite diameter and capacity play a central role in potential theory, harmonic analysis, and other areas of mathematics.

E.11 Transfinite Diameter. Let E be a compact subset of $\mathbb{C}$. Let

$$\Delta_n(E) := \max_{z_1,\dots,z_n\in E} \prod_{\substack{1\le i,j\le n\\ i\ne j}} |z_i - z_j|\,.$$

The points z_i at which the above maximum are obtained are called nth *Fekete points* for E. If the points z_i are the nth Fekete points for E, then the polynomial

$$q_n(z) := \prod_{i=1}^{n} (z - z_i)$$

is called an nth (monic) *Fekete polynomial* for E. The *transfinite diameter* or *logarithmic capacity* of E is defined by

$$\operatorname{cap}(E) := \lim_{n\to\infty} (\Delta_n(E))^{\frac{1}{n(n-1)}}\,,$$

where the limit exists by part c] (below).

a] Let $z_1, z_2, \dots, z_n$ be nth Fekete points for E. Then

$$(\Delta_n(E))^{1/2} = \operatorname{abs}\begin{vmatrix} 1 & z_1 & \dots & z_1^{n-1}\\ 1 & z_2 & \dots & z_2^{n-1}\\ \vdots & \vdots & \ddots & \vdots\\ 1 & z_n & \dots & z_n^{n-1}\end{vmatrix}\,.$$

Hint: See E.2 b] (Vandermonde determinant) of Section 3.2. □

b] Let $q_n(z) := \prod_{i=1}^{n}(z - z_i)$ be an nth Fekete polynomial for E. Let

$$m_n := \min_{i=1,\dots,n} |q_n'(z_i)| \quad\text{and}\quad M_n := \|q_n\|_E\,.$$

Then

$$M_n \le \left(\frac{\Delta_{n+1}(E)}{\Delta_n(E)}\right)^{1/2} \le m_{n+1} \le (\Delta_{n+1}(E))^{\frac{1}{n+1}}\,.$$

Outline. We have

$$|q_n(z)|^2\Delta_n(E) = \prod_{i=1}^{n} |z - z_i|^2 \prod_{\substack{1\le i,j\le n\\ i\ne j}} |z_i - z_j| \le \Delta_{n+1}(E)$$

and

$$\begin{aligned}\Delta_{n+1}(E) = &\prod_{\substack{1\le i,j\le n+1\\ i\ne k, j\ne k, i\ne j}} |z_i - z_j| \prod_{\substack{1\le i\le n+1\\ i\ne k}} |z_k - z_i|^2\\ \le &\Delta_n(E) \prod_{\substack{1\le i\le n+1\\ i\ne k}} |z_k - z_i|^2\,.\end{aligned}$$

From the first line above,

$$M_n^2 \Delta_n(E) \le \Delta_{n+1}(E)\,.$$

From the second line above,

$$\Delta_{n+1}(E) \le \Delta_n(E) m_{n+1}^2\,.$$

□

c] Show that $(\Delta_n(E))^{\frac{1}{n(n-1)}}$ is decreasing, so the limit exists in the definition of $\operatorname{cap}(E)$.

d] The Fekete points lie on the boundary of E. So $\operatorname{cap}(E) = \operatorname{cap}(\partial(E))$.

Hint: Use the maximum principle (see E.1 d] of Section 1.2). □

e] If $E \subset F$, where $F \subset \mathbb{C}$ is also compact, then $\operatorname{cap}(E) \le \operatorname{cap}(F)$.

f] Chebyshev Constants. Let

$$\mathcal{M}_n := \left\{ p \in \mathcal{P}_n^c : \; p(z) = \prod_{j=1}^{n} (z - z_j)\,, \quad z_j \in \mathbb{C} \right\}$$

and

$$\widetilde{\mathcal{M}}_n := \left\{ p \in \mathcal{P}_n^c : \; p(z) = \prod_{j=1}^{n} (z - z_j)\,, \quad z_j \in E \right\}.$$

Let

$$\mu_n(E) := \inf\{\|p\|_E : p \in \mathcal{M}_n\}$$

and

$$\widetilde{\mu}_n(E) := \inf\{\|p\|_E : p \in \widetilde{\mathcal{M}}_n\}\,.$$

Show that the infimum in the definition of $\mu_n(E)$ and $\widetilde{\mu}_n(E)$ is actually minimum. Show also that

$$\mu_{n+m}(E) \le \mu_n(E)\mu_m(E)$$

and

$$\widetilde{\mu}_{n+m}(E) \le \widetilde{\mu}_n(E)\widetilde{\mu}_m(E)$$

for any two nonnegative integers n and m. Finally show that the above inequalities imply that

$$\mu(E) := \lim_{n\to\infty} (\mu_n(E))^{1/n} \quad \text{and} \quad \widetilde{\mu}(E) := \lim_{n\to\infty} (\widetilde{\mu}_n(E))^{1/n}$$

exist.

The numbers $\mu(E)$ and $\widetilde{\mu}(E)$ are called the *Chebyshev constant* and *modified Chebyshev constant*, respectively, associated with E. Obviously $\mu(E) \leq \widetilde{\mu}(E)$.

g] Transfinite Diameter and Chebyshev Constants Are the Same:

$$\operatorname{cap}(E) = \mu(E) = \widetilde{\mu}(E)\,.$$

Proof. Without loss of generality we may assume that E contains infinitely many points. Part b] yields $\widetilde{\mu}(E) \leq \operatorname{cap}(E)$. Therefore, since $\mu(E) \leq \widetilde{\mu}(E)$, it is sufficient to prove that $\operatorname{cap}(E) \leq \mu(E)$. Note that if $p \in \mathcal{M}_n$ and $z_1, z_2, \ldots, z_{n+1} \in E$ are the $(n+1)$th Fekete points for E, then

$$(\Delta_{n+1}(E))^{1/2} = \operatorname{abs} \begin{vmatrix} 1 & z_1 & \cdots & z_1^{n-1} & p(z_1) \\ 1 & z_2 & \cdots & z_2^{n-1} & p(z_2) \\ \vdots & \vdots & \ddots & \vdots & \vdots \\ 1 & z_{n+1} & \cdots & z_{n+1}^{n-1} & p(z_{n+1}) \end{vmatrix}.$$

Expanding the above determinant with respect to its last column, we obtain

$$\begin{aligned}(\Delta_{n+1}(E))^{1/2} &\leq (\Delta_n(E))^{1/2} \sum_{j=1}^{n+1} |p(z_j)| \\ &\leq (n+1)(\Delta_n(E))^{1/2} \|p\|_E\,,\end{aligned}$$

so

$$(\Delta_{n+1}(E))^{1/2} \leq (n+1)(\Delta_n(E))^{1/2} \mu_n(E)\,.$$

For the sake of brevity let

$$c_n := \left((n+1)^2 (\mu_n(E))^2\right)^{1/n} \quad \text{and} \quad d_n := (\Delta_n(E))^{\frac{2}{n(n-1)}}\,.$$

Then

$$d_{n+1}^{n+1} \leq c_n d_n^{n-1}\,.$$

Since E contains infinitely many points, $c_n > 0$ and $d_n > 0$ hold for each $n = 2, 3, \ldots$. Multiplying the above inequalities for $n = 1, 2, \ldots, k$, we obtain after simplification that

$$(d_2 d_3 \cdots d_{k+1})^{\frac{1}{k-1}} (d_{k+1})^{\frac{k}{k-1}} \leq (d_2)^{\frac{2}{k-1}} (c_2 c_3 \cdots c_k)^{\frac{1}{k-1}}\,.$$

Since $\lim_{k\to\infty} d_k = \operatorname{cap}(E)$ and $\lim_{k\to\infty} c_k = (\mu(E))^2$, we conclude

$$(\operatorname{cap}(E))^2 \leq (\mu(E))^2\,,$$

which finishes the proof. □

h] Show that $\text{cap}([a,b]) = \frac{1}{4}(b-a)$.

i] Show that $\text{cap}(\overline{D}_\rho) = \rho$, where

$$\overline{D}_\rho := \{z \in \mathbb{C} : |z| \le \rho\}.$$

j] Show that $\text{cap}(A_\alpha) = \sin(\alpha/4)$, where A_α is an arc of the unit circle C of length α, $0 \le \alpha \le 2\pi$.

Hint: Without loss of generality we may assume that the arc A_α is symmetric with respect to the x-axis and $1 \in A_\alpha$. Now use part h] and the transformation $x = \frac{1}{2}(z + z^{-1})$. □

2.2 Orthogonal Functions

The most basic properties of orthogonal functions are explored in this section. The following section specializes the discussion to polynomials.

In this section the functions are complex-valued and the vector spaces are over the complex numbers. All the results have obvious real analogs and in many later applications we will restrict to these corresponding real cases.

An *inner product* on a vector space V is a function $\langle \cdot, \cdot \rangle$ from $V \times V$ to $\mathbb{C}$ that satisfies, for all $f, g, h \in V$ and $\alpha, \beta \in \mathbb{C}$,

(2.2.1) $\quad \langle f, f \rangle > 0$ unless $f = 0$ $\quad$ (*positivity*)

(2.2.2) $\quad \langle f, g \rangle = \overline{\langle g, f \rangle}$ $\quad$ (*conjugate symmetry*)

(2.2.3) $\quad \langle \alpha f + \beta g, h \rangle = \alpha \langle f, h \rangle + \beta \langle g, h \rangle$ $\quad$ (*linearity*).

A vector space V equipped with an inner product is called an *inner product space*. It is a normed linear space with the norm $\|\cdot\| := \langle \cdot, \cdot \rangle^{1/2}$.

The canonical example for us will be the space $C[a,b]$ of all complex-valued continuous functions on $[a,b]$ with the inner product

$$\langle f, g \rangle := \int_a^b f(x)\overline{g(x)}w(x)\,dx, \tag{2.2.4}$$

where $w(x)$ is a nonnegative integrable function on $[a,b]$ that is positive except possibly on a set of measure zero. It is a normed linear space with the norm

$$\|f\|_{L_2(w)} := \langle f, f \rangle^{1/2} = \left(\int_a^b |f(x)|^2 w(x)\,dx \right)^{1/2}. \tag{2.2.5}$$

More generally, if (X, μ) is a measure space (with μ nonnegative), then

$$\langle f, g \rangle := \int_X f(x)\overline{g(x)}\, d\mu(x) \tag{2.2.6}$$

is an inner product on the space $L_2(\mu)$ of square integrable functions. More precisely, $L_2(\mu)$ denotes the space of equivalence classes of measurable functions for which

$$\|f\|_{L_2(\mu)} := \langle f, f \rangle^{1/2} = \left\{ \int_X |f(x)|^2\, d\mu(x) \right\}^{1/2}$$

is finite. The equivalence classes are defined by the equivalence relation $f \sim g$ if $f = g$ μ-almost everywhere on X.

If V is a vector space equipped with an inner product $\langle \cdot, \cdot \rangle$, then a metric ρ can be defined on V by $\rho(f, g) := \langle f - g, f - g \rangle^{1/2}$. The fact that this ρ is a metric on V is an immediate consequence of (2.2.1) and Theorem 2.2.1 b]. If this metric space (V, ρ) is complete (that is, if every Cauchy sequence in (V, ρ) converges to some $x \in V$), then V is called a *Hilbert space.*

It can be shown that $L_2(\mu)$ is a Hilbert space for every measure space (X, μ) (see Rudin [87]), while $C[a, b]$ equipped with the inner product (2.2.4), where $w(x) \equiv 1$, is not a Hilbert space (see E.1).

When we write $L_2[a, b]$ we always mean $L_2(\mu)$ where μ is the Lebesgue measure on $X = [a, b]$. The fact that the inner product gives a norm is part of the next theorem.

Theorem 2.2.1. *If $(V, \langle \cdot, \cdot \rangle)$ is an inner product space equipped with the norm $\| \cdot \| := \langle \cdot, \cdot \rangle^{1/2}$, then for all $f, g \in V$,*

a] $|\langle f, g \rangle| \le \|f\|\, \|g\|$ *Cauchy-Schwarz inequality*

b] $\|f + g\| \le \|f\| + \|g\|$ *triangle inequality*

c] $\|f + g\|^2 + \|f - g\|^2 = 2\, \|f\|^2 + 2\, \|g\|^2$ *parallelogram law.*

Proof. Let $f, g \in V$ be arbitrary. To prove the Cauchy-Schwarz inequality, without loss of generality we may assume $\langle g, g \rangle = 1$ and we may assume $\langle f, g \rangle$ is real (why?). Let $\zeta := \langle f, g \rangle$ and note that by (2.2.1) and (2.2.3),

$$\begin{aligned} 0 \le \langle f - \zeta g, f - \zeta g \rangle &= \langle f, f \rangle - 2\zeta \langle f, g \rangle + \zeta^2 \langle g, g \rangle \\ &= \|f\|^2 - \langle f, g \rangle^2 , \end{aligned}$$

which finishes the proof of part a].

Using the Cauchy-Schwarz inequality, we obtain

$$\begin{aligned} \|f + g\|^2 = \langle f + g, f + g \rangle &= \langle f, f \rangle + 2\mathrm{Re}(\langle f, g \rangle) + \langle g, g \rangle \\ &\le \|f\|^2 + 2\, \|f\|\, \|g\| + \|g\|^2 \\ &\le (\|f\| + \|g\|)^2 , \end{aligned}$$

which is the triangle inequality.

The parallelogram law follows from

$$\|f+g\|^2+\|f-g\|^2$$
$$=\langle f,f\rangle+2\,\mathrm{Re}(\langle f,g\rangle)+\langle g,g\rangle+\langle f,f\rangle-2\,\mathrm{Re}(\langle f,g\rangle)+\langle g,g\rangle\,.$$

□

For the space $L_2(\mu)$ of all square integrable functions, the Cauchy-Schwarz inequality becomes

$$\left|\int_a^b f\overline{g}\,d\mu\right|\le\left(\int_a^b|f|^2\,d\mu(x)\right)^{1/2}\left(\int_a^b|g|^2\,d\mu\right)^{1/2}.$$

Applying this with f and g replaced by $|f|$ and $|g|$, we obtain

$$\int_a^b|fg|\,d\mu\le\left(\int_a^b|f|^2\,d\mu(x)\right)^{1/2}\left(\int_a^b|g|^2\,d\mu\right)^{1/2}. \tag{2.2.7}$$

A collection of vectors $\{f_\alpha:\alpha\in\mathcal{A}\}$ in an inner product space $(V,\langle\cdot,\cdot\rangle)$ is said to be *orthogonal* if

$$\langle f_\alpha,f_\beta\rangle=0\,,\qquad \alpha,\beta\in\mathcal{A},\ \ \alpha\neq\beta\,. \tag{2.2.8}$$

If $\langle f_\alpha,f_\beta\rangle=0$, then we write $f_\alpha\bot f_\beta$. The collection is called *orthonormal* if, in addition to being orthogonal,

$$\langle f_\alpha,f_\alpha\rangle=1\,,\qquad \alpha\in\mathcal{A}\,. \tag{2.2.9}$$

An orthogonal collection $\{f_\alpha:\alpha\in\mathcal{A}\}$ of nonzero vectors in an inner product space can always be orthonormalized as $\{\|f_\alpha\|^{-1}f_\alpha:\alpha\in\mathcal{A}\}$. The vector space over $\mathbb{C}$ generated by $\{f_\alpha:\alpha\in\mathcal{A}\}$ is denoted by

$$\mathrm{span}\{f_\alpha:\alpha\in\mathcal{A}\}\,.$$

So $\mathrm{span}\{f_\alpha:\alpha\in\mathcal{A}\}$ is just the set of all finite linear combinations

$$\left\{\sum_{i=1}^n c_if_{\alpha_i}:\alpha_i\in\mathcal{A},\ \ c_i\in\mathbb{C},\ \ n\in\mathbb{N}\right\}.$$

Any linearly independent collection of vectors can be orthonormalized, as the next theorem shows.

Theorem 2.2.2 (Gram-Schmidt). *Let $(V, \langle \cdot, \cdot \rangle)$ be an inner product space with norm $\| \cdot \| := \langle \cdot, \cdot \rangle^{1/2}$. Suppose $\{f_i\}_{i=1}^{\infty}$ is a linearly independent collection of vectors in V. Let*

$$g_1 := \frac{f_1}{\|f_1\|}$$

and (inductively) let

$$u_n := f_n - \sum_{k=1}^{n-1} \langle f_n, g_k \rangle g_k \qquad \text{and} \qquad g_n := \frac{u_n}{\|u_n\|}.$$

Then $\{g_n\}_{n=1}^{\infty}$ is an orthonormal collection, and for each n,

$$\operatorname{span}\{g_1, g_2, \dots, g_n\} = \operatorname{span}\{f_1, f_2, \dots, f_n\}.$$

Proof. This can be proved easily by induction where the inductive step is: for $m < n$,

$$\begin{aligned} \langle u_n, g_m \rangle &= \langle f_n, g_m \rangle - \sum_{k=1}^{n-1} \langle f_n, g_k \rangle \langle g_k, g_m \rangle \\ &= \langle f_n, g_m \rangle - \sum_{k=1}^{n-1} \langle f_n, g_k \rangle \delta_{k,m} = 0. \end{aligned}$$

□

The key approximation theoretic property orthonormal sets have is encapsulated in the following result:

Theorem 2.2.3 (Best Approximation by Linear Combinations). *Let $(V, \langle \cdot, \cdot \rangle)$ be an inner product space with norm $\| \cdot \| := \langle \cdot, \cdot \rangle^{1/2}$. Suppose $\{f_1, \dots, f_n\}$ is an orthonormal collection of vectors in V. Let $f \in V$. Then*

$$\min_{c_i \in \mathbb{C}} \left\| \sum_{i=1}^{n} c_i f_i - f \right\|$$

is attained if and only if

$$c_i = \langle f, f_i \rangle, \qquad i = 1, 2, \dots, n.$$

In other words, the sum $\sum_{i=1}^{n} \langle f, f_i \rangle f_i$ is the best approximation to f from $\operatorname{span}\{f_1, \dots, f_n\}$ in the norm $\langle \cdot, \cdot \rangle^{1/2}$.

Proof. Fix $f \in V$, and let c_i be as above. Let

$$g := \sum_{i=1}^{n} c_i f_i$$

and let $h \in \operatorname{span}\{f_1, \dots, f_n\}$. Note that

$$(g - f) \perp f_i\,, \qquad i = 1, 2, \dots, n$$

since by orthonormality

$$\langle g - f, f_i \rangle = \sum_{j=1}^{n} c_j \langle f_j, f_i \rangle - c_i = 0\,.$$

Thus

$$(g - f) \perp (h - g)$$

and so

$$\begin{aligned}
\|h - f\|^2 &= \|(h - g) + (g - f)\|^2 \\
&= \|h - g\|^2 + 2\mathrm{Re}(\langle h - g, g - f \rangle) + \|g - f\|^2 \\
&= \|h - g\|^2 + \|g - f\|^2 \\
&\geq \|g - f\|^2
\end{aligned}$$

with strict inequality unless $h = g$. This finishes the proof. □

Note that the above theorem gives the following corollary:

Corollary 2.2.4. *If $\{f_1, \dots, f_n\}$ is an orthonormal collection, then every*

$$g \in \operatorname{span}\{f_1, \dots, f_n\}$$

can be written as

$$g = \sum_{i=1}^{n} \langle g, f_i \rangle f_i\,.$$

Comments, Exercises, and Examples.

The theory of orthogonal functions, and in particular orthogonal polynomials, is old and far-reaching. As we will see in the next section, the names associated with the classical orthogonal polynomials including Chebyshev, Laguerre, Legendre, Hermite, Jacobi, and Stieltjes, are the "who's who" of nineteenth century analysis. Various aspects of this beautiful body of theory are explored in the exercises of this and the next section.

Much of this material is available in G. Szegő's [75] classical treatise "Orthogonal Polynomials." Of course, orthogonal polynomials are intimately connected to Fourier series and parts of harmonic and functional analysis generally. The standard functional analysis in the following exercises is available in many sources. See, for example, Rudin [73, 87].

E.1 $C[0,1]$ Is Not a Hilbert Space. Construct a sequence of continuous functions $(f_n)_{n=1}^{\infty}$ on $[0,1]$ for which

$$\|f_n - f\|_{L_2[0,1]} \to 0$$

with some $f \notin C[0,1]$ (in the sense that f cannot be modified on a set of measure zero to be in $C[0,1]$).

So $C[0,1]$ equipped with the inner product (2.2.4), where $[a,b] = [0,1]$ and $w(x)$ is identically 1, is not a Hilbert space. It can be shown that there is no way of putting a norm on $C[0,1]$ that preserves the uniform topology and makes $C[0,1]$ into a Hilbert space, essentially because $C[0,1]$ is not reflexive (see Rudin [73], Chapter 4). This, in fact, shows that $C[0,1]$ is not isomorphic to $L_p[0,1]$ for any $p \in (1,\infty)$. For the definition of $L_p[0,1]$, see E.7.

E.2 On $L_2(w)$. Consider

$$\langle f, g\rangle = \int_a^b f(x)g(x)w(x)\,dx\,.$$

What conditions on w guarantee that $\langle f, g\rangle$ is an inner product on $C[a,b]$?

E.3 Cauchy-Schwarz Inequality for Sequences. Show that

$$\left|\sum_{i=1}^{n} \alpha_i \beta_i\right|^2 \le \left(\sum_{i=1}^{n} |\alpha_i|^2\right)\left(\sum_{i=1}^{n} |\beta_i|^2\right)$$

for all $\alpha_1, \ldots, \alpha_n, \beta_1, \ldots, \beta_n \in \mathbb{C}$. Equality holds if and only if there exists a $\gamma \in \mathbb{C}$ so that either $\alpha_i = \gamma\overline{\beta}_i$ for each i or $\beta_i = \gamma\overline{\alpha}_i$ for each i.

Hint: $\mathbb{C}^n$ is a Hilbert space with inner product

$$\langle(\alpha_1, \alpha_2, \ldots, \alpha_n), (\beta_1, \beta_2, \ldots, \beta_n)\rangle = \sum_{i=1}^{n} \alpha_i \overline{\beta}_i\,.$$

□

E.4 Bessel's Inequality. Let $(V, \langle\cdot,\cdot\rangle)$ be an inner product space with norm $\|\cdot\| := \langle\cdot,\cdot\rangle^{1/2}$. Suppose $\{f_i\}_{i=1}^{\infty}$ is a countable collection of orthonormal vectors in V.

a] Show that

$$\sum_{i=1}^{\infty} |\langle f_i, f\rangle|^2 \le \|f\|^2\,.$$

Hint: With $h := 0$ in the last expression of the proof of Theorem 2.2.3

$$\|f\|^2 = \|g\|^2 + \|g - f\|^2\,.$$

□

b] Suppose

$$f = \sum_{i=1}^{\infty} \langle f_i, f \rangle f_i$$

in the sense that the partial sums of the right-hand side converge to f in the norm $\|\cdot\|$. Show that

$$\|f\|^2 = \sum_i |\langle f_i, f \rangle|^2 .$$

E.5 The Kernel Function. Let $\{p_i\}_{i=0}^n$ be a collection of orthonormal functions in $L_2[a,b]$ with respect to the inner product defined by (2.2.6), where $X := [a,b]$. Define the *kernel function* by

$$K_n(x_0, x) := \overline{p_0(x_0)}p_0(x) + \overline{p_1(x_0)}p_1(x) + \cdots + \overline{p_n(x_0)}p_n(x) .$$

a] Reproducing Property. If $q \in \operatorname{span}\{p_0, \dots, p_n\}$, then

$$\int_a^b K_n(t,x)q(t)\, d\mu(t) = q(x) .$$

Hint: Expand q in terms of $p_0, \dots, p_n$ as in Corollary 2.2.4. □

b] $(K_n(x_0,x_0))^{-1/2}K_n(x_0,x)$ solves the following maximization problem:

$$\max\left\{ |q(x_0)| : q \in \operatorname{span}\{p_0, p_1, \dots, p_n\} \text{ and } \int_a^b |q(x)|^2\, d\mu(x) = 1 \right\} .$$

Outline. Write $q = \sum_{i=0}^n c_i p_i$. Then, as in E.4 b],

$$\|q\|_{L_2(\mu)}^2 = |c_0|^2 + |c_1|^2 + \cdots + |c_n|^2 = 1 .$$

The Cauchy-Schwarz inequality of E.3 yields that

$$|q(x_0)|^2 \le \left(\sum_{i=0}^n |c_i|^2\right)\left(\sum_{i=0}^n |p_i(x_0)|^2\right) = K_n(x_0,x_0) .$$

However, if

$$c_i = \frac{\overline{p_i(x_0)}}{\left(\sum_{j=0}^n |p_j(x_0)|^2\right)^{1/2}} ,$$

so

$$q(x) = \frac{K_n(x_0,x)}{(K_n(x_0,x_0))^{1/2}} ,$$

then equality holds in the above inequality. □

c] Show, as in a], that if $q \in \operatorname{span}\{p_0, \dots, p_n\}$ and $p_0, \dots, p_n$ are m times differentiable at x_0, then

$$|q^{(m)}(x_0)| \le \left(\sum_{k=0}^m |p_k^{(m)}(x_0)|^2\right)^{1/2} \|q\|_{L_2(\mu)} .$$

When does equality hold?

E.6 Completeness. Let $\{f_\alpha : \alpha \in \mathcal{A}\}$ be an orthonormal collection in a Hilbert space H. The collection $\{f_\alpha : \alpha \in \mathcal{A}\}$ is called a *maximal orthonormal set* in H if there is no $f \neq 0$ so that $\langle f, f_\alpha \rangle = 0$ for every $\alpha \in \mathcal{A}$.

The following statements are equivalent:

(1) The set of all finite linear combinations of $f_\alpha, \alpha \in \mathcal{A}$, is dense in H.

(2) $\|f\|^2 = \sum_{\alpha \in \mathcal{A}} |\langle f_\alpha, f \rangle|^2$ for all $f \in H$.

(3) $\langle f, g \rangle = \sum_{\alpha \in \mathcal{A}} \langle f_\alpha, f \rangle \overline{\langle f_\alpha, g \rangle}$ for all $f, g \in H$.

(4) $\{f_\alpha : \alpha \in \mathcal{A}\}$ is a maximal orthonormal set in H.

If any of the above holds, then the orthonormal collection is called a *complete orthonormal system*. (See, for example, Rudin [87].)

a] Deduce (1) $\Rightarrow$ (2) from Theorem 2.2.3.

b] Deduce (2) $\Rightarrow$ (3) from the simple identity

$$4\langle f, g \rangle = \|f + g\|^2 - \|f - g\|^2 + i\|f + ig\|^2 - i\|f - ig\|^2 .$$

The above identity is called *polarization.*

c] Prove (3) $\Rightarrow$ (4).

d] Prove (4) $\Rightarrow$ (1) by contradiction.

Equality (3) is called *Parseval's identity.*

The remaining exercises assumes some familiarity with measure theory.

E.7 Basic Theory of L_p Spaces. Let (X, μ) be a measure space (μ is nonnegative) and $p \in (0, \infty]$. The space $L_p(\mu)$ is defined as the collection of equivalence classes of measurable functions for which $\|f\|_{L_p(\mu)} < \infty$, where

$$\|f\|_{L_p(\mu)} := \left(\int_X |f|^p \, d\mu \right)^{1/p}, \qquad p \in (0, \infty)$$

and

$$\|f\|_{L_\infty(\mu)} := \sup\{\alpha \in \mathbb{R} : \mu(\{x \in X : |f(x)| > \alpha\}) > 0\} < \infty .$$

In any of the cases the equivalence classes are defined by the equivalence relation $f \sim g$ if $f = g$ μ-almost everywhere on X. When we write $L_p[a, b]$ we always mean $L_p(\mu)$, where μ is the Lebesgue measure on $X = [a, b]$. The notations $L_p(a, b), L_p[a, b)$, and $L_p(a, b]$ are also used analogously to $L_p[a, b]$.

a] Hölder's Inequality. Suppose $1 \le p < q \le \infty$ and $p^{-1} + q^{-1} = 1$. Show that

$$\left| \int_X fg \, d\mu \right| \le \|f\|_{L_p(\mu)} \|g\|_{L_q(\mu)}$$

for every $f \in L_p(\mu)$ and $g \in L_q(\mu)$.

If $1 < p, q < \infty$, then equality holds if and only if $\alpha|f|^p = \beta|g|^q$ μ-almost everywhere on X for some $\alpha, \beta \in \mathbb{R}$ with $\alpha^2 + \beta^2 > 0$, and there is a $c \in \mathbb{C}$ with $|c| = 1$ so that cfg is nonnegative μ-almost everywhere on X.

Hölder's inequality was proved by Rogers [1888] before Hölder [1889] proved it independently.

Hint: If the right-hand side is 0, then the inequality is obvious. If it is different from 0, then let

$$F := \frac{|f|}{\|f\|_{L_p(\mu)}} \quad \text{and} \quad G := \frac{|g|}{\|g\|_{L_q(\mu)}} .$$

If $x \in X$ is such that $0 < F(x) < \infty$ and $0 < G(x) < \infty$, then there are real numbers s and t such that

$$F(x) = e^{s/p} \quad \text{and} \quad G(x) = e^{t/q} .$$

Use the convexity of the exponential function to show that

$$e^{s/p + t/q} \le p^{-1}e^s + q^{-1}e^t .$$

Apply this with the above choices of s and t, and integrate both sides on X with respect to μ. □

b] Minkowski's Inequality for $p \in [1, \infty]$. Let $p \in [1, \infty]$. Show that

$$\|f + g\|_{L_p(\mu)} \le \|f\|_{L_p(\mu)} + \|g\|_{L_p(\mu)}$$

for every $f, g \in L_p(\mu)$.

If $1 < p, q < \infty$, then equality holds if and only if $\alpha f = \beta g$ μ-almost everywhere on X for some $\alpha, \beta \in \mathbb{R}$ with $\alpha^2 + \beta^2 > 0$.

Hint: The cases $p = 1$ and $p = \infty$ are straightforward. Let $p \in (1, \infty)$. Then

$$|f + g|^p \le |f| \, |f + g|^{p-1} + |g| \, |f + g|^{p-1}$$

and apply Hölder's inequality (part a]) to each term separately. □

By part b], $L_p(\mu)$ is a vector space and $\|\cdot\|_{L_p(\mu)}$ is a norm on $L_p(\mu)$ whenever $p \in [1,\infty]$. If $p \in (0,1)$, then $\|\cdot\|_{L_p(\mu)}$ is still called a norm in the literature, however, for $p \in (0,1)$ the subadditive property, in general, fails. In fact, if $p \in (0,1)$, then $\|\cdot\|_{L_p[a,b]}$ is superadditive for Riemann integrable functions in $L_p[a,b]$; see Pólya and Szegő [76].

c] Assume $\mu(X) < \infty$. Show that $L_q(\mu) \subset L_p(\mu)$ for every $0 < p < q \leq \infty$. If $\mu(X) \leq 1$, then prove that

$$\|f\|_{L_p(\mu)} \leq \|f\|_{L_q(\mu)}$$

for every measurable function f.

d] Assume $f \in L_q(\mu)$ for some $q > 0$. Show that

$$\lim_{p\to\infty} \|f\|_{L_p(\mu)} = \|f\|_{L_\infty(\mu)}\,.$$

e] **Riesz-Fischer Theorem.** *Show that if* $1 \leq p \leq \infty$, *then* $(L_p(\mu), \rho)$ *is a complete metric space, where*

$$\rho(f,g) := \|f-g\|_{L_p(\mu)}\,.$$

Hint: Use the monotone convergence theorem and Minkowski's inequality (part b]); see Rudin [87] for details. □

If $p \in [1,\infty]$, then $q \in [1,\infty]$ defined by $p^{-1} + q^{-1} = 1$ is called *conjugate* to p.

f] **Bounded Linear Functionals on $L_p(\mu)$.** Let $1 \leq p < \infty$ and $g \in L_q(\mu)$, where q is the conjugate exponent to p. Show that

$$\Phi_g(f) := \int_X fg\,d\mu$$

is a bounded linear functional on $L_p(\mu)$.

Hint: Use Hölder's inequality (part a]). □

g] **Riesz Representation Theorem.** *Suppose* $1 \leq p < \infty$, μ *is* (σ-) *finite and* Φ *is a bounded linear functional on* $L_p(\mu)$. *Then there is a unique* $g \in L_q(\mu)$, *where* q *is the conjugate exponent to* p, *so that*

$$\Phi(f) = \int_X fg\,d\mu\,, \qquad f \in L_p(\mu)\,.$$

Moreover, if Φ *and* g *are related as above, then*

$$\|\Phi\| := \max\{\Phi(f) : f \in L_p(\mu),\ \|f\|_{L_p(\mu)} = 1\} = \|g\|_{L_q(\mu)}\,.$$

Proof. See, for example, Rudin [87] or Royden [88]. □

If X is a locally compact Hausdorff space, then the characterization of bounded linear functionals on the space $C_c(X)$ of continuous functions with compact support equipped with the uniform norm is also known as the Riesz representation theorem, and its proof may be found in, for example, Rudin [87].

h] Orthogonality in $L_p(\mu)$. Suppose $1 \le p < \infty, \mu$ is (σ-)finite, Y is a finite-dimensional subspace of $L_p(\mu)$. The function $f \in L_p(\mu)$ is said to be *orthogonal* to Y in $L_p(\mu)$, written $f \perp Y$, if

$$\|f\|_{L_p(\mu)} \le \|f + h\|_{L_p(\mu)}$$

for every $h \in Y$. Show that an element $f \in L_p(\mu)$ is orthogonal to Y if and only if

$$\int_X |f|^{p-1}\overline{\text{sign}}(f)h \, d\mu = 0$$

for every $h \in Y$, where

$$\overline{\text{sign}}(f(x)) := \begin{cases} \dfrac{\overline{f(x)}}{|f(x)|} & \text{if } f(x) \ne 0 \\ 0 & \text{if } f(x) = 0 \, . \end{cases}$$

Outline. Suppose that the integral vanishes for every $h \in Y$. Let q be the conjugate exponent to p. defined by $p^{-1} + q^{-1} = 1$. Observe that

$$g := |f|^{p-1}\,\overline{\text{sign}}(f) \in L_q(\mu)$$

and

$$\int_X |g|^q d\mu = \int_X |f|^p \, d\mu \, .$$

Without loss of generality we may assume that $\|f\|_{L_p(\mu)} = 1$. Then for every $h \in Y$, Hölder's inequality yields that

$$\begin{aligned} \|f\|_{L_p(\mu)} = 1 = \int_X fg \, d\mu &= \int_X (f+h)g \, d\mu \\ &\le \|f+h\|_{L_p(\mu)}\|g\|_{L_q(\mu)} = \|f+h\|_{L_p(\mu)} \, , \end{aligned}$$

proving that $f \perp Y$. (Observe that this argument is also valid for $p = 1$.)

Suppose now $f \perp Y$. Without loss of generality we may assume that $f \notin Y$. By a standard corollary to the Hahn-Banach theorem (see, for example, Rudin [87]), there exists a linear functional M on $L_p(\mu)$ such that $M(f) = 1$, $M(h) = 0$ for every $h \in Y$, and $\|M\| = \|f\|^{-1}_{L_p(\mu)}$. This M is then representable by some element $g \in L_q(\mu)$, that is,

$$M(f) = \int_X fg \, d\mu, \qquad \|g\|_{L_p(\mu)} = \|f\|^{-1}_{L_p(\mu)}$$

(see part g]). Therefore

$$\int_X fg\,d\mu = \|f\|_{L_p(\mu)}\,\|g\|_{L_q(\mu)}\,,$$

and by the conditions for equality to hold in Hölder's inequality,

$$g(x)f(x) \geq 0 \qquad a.e.\ [\mu]\ on\ X$$

and

$$|g(x)|^q = \lambda|f(x)|^p \qquad a.e.\ [\mu]\ on\ X$$

for a suitable constant $\lambda > 0$. Hence

$$g(x) = \lambda|f(x)|^{p-1}\,\overline{\text{sign}}(f(x))\,,$$

and so $M(h) = 0$, $h \in Y$, implies

$$\int_X |f|^{p-1}\,\overline{\text{sign}}(f)h\,d\mu = 0$$

for every $h \in Y$. □

The statement of part h] remains valid for closed subspaces Y instead of finite-dimensional subspaces, see, for example, Shapiro [71].

i] Minkowski's Inequality for $p \in (0,1)$. Show that

$$\|f+g\|_{L_p(\mu)} \leq 2^{1/p-1}\left(\|f\|_{L_p(\mu)} + \|g\|_{L_p(\mu)}\right)$$

for every $f, g \in L_p(\mu)$ and $p \in (0,1)$.

Hint: Verify that

$$\begin{aligned}\|f+g\|_{L_p(\mu)} &\leq \left(\int_X (|f|+|g|)^p\,d\mu\right)^{1/p}\\ &\leq \left(\int_X |f|^p\,d\mu + \int_X |g|^p\,d\mu\right)^{1/p}\\ &\leq 2^{1/p-1}\left(\left(\int_X |f|^p\,d\mu\right)^{1/p} + \left(\int_X |g|^p\,d\mu\right)^{1/p}\right)\end{aligned}$$

whenever $f, g \in L_p(\mu)$ and $p \in (0,1)$. □

Further properties of $L_p(\mu)$ spaces may be found in Rudin [87].

E.8 Fourier Series.

a] Show that

$$\left\{ \frac{e^{in\theta}}{\sqrt{2\pi}} : n \in \mathbb{Z} \right\}$$

is a maximal orthonormal collection in $L_2[-\pi, \pi]$.

Hint: The orthonormality is obvious. In order to show the maximality, first note that $L_2[-\pi, \pi]$ is a Hilbert space by the Riesz-Fischer theorem (E.7 e]). Hence, by E.6, it is sufficient to show that the set $\mathcal{T}^c$ of all complex trigonometric polynomials is dense in $L_2[-\pi, \pi]$. By the Stone-Weierstrass theorem (E.2 of Section 4.1) $\mathcal{T}^c$ is dense in $C^*[-\pi, \pi]$, where $C^*[-\pi, \pi]$ is the space of all complex-valued 2π-periodic continuous functions on $\mathbb{R}$ equipped with the uniform norm on $\mathbb{R}$. Finally, it is a standard measure theoretic argument to show that $C^*[-\pi, \pi]$ is dense in $L_2[-\pi, \pi]$; see, for example, Rudin [87]. □

The kth *Fourier coefficient* $\widehat{f}(k)$ of a function $f \in L_1[-\pi, \pi]$ is defined by

$$\widehat{f}(k) := \frac{1}{2\pi} \int_{-\pi}^{\pi} f(\theta) e^{-ik\theta}\, d\theta\,.$$

The (formal) *Fourier series* of a function $f \in L_1[-\pi, \pi]$ is defined by

$$f \sim \sum_{k=-\infty}^{\infty} \widehat{f}(k) e^{ik\theta}\,.$$

The functions

$$S_n(\theta) := \sum_{k=-n}^{n} \widehat{f}(k) e^{ik\theta}$$

are called the nth *partial sums* of the Fourier series of f.

b] Show that if $f \in L_2[-\pi, \pi]$, then

$$\sum_{k=-\infty}^{\infty} |\widehat{f}(k)|^2 = \|f\|^2_{L_2[-\pi,\pi]} < \infty\,.$$

Hint: Use part a], E.6, and E.7 e]. □

c] Show that if $f \in L_2[-\pi, \pi]$, then

$$\lim_{n \to \infty} \|f - S_n\|_{L_2[-\pi,\pi]} = 0\,,$$

so f is the $L_2[-\pi, \pi]$ limit of the partial sums of its formal Fourier series.

Hint: Use part b]. □

Carleson, in 1966, solved Luzin's problem by showing that $S_n \to f$ almost everywhere on $[-\pi, \pi]$ for every $f \in L_2[-\pi, \pi]$. Earlier, Kolmogorov showed that there is a function $f \in L_1[-\pi, \pi]$ so that S_n diverges everywhere on $[-\pi, \pi]$.

d] Isometry of $L_2[-\pi, \pi]$ and ℓ_2^c. Let

$$\ell_2^c := \left\{ \mathbf{x} = (x_k)_{k=-\infty}^{\infty} : x_k \in \mathbb{C}, \quad \sum_{k=-\infty}^{\infty} |x_k|^2 < \infty \right\}$$

and

$$\|\mathbf{x}\|_{\ell_2^c} := \left(\sum_{k=-\infty}^{\infty} |x_k|^2 \right)^{1/2}, \qquad \mathbf{x} = (x_k)_{-\infty}^{\infty}, \quad x_k \in \mathbb{C}.$$

Show that the function $I : L_2[-\pi, \pi] \to \ell_2^c$ defined by

$$I(f) := \widehat{f} := (\widehat{f}(k))_{-\infty}^{\infty}$$

is one-to-one and onto, and

$$\|I(f)\|_{\ell_2^c} = \|f\|_{L_2[-\pi,\pi]}.$$

Hint: Use part a] to show that I is one-to-one. Use the Riesz-Fischer theorem (E.7 e]) to show that I is onto. The norm-preserving property is the content of part b]. □

Part d] shows that the structure of $L_2[-\pi, \pi]$ is the same as that of ℓ_2^c. Hence $L_2[-\pi, \pi]$ is a separable Hilbert space, that is, it has a countable dense subset. So if $\epsilon > 0$ is fixed, then any collection $\{f_\alpha : \alpha \in \mathcal{A}\}$ from $L_2[-\pi, \pi]$ for which

$$\|f_\alpha - f_\beta\|_{L_2[-\pi,\pi]} \geq \epsilon, \qquad \alpha, \beta \in \mathcal{A}, \ \alpha \neq \beta$$

must be countable.

e] The Riemann-Lebesgue Lemma. *If $f \in L_1[-\pi, \pi]$, then $\widehat{f}(k) \to 0$ as $k \to \infty$.*

Hint: First prove it for step functions, then extend the result to every $f \in L_1[-\pi, \pi]$ by using the fact that step functions form a dense set in $L_1[-\pi, \pi]$. □

f] Show that

$$\sum_{n=1}^{\infty} \frac{1}{n^2} = \frac{\pi^2}{6}, \qquad \sum_{n=1}^{\infty} \frac{1}{n^4} = \frac{\pi^4}{90}, \quad \text{and} \quad \sum_{n=1}^{\infty} \frac{1}{n^{2k}} = r_k \pi^{2k}$$

for every $k = 3, 4, \cdots$, where r_k is a rational number.

Hint: Let f be the 2π periodic function defined by

$$f(\theta) := \left(\frac{\pi - \theta}{2} \right)^k, \qquad \theta \in [0, 2\pi).$$

Apply part b]. □

E.9 Denseness of Polynomials in $L_2(\mu)$ on $\mathbb{R}$.

a] Let μ be a finite Borel measure on $[a,b]$ and $f \in L_1(\mu)$. Show that if

$$\int_a^b f(x)e^{itx}\,d\mu(x) = 0\,, \qquad t = \frac{2\pi k}{b-a}\,, \quad k = 0, \pm 1, \pm 2, \ldots\,,$$

then $f(x) = 0$ a.e. $[\mu]$ on $[a,b]$.

Outline. Use the fact that the set $\mathcal{T}^c$ of all complex trigonometric polynomials is dense in $C^*[-\pi,\pi]$ (see the hint given for E.8 a]) and standard measure theoretic arguments to show that the assumption of part a] implies

$$\int_a^b f(x)g(x)\,d\mu(x) = 0$$

for every bounded measurable function g defined on $[a,b]$. Now, choosing

$$g(x) := \overline{\text{sign}}(f(x)) := \begin{cases} \dfrac{\overline{f(x)}}{|f(x)|} & \text{if } f(x) \neq 0 \\ 0 & \text{if } f(x) = 0\,, \end{cases}$$

we obtain

$$\int_{-\pi}^{\pi} |f(x)|\,d\mu(x) = 0\,,$$

and the result follows. □

b] Let μ be a finite Borel measure on $\mathbb{R}$ and $f \in L_1(\mu)$. Show that if

$$\int_{\mathbb{R}} f(x)e^{itx}\,d\mu(x) = 0\,, \qquad t \in \mathbb{R}\,,$$

then $f(x) = 0$ a.e. $[\mu]$ on $\mathbb{R}$.

Hint: Use part a] to show that the assumption of part b] implies

$$\int_{\mathbb{R}} f(x)g(x)\,d\mu(x) = 0$$

for every bounded measurable function defined on $\mathbb{R}$ (first assume that g has compact support, then eliminate this assumption). Finish the proof as in part a]. □

c] Let μ be a Borel measure on $\mathbb{R}$ satisfying

$$\int_{\mathbb{R}} e^{r|x|}\,d\mu(x) < \infty$$

with some $r > 0$. Show that the set $\mathcal{P}^c$ of all complex algebraic polynomials is dense in $L_2(\mu)$.

Outline. First observe that the assumption on μ implies $\mathcal{P}^c \subset L_2(\mu)$. The fact that $L_2(\mu)$ is a Hilbert space (see E.7 e]), Theorem 2.2.2 (Gram-Schmidt), and E.6 imply that it is sufficient to prove that if $f \in L_2(\mu)$ and

$$\int_{\mathbb{R}} f(x)x^k \, d\mu(x) = 0\,, \qquad k = 0, 1, 2, \dots\,,$$

then $f(x) = 0$ a.e. $[\mu]$ on $\mathbb{R}$. Assume that $f \in L_2(\mu)$ satisfies the above orthogonality relation. Use Theorem 2.2.1 a] (Cauchy-Schwarz inequality) to show that

$$F(t) := \int_{\mathbb{R}} f(x)e^{-itx} \, d\mu(x)$$

is well-defined on $\mathbb{R}$. For every $t_0 \in \mathbb{R}$, we have

$$\begin{aligned} f(x)e^{-itx} &= f(x)e^{-it_0x}e^{-i(t-t_0)x} \\ &= \sum_{k=0}^{\infty} (-i)^k \frac{(t-t_0)^k}{k!} f(x)e^{-it_0x}x^k\,. \end{aligned}$$

Note that if $|t - t_0| \le r/2$, then the integral of the right-hand side with respect to $\mu(x)$ on $\mathbb{R}$ can be calculated by integrating term by term since

$$\begin{aligned} &\sum_{k=0}^{\infty} \int_{\mathbb{R}} \left| (-i)^k \frac{(t-t_0)^k}{k!} f(x)e^{-it_0x}x^k \right| d\mu(x) \\ &\quad \le \sum_{k=0}^{\infty} \int_{\mathbb{R}} \frac{|t-t_0|^k}{k!} |f(x)||x|^k \, d\mu(x) \\ &\quad = \int_{\mathbb{R}} |f(x)|e^{|t-t_0||x|} \, d\mu(x) \\ &\quad \le \left| \int_{\mathbb{R}} |f(x)|^2 \, d\mu(x) \int_{\mathbb{R}} e^{2|t-t_0||x|} \, d\mu(x) \right|^{1/2} < \infty\,. \end{aligned}$$

Therefore, if $|t - t_0| \le r/2$, then

$$F(t) = \sum_{k=0}^{\infty} (-i)^k \frac{(t-t_0)^k}{k!} \int_{\mathbb{R}} f(x)e^{-it_0x}x^k \, d\mu(x)\,.$$

This means that F has a Taylor series expansion about every $t_0 \in \mathbb{R}$ with radius of convergence at least $r/2$. Also, with the choice $t_0 = 0$, by the assumed orthogonality relations, we have $F(t) = 0$ whenever $|t| \le r/2$. We can now deduce that $F(t) = 0$ for every $t \in \mathbb{R}$. Hence it follows from part b] that $f(x) = 0$ a.e. $[\mu]$ on $\mathbb{R}$. □

2.3 Orthogonal Polynomials

The classical orthogonal polynomials arise on orthogonalizing the sequence

$$(1, x, x^2, \dots)$$

with respect to various particularly nice weights, $w(x)$, on an interval, which, after a linear transformation, may be taken to be one of $[-1,1]$, $[0,\infty)$, or $(-\infty,\infty)$. The main examples we consider are the *Jacobi polynomials*

$$P_n^{(\alpha,\beta)}(x)\,, \text{ where } w(x) := (1-x)^\alpha(1+x)^\beta \text{ on } [-1,1]\,, \quad \alpha,\beta > -1\,. \tag{2.3.1}$$

When $\alpha = \beta = -1/2$ the Jacobi polynomials are the *Chebyshev polynomials of the first kind,*

$$T_n(x)\,, \text{ where } w(x) := (1-x^2)^{-1/2} \text{ on } [-1,1]\,. \tag{2.3.2}$$

When $\alpha = \beta = 1/2$ they are the *Chebyshev polynomials of the second kind,*

$$U_n(x)\,, \text{ where } w(x) := (1-x^2)^{1/2} \text{ on } [-1,1]\,. \tag{2.3.3}$$

Another special case of importance is $\alpha = \beta = 0$, which gives the *Legendre polynomials,*

$$P_n(x)\,, \text{ where } w(x) = 1 \text{ on } [-1,1]\,. \tag{2.3.4}$$

The *Laguerre polynomials* are

$$L_n(x)\,, \text{ where } w(x) := e^{-x} \text{ on } [0,\infty)\,. \tag{2.3.5}$$

The *Hermite polynomials* are

$$H_n(x)\,, \text{ where } w(x) := e^{-x^2} \text{ on } (-\infty,\infty)\,. \tag{2.3.6}$$

The above notation is traditionally used to denote orthogonal polynomials with a standard normalization; see the exercises. It is not usually the case that this normalization gives orthonormality. All of these much studied polynomials arise naturally, as do all the special functions, in the study of differential equations. We catalog some of the special properties of these classical orthogonal polynomials in the exercises.

In general, a nondecreasing bounded function α (typically the distribution function of a finite measure) defined on $\mathbb{R}$ is called an *m-distribution* if it takes infinitely many distinct values, and its *moments*, that is, the improper Stieltjes integrals

$$\int_{-\infty}^{\infty} x^n \, d\alpha(x) = \lim_{\substack{\omega_1 \to -\infty \\ \omega_2 \to +\infty}} \int_{\omega_1}^{\omega_2} x^n \, d\alpha(x) \,,$$

exist and are finite for $n = 0, 1, \dots$.

Theorem 2.3.1 (Existence and Uniqueness of Orthonormal Polynomials). *For every m-distribution α there is a unique sequence of polynomials $(p_n)_{n=0}^{\infty}$ with the following properties:*

(*i*) $$p_n(x) = \gamma_n x^n + r_{n-1}(x) \,, \qquad \gamma_n > 0 \,, \quad r_{n-1} \in \mathcal{P}_{n-1} \,,$$

(*ii*) $$\int_{\mathbb{R}} p_n(x) p_m(x) \, d\alpha(x) = \delta_{n,m} = \begin{cases} 1 & \text{for } n = m \\ 0 & \text{for } n \neq m \,. \end{cases}$$

Proof. The result follows from Theorem 2.2.2 (Gram-Schmidt). Note that the defining property of an m-distribution α ensures that

$$\langle p, q \rangle := \int_{\mathbb{R}} pq \, d\alpha$$

is an inner product on $\mathcal{P}_n^c$. □

The sequence $(p_n)_{n=0}^{\infty}$ defined by Theorem 2.3.1 is called the *sequence of orthonormal polynomials* associated with an m-distribution α. The sequence $(q_n)_{n=0}^{\infty}$ is called a *sequence of orthogonal polynomials* associated with an m-distribution α if

$$q_n = c_n p_n \,, \quad 0 \neq c_n \in \mathbb{C} \,, \quad n = 0, 1, \dots \,,$$

where $(p_n)_{n=0}^{\infty}$ is the sequence of orthonormal polynomials associated with α. The support $\operatorname{supp}(\alpha)$ of an m-distribution α is defined as the closure of the set

$$\{x \in \mathbb{R} : \alpha \text{ is increasing at } x\} \,.$$

If α is absolutely continuous on $\mathbb{R}$, then

$$d\alpha(x) = w(x) \, dx \quad \text{with some} \quad 0 \leq w \in L_1(\infty, \infty)$$

in which case α may be identified as a nonnegative weight function $w \in L_1(-\infty, \infty)$ whose integral takes infinitely many distinct values. If (a, b) is an interval and $w \in L_1[a, b]$ has an integral that takes infinitely many distinct values, then the sequence of orthogonal (orthonormal) polynomials associated with

$$\widetilde{w}(x) = \begin{cases} w(x) & \text{if } x \in (a, b) \\ 0 & \text{if } x \notin (a, b) \end{cases}$$

is said to be *orthogonal (orthonormal) with respect to the weight w.*

One thing distinguishing orthogonal polynomials from general orthogonal systems is the existence of a three-term recursion.

Theorem 2.3.2 (Three-Term Recursion). *Suppose $(p_n)_{n=0}^\infty$ is a sequence of orthonormal polynomials with respect to an m-distribution α. Then*

$$xp_n(x) = a_n p_{n+1}(x) + b_n p_n(x) + a_{n-1} p_{n-1}(x)\,, \qquad n = 0, 1, \dots\,,$$

where

$$p_{-1} := 0\,, \quad a_{-1} = 0\,, \quad a_n = \frac{\gamma_n}{\gamma_{n+1}} > 0\,, \quad b_n \in \mathbb{R}\,, \quad n = 0, 1, \dots$$

(γ_n is the leading coefficient of p_n).

This theorem has a converse due to Favard [35]; see E.12.

Proof. Since $xp_n(x) \in \mathcal{P}_{n+1}$, we may write

$$xp_n(x) = \sum_{k=0}^{n+1} d_k p_k(x)\,, \qquad d_k \in \mathbb{R}\,. \tag{2.3.7}$$

For notational convenience, let

$$\langle p, q \rangle := \int_{\mathbb{R}} p(x) q(x)\, d\alpha(x)$$

for any two polynomials p and q. Since $\langle p_n, q \rangle = 0$ for every $q \in \mathcal{P}_{n-1}$, we have

$$\langle xp_n(x), q(x) \rangle = \langle p_n(x), xq(x) \rangle = 0$$

for every $q \in \mathcal{P}_{n-2}$. In particular,

$$\langle xp_n(x), p_k(x) \rangle = 0\,, \qquad k = 0, 1, \dots, n-2\,.$$

On the other hand, using (2.3.7) and the orthonormality of $(p_n)_{n=0}^\infty$, we obtain

$$\langle xp_n(x), p_k(x) \rangle = d_k \langle p_k, p_k \rangle = d_k\,.$$

Hence $d_k = 0$ for each $k = 0, 1, \dots, n-2$ and

$$xp_n(x) = d_{n+1} p_{n+1}(x) + d_n p_n(x) + d_{n-1} p_{n-1}(x)\,. \tag{2.3.8}$$

Here the lead coefficient of the left-hand side polynomial is γ_n, while the lead coefficient of the right-hand side polynomial is $d_{n+1}\gamma_{n+1}$, so

$$a_n := d_{n+1} = \gamma_n / \gamma_{n+1}\,.$$

In order to show that $a_{n-1} := d_{n-1} = \gamma_{n-1}/\gamma_n$, note that (2.3.8) and the orthonormality of $(p_n)_{n=0}^\infty$ imply

$$\begin{aligned}
0 &= \langle p_{n+1}, p_{n-1}\rangle \\
&= \frac{1}{d_{n+1}}\langle xp_n(x), p_{n-1}(x)\rangle - \frac{d_n}{d_{n+1}}\langle p_n, p_{n-1}\rangle - \frac{d_{n-1}}{d_{n+1}}\langle p_{n-1}, p_{n-1}\rangle \\
&= \frac{1}{d_{n+1}}\langle p_n(x), \gamma_{n-1}x^n\rangle - \frac{d_{n-1}}{d_{n+1}} \\
&= \frac{1}{d_{n+1}}\frac{\gamma_{n-1}}{\gamma_n} - \frac{d_{n-1}}{d_{n+1}}.
\end{aligned}$$

Hence

$$a_{n-1} := d_{n-1} = \frac{\gamma_{n-1}}{\gamma_n}.$$

□

Theorem 2.3.3 (Christoffel-Darboux Formula). *With the notation of the previous theorem,*

$$\sum_{k=0}^{n} p_k(x)p_k(y) = \frac{\gamma_n}{\gamma_{n+1}}\left(\frac{p_{n+1}(x)p_n(y) - p_n(x)p_{n+1}(y)}{x-y}\right)$$

for all $x \neq y \in \mathbb{C}$.

Proof. Theorem 2.3.2 (three-term recursion) yields that

$$\begin{aligned}
\Delta_k :&= p_{k+1}(x)p_k(y) - p_k(x)p_{k+1}(y) \\
&= \frac{1}{a_k}(x-y)p_k(x)p_k(y) + \frac{a_{k-1}}{a_k}\left(p_k(x)p_{k-1}(y) - p_{k-1}(x)p_k(y)\right).
\end{aligned}$$

So

$$a_k\frac{\Delta_k}{x-y} = p_k(x)p_k(y) + a_{k-1}\frac{\Delta_{k-1}}{x-y},$$

and we sum the above from 0 to n to get the desired formula. □

Corollary 2.3.4. *In the notation of Theorem 2.3.2*

$$\sum_{k=0}^{n} p_k^2(x) = \frac{\gamma_n}{\gamma_{n+1}}\left(p'_{n+1}(x)p_n(x) - p'_n(x)p_{n+1}(x)\right).$$

Proof. Let $y \to x$ in Theorem 2.3.3. □

We can deduce quite easily from this that orthogonal polynomials associated with an m-distribution α have real interlacing zeros lying in the interior of the smallest interval containing $\text{supp}(\alpha)$; see E.1 and E.2.

Comments, Exercises, and Examples.

Askey, in comments following an outline of the history of orthogonal polynomials by Szegő [82, vol. III], writes:

"The classical orthogonal polynomials are mostly attributed to someone other than the person who introduced them. Szegő refers to Abel and Lagrange and Tschebyscheff in [75, chapter 5] for work on the Laguerre polynomials $L_n^0(x)$. Abel's work was published posthumously in 1881. Probably the first published work on these polynomials that uses their orthonormality was by Murphy (1833). Hermite polynomials were studied extensively by Laplace in connection with work on probability theory. Hermite's real contribution to these polynomials was to introduce Hermite polynomials in several variables. Lagrange came across the recurrence relation for Legendre polynomials."

Perhaps this is not very surprising given the many diverse ways in which these polynomials can arise.

There are many sources for the basic properties of orthogonal polynomials. In particular, Askey and Ismail [84], Chihara [78], Erdélyi et al. [53], Freud [71], Nevai [79b], [86], Szegő [75], and, in tabular form, Abramowitz and Stegun [65] are such sources. Exercises include a treatment of the elementary properties of the most familiar orthogonal polynomials. The connections linking orthogonal polynomials, the moment problem, and Favard's converse theorem to the three-term recursion are also examined in the exercises.

E.1 Simple Real Zeros. Let $(p_n)_{n=0}^{\infty}$ be the sequence of orthonormal polynomials associated with an m-distribution α. Show that each p_n has exactly n simple real zeros lying in the interior of the smallest interval containing $\operatorname{supp}(\alpha)$.

Hint: Suppose the statement is false. Then p_n has at most $n-1$ sign changes on $[a,b]$, hence there exists $0 \neq q \in \mathcal{P}_{n-1}$ so that

$$p_n(x)q(x) \geq 0\,, \qquad x \in [a,b]\,.$$

Show that this contradicts the orthogonality relation

$$0 = \int_{\mathbb{R}} p_n(x)q(x)\,d\alpha(x) = \int_a^b p_n(x)q(x)\,d\alpha(x)\,.$$

□

E.2 Interlacing of Zeros. Let $(p_n)_{n=0}^{\infty}$ be the sequence of orthonormal polynomials associated with an m-distribution α. Then the zeros of p_n and p_{n+1} strictly interlace. That is, there is exactly one zero of p_n strictly between any two consecutive zeros of p_{n+1}.

Hint: From Corollary 2.3.4,

$$p'_{n+1}(x)p_n(x) - p'_n(x)p_{n+1}(x)$$

is positive on $\mathbb{R}$. Since p_{n+1} has $n+1$ simple real zeros (see E.1), we see that if γ and δ are two consecutive zeros of p_{n+1}, then

$$\operatorname{sign}(p'_{n+1}(\gamma)) = -\operatorname{sign}(p'_{n+1}(\delta)),$$

and hence

$$\operatorname{sign}(p_n(\gamma)) = -\operatorname{sign}(p_n(\delta)).$$

□

E.3 Orthogonality of $(K_n(x_0, x))_{n=0}^\infty$. Let $(p_n)_{n=0}^\infty$ be the sequence of orthonormal polynomials associated with an m-distribution α. Let

$$x_0 < \min \operatorname{supp}(\alpha) \quad \text{or} \quad x_0 > \max \operatorname{supp}(\alpha).$$

Let $(K_n(x_0, x))_{n=0}^\infty$ be the sequence of associated kernel functions (as in E.5 of Section 2.2). Show that

$$\int_{\mathbb{R}} K_n(x_0, x) K_m(x_0, x) |x - x_0| \, d\alpha(x) = 0,$$

for any two nonnegative integers $n \neq m$.

E.4 Hypergeometric Functions. We introduce the following standard notation: the *rising factorial (or Pochammer symbol)*

$$(a)_n := a(a+1)\cdots(a+n-1), \qquad (a)_0 := 1$$

for $a \in \mathbb{C}$ and $n = 1, 2, \ldots$; the *binomial coefficient*

$$\binom{a}{n} := \frac{a(a-1)\cdots(a-n+1)}{n!}, \qquad \binom{a}{0} := 1$$

for $a \in \mathbb{C}$ and $n = 1, 2, \ldots$; and the *Gaussian hypergeometric series*

$${}_2F_1(a, b; c; z) := F(a, b; c; z) := \sum_{n=0}^\infty \frac{(a)_n (b)_n}{(c)_n} \frac{z^n}{n!}$$

for $a, b, c \in \mathbb{C}$.

a] For $\operatorname{Re}(c) > \operatorname{Re}(b) > 0$,

$$F(a, b; c; z) = \frac{\Gamma(c)}{\Gamma(b)\Gamma(c-b)} \int_0^1 t^{b-1}(1-t)^{c-b-1}(1-tz)^{-a} \, dt,$$

where Γ is, as usual, the *gamma function* defined by

$$\Gamma(z) := \int_0^\infty t^{z-1} e^{-t}\, dt\,, \qquad \mathrm{Re}(z) > 0\,.$$

Proof. See, for example, Szegő [75]. □

b] Hypergeometric Differential Equation. The function $y = F(a, b\,; c\,; z)$ satisfies

$$z(1-z)\,\frac{d^2y}{dz^2} + [c - (a+b+1)z]\,\frac{dy}{dz} - aby = 0\,.$$

Proof. See, for example, Szegő [75]. □

In E.5, E.6, and E.7 we catalog some of the basic properties of some of the classical orthogonal polynomials. Proofs are available in Szegő [75], for example.

E.5 Jacobi Polynomials.

a] Rodrigues' Formula. Let

$$P_n^{(\alpha,\beta)}(x) := (-1)^n \frac{2^{-n}}{n!}(1-x)^{-\alpha}(1+x)^{-\beta}\frac{d^n}{dx^n}\left[(1-x)^\alpha(1+x)^\beta(1-x^2)^n\right].$$

Then $(P_n^{(\alpha,\beta)})_{n=0}^\infty$ is a sequence of orthogonal polynomials on $[-1,1]$ associated with the weight function

$$w(x) := (1-x)^\alpha(1+x)^\beta\,, \qquad -1 < \alpha, \beta < \infty\,.$$

That is,

$$P_n^{(\alpha,\beta)} \in \mathcal{P}_n \quad \text{and} \quad \int_{-1}^1 P_n^{(\alpha,\beta)} P_m^{(\alpha,\beta)} (1-x)^\alpha(1+x)^\beta\, dx = 0$$

for any two nonnegative integers $n \neq m$.

In the rest of the exercise, the polynomials $P_n^{(\alpha,\beta)}$ are as in part a].

b] Normalization. We have

$$P_n^{(\alpha,\beta)}(1) = \binom{n+\alpha}{n} = \frac{(\alpha+1)_n}{n!}$$

and

$$\int_{-1}^1 (P_n^{(\alpha,\beta)}(x))^2(1-x)^\alpha(1+x)^\beta\, dx$$
$$= \frac{2^{\alpha+\beta+1}}{2n+\alpha+\beta+1}\,\frac{\Gamma(n+\alpha+1)\Gamma(n+\beta+1)}{\Gamma(n+1)\Gamma(n+\alpha+\beta+1)}\,.$$

c] Explicit Form.

$$
\begin{aligned}
P_n^{(\alpha,\beta)}(x) &= \frac{1}{2^n}\sum_{m=0}^{n}\binom{n+\alpha}{m}\binom{n+\beta}{n-m}(x-1)^{n-m}(x+1)^m \\
&= \sum_{m=0}^{n}\binom{\alpha+n}{n-m}\binom{\alpha+\beta+n+m}{m}\left(\frac{x-1}{2}\right)^m \\
&= \binom{n+\alpha}{n}\,{}_2F_1\left(-n, n+\alpha+\beta+1\,;\alpha+1\,;\frac{1-x}{2}\right).
\end{aligned}
$$

d] Differential Equation. The function $y = P_n^{(\alpha,\beta)}(x)$ satisfies

$$(1-x^2)\frac{d^2y}{dx^2} + [\beta-\alpha-(\alpha+\beta+2)x]\frac{dy}{dx} + n(n+\alpha+\beta+1)y = 0\,.$$

e] Recurrence Relation. The sequence $(P_n^{(\alpha,\beta)}(x))_{n=0}^{\infty}$ satisfies

$$D_n P_{n+1}^{(\alpha,\beta)}(x) = (A_n + B_n x)P_n^{(\alpha,\beta)}(x) - C_n P_{n-1}^{(\alpha,\beta)}(x)\,,$$

where

$$P_0^{(\alpha,\beta)} = 1 \qquad \text{and} \qquad P_1^{(\alpha,\beta)}(x) = \tfrac{1}{2}[\alpha-\beta+(\alpha+\beta+2)x]$$

and

$$
\begin{aligned}
D_n &= 2(n+1)(n+\alpha+\beta+1)(2n+\alpha+\beta) \\
A_n &= (2n+\alpha+\beta+1)(\alpha^2-\beta^2) \\
B_n &= (2n+\alpha+\beta+2)(2n+\alpha+\beta+1)(2n+\alpha+\beta) \\
C_n &= 2(n+\alpha)(n+\beta)(2n+\alpha+\beta+2)\,.
\end{aligned}
$$

f] Generating Function.

$$\sum_{n=0}^{\infty} P_n^{(\alpha,\beta)}(x)z^n = \frac{2^{\alpha+\beta}}{R(1-z+R)^{\alpha}(1+z+R)^{\beta}}\,,$$

where $R = \sqrt{1-2xz+z^2}$.

There are various special cases, some of which we have previously defined. The Legendre polynomials P_n are defined by

$$P_n := P_n^{(0,0)}\,, \qquad n = 0, 1, \ldots\,.$$

The Chebyshev polynomials T_n defined in Section 2.1 satisfy

$$T_n = \frac{4^n}{\binom{2n}{n}} P_n^{(-1/2,-1/2)}, \qquad n = 0, 1, \ldots.$$

The *ultraspherical* (or *Gegenbauer*) *polynomials* $C_n^{(\alpha)}$ are defined by

$$C_n^{(\alpha)} := \frac{\Gamma(2\alpha+n)\Gamma(\alpha+\frac{1}{2})}{\Gamma(2\alpha)\Gamma(\alpha+n+\frac{1}{2})} P_n^{(\alpha-1/2,\alpha-1/2)}, \qquad n = 0, 1, \ldots.$$

In terms of $C_n^{(\alpha)}$, the Chebyshev polynomials of the first and second kind are given by

$$T_n = \frac{n}{2} C_n^{(0)} \quad \text{and} \quad U_n = C_n^{(1)}, \qquad n = 0, 1, \ldots.$$

E.6 Hermite Polynomials.

a] Rodrigues' Formula. Let

$$H_n(x) := \frac{(-1)^n}{\exp(-x^2)} \frac{d^n}{dx^n} \exp(-x^2).$$

Then $(H_n)_{n=0}^{\infty}$ is a sequence of orthogonal polynomials on $(-\infty, \infty)$ associated with the weight function

$$w(x) := \exp(-x^2).$$

That is,

$$H_n \in \mathcal{P}_n \quad \text{and} \quad \int_{\mathbb{R}} H_n(x) H_m(x) \exp(-x^2)\, dx = 0$$

for any two nonnegative integers $n \neq m$.

In the rest of the exercise, the polynomials H_n are as in part a].

b] Normalization. We have

$$\int_{-\infty}^{\infty} (H_n(x))^2 \exp(-x^2)\, dx = \sqrt{\pi}\, 2^n n!$$

and

$$H_{2n+1}(0) = 0, \qquad H_{2n}(0) = (-1)^n \frac{(2n)!}{n!}.$$

c] Explicit Form.

$$H_n(x) = n! \sum_{m=0}^{\lfloor n/2 \rfloor} \frac{(-1)^m (2x)^{n-2m}}{m!(n-2m)!}.$$

d] Differential Equation. The function $y = H_n(x)$ satisfies

$$\frac{d^2y}{dx^2} - 2x\,\frac{dy}{dx} + 2ny = 0\,.$$

e] Recurrence Relation. The sequence $(H_n(x))_{n=0}^{\infty}$ satisfies

$$H_{n+1}(x) = 2xH_n(x) - 2nH_{n-1}(x)$$

with

$$H_0(x) = 1 \qquad \text{and} \qquad H_1(x) = 2x\,.$$

f] Generating Function.

$$\sum_{n=0}^{\infty} H_n(x)\frac{z^n}{n!} = \exp(2xz - z^2)\,.$$

E.7 Laguerre Polynomials. Let $\alpha \in (-1, \infty)$.

a] Rodrigues' Formula. Let

$$L_n^{(\alpha)}(x) := \frac{1}{n!e^{-x}x^{\alpha}}\frac{d^n}{dx^n}(e^{-x}x^{\alpha+n})\,.$$

Then $(L_n^{(\alpha)})_{n=0}^{\infty}$ is a sequence of orthogonal polynomials on $[0, \infty)$ associated with the weight function

$$w(x) := x^{\alpha}\exp(-x)\,.$$

That is,

$$L_n^{(\alpha)} \in \mathcal{P}_n \quad \text{and} \quad \int_0^{\infty} L_n^{(\alpha)}(x)L_m^{(\alpha)}(x)x^{\alpha}\exp(-x)\,dx = 0$$

for any two nonnegative integers $n \neq m$.

In the rest of the exercise, the polynomials $L_n^{(\alpha)}$ are defined as in part a].

b] Normalization. We have

$$\int_0^{\infty} (L_n^{(\alpha)}(x))^2 x^{\alpha}e^{-x}\,dx = \frac{\Gamma(\alpha+n+1)}{n!}$$

and

$$L_n^{(\alpha)}(0) = \binom{n+\alpha}{n}\,.$$

c] Explicit Form.

$$L_n^{(\alpha)}(x) = \sum_{m=0}^{n} \frac{(-1)^m}{m!}\binom{n+\alpha}{n-m}x^m\,.$$

d] Differential Equation. The function $y = L_n^{(\alpha)}(x)$ satisfies

$$x\,\frac{d^2y}{dx^2} + (\alpha + 1 - x)\,\frac{dy}{dx} + ny = 0\,.$$

e] Recurrence Relation. The sequence $(L_n^{(\alpha)}(x))_{n=0}^{\infty}$ satisfies

$$(n+1)L_{n+1}^{(\alpha)}(x) = [(2n + \alpha + 1) - x]L_n^{(\alpha)}(x) - (n + \alpha)L_{n-1}^{(\alpha)}(x)$$

with

$$L_0^{(\alpha)} = 1 \qquad \text{and} \qquad L_1^{(\alpha)}(x) = -x + \alpha + 1\,.$$

f] Generating Function.

$$\sum_{n=0}^{\infty} L_n^{(\alpha)}(x)z^n = \exp\left(\frac{xz}{z-1}\right)(1-z)^{-\alpha-1}\,.$$

E.8 Christoffel Numbers and Gauss-Jacobi Quadrature. Let $(p_n)_{n=0}^{\infty}$ be the sequence of orthonormal polynomials associated with an m-distribution α. Let $x_{\nu,n}$, $\nu = 1, 2, \dots, n$, denote the zeros of p_n. Let

$$\lambda_{\nu,n} := \frac{1}{p_n'(x_{\nu,n})} \int_{\mathbb{R}} \frac{p_n(x)}{x - x_{\nu,n}}\,d\alpha(x)\,, \qquad \nu = 1, 2, \dots, n\,.$$

The numbers $\lambda_{\nu,n}$ are called the *Christoffel* or *Cotes numbers.*

a] Show that, for any $q \in \mathcal{P}_{2n-1}$,

$$\int_{\mathbb{R}} q(x)d\alpha(x) = \sum_{\nu=1}^{n} \lambda_{\nu,n} q(x_{\nu,n})\,.$$

Hint: First show the equality for every $q \in \mathcal{P}_{n-1}$ by using the Lagrange interpolation formula (E.6 of Section 1.1). If $q \in \mathcal{P}_{2n-1}$, then $q = sp_n + r$ with some $s, r \in \mathcal{P}_{n-1}$, where s is orthogonal to p_n. □

b] Show that $\lambda_{\nu,n} > 0$ for every $\nu = 1, 2, \dots, n$.

Hint: Use part a] to show that

$$\lambda_{\nu,n} = \frac{1}{(p_n'(x_{\nu,n}))^2} \int_{\mathbb{R}} \left(\frac{p_n(x)}{x - x_{\nu,n}}\right)^2 d\alpha(x)\,.$$

□

c] Suppose $[a,b]$ is a finite interval containing $\operatorname{supp}(\alpha)$. Let $f \in C[a,b]$. Show that

$$\left| \int_a^b f(x)\, d\alpha(x) - \sum_{\nu=1}^{n} \lambda_{\nu,n} f(x_{\nu,n}) \right| \le 2(\alpha(b+) - \alpha(a-)) \min_{p \in \mathcal{P}_{2n-1}} \|f - p\|_{[a,b]} \,.$$

Hint: Use parts a] and b] together with the observation

$$\sum_{\nu=1}^{n} \lambda_{\nu,n} = \int_a^b d\alpha(x) = \alpha(b+) - \alpha(a-)\,.$$

□

d] Suppose $\operatorname{supp}(\alpha) \subset [a,b]$, where $a, b \in \mathbb{R}$. Show that

$$\sum_{\nu=1}^{n} \lambda_{\nu,n} f(x_{\nu,n}) \underset{n \to \infty}{\longrightarrow} \int_a^b f(x) d\alpha(x)$$

for every Riemann-Stieltjes integrable function on $[a,b]$ with respect to α.

Hint: First show that f is Riemann-Stieltjes integrable on $[a,b]$ with respect to α if and only if for every $\epsilon > 0$ there are $g_1, g_2 \in C[a,b]$ so that

$$g_1(x) \le f(x) \le g_2(x)\,, \qquad x \in [a,b]$$

and

$$\int_a^b (g_2(x) - g_1(x))\, d\alpha(x) < \epsilon\,.$$

Finish the proof by part c] and the Weierstrass approximation theorem (see E.1 of Section 4.1). □

e] Suppose $\operatorname{supp}(\alpha)$ is compact. Let

$$Z := \{x_{\nu,n} : \nu = 1, 2, \dots, n, \;\; n = 1, 2, \dots\}\,.$$

Show that $\operatorname{supp}(\alpha) \subset \overline{Z}$, where $\overline{Z}$ denotes the closure of Z.

Hint: Use part d]. □

f] Show by an example that $\operatorname{supp}(\alpha) \ne \overline{Z}$ is possible.

E.9 Characterization of Compact Support. Using the notation of Theorem 2.3.2 and E.8, show that the following statements are equivalent:

(1) $\operatorname{supp}(\alpha)$ is compact.

(2) $\sup_{n \in \mathbb{N}} \{|a_n| + |b_n|\} < \infty$.

(3) The set $Z := \{x_{\nu,n} : \nu = 1, 2, \dots, n, \;\; n = 1, 2, \dots\}$ is bounded.

Outline. (1) $\Rightarrow$ (2). Note that the orthogonality of $\{p_n\}_{n=0}^{\infty}$ implies

$$a_n = \int_{\mathbb{R}} x p_{n-1}(x) p_n(x)\, d\alpha(x) \qquad \text{and} \qquad b_n = \int_{\mathbb{R}} x p_n^2(x)\, d\alpha(x)\,.$$

So $\operatorname{supp}(\alpha) \subset [-K, K]$, the Cauchy-Schwarz inequality, and the orthonormality of $(p_n)_{n=0}^{\infty}$ yield

$$\begin{aligned} |a_n| &\le K \int_{-K}^{K} |p_{n-1}(x) p_n(x)|\, d\alpha(x) \\ &\le K \left(\int_{-K}^{K} p_{n-1}^2(x)\, d\alpha(x) \right)^{1/2} \left(\int_{-K}^{K} p_n^2(x)\, d\alpha(x) \right)^{1/2} \le K \end{aligned}$$

and

$$|b_n| \le K \int_{-K}^{K} p_n^2(x) d\alpha(x) = K\,.$$

(2) $\Rightarrow$ (3). Use Theorem 2.3.2 (three-term recursion) to show that

$$x_{\nu,n} \sum_{k=0}^{n-1} p_k^2(x_{\nu,n}) \le 2 \sum_{k=0}^{n-1} a_{k+1} p_k(x_{\nu,n}) p_{k+1}(x_{\nu,n}) + \sum_{k=0}^{n-1} b_k p_k^2(x_{\nu,n})\,.$$

Hence

$$|x_{\nu,n}| \sum_{k=0}^{n-1} p_k^2(x_{\nu,n}) \le \left(2 \max_{0 \le k \le n-1} |a_k| + \max_{0 \le k \le n-1} |b_k| \right) \sum_{k=0}^{n-1} p_k^2(x_{\nu,n})\,.$$

(3) $\Rightarrow$ (1). If $Z \subset [-K, K]$, then by E.8 a]

$$\int_{\mathbb{R}} x^{2n-2}\, d\alpha(x) = \sum_{\nu=1}^{n} \lambda_{\nu,n} x_{\nu,n}^{2n-2} \le K^{2n-2} \int_{\mathbb{R}} d\alpha(x)\,,$$

which implies $\operatorname{supp}(\alpha) \subset [-K, K]$. □

E.10 A Condition for $\operatorname{supp}(\alpha) \subset [0, \infty)$**.** Let $(p_n)_{n=0}^{\infty}$ be the sequence of orthonormal polynomials associated with an m-distribution α. Suppose $\operatorname{supp}(\alpha)$ is compact and

$$p_n(0) p_{n+1}(0) < 0\,, \qquad n = 0, 1, \ldots\,.$$

Show that $\operatorname{supp}(\alpha) \subset [0, \infty)$.

Hint: Use the interlacing property of the zeros of p_n (E.2) to show that

$$Z := \{x_{\nu,n} : \nu = 1, 2, \ldots, n, \;\; n = 1, 2, \ldots\} \subset [0, \infty).$$

Now use E.8 e] to obtain $\operatorname{supp}(\alpha) \subset [0, \infty)$. □

E.11 The Solvability of the Moment Problem. Let $(\mu_n)_{n=0}^{\infty}$ be a sequence of real numbers. We would like to characterize those sequences $(\mu_n)_{n=0}^{\infty}$ for which there exists an m-distribution α so that

$$\int_{\mathbb{R}} x^n \, d\alpha(x) = \mu_n \,, \qquad n = 0, 1, \ldots \,.$$

Let

$$\mu(p) := \sum_{k=0}^{n} a_k \mu_k$$

for every $p \in \mathcal{P}_n$ of the form $p(x) = \sum_{k=0}^{n} a_k x^k$.

A polynomial p is called *nonnegative* if it takes nonnegative values on the real line. The sequence $(\mu_n)_{n=0}^{\infty}$ is called *positive definite* if

$$\mu(p) := \sum_{k=0}^{n} a_k \mu_k > 0 \,, \qquad n = 0, 1, \ldots$$

holds for every nonnegative polynomial $p \in \mathcal{P}_n$ of the form

$$p(x) = \sum_{k=0}^{n} a_k x^k \,.$$

The aim of this exercise is to outline the proof of Hamburger's characterization of the solvability of the moment problem by the positive definiteness of the sequence of moments. See part o].

a] Show that if there exists an m-distribution α so that

$$\int_{\mathbb{R}} x^n \, d\alpha(x) = \mu_n \,, \qquad n = 0, 1, \ldots \,,$$

then $(\mu_n)_{n=0}^{\infty}$ is positive definite.

Hint: An m-distribution α is increasing at infinitely many points. □

b] Show that $(\mu_n)_{n=0}^{\infty}$ is positive definite if and only if $\mu(p^2) > 0$ holds for every $0 \neq p \in \mathcal{P}_n$, $n = 0, 1, \ldots$.

Hint: Use E.3 of Section 2.4. □

c] Show that $(\mu_n)_{n=0}^{\infty}$ is positive definite if and only if

$$\begin{vmatrix} \mu_0 & \mu_1 & \cdots & \mu_n \\ \mu_1 & \mu_2 & \cdots & \mu_{n+1} \\ \vdots & \vdots & \ddots & \vdots \\ \mu_n & \mu_{n+1} & \cdots & \mu_{2n} \end{vmatrix} > 0 \,, \qquad n = 0, 1, \ldots \,.$$

Hint: Use part b] and the law of inertia of Sylvester. See, for example, van der Waerden [50]. □

d] Helly's Selection Theorem. *Suppose the functions f_n, $n = 1, 2, \dots$, are nondecreasing on $\mathbb{R}$, and*

$$\sup_{n \in \mathbb{N}} \|f_n\|_{\mathbb{R}} < \infty \, .$$

Then there exists a subsequence of $(f_n)_{n=1}^{\infty}$ that converges for every $x \in \mathbb{R}$. That is, we can select a pointwise convergent subsequence.

Hint: See, for example, Freud [71]. □

e] Helly's Convergence Theorem. *Let $[a, b]$ be a finite interval. Suppose the functions α_n, $n = 1, 2, \dots$, are nondecreasing on $[a, b]$ and*

$$\sup_{n \in \mathbb{N}} \|\alpha_n\|_{[a,b]} < \infty \, .$$

Suppose also that $(\alpha_n(x))_{n=1}^{\infty}$ converges to $\alpha(x)$ for every $x \in [a, b]$. Then

$$\lim_{n \to \infty} \int_a^b f(x) \, d\alpha_n(x) = \int_a^b f(x) \, d\alpha(x)$$

for every $f \in C[a, b]$.

Hint: See, for example, Freud [71]. □

In the rest of the exercise (except for the last part) we assume that $(\mu_n)_{n=0}^{\infty}$ is positive definite. Our goal is to prove the converse of part a]. Let

$$p_n^*(x) := \begin{vmatrix} \mu_0 & \mu_1 & \dots & \mu_{n-1} & 1 \\ \mu_1 & \mu_2 & \dots & \mu_n & x \\ \vdots & \vdots & \ddots & \vdots & \vdots \\ \mu_n & \mu_{n+1} & \dots & \mu_{2n-1} & x^n \end{vmatrix} .$$

f] Show that

$$\mu(p_n^* q) = 0 \, , \qquad q \in \mathcal{P}_{n-1} \, .$$

g] Show that each p_n^* has n simple real zeros.

Hint: Use part f]. □

Let $x_{1,n} > x_{2,n} > \dots > x_{n,n}$ be the zeros of p_n^*. Let

$$l_{\nu,n}(x) := \frac{p_n^*(x)}{p_n^{*\prime}(x_{\nu,n})(x - x_{\nu,n})} \, , \qquad \nu = 1, 2, \dots, n \, , \quad n = 1, 2, \dots$$

(see E.6 of Section 1.1), and let

$$\lambda_{\nu,n} := \mu(l_{\nu,n}) \, .$$

h] Show that

$$\mu(q) = \sum_{\nu=1}^{n} \lambda_{\nu,n} q(x_{\nu,n})$$

for every $q \in \mathcal{P}_{2n-1}$.

Hint: Use part f]. □

i] Show that

$$\lambda_{\nu,n} = \mu(l_{\nu,n}^2) > 0\,, \qquad \nu = 1, 2, \ldots, n\,, \quad n = 1, 2, \ldots\,.$$

Hint: Use part h]. □

For $x \in \mathbb{R}$, let

$$\alpha_n(x) := \sum_{\{\nu :\, x_{\nu,n} \le x\}} \lambda_{\nu,n}\,, \qquad n = 1, 2, \ldots\,.$$

j] Show that $0 \le \alpha_n(x) \le \mu_0$ on $\mathbb{R}$ for each n, and there is a subsequence of $(\alpha_n)_{n=1}^{\infty}$ that converges pointwise to a nondecreasing real-valued function α on $\mathbb{R}$.

Hint: Use parts h], i], and d]. □

k] Show that for every finite interval $[a, b]$,

$$\lim_{k \to \infty} \int_a^b x^m \, d\alpha_{n_k}(x) = \int_a^b x^m \, d\alpha(x)\,, \qquad m = 0, 1, 2, \ldots\,,$$

where α is defined in part j].

Hint: Use part e]. □

l] Let m be a fixed nonnegative integer and let $r := \lfloor m/2 \rfloor + 1$. Show that if $n_k \ge r + 1$, $a \le -1$ and $b \ge 1$, then

$$\left| \int_{-\infty}^{a} x^m \, d\alpha_{n_k}(x) + \int_b^{\infty} x^m \, d\alpha_{n_k}(x) \right| \le \left(\frac{1}{|a|} + \frac{1}{|b|} \right) \int_{\mathbb{R}} x^{2r} \, d\alpha_{n_k}(x) = \left(\frac{1}{|a|} + \frac{1}{|b|} \right) \mu_{2r}\,.$$

Hint: Use part h]. □

m] Show that

$$\int_{\mathbb{R}} x^m \, d\alpha(x) = \mu_m\,, \qquad m = 0, 1, 2, \ldots$$

where α is defined in part j].

Hint: Use parts k] and l]. □

n] Show that α defined in part j] is an m-distribution.

o] There exists an m-distribution α so that

$$\mu_n = \int_{\mathbb{R}} x^n \, d\alpha(x)$$

if and only if $(\mu_n)_{n=0}^{\infty}$ is positive definite, that is, if and only if

$$\begin{vmatrix} \mu_0 & \mu_1 & \cdots & \mu_n \\ \mu_1 & \mu_2 & \cdots & \mu_{n+1} \\ \vdots & \vdots & \ddots & \vdots \\ \mu_n & \mu_{n+1} & \cdots & \mu_{2n} \end{vmatrix} > 0 \,, \qquad n = 0, 1, \ldots \,.$$

Hint: Combine parts a], c], m], and n]. □

Necessary and sufficient conditions for the uniqueness of the solution of the moment problem are given in Freud [71], for example.

E.12 Favard's Theorem. *Given $(a_n)_{n=0}^{\infty} \subset (0, \infty)$ and $(b_n)_{n=0}^{\infty} \subset \mathbb{R}$, the polynomials $p_n \in \mathcal{P}_n$ are defined by*

$$\begin{aligned} &xp_n(x) = a_n p_{n-1}(x) + b_n p_n(x) + a_{n+1} p_{n+1}(x) \,, \\ &p_{-1} = 0 \,, \qquad p_0 = \gamma_0 > 0 \,. \end{aligned}$$

Then there exists an m-distribution α such that

$$\int_{\mathbb{R}} p_n(x) p_m(x) \, d\alpha(x) = 0$$

for any two distinct nonnegative integers n and m. In other words, the converse of Theorem 2.3.2 is true.

In order to prove Favard's theorem, proceed as follows:

a] Show that the polynomials p_n are of the form

$$p_n(x) = \gamma_n x^n + r(x) \,, \qquad \gamma_n > 0 \,, \quad r \in \mathcal{P}_{n-1} \,.$$

b] Let $\widetilde{p}_n := \gamma_n^{-1} p_n$, $n = 0, 1, \ldots$. The sequence $(\mu_n)_{n=0}^{\infty}$ is defined as follows. Let

$$\mu_0 := 1 \,, \quad \mu(q) := c \quad \text{if} \quad q = c \,, \quad c \in \mathbb{R} \,.$$

If $\mu_0, \mu_1, \ldots, \mu_n$ have already been defined, then let

$$\mu(q) := \sum_{k=0}^{n} c_k \mu_k \quad \text{whenever} \quad q(x) = \sum_{k=0}^{n} c_k x^k \,, \quad c_k \in \mathbb{R}$$

and let

$$\mu_{n+1} := \mu(x^{n+1} - \widetilde{p}_{n+1}(x))\,.$$

Show that

$$\mu(\widetilde{p}_n\widetilde{p}_m) = 0\,, \qquad m = 0, 1, \ldots, n-1\,, \quad n = 0, 1, \ldots$$

and

$$\mu(\widetilde{p}_n^2) > 0\,, \qquad n = 0, 1, \ldots\,.$$

Hint: It is sufficient to prove that

$$\mu(\widetilde{p}_n(x)x^m) = 0\,, \qquad m = 0, 1, \ldots, n-1\,, \quad n = 0, 1, \ldots$$

and

$$\mu(\widetilde{p}_n(x)x^n) > 0\,, \qquad n = 0, 1, \ldots\,.$$

These can be obtained from the definition of μ and from Theorem 2.3.2 (three-term recursion) by induction on n. □

c] Show that every $q \in \mathcal{P}_n$ is of the form

$$q = \sum_{k=0}^{n} c_k \widetilde{p}_k\,, \qquad c_k \in \mathbb{R}$$

and if $q \neq 0$, then

$$\mu(q^2) = \sum_{k=0}^{n} c_k^2 \mu(\widetilde{p}_k^2) > 0\,.$$

d] Show that $(\mu_n)_{n=0}^{\infty}$ is positive definite in the sense of E.11.

Hint: Use the previous part and E.11 b]. □

e] Prove Favard's theorem.

Hint: Use part o] of E.11, parts d] and b] of this exercise, and the definition of μ. □

E.13 Christoffel Function. Let α be an m-distribution. For a fixed $n \in \mathbb{N}$, the function

$$\lambda_n(z) = \inf\left\{\int_{\mathbb{R}} q^2(x)\,d\alpha(x) : q \in \mathcal{P}_{n-1}\,,\ |q(z)| = 1\right\}, \qquad z \in \mathbb{C}$$

is called the nth *Christoffel function* associated with α.

a] Show that

$$\lambda_n(z) := \left(\sum_{k=0}^{n-1} |p_k(z)|^2\right)^{-1},$$

where $(p_n)_{n=0}^{\infty}$ is the sequence of orthonormal polynomials associated with α.

Show also that the infimum in the definition of $\lambda_n(z)$ is actually a minimum, and it is attained if and only if

$$q(x) = \frac{\sum_{k=0}^{n-1} \overline{p_k(z)} p_k(x)}{\sum_{k=0}^{n-1} |p_k(z)|^2} .$$

Hint: Write

$$q = \sum_{k=0}^{n-1} c_k p_k , \qquad c_k \in \mathbb{C}$$

and observe that the orthonormality of $(p_n)_{n=0}^{\infty}$ implies

$$\int_{\mathbb{R}} q^2(x)\, d\alpha(x) = \sum_{k=0}^{n-1} |c_k|^2 .$$

Now use the Cauchy-Schwarz inequality (E.3 of Section 2.2) to find the maximum of $|q(z)|$ for polynomials $q \in \mathcal{P}_{n-1}^c$ satisfying

$$\int_{\mathbb{R}} q^2(x)\, d\alpha(x) \le 1$$

where $z \in \mathbb{C}$ is fixed.

b] Let $\lambda_{\nu,n}$, $\nu = 1, 2, \ldots, n$, be the Christoffel numbers associated with an m-distribution α, that is, the coefficients in the Gauss-Jacobi quadrature formula, as in E.8. Show that

$$\lambda_{\nu,n} = \lambda_n(x_{\nu,n}) , \qquad \nu = 1, 2, \ldots, n ,$$

that is, the Christoffel numbers are the values of the Christoffel function at the zeros of the nth orthonormal polynomial p_n.

Hint: Use parts E.8 a] and E.8 b]. □

c] Let $x \in \mathbb{R}$ be fixed. Show that

$$\sum_{n=0}^{\infty} p_n^2(x) < \infty$$

if and only if x is a mass point of α, that is, $\alpha(x-) < \alpha(x+)$, in which case

$$\sum_{n=0}^{\infty} p_n^2(x) = (\alpha(x+) - \alpha(x-))^{-1} .$$

Hint: Use part a] and the Weierstrass approximation theorem. See E.1 of Section 4.1. □

E.14 The Markov-Stieltjes Inequality. Let α be an m-distribution with associated orthonormal polynomials $(p_n)_{n=0}^{\infty}$. Let $x_{1,n} > x_{2,n} > \cdots > x_{n,n}$ denote the zeros of p_n. Let $x_{0,n} := \infty$ and $x_{n+1,n} := -\infty$. As in E.8 let $\lambda_{\nu,n}$, $\nu = 1, 2, \ldots, n$, be the Christoffel numbers associated with α . Show that

$$\lambda_{\nu,n} \leq \int_{x_{\nu+1,n}}^{x_{\nu-1,n}} d\alpha(x)\,, \qquad \nu = 1, 2, \ldots, n$$

and

$$\int_{x_{\nu,n}}^{x_{\nu-1,n}} d\alpha(x) \leq \lambda_{\nu,n} + \lambda_{\nu-1,n}\,, \qquad \nu = 2, 3, \ldots, n\,.$$

Hint: Let $1 \leq k \leq n$ be fixed. Use E.7 of Section 1.1 (Hermite interpolation) to find polynomials $P \in \mathcal{P}_{2n-1}$ and $Q \in \mathcal{P}_{2n-1}$ with the following properties:

(1) $P(x_{j,n}) = Q(x_{j,n}) = 1\,, \qquad j = 1, 2, \ldots, k-1\,,$

(2) $P(x_{k,n}) = 0\,, \qquad Q(x_{k,n}) = 1\,,$

(3) $P(x_{j,n}) = Q(x_{j,n}) = 0\,, \qquad j = k+1, k+2, \ldots, n\,,$

(4) $P(x) \leq \chi_{(-\infty,\, x_{k,n}]}(x) \leq Q(x)\,, \qquad x \in \mathbb{R}\,,$

where

$$\chi_{(-\infty,\, x_{k,n}]}(x) := \begin{cases} 1 & \text{if} \quad -\infty < x \leq x_{k,n} \\ 0 & \text{if} \quad x_{k,n} < x < \infty\,. \end{cases}$$

Now apply E.8 (Gauss-Jacobi quadrature formula) to P and Q. □

E.15 Orthonormal Polynomials as Determinants. Suppose α is an m-distribution with moments

$$\mu_n = \int_{\mathbb{R}} x^n \, d\alpha(x)\,, \qquad n = 0, 1, \ldots\,.$$

Let

$$\Delta_n := \begin{vmatrix} \mu_0 & \mu_1 & \cdots & \mu_n \\ \mu_1 & \mu_2 & \cdots & \mu_{n+1} \\ \vdots & \vdots & \ddots & \vdots \\ \mu_n & \mu_{n+1} & \cdots & \mu_{2n} \end{vmatrix}\,, \qquad n = 0, 1, \ldots\,.$$

a] Show that $\Delta_n > 0$, $n = 0, 1, 2, \ldots$.

b] Show that the orthonormal polynomials p_n associated with α are of the form

$$p_n(x) = (\Delta_n \Delta_{n-1})^{-1/2} \begin{vmatrix} \mu_0 & \mu_1 & \cdots & \mu_{n-1} & 1 \\ \mu_1 & \mu_2 & \cdots & \mu_n & x \\ \vdots & \vdots & \ddots & \vdots & \vdots \\ \mu_n & \mu_{n+1} & \cdots & \mu_{2n-1} & x^n \end{vmatrix}\,.$$

c] Let $(a_n)_{n=0}^\infty \subset (0,\infty)$ and $(b_n)_{n=0}^\infty \subset \mathbb{R}$ be the coefficients in the three-term recursion for the sequence $(p_n)_{n=0}^\infty$ of orthonormal polynomials associated with α as in Theorem 2.3.2. Show that the monic orthogonal polynomials $\widetilde{p}_n := \gamma_n^{-1} p_n$ are of the form

$$\widetilde{p}_n(x) = \det\,(xI_n - J_n)$$

where J_n is the tridiagonal n by n *Jacobi matrix*

$$J_n := \begin{pmatrix} b_0 & a_1 & & & \\ a_1 & b_1 & a_2 & & \\ & a_2 & b_2 & a_3 & \\ & & \ddots & \ddots & \ddots \\ & & & a_n & b_n \end{pmatrix}$$

and I_n is the n by n unit matrix.

E.16 The Support of α. Let $(a_n)_{n=0}^\infty \subset (0,\infty)$ and $(b_n)_{n=0}^\infty \subset \mathbb{R}$ be the coefficient sequences in the three-term recursion for the sequence of $(p_n)_{n=0}^\infty$ of orthonormal polynomials associated with an m-distribution α as in Theorem 2.3.2.

a] Show that if $\text{supp}(\alpha) \subset [b-a, b+a]$ with some $a > 0$ and $b \in \mathbb{R}$, then

$$a_n \le a \quad \text{and} \quad |b_n - b| \le a\,, \qquad n = 0, 1, \ldots\,.$$

Hint: Use the orthonormality of $(p_n)_{n=0}^\infty$ to show that

$$a_n = \int_{b-a}^{b+a} (x-b)p_{n-1}(x)p_n(x)\,d\alpha(x)$$

and

$$b_n - b = \int_{b-a}^{b+a} (x-b)p_n^2(x)\,d\alpha(x)\,.$$

Now apply the Cauchy-Schwarz inequality, and use the orthonormality of $(p_n)_{n=0}^\infty$ again. □

b] Show that

$$\text{supp}(\alpha) \subset [-K, K]$$

where

$$K := 2\sup\{a_n : n \in \mathbb{N}\} + \sup\{|b_n| : n \in \mathbb{N}\}$$

(the suprema are taken over all nonnegative integers).

Hint: Suppose $K < \infty$; otherwise there is nothing to prove. Combine E.9, the inequality in the hint to the direction (2) ⇒ (3) of E.9, and E.8 e]. □

c] Blumenthal's Theorem. *Assume that*

$$\lim_{n\to\infty} a_n =: \frac{a}{2} > 0 \qquad \text{and} \qquad \lim_{n\to\infty} b_n =: b \in \mathbb{R}\,.$$

Then

$$\operatorname{supp}(\alpha) = [b-a, b+a] \cup F$$

where $F \subset \mathbb{R} \setminus [b-a, b+a]$ is a countable bounded set for which

$$F \setminus (b-a-\epsilon, b+a+\epsilon)$$

is finite for every $\epsilon > 0$.

Proof. See Nevai [79b] or Máté, Nevai, and Van Assche [91]. □

d] Rakhmanov's Theorem. *Suppose $\operatorname{supp}(\alpha) \subset [b-a, b+a]$ with some $a > 0$ and $b \in \mathbb{R}$. Suppose also that $\alpha'(x) > 0$ a.e. in $[b-a, b+a]$. Then*

$$\lim_{n\to\infty} a_n = \frac{a}{2} \quad \text{and} \quad \lim_{n\to\infty} b_n = b\,.$$

Proof. See, for example, Máté, Nevai, and Totik [85], or Nevai [91]. □

There is an analogous theory of orthogonal polynomials on the unit circle initiated by Geronimus, Shohat, and Achiezer. An important contribution, called Szegő theory, may be found in Freud [71].

E.17 A Theorem of Stieltjes [14]. Let w be a positive continuous weight function on $[-a, a]$. Denote the nth moment by

$$\mu_n := \int_{-a}^{a} x^n w(x)\, dx\,.$$

Let $\widetilde{p}_n \in \mathcal{P}_n$ denote the nth *monic* orthogonal polynomial on $[-a, a]$ associated with the weight w. Then $(\widetilde{p}_n)_{n=0}^{\infty}$ satisfies a three-term recursion

$$\widetilde{p}_n(x) = (x - A_n)\widetilde{p}_{n-1}(x) - B_n \widetilde{p}_{n-2}(x)$$

with $\widetilde{p}_0(x) = 1$ and $\widetilde{p}_1(x) := x - A_1$; see Theorem 2.3.2.

Suppose the sequence of polynomials $(q_n)_{n=0}^{\infty}$ satisfies the same recursion commencing with $q_0(x) := 0$ and $q_1(x) := B_1$. Stieltjes' theorem (see, for example, Cheney [66]) states the following.

Theorem. *For any $x \notin [-a, a]$,*

$$\int_{-a}^{a} \frac{w(t)\, dt}{x-t} = \frac{\mu_0}{x} + \frac{\mu_1}{x^2} + \cdots$$

$$= \cfrac{B_1}{x - A_1 - \cfrac{B_2}{x - A_2 - \cfrac{B_3}{x - A_3 - \cdots}}}\,.$$

Furthermore, the nth convergent $q_n/\widetilde{p}_n$ satisfies

$$\frac{q_n(x)}{\widetilde{p}_n(x)} = \cfrac{B_1}{x - A_1 - \cfrac{B_2}{\ddots \; \cfrac{B_n}{x - A_n}}}.$$

E.18 Completeness of Orthogonal Polynomials. Let $(p_n)_{n=0}^{\infty}$ be the sequence of orthonormal polynomials associated with an m-distribution α. If $\text{supp}(\alpha) \subset [a, b]$, where $[a, b]$ is a finite interval, then $(p_n)_{n=0}^{\infty}$ is a maximal orthogonal collection in $L_2[a, b]$.

Hint: Use the Weierstrass approximation theorem (E.1 of Section 4.1) on $[a, b]$. □

E.19 Bounds for Jacobi Polynomials. For all Jacobi weight functions $w(x) = (1-x)^\alpha(1+x)^\beta$ with $\alpha \geq -1/2$ and $\beta \geq -1/2$, the inequalities

$$\max_{x\in[-1,1]} \frac{p_n^2(x)}{\sum_{k=0}^{n} p_k^2(x)} \leq \frac{4\left(2+\sqrt{\alpha^2+\beta^2}\right)}{2n+\alpha+\beta+2}$$

and

$$\max_{x\in[-1,1]} \sqrt{1-x^2}\,w(x)p_n^2(x) \leq \frac{2e\left(2+\sqrt{\alpha^2+\beta^2}\right)}{\pi}$$

hold, where $(p_n)_{n=0}^{\infty}$ is the sequence of orthonormal Jacobi polynomials associated with the weight function w.

Proof. See Erdélyi, Magnus, and Nevai [94]. □

2.4 Polynomials with Nonnegative Coefficients

A quadratic polynomial $x^2 + \alpha x + \beta$ with real coefficients has both roots in the halfplane $\{z \in \mathbb{C} : \text{Re}(z) \leq 0\}$ if and only if $\beta \geq 0$ and $\alpha \geq 0$. This easy consequence of the quadratic formula gives the following lemma:

Lemma 2.4.1. *If $p \in \mathcal{P}_n$ has all its zeros in $\{z \in \mathbb{C} : \text{Re}(z) \leq 0\}$, then either p or $-p$ has all nonnegative coefficients.*

The converse of this is far from true. Indeed, the following result of Meissner holds (see Pólya and Szegő [76]). We denote by $\mathcal{P}_n^+$ the set of polynomials in $\mathcal{P}_n$, that have all nonnegative coefficients.

Theorem 2.4.2. *If $p \in \mathcal{P}_n^c$ and $p(x) > 0$ for $x > 0$, then $p = s/t$, where s and t are both polynomials with all nonnegative coefficients.*

Since a polynomial p that is real-valued on the positive real axis has real coefficients, and since $p(x) > 0$ for all $x > 0$ implies that the leading coefficient of p is positive, Theorem 2.4.2 will follow immediately from the next lemma.

Lemma 2.4.3. *Suppose $\alpha, \beta \in \mathbb{R}$ and suppose $x^2 - \alpha x + \beta$ has no nonnegative root. Then $x^2 - \alpha x + \beta = p(x)/q(x)$, where $p, q \in \mathcal{P}_m$ both have all nonnegative coefficients, and where*

$$m \le 10\left(4 - \frac{\alpha^2}{\beta}\right)^{-1/2}.$$

Proof. The quadratic polynomial $x^2 - \alpha x + \beta$ has no positive root if and only if $\alpha^2 < 4\beta$. We set $c := \alpha^2/\beta$ and note that $c < 4$. Consider

$$\begin{aligned}(x^2 - \alpha x + \beta)(x^2 + \alpha x + \beta) &= x^4 + (2\beta - \alpha^2)x^2 + \beta^2 \\ &= x^4 + \beta(2 - c)x^2 + \beta^2 .\end{aligned}$$

If $c \le 2$ we have the desired factorization. If $c > 2$, consider

$$\begin{aligned}&(x^4 + \beta(2 - c)x^2 + \beta^2)(x^4 - \beta(2 - c)x^2 + \beta^2) \\ &\qquad = x^8 + \beta^2(2 - (2 - c)^2)x^4 + \beta^4 .\end{aligned}$$

If $2 - (2 - c)^2 > 0$ we are finished. In general, we proceed as follows: Let

$$\begin{aligned}P_n(x) :&= x^{2^{n+1}} + \beta^{2^{n-1}}(2 - (2 - (2 - \cdots 2 - (2 - c)^2)^2 \cdots)^2)x^{2^n} + \beta^{2^n} \\ &= x^{2^{n+1}} + \beta^{2^{n-1}} c_n x^{2^n} + \beta^{2^n} ,\end{aligned}$$

where c_n has n nested terms. Let

$$Q_n(x) := x^{2^{n+1}} - \beta^{2^{n-1}} c_n x^{2^n} + \beta^{2^n} .$$

Note that, since $c_{n+1} = 2 - c_n^2$

$$\begin{aligned}P_n(x)Q_n(x) &= x^{2^{(n+1)+1}} - \beta^{2^n} c_n^2 x^{2^{n+1}} + 2\beta^{2^n} x^{2^{n+1}} + \beta^{2^{n+1}} \\ &= x^{2^{n+2}} + \beta^{2^n} c_{n+1} x^{2^{n+1}} + \beta^{2^{n+1}} \\ &= P_{n+1}(x) .\end{aligned}$$

Consider the smallest n (if it exists) such that c_n is nonnegative. Then

$$(x^2 - \alpha x + \beta)(x^2 + \alpha x + \beta) = P_1(x)$$

and

$$P_1 Q_1 Q_2 \cdots Q_{n-1} = P_n \,,$$

where $Q_1 Q_2 \cdots Q_{n-1} \in \mathcal{P}^+_{2^{n+1}-4}$ since each $c_k < 0$ for $k < n$, and where $P_n \in P^+_{2^{n+1}}$ since $c_n \geq 0$. Thus, we have the desired representation

$$x^2 - \alpha x + \beta = \frac{P_n(x)}{(x^2 + \alpha x + \beta)(Q_1 Q_2 \cdots Q_{n-1})(x)}\,,$$

where n is the smallest integer such that $c_n > 0$.

Now suppose $c_1, \ldots, c_{n-1}, c_n$ are all nonpositive. Then

$$c_k = -\sqrt{2 - c_{k+1}}\,, \qquad k = 1, 2, \ldots, n-1$$

and $c_1 = 2 - c$ imply

$$c > 2 + (2 + \cdots (2 + (2 + 2^{1/2})^{1/2})^{1/2} \cdots)^{1/2} =: \delta_n \,, \tag{2.4.1}$$

where the above formula contains n iterations. Since, by assumption, $c < 4$, and since $\delta_n \to 4$ as $n \to \infty$, it is clear that (2.4.1) is not satisfied for some n, and eventually some c_n is greater than zero.

The estimate on the degree requires analyzing the rate of convergence of $(\delta_n)_{n=0}^{\infty}$. Since $\delta_n = 2 + \sqrt{\delta_{n-1}}$, we have

$$4 - \delta_n = 2 - \sqrt{\delta_{n-1}} = \frac{4 - \delta_{n-1}}{2 + \sqrt{\delta_{n-1}}} \leq \frac{4 - \delta_{n-1}}{2}\,.$$

By repeated applications of the above,

$$4 - \delta_n \leq \frac{4 - \delta_0}{2^n} = \frac{1}{2^{n-1}}\,.$$

Now we can improve the above estimate as follows: We have

$$\begin{aligned}
4 - \delta_n =& \frac{4 - \delta_{n-1}}{2 + \sqrt{\delta_{n-1}}} = \frac{4 - \delta_{n-2}}{(2 + \sqrt{\delta_{n-1}})(2 + \sqrt{\delta_{n-2}})} \\
=& \frac{4 - \delta_0}{(2 + \sqrt{\delta_{n-1}})(2 + \sqrt{\delta_{n-2}}) \cdots (2 + \sqrt{\delta_0})} \\
\leq& \frac{2}{(2 + \sqrt{4 - 2^{2-n}})(2 + \sqrt{4 - 2^{3-n}}) \cdots (2 + \sqrt{4 - 2^{(n+1)-n}})} \\
\leq& \frac{2}{(2 + 2 - 2^{1-n})(2 + 2 - 2^{2-n}) \cdots (2 + 2 - 2^{n-n})} \\
\leq& 2 \cdot 4^{-n} \prod_{j=2}^{n+1} \frac{1}{1 - 2^{-j}} \leq 2 \cdot 4^{-n} \prod_{j=2}^{n+1} (1 + 2 \cdot 2^{-j}) \leq 2e4^{-n}\,.
\end{aligned}$$

So if

$$m := 2^{n+1} \geq \frac{2\sqrt{2e}}{\sqrt{4-c}},$$

then

$$4 - \delta_n \leq 2e4^{-n} \leq 4 - c,$$

that is, $\delta_n \geq c$. □

We note that in the above proof a little additional effort yields a slightly better constant than 10.

Let $k \in \mathbb{N}$ and $\epsilon \in (0, \pi)$. It follows easily from Lemma 2.3.4 that if $p \in \mathcal{P}_k$ has no zeros in the cone

$$\{z \in \mathbb{C} : |\arg(z)| < \epsilon\},$$

then there are $s, t \in \mathcal{P}_m^+$ with $m \leq \frac{5}{4}\pi k \epsilon^{-1}$ so that $p = s/t$; see E.1 d]. The essential sharpness of this upper bound is shown by E.1 e]. An easier proof of Theorem 2.4.2 that gives a weaker bound for the degree of the numerator and denominator in the representation is given by E.1 f].

A similar sort of representation theorem due to Bernstein [15] is the following:

Theorem 2.4.4. *If $p \in \mathcal{P}_n$ and $p(x) > 0$ for $x \in (-1, 1)$, then there is a representation*

$$p(x) = \sum_{j=0}^{d} a_j (1-x)^j (1+x)^{d-j}$$

with each $a_j \geq 0$. (The smallest $d := d(p)$ for which such a representation exists is called the Lorentz degree of p.)

It suffices to prove this result for quadratic polynomials; this is left as an exercise; see E.1 f].

The proof of the following interesting result of Barnard et al. [91] is surprisingly complicated, and we do not reproduce it here.

Theorem 2.4.5. *Suppose that $p \in \mathcal{P}_n$ has all nonnegative coefficients. Suppose that the zeros of p are $z_1, z_1, \ldots, z_n \in \mathbb{C}$. For $\tau \geq 0$, let*

$$p_\tau(z) = \prod_{\substack{j=1 \\ |\arg(z_j)| > \tau}}^{n} \left(1 - \frac{z}{z_j}\right),$$

where $\arg(z)$ is defined so that $\arg(z) \in [-\pi, \pi)$. Then $p_\tau(z)$ has all nonnegative coefficients.

It follows from this result that if $p \in \mathcal{P}_n$ has all nonnegative coefficients and if $q(x) = x^2 + \alpha x + \beta$ is a quadratic polynomial with zeros forming a pair of conjugate zeros of p that have least angular distance from the positive x-axis, then p/q also has all nonnegative coefficients.

Comments, Exercises, and Examples.

Polynomials with all nonnegative coefficients have a number of distinguishing properties that are explored in the exercises. For example, only analytic functions with all nonnegative coefficients can be approximated uniformly on $[0,1]$ by such polynomials; see E.2. So a Weierstrass-type theorem does not hold for these polynomials. This is quite different from approximation by polynomials of the form

$$\sum a_{i,j}(x+1)^i(1-x)^j\,, \qquad a_{i,j} \geq 0\,. \tag{2.4.2}$$

Since every polynomial that is strictly positive on $(-1,1)$ has such a representation (E.1 b]), it follows from the Weierstrass approximation theorem that all nonnegative functions from $C[-1,1]$ are in the uniform closure.

It follows from Theorem 2.4.2 and the Weierstrass approximation theorem (see E.1 of Section 4.1) that fractions of polynomials with all nonnegative coefficients form a dense set in the uniform norm on $[0,1]$ in the set of nonnegative continuous functions on a finite closed interval $[0,1]$. Hence they have a much larger uniform closure on $[0,1]$ than that of the polynomials with all nonnegative coefficients.

Various inequalities for polynomials of the form (2.4.2) are considered in Appendix 5.

E.1 Remarks on Theorem 2.4.2.

a] Suppose $\alpha, \beta \in \mathbb{R}, \epsilon \in (0,\pi)$, and suppose $x^2+\alpha x+\beta$ has no zeros in the cone

$$\{z \in \mathbb{C} : |\arg(z)| < \epsilon\}\,.$$

Show that there are $p, q \in \mathcal{P}_m^+$ with $m \leq \frac{5}{2}\pi\epsilon^{-1}$ such that

$$x^2+\alpha x+\beta = \frac{p(x)}{q(x)}\,.$$

Hint: This is a reformulation of Theorem 2.4.3 by introducing the angle between the positive x-axis and the zero of the quadratic polynomial. □

b] Let $n \in \mathbb{N}$. Show that if $p \in \mathcal{P}_n^+$, then p has no zeros in the cone

$$\{z \in \mathbb{C} : |\arg(z)| < \pi/(2n)\}\,.$$

This is sharp, as the example $p(x) := x^n + 1$ shows.

c] Show that the result of part a] is sharp up to the constant $\frac{5}{2}$.

Hint: Let $n \in \mathbb{N}$. Consider

$$x^2+\alpha x+\beta = \left(x - \exp\left(\tfrac{2\pi i}{2n}\right)\right)\left(x - \exp\left(\tfrac{-2\pi i}{2n}\right)\right).$$

Show that if there are $p, q \in \mathcal{P}_m^+$ so that $x^2 + \alpha x + \beta = p(x)/q(x)$, then $m \geq n - 1$. □

d] Let $\epsilon \in (0, \pi)$ and $k \in \mathbb{N}$. Suppose $p \in \mathcal{P}_k$ has no zeros in the cone

$$\{z \in \mathbb{C} : |\arg(z)| < \epsilon\} .$$

Show that there are $s, t \in \mathcal{P}_m^+$ with $m \leq \frac{5}{4}\pi k \epsilon^{-1}$ so that $p = s/t$.

e] Let $\epsilon \in (0, \pi)$ and $k \in \mathbb{N}$. Let

$$f_k(x) := ((x - z_0)(x - \overline{z}_0))^k ,$$

where $\arg(z_0) = \epsilon$. Assume that $f_k = s/t$, where $s, t \in \mathcal{P}_m^+$. Show that

$$m \geq (\log 2)k\epsilon^{-1} .$$

Hint: First observe that

$$s(y) \leq s(y + y\delta m^{-1}) \leq e^{\delta} s(y)$$

for every $s \in \mathcal{P}_m^+$, $y \in (0, \infty)$, and $\delta \in (0, \infty)$. Therefore

$$f_k(y + y\delta m^{-1}) \leq e^{\delta} f_k(y)$$

for every $y \in (0, \infty)$ and $\delta \in (0, \infty)$. Now let $y \geq |z_0|$ be chosen so that $|y - z_0| = \epsilon$. Applying the above inequality with this y and $\delta := m\alpha$, we obtain

$$f_k(y + \epsilon y) \leq e^{m\epsilon} f_k(y) ,$$

hence $2^k \leq e^{m\epsilon}$, that is, $k \log 2 \leq m\epsilon$. □

f] Prove that if $r \in \mathcal{P}_n$ and $r(x) > 0$ for all $x > 0$, then there is an integer $d \geq n$ such that

$$r(x) = \frac{q(x)}{(1 + x)^{d-n}} ,$$

where $q \in \mathcal{P}_d^+$.

Hint: Let $\alpha, \beta \in \mathbb{R}$ and $\alpha^2 < 4\beta$. Consider

$$(x^2 - \alpha x + \beta)(1 + x)^d = \sum_{j=0}^{d+2} c_j x^j$$

and compute c_j explicitly. □

g] If $p \in \mathcal{P}_n$ and $p(x) > 0$ for all $x \in (-1, 1)$, then it is of the form

$$p(x) = \sum_{j=0}^{d} a_j (1 - x)^j (1 + x)^{d-j} , \qquad a_j \geq 0$$

for some $d \geq n$.

Hint: Apply a] to

$$r(u) := (1 + u)^n p\left(\frac{1 - u}{1 + u}\right) , \qquad u := \frac{1 - x}{1 + x} .$$

□

E.2 Polynomials with Nonnegative Coefficients.

a] If $p \in \mathcal{P}_n^+$, then for $x > 0$

$$0 \le p'(x) < \frac{n}{x} p(x) .$$

b] If $(p_n)_{n=1}^\infty$ is a sequence of polynomials with $p_n \in \mathcal{P}^+ := \cup_{k=0}^\infty \mathcal{P}_k^+$ and $(p_n)_{n=1}^\infty$ converges to f uniformly on $[0,1]$, then f is the restriction to $[0,1]$ of a function analytic in $D := \{z \in \mathbb{C} : |z| < 1\}$ of the form

$$f(z) = \sum_{n=0}^\infty a_n z^n , \qquad a_n \ge 0 .$$

Hint: Since $(p_n(1))_{n=1}^\infty$ converges and each p_n has nonnegative coefficients, there is a constant C such that

$$\|p_n\|_D \le p_n(1) \le C , \qquad n = 1, 2, \ldots .$$

Now Montel's theorem (see, for example, Ash [71]) implies that $(p_n)_{n=1}^\infty$ has a locally uniformly convergent subsequence on D. Deduce that this subsequence converges to an f with nonnegative coefficients. □

E.3 Nonnegative-Valued Polynomials and Sums of Squares.

a] Suppose $p \in \mathcal{P}_{2n}$ is nonnegative on $\mathbb{R}$. Then there exist $s, t \in \mathcal{P}_n$ such that

$$p(x) = s^2(x) + t^2(x) .$$

Hint: If $p \in \mathcal{P}_2$ and p is nonnegative, then for some real numbers α and β,

$$p(x) = (x - \alpha)^2 + \beta^2 .$$

Now use the identity

$$(a^2 + b^2)(c^2 + d^2) = (ac + bd)^2 + (ad - bc)^2 .$$

□

b] If $p \in \mathcal{P}_{2n}$ is nonnegative for $x \ge 0$, then there exist $s, t, u, v \in \mathcal{P}_n$ such that

$$p(x) = s^2(x) + t^2(x) + xu^2(x) + xv^2(x) .$$

c] Suppose $t \in \mathcal{T}_n$ is nonnegative on $\mathbb{R}$. Show that there exists a $q \in \mathcal{P}_n^c$ such that

$$t(\theta) = |q(e^{i\theta})|^2 , \qquad \theta \in \mathbb{R} .$$

Show also that if, in addition, $t \in \mathcal{T}_n$ is even, then there exists a $q \in \mathcal{P}_n$ such that the above holds.

d] If $p \in \mathcal{P}_n$ is nonnegative on $[-1,1]$, then there exist $s,\, t \in \mathcal{P}_n$ such that

$$p(x) = s^2(x) + (1-x^2)t^2(x)\,.$$

Hint: Write, by c],

$$p(\cos\theta) = |s(\cos\theta) + it(\cos\theta)\sin\theta|^2\,.$$

□

The above exercise follows Pólya and Szegő [76]; see also E.1 of Section 7.2 where this result is extended.

The following two exercises discuss results proved in Erdélyi and Szabados [88], [89b], and Erdélyi [91c].

E.4 Lorentz Degree of Polynomials. Given a polynomial $p \in \mathcal{P}_n$, let $d = d(p)$ be the minimal nonnegative integer for which the polynomial p is of the form

$$p(x) = \pm \sum_{j=0}^{d} a_j(1-x)^j(x+1)^{d-j}\,, \quad a_j \geq 0\,.$$

If there is no such d, then let $d(p) := \infty$. We call $d := d(p)$ the *Lorentz degree* of the polynomial p.

a] Let $p \in \mathcal{P}_n \setminus \mathcal{P}_{n-1}$ be of the form

$$p(x) = \sum_{j=0}^{n} b_j(1-x)^j(x+1)^{n-j}\,, \qquad b_j \in \mathbb{R}\,.$$

For $m \geq n$, let the numbers $b_{j,m}$, $j = 0, 1, \ldots, m$, be defined by

$$\begin{aligned} p(x) &= \left(\sum_{j=0}^{n} b_j(1-x)^j(x+1)^{n-j}\right)\left(\frac{1-x}{2} + \frac{x+1}{2}\right)^{m-n} \\ &= \sum_{j=0}^{m} b_{j,m}(1-x)^j(x+1)^{m-j}\,. \end{aligned}$$

Show that if $d(p)$ is finite, then it is the smallest value of m for which each $b_{j,m}$ is nonnegative or each $b_{j,m}$ is nonpositive.

b] Show that if $p \in \mathcal{P}_1 \setminus \mathcal{P}_0$ has no zeros in $(-1,1)$, then $d(p) = 1$.

c] Show that if $p \in \mathcal{P}_2 \setminus \mathcal{P}_1$ has no zeros in the open unit disk, then $d(p) = 2$.

d] Show that if the zeros of a polynomial $p \in \mathcal{P}_2 \setminus \mathcal{P}_1$ lie on the ellipse

$$B_\epsilon := \left\{z = x + iy : \; y^2 = \epsilon^2(1-x^2), \;\; x \in (-1,1)\right\}$$

with $\epsilon \in (0,1]$, then

$$\epsilon^{-2} \leq d(p) < 2\epsilon^{-2} + 1.$$

e] Let $\epsilon \in (0,1)$ be such that ϵ^{-2} is an integer. Let

$$p_1(x) := x^2 + 2\frac{3\epsilon^2 - 1}{1-\epsilon^2}x + \frac{8\epsilon^4 - 5\epsilon^2 + 1}{1-\epsilon^2}.$$

Show that p_1 has its zeros on B_ϵ defined in d], and $d(p_1) = \epsilon^{-2}$.

f] Let $\epsilon \in (0,1)$ be such that $2\epsilon^{-2}$ is an integer. Let

$$p_2(x) := x^2 - 2\frac{2 - 3\epsilon^2}{(1-\epsilon^2)(2-\epsilon^2)}x + \frac{-\epsilon^8 - 5\epsilon^6 + \epsilon^4 - 8\epsilon^2 + 4}{(1-\epsilon^2)(2-\epsilon^2)^2}.$$

Show that p_2 has its zeros on B_ϵ defined in part d], and $d(p_2) = 2\epsilon^{-2}$.

g] Show that $d(pq) \le d(p) + d(q)$ for any two polynomials p and q.

h] Let $\epsilon \in (0,1]$. Show that if $p \in \mathcal{P}_n$ has no zeros in

$$D_\epsilon := \left\{z = x + iy : y^2 < \epsilon^2(1-x^2),\ \ x \in (-1,1),\ \ y \in \mathbb{R}\right\},$$

then

$$d(p) < 2n\epsilon^{-2} + n < 3n\epsilon^{-2}.$$

i] Let p be a polynomial. Show that $d(p) < \infty$ if and only if $p = 0$ or p has no zeros in $(-1,1)$.

j] Show that

$$\left|p\left(y + \tfrac{i}{2}\sqrt{\tfrac{(1-y^2)n}{d}}\right)\right| \le 2^n \left|p\left(y - \tfrac{n}{4d}\right)\right|$$

for every $p \in \mathcal{B}_d(-1,1)$, $1 \le n \le d$, and $y \in [0,1)$ (i is the imaginary unit).

Hint: Modify the proof of Lemma A.5.4. □

k] Let $b \in [0,1]$. Show that

$$p'(b) \le d\,p(b)$$

for every $p \in \mathcal{B}_d(-1,1)$, positive in $(-1,1)$.

Hint: If $q_{j,d}(x) := (1-x)^j(x+1)^{d-j}$, then

$$q'_{j,d}(b) = q_{j,d}(b)\left(\frac{d-j}{1+b} - \frac{j}{1-b}\right) \le d\,q_{j,d}(b), \quad j = 0,1,\dots,d$$

for every $b \in [0,1]$. □

l] Show that $d(p) \ge \frac{1}{17}n\epsilon^{-2}$ whenever

$$p(x) = ((x - z_0)(x - \bar{z}_0))^n, \quad z_0 \in B_\epsilon, \quad \epsilon \in (0,1],$$

where B_ϵ is defined in part d].

Proof. Let $z_0 = y + i\epsilon(1-y^2)^{1/2}$, $y \in (-1,1)$. Without loss of generality it may be assumed that $0 \le y < 1$. Distinguish two cases.

Case 1: $1 - 2\epsilon^2 \le y < 1$. By part k],

$$d(p) \ge \frac{p'(1)}{p(1)} = \frac{2n(1-y)}{(1-y)^2 + \epsilon^2(1-y^2)} = \frac{2n}{(1-y) + \epsilon^2(1+y)} > \frac{n}{2\epsilon^2}\,.$$

Case 2: $0 \le y < 1 - 2\epsilon^2$. Applying part k] with

$$b := y + \epsilon(1-y^2)^{1/2} \in [0,1]$$

deduce that

$$d(p) \ge \frac{p'(b)}{p(b)} = \frac{2n(b-y)}{(b-y)^2 + \epsilon^2(1-y^2)} = \frac{n}{\epsilon(1-y^2)^{1/2}}\,. \tag{2.4.3}$$

Use part j] to obtain

$$\left|(1-y^2)\epsilon^2 - \frac{(1-y^2)n}{4d}\right|^n \le 2^n\left((1-y^2)\epsilon^2 + \frac{n^2}{16d^2}\right)^n, \tag{2.4.4}$$

where $d := d(p)$ and $n = \frac{1}{2}\deg(p)$. If

$$(1-y^2)\epsilon^2 \ge \frac{(1-y^2)n}{8d}\,,$$

then there is nothing to prove. Therefore assume that

$$(1-y^2)\epsilon^2 \le \frac{(1-y^2)n}{8d}\,. \tag{2.4.5}$$

Now (2.4.3) to (2.4.5) yield

$$\left(\frac{(1-y^2)n}{8d}\right)^n \le 2^n\left((1-y^2)\epsilon^2 + \frac{n^2}{16d^2}\right)^n$$

and so

$$\frac{(1-y^2)n}{8d} \le 2\left((1-y^2)\epsilon^2 + \frac{n^2}{16d^2}\right).$$

Since, by (2.4.3), $n^2d^{-2} \le (1-y^2)\epsilon^2$, the above inequality implies

$$\frac{(1-y^2)n}{8d} \le 2\,\frac{17}{16}(1-y^2)\epsilon^2$$

and so $d \ge \frac{1}{17}n\epsilon^{-2}$. □

m] Show that $p \in \mathcal{P}_n \setminus \mathcal{P}_{n-1}$ and $d(p) = n$ imply that the zeros $z_1, z_2, \dots, z_n$ of p satisfy $|z_1 z_2 \cdots z_n| \geq 1$.

n] Show that $d(pq) < \max\{d(p), d(q)\}$ can happen.

Hint: Let

$$p(x) := (1-x)^2 - 2(1-x^2) + 4(x+1)^2 \quad \text{and} \quad q(x) := (x+1) + \tfrac{1}{2}(1-x)\,.$$

Show that $d(p) = 4$, $d(q) = 1$, and $d(pq) = 3$. □

o] Show that if $p \in \mathcal{P}_k \setminus \mathcal{P}_{k-1}$ has no zeros in $(-1,1)$, $z \in \mathbb{C}$, $|z| > 1$, and

$$p(x) = ((x-z)(x-\overline{z}))^m q(x)\,,$$

then $d(p) = \deg(p) = k + 2m$ if m is sufficiently large. This shows that polynomials p with the property $d(p) = \deg(p)$ can have arbitrary many prescribed zeros in $\mathbb{C} \setminus (-1,1)$.

E.5 Lorentz Degree of Trigonometric Polynomials. Given $\omega \in (0, \pi]$ and a real trigonometric polynomial $t \in \mathcal{T}_n$, let $d = d_\omega(t)$ be the minimal nonnegative integer for which t is of the form

$$t(\theta) = \pm \sum_{j=0}^{2d} a_j \sin^j \frac{\omega - \theta}{2} \sin^{2d-j} \frac{\theta + \omega}{2}\,, \qquad a_j \geq 0\,.$$

If there is no such d, then let $d_\omega(t) := \infty$. We call $d = d_\omega(t)$ the *Lorentz degree* of t.

a] Let $t \in \mathcal{T}_n \setminus \mathcal{T}_{n-1}$ be of the form

$$t(\theta) = \sum_{j=0}^{2n} b_j \sin^j \frac{\omega - \theta}{2} \sin^{2n-j} \frac{\theta + \omega}{2}\,.$$

For $m \geq n$ let the numbers $b_{j,m}$, $j = 0, 1, \dots, 2m$, be defined by

$$\begin{aligned} t(\theta) &= \left(\sum_{j=0}^{2n} b_j \sin^j \frac{\omega - \theta}{2} \sin^{2n-j} \frac{\theta + \omega}{2} \right) \\ &\times \left(\frac{1}{\sin^2 \omega} \left(\sin^2 \frac{\omega - \theta}{2} + 2 \cos \omega \sin \frac{\omega - \theta}{2} \sin \frac{\theta + \omega}{2} + \sin^2 \frac{\theta + \omega}{2} \right) \right)^{m-n} \\ &= \sum_{j=0}^{2m} b_{j,m} \sin^j \frac{\omega - \theta}{2} \sin^{2m-j} \frac{\theta + \omega}{2}\,. \end{aligned}$$

Show that if $d_\omega(t)$ is finite, then it is the smallest value of m for which each $b_{j,m}$ is nonnegative or each $b_{j,m}$ is nonpositive.

Hint: The second factor in the representation of t is identically 1. □

We introduce the notation

$$G := \{z = x + iy : -\pi \le x < \pi, \ y \in \mathbb{R}\}$$

and

$$G_\omega := \{z = x + iy : \cos\omega \cosh y \ge \cos x, \ -\pi \le x < \pi, \ y \in \mathbb{R}\}.$$

b] Let $t \in \mathcal{T}_1 \setminus \mathcal{T}_0$. Show that $d_\omega(t) = 1$ if and only if t has its zeros in G_ω.

c] Assume $0 < \omega < \pi/2$, $t \in \mathcal{T}_1$, and $t(z) = 0$ for some

$$z := x + iy \in G \setminus (G_\omega \cup (-\omega, \omega)).$$

Show that

$$d_\omega(t) < \max \frac{4\sin(\omega \pm x)(\sin\omega \cosh y \mp \sin x)}{\cos\omega \sinh^2 y} - 1,$$

where the maximum is to be taken over both sets of signs.

d] Suppose $\pi/2 \le \omega \le \pi$, and $t \in \mathcal{T}_1$, $t(z) = 0$ for some $z \in G \setminus G_\omega$. Show that $d_\omega(t) = \infty$.

e] Show that $d_\omega(t_1 t_2) \le d_\omega(t_1) + d_\omega(t_2)$ for any two trigonometric polynomials t_1 and t_2.

f] Let $0 < \omega < \pi/2$ and $0 < \epsilon < \infty$. Show that if $t \in \mathcal{T}_n$ has no zeros in

$$E_{\omega,\epsilon} := \{z = x + iy : y^2 < \epsilon^2(\omega^2 - x^2), \ x \in (-\omega, \omega), \ y \in \mathbb{R}\},$$

then

$$d_\omega(t) \le n\left(\frac{4}{\cos\omega}\epsilon^{-2} + 2\tan\omega + 1\right).$$

g] Let p be a trigonometric polynomial and $0 < \omega < \pi/2$. Show that $d_\omega(t) < \infty$ if and only if $t = 0$ or t has no zeros in $(-\omega, \omega)$. (Note that part d] shows that this conclusion fails to hold when $\pi/2 \le \omega \le \pi$.)

h] Show that there is an absolute constant $c > 0$ (independent of n, ω and z_0) so that $d_\omega(t) \ge cn\epsilon^{-2}$ whenever

$$t(\theta) = \left(\sin\frac{\theta - z_0}{2} \sin\frac{\theta - \overline{z}_0}{2}\right)^n$$

with $z_0 \in \partial E_{\omega,\epsilon} \setminus \{-\omega, \omega\}$, $0 < \epsilon < \infty$, where $\partial E_{\omega,\epsilon}$ denotes the boundary of $E_{\omega,\epsilon}$ defined in part f].

3

Chebyshev and Descartes Systems

Overview

A *Chebyshev space* is a finite-dimensional subspace of $C(A)$ of dimension $n+1$ that has the property that any element that vanishes at $n+1$ points vanishes identically. Such spaces, whose prototype is the space $\mathcal{P}_n$ of real algebraic polynomials of degree at most n, share with the polynomials many basic properties. The first section is an introduction to these Chebyshev spaces. A basis for a Chebyshev space is called a *Chebyshev system.* Two special families of Chebyshev systems, namely, *Markov systems* and *Descartes systems*, are examined in the second section. The third section examines the *Chebyshev "polynomials"* associated with Chebyshev spaces. These associated Chebyshev polynomials, which equioscillate like the usual Chebyshev polynomials, are extremal for various problems in the supremum norm. The fourth section studies particular Descartes systems

$$(x^{\lambda_0}, x^{\lambda_1}, \dots)\,, \qquad \lambda_0 < \lambda_1 < \dots$$

on $(0, \infty)$ in detail. These systems, which we call *Müntz systems*, can be very explicitly orthonormalized on $[0, 1]$, and this orthogonalization is also examined. The final section constructs Chebyshev "polynomials" associated with the Chebyshev spaces

$$\operatorname{span}\left\{1, \frac{1}{x-a_1}, \dots, \frac{1}{x-a_n}\right\}, \qquad a_i \in \mathbb{R} \setminus [-1, 1]$$

on $[-1, 1]$ and explores their various properties.

3.1 Chebyshev Systems

From an approximation theoretic point of view an essential property that polynomials of degree at most n have is that they can uniquely interpolate at $n+1$ points. This is equivalent to the fact that a polynomial of degree at most n that vanishes at $n+1$ points vanishes identically. Any $(n+1)$-dimensional vector space of continuous functions with this property is called a *Chebyshev space* or sometimes a *Haar space*. Many basic approximation properties extend to these spaces. The precise definition is the following.

Definition 3.1.1 (Chebyshev System). Let A be a Hausdorff space. The sequence $(f_0, \dots, f_n)$ is called a *real* (or *complex*) *Chebyshev system* or *Haar system* of dimension $n+1$ on A if $f_0, \dots, f_n$ are real- (or complex-) valued continuous functions on A, $\operatorname{span}\{f_0, \dots, f_n\}$ over $\mathbb{R}$ (or $\mathbb{C}$) is an $(n+1)$-dimensional subspace of $C(A)$, and any element of $\operatorname{span}\{f_0, \dots, f_n\}$ that has $n+1$ distinct zeros in A is identically zero.

If $(f_0, \dots, f_n)$ is a Chebyshev system on A, then $\operatorname{span}\{f_0, \dots, f_n\}$ is called a *Chebyshev space* or *Haar space* on A.

Chebyshev systems and spaces will be assumed to be real, unless we explicitly specify otherwise. If $A \subset \mathbb{R}$, then the topology on A is always meant to be the usual metric topology.

Implicit in the definition is that A contains at least $n+1$ points. Being a Chebyshev system is a property of the space spanned by the elements of the system, so every basis of a Chebyshev space is a Chebyshev system.

A point $x_0 \in (a, b)$ is called a *double zero* of an $f \in C[a, b]$ if $f(x_0) = 0$ and $f(x_0 - \epsilon)f(x_0 + \epsilon) > 0$ for all sufficiently small $\epsilon > 0$ (in other words, if f vanishes without changing sign at x_0). It is easy to see that if $(f_0, \dots, f_n)$ is a Chebyshev system on $[a, b] \subset \mathbb{R}$, then every $0 \neq f \in \operatorname{span}\{f_0, \dots, f_n\}$ has at most n zeros even if each double zero is counted twice; see E.10. Chebyshev spaces are defined via zero counting, and many of the theorems in the theory of Chebyshev spaces are proved by zero counting arguments. So it is important to make the agreement that, unless it is stated explicitly otherwise, we count the zeros of an element f from a Chebyshev space on $[a, b]$ so that each double zero of f is counted twice.

The following simple equivalences hold:

Proposition 3.1.2 (Equivalences). *Let $f_0, \dots, f_n$ be real- (or complex-) valued continuous functions on a Hausdorff space A (containing at least $n+1$ points). Then the following are equivalent:*

a] *Every $0 \neq p \in \operatorname{span}\{f_0, \dots, f_n\}$ has at most n distinct zeros in A.*

b] *If $x_0, \dots, x_n$ are distinct elements of A and $y_0, \dots, y_n$ are real (or complex) numbers, then there exists a unique $p \in \operatorname{span}\{f_0, \dots, f_n\}$ such that*

$$p(x_i) = y_i\,, \qquad i = 1, 2, \dots, n\,.$$

c] *If $x_0, \dots, x_n$ are distinct points of A, then*

$$D(x_0, \dots, x_n) := \begin{vmatrix} f_0(x_0) & \dots & f_n(x_0) \\ \vdots & \ddots & \vdots \\ f_0(x_n) & \dots & f_n(x_n) \end{vmatrix} \neq 0\,.$$

Proof. These equivalences are all elementary facts in linear algebra. □

On an interval there is a sign regularity to the determinants in c].

Proposition 3.1.3. *Suppose $(f_0, \dots, f_n)$ is a (real) Chebyshev system on $[a, b] \subset \mathbb{R}$. Then there exists a $\delta := -1$ or $\delta := 1$ such that*

$$\delta \begin{vmatrix} f_0(x_0) & \dots & f_n(x_0) \\ \vdots & \ddots & \vdots \\ f_0(x_n) & \dots & f_n(x_n) \end{vmatrix} > 0$$

for any $a \le x_0 < x_1 < \cdots < x_n \le b$.

Proof. This follows immediately from part c] of the previous proposition and continuity considerations. That is, if $D(x_0, \dots, x_n) < 0$ while $D(y_0, \dots, y_n) > 0$, then for some $\lambda \in (0, 1)$

$$D(\lambda x_0 + (1 - \lambda) y_0, \dots, \lambda x_n + (1 - \lambda) y_n) = 0\,,$$

which is impossible. □

The intimate relationship between Chebyshev systems and best approximation in the uniform norm is indicated by the next result. In order to state it we need to introduce the notion of an alternation sequence.

Definition 3.1.4 (Alternation Sequence). Let $A \subset \mathbb{R}$ and let

$$x_0 < x_1 < \cdots < x_n$$

be $n + 1$ points of A. Then $(x_0, x_1, \dots, x_n)$ is said to be an *alternation sequence* of *length* $n + 1$ for a real valued $f \in C(A)$ if

$$|f(x_i)| = \|f\|_A\,, \qquad i = 0, 1, \dots, n$$

and

$$\operatorname{sign}(f(x_{i+1})) = -\operatorname{sign}(f(x_i))\,, \qquad i = 0, 1, \dots, n-1\,.$$

Definition 3.1.5 (Best Approximation). Suppose that U is a (finite-dimensional) subspace of a normed space $(V, \|\cdot\|)$. If $g \in V$ and $p \in U$ satisfy

$$\|g - p\| = \inf_{h \in U} \|g - h\|,$$

then p is said to be a *best approximation* to g from U.

As a result of the finite dimensionality of the subspace U, at least one best approximation to any $g \in V$ from U exists. This is straightforward since

$$T := \{p \in U : \|p\| \leq \|g\| + 1\}$$

is a compact subset of U, so any sequence (p_j) of approximations to g from U satisfying

$$\|g - p_j\| \leq j^{-1} + \inf_{h \in U} \|g - h\|$$

has a convergent subsequence with limit in U. This limit is then a best approximation to g from U.

Theorem 3.1.6 (Alternation of Best Approximations). *Suppose $(f_0, \ldots, f_n)$ is a Chebyshev system on $[a, b] \subset \mathbb{R}$. Let A be a closed subset of $[a, b]$ containing at least $n + 2$ distinct points. Then $p \in H_n := \operatorname{span}\{f_0, \ldots, f_n\}$ is a best approximation to $g \in C(A)$ from H_n in the uniform norm on A if and only if there exists an alternation sequence of length $n + 2$ for $g - p$ on A.*

Proof. The proof of the *only if* part of the theorem is mostly an example of a standard type of perturbation argument that will recur later.

The perturbation argument goes as follows. Suppose p is a best approximation of required type and suppose a alternation sequence of maximal length for $g - p$ is

$$(x_0 < x_1, < \cdots < x_m)$$

where $x_i \in A$ and where m is strictly less than $n + 1$. Suppose, without loss of generality, that

$$g(x_0) - p(x_0) > 0$$

(otherwise multiply by -1). Now let

$$Y := \{x \in A : |g(x) - p(x)| = \|g - p\|_A\}.$$

Note that Y is compact. Since $(x_0 < x_1 < \ldots < x_m)$ is an alternation sequence of maximal length, we can divide Y into $m + 1$ disjoint compact subsets $Y_0, Y_1, \ldots, Y_m$ with

$$x_0 \in Y_0, \quad x_1 \in Y_1, \ldots, \quad x_m \in Y_m$$

so that

$$\operatorname{sign}(g(x) - p(x)) = -\operatorname{sign}(g(y) - p(y)) \neq 0\,, \qquad x \in Y_i\,,\ \ y \in Y_{i+1}\,.$$

Now choose a $p^* \in \operatorname{span}\{f_0, \ldots, f_n\}$ satisfying

$$\operatorname{sign}_{x \in Y_i}(p^*(x)) = (-1)^i\,, \qquad i = 0, 1, \ldots, m\,.$$

This can be done by choosing points z_i with

$$\max Y_{i-1} < z_i < \min Y_i\,, \qquad i = 1, 2, \ldots, m$$

and then applying E.11. We now claim that, for $\delta > 0$ sufficiently small,

$$\|g - (p + \delta p^*)\|_A < \|g - p\|_A\,, \tag{3.1.1}$$

which contradicts the fact that p is a best approximation, and so there must exist an alternation set of length $n + 2$ for $g - p$ on A. To verify (3.1.1) we proceed as follows:

For each $i = 0, 1, \ldots, m$ choose an open set $O_i \subset [a, b]$ (in the usual metric topology relative to $[a, b]$) containing Y_i so that for every $x \in \overline{O}_i$,

$$\operatorname{sign}(g(x) - p(x)) = \operatorname{sign}(p^*(x)) \tag{3.1.2}$$

and

$$|g(x) - p(x)| \geq \tfrac{1}{2}\|g - p\|_A\,. \tag{3.1.3}$$

Now pick a $\delta_1 > 0$ such that for every $x \in B := A \setminus \bigcup_{i=0}^m O_i$ and $\delta \in (0, \delta_1)$,

$$|g(x) - (p(x) + \delta p^*(x))| < \|g - p\|_A\,,$$

which can be done since B is compact and by construction

$$\|g - p\|_B < \|g - p\|_A\,.$$

Note that (3.1.2) and (3.1.3) allow us to pick a $\delta_2 > 0$ such that for $x \in \bigcup_{i=0}^m \overline{O}_i$ and $\delta \in (0, \delta_2)$,

$$|g(x) - (p(x) + \delta p^*(x))| < |g(x) - p(x)|\,.$$

This verifies (3.1.1) and finishes the direct half of the theorem.

The proof of the converse is simple. Suppose there is an alternation sequence of length $n + 2$ for $g - p$ on A, and suppose there exists a p^* with

$$\|g - p^*\|_A < \|g - p\|_A.$$

Then $p^* - p$ has at least $n+1$ zeros on $[a, b]$, one between any two consecutive alternation points for $g - p$ on A, and hence it vanishes identically. This contradiction finishes the proof. □

In the setting of Theorem 3.1.6 the best approximation is unique; see E.5.

Comments, Exercises, and Examples.

The terminology is not entirely standard in the literature with Chebyshev systems often referred to as Haar systems on intervals of $\mathbb{R}$. There are various proofs of Theorem 3.1.6, of which ours is by no means the most elegant (see Cheney [66], for example). The point of this proof is that it easily modifies to deal with characterizations of extremal functions for various other extremal problems. Many good books cover this standard material. See, for example, Cheney [66], Lorentz [86a], or Pinkus [89]; see also Appendix 3. An extensive treatment of Chebyshev systems is available in Karlin and Studden [66] or Nürnberger [89], where E.3 can be found. E.4 shows that real Chebyshev systems are intrinsically one-dimensional.

E.1 Examples of Chebyshev systems.

a] Suppose $0 = \lambda_0 < \lambda_1 < \cdots < \lambda_n$. Show that

$$(x^{\lambda_0}, x^{\lambda_1}, \dots, x^{\lambda_n})$$

is a Chebyshev system on $[0, \infty)$.

b] Suppose $\lambda_0 < \lambda_1 < \cdots < \lambda_n$. Show that

$$(x^{\lambda_0}, x^{\lambda_1}, \dots, x^{\lambda_n})$$

is a Chebyshev system on $(0, \infty)$.

c] Suppose $\lambda_0 < \lambda_1 < \cdots < \lambda_n$. Show that

$$(x^{\lambda_0},\, x^{\lambda_0}\log x\,,\, x^{\lambda_1},\, x^{\lambda_1}\log x\,, \dots ,\, x^{\lambda_n},\, x^{\lambda_n}\log x)$$

is a Chebyshev system on $(0, \infty)$.

d] Suppose $\lambda_0 < \lambda_1 < \cdots < \lambda_n$. Show that

$$\left(\frac{1}{x-\lambda_0}, \frac{1}{x-\lambda_1}, \dots, \frac{1}{x-\lambda_n}\right)$$

is a Chebyshev system on $(-\infty, \infty) \setminus \{\lambda_0, \lambda_1, \dots, \lambda_n\}$.

e] Suppose $\lambda_0 < \lambda_1 < \cdots < \lambda_n$. Show that

$$(e^{\lambda_0 x},\, e^{\lambda_1 x},\, \dots ,\, e^{\lambda_n x})$$

is a Chebyshev system on $(-\infty, \infty)$.

f] Show that

$$(1,\, \cos\theta,\, \sin\theta,\, \cos 2\theta,\, \sin 2\theta,\, \dots ,\, \cos n\theta,\, \sin n\theta)$$

is a Chebyshev system on $[0, 2\pi)$.

g] Show that

$$(1,\, \cos\theta,\, \cos 2\theta,\, \dots ,\, \cos n\theta)$$

is a Chebyshev system on $[0, \pi)$.

E.2 More Examples.

a] If $(f_0, \dots, f_n)$ is a Chebyshev system on A, then it is also a Chebyshev system on any subset B of A containing at least $n+1$ points.

b] If $(f_0, \dots, f_n)$ is a Chebyshev system on A and $g \in C(A)$ is strictly positive on A, then $(gf_0, \dots, gf_n)$ is also a Chebyshev system on A.

c] If $(f_0, \dots, f_n)$ is a Chebyshev system on $[0,1]$, then

$$\left(1, \int_0^x f_0(t)\,dt, \dots, \int_0^x f_n(t)\,dt\right)$$

is also a Chebyshev system on $[0,1]$.

See E.8 of Section 3.2, which treats the effect of differentiation on a Chebyshev system.

E.3 Extended Complete Chebyshev Systems. Let $(g_0, \dots, g_n)$ be a sequence of functions in $C^n[a,b]$. Define the *Wronskian determinant*

$$W(g_0, \dots, g_m)(t) := \begin{vmatrix} g_0(t) & g_1(t) & \dots & g_m(t) \\ g_0'(t) & g_1'(t) & \dots & g_m'(t) \\ \vdots & \vdots & \ddots & \vdots \\ g_0^{(m)}(t) & g_1^{(m)}(t) & \dots & g_m^{(m)}(t) \end{vmatrix}.$$

We say that $(g_0, \dots, g_n)$ is an *extended complete Chebyshev system (ECT system)* on $[a,b]$ if

$$W(g_0, \dots, g_m)(t) > 0, \quad m = 0, 1, \dots, n, \quad t \in [a,b].$$

a] Let $\operatorname{span}\{g_0, \dots, g_n\}$ be an $(n+1)$-dimensional subspace of $C^n[a,b]$. Show that the following statements are equivalent:

(i) For every $m = 0, 1, \dots, n$, $0 \neq f \in \operatorname{span}\{g_0, \dots, g_m\}$ has at most m zeros in $[a,b]$ counting multiplicities ($x_0 \in [a,b]$ is a zero of f with multiplicity k if $f(x_0) = f'(x_0) = \cdots f^{(k-1)}(x_0) = 0$ and $f^{(k)}(x_0) \neq 0$).

(ii) For each $i = 0, 1, \dots, n$, there exists a choice of $\delta_i := 1$ or $\delta_i := -1$ such that

$$(\delta_0 g_0, \delta_1 g_1, \dots, \delta_n g_n)$$

is an ECT system on $[a,b]$.

In particular, every ECT system on $[a,b]$ is a Chebyshev system on $[a,b]$.

Proof. For details see Karlin and Studden [66]. □

b] Characterization Theorem. *The following statements are equivalent:*

(i) $(g_0, \dots, g_n)$ *is an ECT system on* $[a, b]$.

(ii) There exist $w_i \in C^{n-i}[a, b]$, $i = 0, 1, \dots, n$, *with* w_i *strictly positive such that*

$$g_0(t) = w_0(t)\,,$$

$$g_1(t) = w_0(t) \int_a^t w_1(t_1)\, dt_1\,,$$

$$\vdots$$

$$g_n(t) = w_0(t) \int_a^t w_1(t_1) \int_a^{t_1} w_2(t_2) \cdots \int_a^{t_{n-1}} w_n(t_n)\, dt_n \cdots dt_2\, dt_1\,.$$

Proof. This is proved by induction on n. See Karlin and Studden [66]. □

c] Suppose $\lambda_0 < \lambda_1 < \cdots < \lambda_n$. Show that $(x^{\lambda_0}, \dots, x^{\lambda_n})$ is an ECT system on $[a, b]$ provided $a > 0$.

E.4 Railway Track Theorem. Real Chebyshev systems exist only on very special subsets of $\mathbb{R}^m$. Indeed, real Chebyshev systems intrinsically live on one-dimensional subsets.

a] Suppose $A \subset \mathbb{R}^m$ contains three distinct arcs that join at a point x_0. Then, for $n \geq 2$, there exists no real Chebyshev system $(f_0, \dots, f_n)$ on A.

Proof. Suppose there exists a real Chebyshev system $(f_0, \dots, f_n)$ on such a set A. Let

$$V(x, y) := D(x, y, x_2, x_3, \dots, x_n)$$

(D is defined in Proposition 3.1.2) which is never zero for distinct points $x, y, x_2, x_3, \dots, x_n$. Choose distinct points $x_0, x_1, \dots, x_n$ on one of the three distinct arcs so that x_0 is adjacent to x_1. Pick the points $z_1 \neq x_0$ and $z_2 \neq x_0$ so that z_1, z_2, and x_1 are on different arcs. Now consider interchanging $x := x_0$ and $y := x_1$ by moving x from x_0 to z_1, y from x_1 to z_2, x from z_1 to x_1, and y from z_2 to x_0. Since $x, y, x_2, x_3, \dots, x_n$ remain distinct, $V(x, y)$ does not vanish in this process. This contradicts the fact that $V(x, y)$ is continuous and $V(x_0, x_1) = -V(x_1, x_0)$. □

The following more general result of Mairhuber [56] also holds:

b] Mairhuber's Theorem. *If* $(f_0, \dots, f_n)$ *is a real Chebyshev space on* A, *then* A *is homeomorphic to a subset of the unit circle.*

E.5 Uniqueness of Best Approximations. Prove that a best approximation from a Chebyshev space satisfying the conditions of Theorem 3.1.6 is unique.

Hint: Suppose f has two best approximations $p_1 \in H_n$ and $p_2 \in H_n$. Then, by the alternation characterization, $p_1 - p_2 \in H_n$ has at least $n+1$ zeros on $[a,b]$ (we count each internal zero without sign change twice). Now E.10 implies that $p_1 - p_2 = 0$. □

E.6 De la Vallée Poussin Theorem. *Suppose H_n is a Chebyshev space of dimension $(n+1)$ on $[a,b]$. If $p \in H_n$ and there exist $n+2$ points*

$$a \leq x_0 < x_1 < \cdots < x_{n+1} \leq b$$

so that

$$\operatorname{sign}(f(x_i) - p(x_i)) = -\operatorname{sign}(f(x_{i+1}) - p(x_{i+1})), \quad i = 0, 1, \ldots, n,$$

then

$$\inf_{p \in H_n} \|f - p\|_{[a,b]} \geq \min_{i=0,\ldots,n+1} |f(x_i) - p(x_i)|.$$

E.7 Haar's Characterization of Chebyshev Spaces. The following pretty theorem is due to Haar (for a proof, see E.3 of Appendix 3):

Theorem. *Let $f_0, \ldots, f_n \in C(A)$ where A is a compact Hausdorff space containing at least $n+1$ points. Then $(f_0, \ldots, f_n)$ is a Chebyshev system on A if and only if every $g \in C(A)$ has a unique best approximation from* $\operatorname{span}\{f_0, \ldots, f_n\}$ *in the uniform norm on A.*

E.8 Best Approximation to x^n. Reprove Theorem 2.1.1 by using the alternation characterization of best approximations.

E.9 Best Rational Approximations. Let $f \in C[a,b]$. Then $p/q \in \mathcal{R}_{n,m}$ is a best approximation to f from $\mathcal{R}_{n,m}$ in $C[a,b]$ if and only if $f - p/q$ has an alternation set of length at least

$$2 + \max\{n + \deg(q), m + \deg(p)\}$$

$[a,b]$. (Here we must assume p/q is written in a reduced form.)

The proof of this is a fairly complicated variant of the proof of Theorem 3.1.6 (see, for example, Cheney [66]).

E.10 Zeros of Functions in Chebyshev Spaces. As before, we call the point $x_0 \in (a,b)$ a double zero of $f \in C[a,b]$ if $f(x_0) = 0$ and

$$f(x_0 - \epsilon) f(x_0 + \epsilon) > 0$$

for all sufficiently small $\epsilon > 0$ (in other words, if f vanishes without changing sign at x_0). Let $(f_0, \ldots, f_n)$ be a Chebyshev system on $[a,b] \subset \mathbb{R}$. Show that every $0 \neq f \in \operatorname{span}\{f_0, \ldots, f_n\}$ has at most n zeros even if each double zero is counted twice.

Hint: Use Proposition 3.1.2 b]. □

E.11 Functions in a Chebyshev Space with Prescribed Sign Changes. Let $(f_0, \dots, f_n)$ be a Chebyshev system on $[a, b]$, and let

$$a < z_1 < z_2 < \cdots < z_m < b, \qquad 0 \le m \le n.$$

Show that there is a function $p^* \in \operatorname{span}\{f_0, \dots, f_n\}$ such that

(i) $p^*(x) = 0$ if and only if $x = z_i$ for some $i = 1, 2, \dots, m$,

(ii) $p^*(x)$ changes sign at each z_i, $i = 1, 2, \dots, m$.

Hint: If $m = n$, then use Proposition 3.1.3 and a continuity argument to show that

$$p^*(x) = \begin{vmatrix} f_0(x) & f_1(x) & \dots & f_n(x) \\ f_0(z_1) & f_1(z_1) & \dots & f_n(z_1) \\ \vdots & \vdots & \ddots & \vdots \\ f_0(z_n) & f_1(z_n) & \dots & f_n(z_n) \end{vmatrix}$$

satisfies the requirements.

If $m < n$, then use the already proved case, a limiting argument, and E.10 to show that there are $p_j \in \operatorname{span}\{f_0, \dots, f_n\}$, $j = 1, 2$, such that

(1) $p_j(x)$ changes sign at x if and only if $x = z_i$, $i = 1, 2, \dots, m$,

(2) $p_1(x) \neq 0$ for every $x \in [a, z_m] \setminus \{z_1, z_2, \dots, z_m\}$,

(3) $p_2(x) \neq 0$ for every $x \in [z_1, b] \setminus \{z_1, z_2, \dots, z_m\}$.

Now show that either $p^* := p_1 + p_2$ or $p^* := p_1 - p_2$ satisfies the requirements. □

E.12 The Dimension of a Chebyshev Space on a Circle. Let $(f_0, \dots, f_n)$ be a Chebyshev system on a circle C. Show that n must be even. Observe that such Chebyshev systems exist.

Hint: Show that for every set of n distinct points $x_1, x_2, \dots, x_n$ on the circle there is a $p \in \operatorname{span}\{f_0, \dots, f_n\}$ such that $p(x) = 0$ if and only if $x \in \{x_1, x_2, \dots, x_n\}$ and $p(x)$ changes sign at each x_i. □

3.2 Descartes Systems

Chebyshev systems capture some of the essential properties of polynomials. There are two additional types of systems that capture some additional properties.

Definition 3.2.1 (Markov System). We say that $(f_0, \dots, f_n)$ is a *Markov system* on a Hausdorff space A if each $f_i \in C(A)$, and $\{f_0, \dots, f_m\}$ is a Chebyshev system for each $m = 0, 1, \dots, n$. (We allow n to tend to $+\infty$, in which case we call the system an *infinite Markov system* on A.)

A Markov system is just a Chebyshev system with each initial segment also a Chebyshev system. Being a Chebyshev system is a property of the space not of the basis. However, the Markov system depends on the basis. For example,

$$(x^{\lambda_0}, x^{\lambda_1}, \ldots), \qquad \lambda_0 < \lambda_1 < \cdots$$

is a Markov system on any $A \subset (0, \infty)$ containing infinitely many points, but not every basis of $\operatorname{span}\{x^{\lambda_0}, x^{\lambda_1}, \ldots\}$ is a Markov system on A (see E.1).

Proposition 3.2.2. *$(f_0, \ldots, f_n)$ is a Markov system on an interval $[a, b]$ if and only if for each $i = 0, 1, \ldots, n$, there exists a choice of $\delta_i := 1$ or $\delta_i := -1$ such that with $g_i := \delta_i f_i$,*

$$D\begin{pmatrix} g_0 & g_1 & \cdots & g_m \\ x_0 & x_1 & \cdots & x_m \end{pmatrix} := \begin{vmatrix} g_0(x_0) & \cdots & g_m(x_0) \\ \vdots & \ddots & \vdots \\ g_0(x_m) & \cdots & g_m(x_m) \end{vmatrix} > 0$$

for every $a \le x_0 < x_1 < \cdots < x_m \le b$ and $m = 0, 1, \ldots, n$.

Proof. This is an easy consequence of Proposition 3.1.3 by induction on n. □

A stronger property that a system on an interval can have is the following:

Definition 3.2.3 (Descartes System). The system $(f_0, \ldots, f_n)$ is said to be a *Descartes system* (or *order complete Chebyshev system*) on an interval I if each $f_i \in C(I)$ and

$$D\begin{pmatrix} f_{i_0} & f_{i_1} & \cdots & f_{i_m} \\ x_0 & x_1 & \cdots & x_m \end{pmatrix} > 0$$

for any $0 \le i_0 < i_1 < \cdots < i_m \le n$ and $x_0 < x_1 < \cdots < x_m$ from I. (Once again we allow n to tend to ∞.)

This again is a property of the basis. It implies that any finite-dimensional subspace generated by some basis elements is a Chebyshev space on I. The canonical example of a Descartes system on $[a, b]$, $a > 0$, is

$$(x^{\lambda_0}, x^{\lambda_1}, \ldots), \qquad \lambda_0 < \lambda_1 < \cdots$$

(see E.2). A Descartes system on I is obviously a Descartes system on any subinterval of I.

The following version of Descartes' rule of signs holds for Descartes systems.

Theorem 3.2.4 (Descartes' Rule of Signs). *If $(f_0, \ldots, f_n)$ is a Descartes system on $[a,b]$, then the number of distinct zeros of any*

$$0 \neq f = \sum_{i=0}^{n} a_i f_i \,, \qquad a_i \in \mathbb{R}$$

is not greater than the number of sign changes in $(a_0, \ldots, a_n)$.

A sign change occurs between a_i and a_{i+k} exactly when $a_i a_{i+k} < 0$ and $a_{i+1} = a_{i+2} = \cdots = a_{i+k-1} = 0$.

Proof. Suppose $(a_0, \ldots, a_n)$ has p sign changes. Then we can partition $\{a_0, \ldots, a_n\}$ into exactly $p+1$ blocks so that each block is of the form

$$a_{n_m+1}, a_{n_m+2}, \ldots, a_{n_{m+1}}, \qquad m = 0, 1, \ldots, p$$

$(n_0 := -1, n_{p+1} := n)$, where all of the coefficients in each of the blocks are of the same sign and not all the coefficients in a block vanish. Now let

$$g_m := \sum_{i=n_m+1}^{n_{m+1}} |a_i| f_i \,, \qquad m = 0, 1, \ldots, p \,.$$

Then, for $a \leq x_0 < x_1 < \cdots < x_p \leq b$,

$$D\begin{pmatrix} g_0 & g_1 & \cdots & g_p \\ x_0 & x_1 & \cdots & x_p \end{pmatrix}$$
$$= \sum_{i_0=n_0+1}^{n_1} \cdots \sum_{i_p=n_p+1}^{n_{p+1}} |a_{i_0}| \cdots |a_{i_p}| D\begin{pmatrix} f_{i_0} & f_{i_1} & \cdots & f_{i_p} \\ x_0 & x_1 & \cdots & x_p \end{pmatrix} > 0$$

since each of the determinants in the sum is positive. Thus $\{g_0, \ldots .g_p\}$ is a $(p+1)$-dimensional Chebyshev system on $[a,b]$, and hence

$$f := \sum_{i=0}^{p} \delta_i g_i \,, \qquad \delta_i = \pm 1$$

has at most p zeros. This finishes the proof. □

A refined version of Descartes' rule of signs for ordinary polynomials is presented in the exercises. The following comparison theorem due to Pinkus and, independently, Smith [78] will be of use later.

Theorem 3.2.5. *Suppose $(f_0, \dots, f_n)$ is a Descartes system on $[a,b]$. Suppose*

$$p = f_\alpha + \sum_{i=1}^{k} a_i f_{\lambda_i}\,, \qquad \text{and} \qquad q = f_\alpha + \sum_{i=1}^{k} b_i f_{\gamma_i}\,,$$

where $0 \le \lambda_1 < \lambda_2 < \cdots < \lambda_k \le n$, $0 \le \gamma_1 < \gamma_2 < \cdots < \gamma_k \le n$,

$$0 \le \gamma_i \le \lambda_i < \alpha\,, \qquad i = 1, 2, \dots, m\,,$$

and

$$\alpha < \lambda_i \le \gamma_i \le n\,, \qquad i = m+1, m+2, \dots, k$$

with strict inequality for at least one index $i = 1, 2, \dots, k$. If

$$p(x_i) = q(x_i) = 0\,, \qquad i = 1, 2, \dots, k\,,$$

where $x_i \in [a,b]$ are distinct, then

$$|p(x)| \le |q(x)|$$

for all $x \in [a,b]$ with strict inequality for $x \neq x_i$.

The proof is left as a guided exercise (see E.4) with some interesting consequences presented in E.5.

Comments, Exercises, and Examples.

Theorem 3.2.4 characterizes Descartes systems; see Karlin and Studden [66, p. 25]. Some caution must be exercised since, as in the previous section, definitions are not entirely standard. We will explore two particular Descartes systems in greater detail later; see E.2 and E.3. For further material, the reader is referred to Karlin and Studden [66], Karlin [68], and Nürnberger [89].

E.1 Distinctions.

a] Given $\lambda_0 < \lambda_1 < \cdots$ and $A \subset (0, \infty)$ with infinitely many points, show that $(x^{\lambda_0}, x^{\lambda_1}, \dots)$ is a Markov system on A, but there is a basis for $\operatorname{span}\{x^{\lambda_0}, x^{\lambda_1}, \dots\}$, which is not a Markov system on A.

b] Find a Markov system that is not a Descartes system.

E.2 Examples of Descartes Systems.

a] Suppose $\lambda_0 < \lambda_1 < \cdots$. Show that the *Müntz system*

$$(x^{\lambda_0}, x^{\lambda_1}, \dots)$$

is a Descartes system on $(0, \infty)$.

Hint: For every $0 \le i_0 < i_1 < \cdots < i_m$, the determinant

$$D\begin{pmatrix} x^{\lambda_{i_0}} & x^{\lambda_{i_1}} & \dots & x^{\lambda_{i_m}} \\ x_0 & x_1 & \dots & x_m \end{pmatrix} = \begin{vmatrix} x_0^{\lambda_{i_0}} & \dots & x_0^{\lambda_{i_m}} \\ \vdots & \ddots & \vdots \\ x_m^{\lambda_{i_0}} & \dots & x_m^{\lambda_{i_m}} \end{vmatrix}$$

is nonzero for any $0 < x_0 < x_1 < \cdots < x_m < \infty$ by Proposition 3.1.2 and E.1 a] of Section 3.1. It only remains to prove that it is positive whenever $0 < x_0 < x_1 < \cdots < x_m < \infty$. Observe that the exponents λ_{i_j} can be varied continuously (for fixed x_i) without changing the sign of the determinant provided no two ever become equal. Now perturb $(\lambda_{i_0}, \lambda_{i_1}, \dots, \lambda_{i_m})$ into $(0, 1, \dots, m)$ and observe that the determinant becomes a Vandermonde determinant, as in part b], which in this case is positive. □

b] Vandermonde Determinant. Show that

$$\begin{vmatrix} 1 & x_0 & \dots & x_0^m \\ 1 & x_1 & \dots & x_1^m \\ \vdots & \vdots & \ddots & \vdots \\ 1 & x_m & \dots & x_m^m \end{vmatrix} = \prod_{0 \le i < j \le m} (x_j - x_i)\,.$$

Hint: The determinant is a polynomial in $x_0, x_1, \dots, x_m$ of degree m in each variable that vanishes whenever $x_i = x_j$. □

c] Suppose $\lambda_0 < \lambda_1 < \cdots$. Show that the *exponential system*

$$(e^{\lambda_0 t}, e^{\lambda_1 t}, \dots)$$

is a Descartes system on $(-\infty, \infty)$.

Hint: Use part a]. □

d] Suppose $0 < \lambda_0 < \lambda_1 < \cdots$. Show that

$$(\sinh \lambda_0 t, \sinh \lambda_1 t, \dots)$$

is a Descartes system on $(0, \infty)$.

Outline. Let $0 \le i_0 < i_1 < \cdots < i_m$ be fixed integers. First we show that

$$(\sinh \lambda_{i_0} t, \sinh \lambda_{i_1} t, \dots, \sinh \lambda_{i_m} t)$$

is a Chebyshev system on $(0, \infty)$. Indeed, let

$$0 \ne f \in \operatorname{span}\{\sinh \lambda_{i_0} t, \sinh \lambda_{i_1} t, \dots, \sinh \lambda_{i_m} t\}\,.$$

Then

$$0 \ne f \in \operatorname{span}\{e^{\pm\lambda_{i_0} t}, e^{\pm\lambda_{i_1} t}, \dots, e^{\pm\lambda_{i_m} t}\}$$

and by E.1 e] of Section 3.1, f has at most $2m$ zeros in $(-\infty, \infty)$. Since f is odd, it has at most m zeros in $(0, \infty)$.

Since for every $0 \le i_0 < i_1 < \cdots < i_m$, $(\sinh \lambda_{i_0} t, \ldots, \sinh \lambda_{i_m} t)$ is a Chebyshev system on $(0, \infty)$, the determinant

$$D\begin{pmatrix} \sinh \lambda_{i_0} t & \sinh \lambda_{i_1} t & \ldots & \sinh \lambda_{i_m} t \\ x_0 & x_1 & \ldots & x_m \end{pmatrix}$$

$$= \begin{vmatrix} \sinh \lambda_{i_0} x_0 & \sinh \lambda_{i_1} x_0 & \ldots & \sinh \lambda_{i_m} x_0 \\ \sinh \lambda_{i_0} x_1 & \sinh \lambda_{i_1} x_1 & \ldots & \sinh \lambda_{i_m} x_1 \\ \vdots & \vdots & \ddots & \vdots \\ \sinh \lambda_{i_0} x_m & \sinh \lambda_{i_1} x_m & \ldots & \sinh \lambda_{i_m} x_m \end{vmatrix}$$

is nonzero for any $0 < x_0 < x_1 < \cdots < x_m < \infty$ Proposition 3.1.2. So it only remains to prove that it is positive whenever $0 < x_0 < x_1 < \cdots < x_m < \infty$. Now let

$$D(\alpha) := D\begin{pmatrix} \sinh \lambda_{i_0} t & \sinh \lambda_{i_1} t & \ldots & \sinh \lambda_{i_m} t \\ \alpha x_0 & \alpha x_1 & \ldots & \alpha x_m \end{pmatrix}$$

$$= \begin{vmatrix} \sinh \lambda_{i_0} \alpha x_0 & \sinh \lambda_{i_1} \alpha x_0 & \ldots & \sinh \lambda_{i_m} \alpha x_0 \\ \sinh \lambda_{i_0} \alpha x_1 & \sinh \lambda_{i_1} \alpha x_1 & \ldots & \sinh \lambda_{i_m} \alpha x_1 \\ \vdots & \vdots & \ddots & \vdots \\ \sinh \lambda_{i_0} \alpha x_m & \sinh \lambda_{i_1} \alpha x_m & \ldots & \sinh \lambda_{i_m} \alpha x_m \end{vmatrix}$$

and

$$D^*(\alpha) := D\begin{pmatrix} \frac{1}{2} e^{\lambda_{i_0} t} & \frac{1}{2} e^{\lambda_{i_1} t} & \ldots & \frac{1}{2} e^{\lambda_{i_m} t} \\ \alpha x_0 & \alpha x_1 & \ldots & \alpha x_m \end{pmatrix}$$

$$= \begin{vmatrix} \frac{1}{2} e^{\lambda_{i_0} \alpha x_0} & \frac{1}{2} e^{\lambda_{i_1} \alpha x_0} & \ldots & \frac{1}{2} e^{\lambda_{i_m} \alpha x_0} \\ \frac{1}{2} e^{\lambda_{i_0} \alpha x_1} & \frac{1}{2} e^{\lambda_{i_1} \alpha x_1} & \ldots & \frac{1}{2} e^{\lambda_{i_m} \alpha x_1} \\ \vdots & \vdots & \ddots & \vdots \\ \frac{1}{2} e^{\lambda_{i_0} \alpha x_m} & \frac{1}{2} e^{\lambda_{i_1} \alpha x_m} & \ldots & \frac{1}{2} e^{\lambda_{i_m} \alpha x_m} \end{vmatrix}$$

where $0 < x_0 < x_1 < \cdots < x_m < \infty$ are fixed. Since

$$(\sinh \lambda_{i_0} t, \sinh \lambda_{i_1} t, \ldots, \sinh \lambda_{i_m} t)$$

and

$$(e^{\lambda_{i_0} t}, e^{\lambda_{i_1} t}, \ldots, e^{\lambda_{i_m} t})$$

are Chebyshev systems on $(0, \infty)$, $D(\alpha)$ and $D^*(\alpha)$ are continuous nonvanishing functions of α on $(0, \infty)$. Now, observe that

$$\lim_{\alpha \to \infty} |D(\alpha)| = \lim_{\alpha \to \infty} |D^*(\alpha)| = \infty \qquad \text{and} \qquad \lim_{\alpha \to \infty} \frac{D(\alpha)}{D^*(\alpha)} = 1\,.$$

By part c],

$$(e^{\lambda_{i_0} t}, e^{\lambda_{i_1} t}, \ldots, e^{\lambda_{i_m} t})$$

is a Descartes system on $(-\infty, \infty)$, hence $D^*(\alpha) > 0$ for every $\alpha > 0$. So the above limit relations imply that $D(\alpha) > 0$ for every large enough α, hence for every $\alpha > 0$. In particular,

$$D(1) = D\begin{pmatrix} \sinh \lambda_{i_0} t & \sinh \lambda_{i_1} t & \dots & \sinh \lambda_{i_m} t \\ x_0 & x_1 & \dots & x_m \end{pmatrix} > 0\,,$$

which finishes the proof. □

e] Suppose $0 < \lambda_0 < \lambda_1 < \cdots$. Show that

$$(\cosh \lambda_0 t, \cosh \lambda_1 t, \dots)$$

is a Descartes system on $(0, \infty)$.

Hint: Proceed as in the outline for part d]. □

E.3 Rational Systems.

a] Cauchy Determinants. Show that

$$\begin{vmatrix} \frac{1}{\alpha_1+\beta_1} & \cdots & \frac{1}{\alpha_1+\beta_m} \\ \vdots & \ddots & \vdots \\ \frac{1}{\alpha_m+\beta_1} & \cdots & \frac{1}{\alpha_m+\beta_m} \end{vmatrix} = \frac{\prod\limits_{1\le i<j\le m} (\alpha_j - \alpha_i)(\beta_j - \beta_i)}{\prod\limits_{1\le i,j\le m} (\alpha_i + \beta_j)}\,.$$

Hint: Multiply both sides above by $\prod\limits_{1\le i,j\le m} (\alpha_i + \beta_j)$ and observe that both sides are polynomials of the same degree, $m-1$, in each variable α_i, β_i. Also both sides vanish exactly when $\alpha_i = \alpha_j$ or $\beta_i = \beta_j$. So up to a constant both sides are the same. Now show that the constant is 1. □

b] Let $\alpha_1 > \alpha_2 > \cdots > b$. Show that

$$\left(\frac{1}{\alpha_1 - x}, \frac{1}{\alpha_2 - x}, \dots\right)$$

is a Descartes system on $[a, b]$.

Let $\alpha_1 < \alpha_2 < \cdots < a$. Show that

$$\left(\frac{1}{x - \alpha_1}, \frac{1}{x - \alpha_2}, \dots\right)$$

is a Descartes system on $[a, b]$ (see also E.6 e]).

E.4 Proof of Theorem 3.2.5. Assume the notation of this Theorem 3.2.5.

a] Let $0 \le \delta_0 < \delta_1 < \cdots < \delta_\mu \le n$ and $a < x_1 < x_2 < \cdots < x_\mu < b$. Show that there exists a unique $p = f_{\delta_\mu} + \sum_{i=0}^{\mu-1} a_i f_{\delta_i}$ such that

(1) $p(x_i) = 0\,, \quad i = 1, 2, \dots, \mu\,.$

Show also that the above p has the following properties:

(2) $p(x)$ changes sign at each x_i,

(3) $p(x) \neq 0$ if $x \notin \{x_1, x_2, \dots, x_\mu\}$,

(4) $a_i a_{i+1} < 0\,, \quad i = 0, 1, \dots, \mu - 1\,, \quad a_\mu := 1\,,$

(5) $p(x) > 0\,, \quad x \in (x_\mu, b]\,.$

Hint: Since $(f_{\delta_0}, \dots, f_{\delta_\mu})$ and $(f_{\delta_0}, \dots, f_{\delta_{\mu-1}})$ are Chebyshev systems, E.11 of Section 3.1 shows that there exists a p of the desired form satisfying (1). Since $(f_{\delta_0}, \dots, f_{\delta_{\mu-1}})$ is a Chebyshev system, this p is unique. Now E.11 of Section 3.1 yields that p satisfies (2) and (3). By Theorem 3.2.4, p satisfies (4). The fact that (5) holds for p follows from expanding the determinant

$$D\begin{pmatrix} f_{\delta_0} & f_{\delta_1} & \cdots & f_{\delta_{\mu-1}} & f_{\delta_\mu} \\ x_1 & x_2 & \cdots & x_\mu & x \end{pmatrix}$$

by Cramer's rule. This determinant is just $cp(x)$ with some $c > 0$ since it vanishes at each x_i, and the coefficient of f_{δ_μ} is positive; see Definition 3.2.3. Also, the above determinant is positive for all $x \in (x_\mu, b]$; see Definition 3.2.3 again. □

b] Prove Theorem 3.2.5.

Outline. For notational simplicity assume that $\alpha = n$ (hence $m = k$); the general case is analogous. Further, we may assume that there is an index j such that

$$\gamma_j < \lambda_j \quad \text{and} \quad \gamma_i = \lambda_i \quad \text{whenever} \quad i \neq j$$

since the result follows from this by a finite number of pairwise comparisons. So we assume

$$p = f_n + a_j f_{\lambda_j} + \sum_{\substack{i=1 \\ i \neq j}}^{k} a_i f_{\lambda_i}$$

and

$$q = f_n + b_j f_{\gamma_j} + \sum_{\substack{i=1 \\ i \neq j}}^{k} b_i f_{\lambda_i}\,,$$

where $0 \leq \lambda_1 < \lambda_2 < \cdots < \lambda_k < n$ and $0 \leq \lambda_{j-1} < \gamma_j < \lambda_j$ for some $1 \leq j \leq k$ (of course, the inequality $\lambda_{j-1} < \gamma_j$ holds only if λ_{j-1} is defined, that is, only if $j \geq 2$). Then

$$p - q = a_j f_{\lambda_j} - b_j f_{\gamma_j} + \sum_{\substack{i=1 \\ i \neq j}}^{k} (a_i - b_i) f_{\lambda_i}$$

has exactly k zeros on $[a, b]$ at $x_1, x_2, \dots, x_k$ because $p - q$ is in a $(k+1)$-dimensional Chebyshev subspace.

By property (5) in part a] applied to p and q, respectively,

$$p(x) > 0 \text{ and } q(x) > 0\,, \qquad x \in (x_k, b]\,.$$

Now by property (4) in part a] applied to $c(p-q)$, where c is chosen so that the lead coefficient of $c(p-q)$ is 1, and by the fact that p and $p-q$ have the same coefficient for f_{λ_j}, the lead coefficient of $p-q$ ($a_k - b_k$ provided $\lambda_k > \lambda_j$) is negative. So property (5) in part a] implies that

$$p(x) - q(x) < 0, \quad x \in (x_k, b]\,.$$

Hence $0 < p(x) < q(x)$, $x \in (x_k, b]$.

Now use property (3) in part a] and the fact that all of p, q, and $p-q$ change sign only at x_i to finish the proof. □

The following extension of part a] will be used later:

c] Suppose

$$0 \le \delta_0 < \delta_1 < \cdots < \delta_\mu \le n\,, \quad a \le x_1 \le x_2 \le \cdots \le x_\mu \le b\,,$$

$a < x_2, x_{\mu-1} < b$, and $x_i < x_{i+2}$, $i = 1, 2, \ldots, \mu - 2$. Show that there exists a unique $p = f_{\delta_\mu} + \sum_{i=0}^{\mu-1} a_i f_{\delta_i}$ (with $a_i \in \mathbb{R}$) such that

(1) $p(x_i) = 0, \quad i = 1, 2, \ldots, \mu\,,$

(2) $p(x)$ changes sign at x_i if and only if $x_i \notin \{a, b, x_{i-1}, x_{i+1}\}\,.$

Show also that

(3) $p(x) \ne 0$ if $x \notin \{x_1, x_2, \ldots, x_\mu\}\,,$

(4) $a_i a_{i+1} \le 0\,, \quad i = 0, 1, \ldots, \mu - 1\,, \quad a_\mu := 1\,,$

(5) $p(x) > 0\,, \quad x \in (x_\mu, b)\,,$

(6) $(-1)^\mu p(x) > 0\,, \quad x \in (a, x_1)\,,$

(7) $(-1)^{\mu - i} p(x) > 0\,, \quad x \in (x_i, x_{i+1})\,, \quad i = 1, 2, \ldots, \mu - 1\,.$

Hint: Use part a] and a limiting argument. The uniqueness follows from E.10 of Section 3.1. □

The next exercise provides a solution to a problem of Lorentz, which is settled in Borosh, Chui, and Smith [77].

E.5 A Problem of Lorentz on Best Approximation to x^λ. Suppose that $[a, b] \subset [0, \infty)$, $n \in \mathbb{N}$, and $p \in (0, \infty]$ are fixed. Let μ be a finite Borel measure on $[a, b]$.

a] Suppose $\lambda_1, \lambda_2, \ldots, \lambda_n$ are arbitrary fixed real numbers if $a > 0$, or fixed real numbers greater than $-1/p$ if $a = 0$. Let $f \in L_p(\mu)$ be fixed. Show that

$$E_{p,\mu}(\lambda_1, \lambda_2, \ldots, \lambda_n; f) := \min_{a_i \in \mathbb{R}} \left\| f(x) - \sum_{i=1}^{n} a_i x^{\lambda_i} \right\|_{L_p(\mu)}$$

exists and is finite.

Outline. Use a standard compactness argument. □

b] Suppose $1 < p < \infty$, the support of μ contains at least $n+1$ points, $\lambda_1, \lambda_2, \ldots, \lambda_n, \lambda$ are arbitrary fixed distinct real numbers if $a > 0$, or fixed real numbers greater than $-1/p$ if $a = 0$. Show that if $(\widetilde{a}_i)_{i=1}^n \subset \mathbb{R}$ satisfies

$$E_{p,\mu}(\lambda_1, \lambda_2, \ldots, \lambda_n; x^\lambda) = \Big\| x^\lambda - \sum_{i=1}^n \widetilde{a}_i x^{\lambda_i} \Big\|_{L_p(\mu)},$$

then

$$f(x) := x^\lambda - \sum_{i=1}^n \widetilde{a}_i x^{\lambda_i}$$

has exactly n sign changes on (a, b).

Hint: Since $(x^{\lambda_1}, \ldots, x^{\lambda_n}, x^\lambda)$ is a Chebyshev system, it is sufficient to prove that f has at least n sign changes on (a, b). Suppose f has at most $n-1$ sign changes on $[a, b]$. Then, since $(x^{\lambda_1}, \ldots, x^{\lambda_n})$ is a Chebyshev system, by E.11 of Section 3.1 there exists an element

$$h \in \operatorname{span}\{x^{\lambda_1}, \ldots, x^{\lambda_n}\}$$

such that

$$|f(x)|^{p-1} \operatorname{sign}(f(x)) h(x) \geq 0$$

on $[a, b]$ with strict inequality at all but n points (at every point where f does not vanish). Using that the support of μ contains at least $n+1$ distinct points, this implies

$$\int_a^b |f|^{p-1} \operatorname{sign}(f) h \, d\mu > 0,$$

which contradicts E.7 h] of Section 2.2. □

c] Suppose $p = \infty$, $\operatorname{supp}(\mu) = [a, b]$, $\lambda_1, \lambda_2, \ldots, \lambda_n, \lambda$ are arbitrary fixed distinct real numbers if $a > 0$, or fixed distinct nonnegative real numbers if $a = 0$. Show that if $(\widetilde{a}_i)_{i=1}^n \subset \mathbb{R}$ satisfies

$$E_{\infty,\mu}(\lambda_1, \lambda_2, \ldots, \lambda_n; x^\lambda) = \Big\| x^\lambda - \sum_{i=1}^n \widetilde{a}_i x^{\lambda_i} \Big\|_{L_\infty(\mu)},$$

then

$$f(x) := x^\lambda - \sum_{i=1}^n \widetilde{a}_i x^{\lambda_i}$$

has exactly n sign changes on (a, b).

Hint: Use Theorem 3.1.6. □

d] Best Approximation to x^λ from Certain Classes of Müntz Polynomials. Let $1 < p \le \infty$. Suppose the support of μ contains at least $n+1$ distinct points if $1 < p < \infty$ or $\text{supp}(\mu) = [a,b]$ if $p = \infty$. Let $\lambda > \gamma_1 > \gamma_2 > \cdots$ be arbitrary fixed real numbers if $a > 0$, or fixed nonnegative real numbers if $a = 0$. Suppose we wish to minimize

$$E_{p,\mu}(\lambda_1, \lambda_2, \ldots, \lambda_n; x^\lambda)$$

for all sets of n distinct real numbers $\lambda_1 > \lambda_2 > \cdots > \lambda_n$ satisfying

$$\{\lambda_1, \lambda_2, \ldots, \lambda_n\} \subset \{\gamma_1, \gamma_2, \ldots\}.$$

Show that the minimum occurs if and only if

$$\{\lambda_1, \lambda_2, \ldots, \lambda_n\} = \{\gamma_1, \gamma_2, \ldots, \gamma_n\}.$$

Hint: Let

$$E_{p,\mu}(\lambda_1, \lambda_2, \ldots, \lambda_n; x^\lambda) = \left\| x^\lambda - \sum_{i=1}^n \widetilde{a}_i x^{\lambda_i} \right\|_{L_p(\mu)},$$

where $\{\lambda_1, \lambda_2, \ldots, \lambda_n\}$ is a set of n distinct real numbers for which the minimum is taken; see part a]. By parts b] and c]

$$f(x) := x^\lambda - \sum_{i=1}^n \widetilde{a}_i x^{\lambda_i}$$

has exactly n sign changes $x_1, x_2, \ldots, x_n$ on (a,b). Let

$$g \in \text{span}\{x^{\gamma_1}, x^{\gamma_2}, \ldots, x^{\gamma_n}\}$$

interpolate x^λ at the points $x_1, x_2, \ldots, x_n$. Now use Theorem 3.2.5 to finish the proof. □

E.6 Strictly Totally Positive Kernels (Karlin [68]). A (continuous) function $K(s,t)$ is an *STP kernel* on $[a,b] \times [c,d]$ if

$$\begin{vmatrix} K(s_0,t_0) & \ldots & K(s_0,t_n) \\ \vdots & \ddots & \vdots \\ K(s_n,t_0) & \ldots & K(s_n,t_n) \end{vmatrix} > 0$$

for all $a \le s_0 < \cdots < s_n \le b, \quad c \le t_0 < \cdots < t_n \le d$, and for all $n > 0$.

a] Observe that E.3 b] implies that

$$K(s,t) = \frac{1}{s+t} \quad \text{is STP on } [a,b] \times [a,b]\,, \quad a > 0\,.$$

Observe also that E.2 b] implies that

$$K(s,t) = e^{st} \quad \text{is STP on } (-\infty,\infty) \times (-\infty,\infty)\,.$$

b] Suppose K is STP on $[a,b] \times [c,d]$, and $(f_0,\dots,f_n)$ is a Chebyshev system on $[a,b]$. Show that if

$$v_i(x) = \int_a^b K(t,x) f_i(t)\,dt\,, \qquad i = 0,1,\dots,n\,,$$

then $(v_0,\dots,v_n)$ is a Chebyshev system on $[c,d]$.

c] **Variation Diminishing Property.** Suppose K is STP on $[a,b] \times [c,d]$ and suppose $f \in C[a,b]$. Let

$$g(x) := \int_a^b K(t,x) f(t)\,dt\,.$$

Then g has no more sign changes on $[c,d]$ than f has on $[a,b]$.

d] The Laplace transform of a function $f \in C[0,\infty) \cap L_1[0,\infty)$

$$L(f)(x) := \int_0^\infty f(t) e^{-tx}\,dt$$

has no more sign changes on $[0,\infty)$ than f does.

Proof. This follows from parts a] and c]. It may also be proved directly by induction as follows. Suppose f has exactly n sign changes on $[0,\infty)$, one at x_0. Then $g(x) := (x_0 - x) f(x)$ has exactly $n-1$ sign changes on $[0,\infty)$. Now observe that

$$e^{-x_0 x} \frac{d}{dx} \left(e^{x_0 x} L(f)(x)\right) = L(g)(x)\,,$$

so $L(f)$ has at most one more sign change on $[0,\infty)$ than $L(g)$ does. □

e] Use part d] and E.2 a] to reprove that

$$\left(\frac{1}{x+\alpha_1}, \frac{1}{x+\alpha_2}, \dots\right), \qquad -a < \alpha_1 < \alpha_2 < \cdots$$

satisfies Descartes' rule of signs on $[a,b]$. □

Proof. Observe that

$$\int_0^\infty e^{-\alpha_i t} e^{-tx}\,dt = \frac{1}{x+\alpha_i}$$

for every $x \in (-\alpha_i, \infty)$. □

E.7 Descartes' Rule of Signs for Polynomials.

a] Prove by induction that $\sum_{k=0}^{n} a_k x^k \in \mathcal{P}_n$ has no more zeros in $(0, \infty)$ (repeated zeros are counted according to their multiplicities) than the number of sign changes in $(a_0, a_1, \dots, a_n)$.

b] Let $\alpha > 0$. Let $p(x) = \sum_{k=0}^{n} a_k x^k$ and $q(x) := (x-\alpha)p(x) = \sum_{k=0}^{n+1} b_k x^k$. Show that if the number of sign changes in $\{a_0, a_1, \dots, a_n\}$ is m, then the number of sign changes in $\{b_0, b_1, \dots, b_{n+1}\}$ is at least $m+1$.

c] Give another proof of a] based on b].

d] In part a] the number of sign changes in $(a_0, a_1, \dots, a_n)$ exceeds the number of positive zeros by an even integer.

Hint: See Pólya and Szegő [76]. □

Refinements of the above exercise are presented in E.6 of Appendix 1, where Cauchy indices are discussed.

The first part of the following exercise is a version of a result from Zielke [79]:

E.8 The Effect of Differentiation on Weak Markov Systems. The system $(f_0, \dots, f_n)$ is called a *weak Chebyshev system* on $[a, b]$ if $f_i \in C[a, b]$ for each i and every $f \in \operatorname{span}\{f_0, \dots, f_n\}$ has at most n sign changes on $[a, b]$ (so the only difference between a Chebyshev system and a weak Chebyshev system is that in the definition of the latter, zeros without sign change are not counted).

Analogously, the system $(g_0, \dots, g_n)$ is called a *weak Markov system* on $[a, b]$ if $g_i \in C[a, b]$ for each i and $(g_0, \dots, g_m)$ is a weak Chebyshev system on $[a, b]$ for every $m = 0, 1, \dots, n$ (so the only difference between a Markov system and a weak Markov system is that in the definition of the latter, zeros without sign change are not counted).

a] Suppose $(1, f_1, \dots, f_n)$ is a weak Markov system of C^1 functions on $[a, b]$. Show that $(f_1', \dots, f_n')$ is a weak Markov system on $[a, b]$.

Outline. Proceed by induction on n. If $n = 0$, then the statement is obvious. Suppose that the statement is true for $n-1$. By the inductive hypothesis $(f_1', \dots, f_{n-1}')$ is a weak Markov system on $[a, b]$, hence Rolle's theorem implies that $(1, f_1, \dots, f_{n-1})$ is a Markov system on $[a, b]$.

Suppose that $g \in \operatorname{span}\{f_1', \dots, f_{n-1}'\}$ is of the form

$$g := \sum_{i=1}^{n} a_i f_i', \qquad a_i \in \mathbb{R}$$

and g has at least n sign changes on $[a, b]$. Then there exist $n+2$ distinct points

$$a < x_1 < x_2 < \cdots < x_{n+2} < b$$

and $\epsilon = \pm 1$ such that

$$F := \epsilon \sum_{i=1}^{n} a_i f_i$$

satisfies

$$(-1)^i (F(x_{i+1}) - F(x_i)) > 0\,, \qquad i = 1, 2, \ldots, n+1\,.$$

Since $(1, f_1, \ldots, f_{n-1})$ is a Markov system on $[a, b]$, by Proposition 3.1.2, there exist functions

$$G_\delta \in \operatorname{span}\{1, f_1, \ldots, f_{n-1}\}$$

such that

$$G_\delta(x_i) = F(x_i) + \delta(-1)^i\,, \qquad i = 2, 3, \ldots, n+1$$

for every $\delta > 0$. Then by the inductive hypothesis, G'_δ has at most $n-2$ sign changes on $[a, b]$. It follows that if $\delta > 0$ is sufficiently small, then

$$(-1)(G_\delta(x_2) - G_\delta(x_1)) < 0$$

and

$$(-1)^{n+1}(G_\delta(x_{n+2}) - G_\delta(x_{n+1})) < 0\,,$$

otherwise G'_δ would have at least n sign changes on $[a, b]$. Now show that for sufficiently small $\delta > 0$

$$F - G_\delta \in \operatorname{span}\{1, f_1, \ldots, f_n\}$$

has at least $n+1$ sign changes, which is a contradiction. □

b] Suppose that $(1, f_1, \ldots, f_n)$ is an ECT system on $[a, b]$ and suppose that each $f_i \in C^n[a, b]$ (see E.3 of Section 3.1). Show that $(f'_1, \ldots, f'_n)$ is also an ECT system on $[a, b]$ with each $f'_i \in C^{n-1}[a, b]$.

Hint: Use the definition given in E.3 of Section 3.1. □

c] Suppose $(1, f_1, \ldots, f_n)$ is a weak Markov system on $[a, b]$ with each $f_i \in C^1[a, b]$. Show that $(1, f_1, \ldots, f_n)$ is a Markov system on $[a, b]$.

Hint: Use Rolle's theorem and part a]. □

3.3 Chebyshev Polynomials in Chebyshev Spaces

Suppose

$$H_n := \operatorname{span}\{f_0, f_1, \dots, f_n\}$$

is a Chebyshev space on $[a, b]$, and A is a compact subset of $[a, b]$ with at least $n + 1$ points. We can define the *generalized Chebyshev polynomial*

$$T_n := T_n\{f_0, f_1, \dots, f_n; A\}$$

for H_n on A by the following three properties:

$$T_n \in \operatorname{span}\{f_0, f_1, \dots, f_n\} \tag{3.3.1}$$

there exists an alternation sequence $(x_0 < x_1 < \cdots < x_n)$ for T_n on A, that is,

$$\operatorname{sign}(T_n(x_{i+1})) = -\operatorname{sign}(T_n(x_i)) = \pm\|T_n\|_A \tag{3.3.2}$$

for $i = 0, 1, \dots, n - 1$, and

$$\|T_n\|_A = 1 \quad \text{with} \quad T_n(\max A) > 0\,. \tag{3.3.3}$$

Of course the existence and uniqueness of such a T_n has to be proved. Note that if together with $\operatorname{span}\{f_0, \dots, f_n\}$, $\operatorname{span}\{f_0, \dots, f_{n-1}\}$ is also a Chebyshev space, then Theorem 3.1.6 implies that

$$T_n = c\left(f_n - \sum_{k=0}^{n-1} a_k f_k\right),$$

where the numbers $a_0, a_1, \dots, a_{n-1} \in \mathbb{R}$ are chosen to minimize

$$\left\| f_n - \sum_{k=0}^{n-1} a_k f_k \right\|_A, \tag{3.3.4}$$

satisfies properties (3.3.1) and (3.3.2), and the normalization constant $c \in \mathbb{R}$ can be chosen so that T_n satisfies property (3.3.3) as well. In E.1 we outline the proof of the existence and uniqueness of a T_n satisfying properties (3.3.1) to (3.3.3) without assuming that $\operatorname{span}\{f_0, \dots, f_{n-1}\}$ is a Chebyshev space.

Note that if $(f_0, \dots, f_n)$ is a Descartes system on $[a, b]$, then the normalization constant (that is, the lead coefficient) c in T_n is positive. This follows from E.4 of Section 3.2.

On intervals, with $f_i(x) := x^i$, the definition (3.3.1) to (3.3.3) gives the usual Chebyshev polynomials; see E.7 of Section 2.1.

The Chebyshev polynomials T_n for H_n on A encode much of the information of how the space H_n behaves with respect to the uniform norm on A. Many extremal problems are solved by the Chebyshev polynomials.

When $(f_0, f_1, \dots)$ is a Markov system on $[a, b]$ we can introduce the sequence $(T_n)_{n=0}^{\infty}$ of *associated Chebyshev polynomials*

$$T_n := T_n\{f_0, f_1, \dots , f_n; [a, b]\}$$

for H_n on $[a, b]$. Then $(T_0, T_1, \dots)$ is a Markov system on $[a, b]$ again with the same span. (One reason for not always choosing this as a canonical basis is that it is never a Descartes system.)

The denseness of Markov spaces in $C[a, b]$ is intimately tied to the location of the zeros of the associated Chebyshev polynomials; see Section 4.1.

An example of an extremal problem solved by the Chebyshev polynomials is the following:

Theorem 3.3.1. *Suppose* $H_n := \operatorname{span}\{f_0, \dots , f_n\}$ *is a Chebyshev space on* $[a, b]$ *with associated Chebyshev polynomial*

$$T_n := T_n\{f_0, f_1, \dots , f_n; [a, b]\}$$

and each f_i *is differentiable at* b. *Then*

$$\max\{|p'(b)| : p \in H_n\,, \ \ \|p\|_{[a,b]} \le 1\,, \ \ p(b) = T_n(b)\}$$

is attained by T_n.

Proof. Suppose $p \in H_n$, $\|p\|_{[a,b]} \le 1$, and $p(b) = T_n(b)$. We need to show that $|p'(b)| \le |T_n'(b)|$. Let $a \le \zeta_0 < \zeta_1 < \cdots < \zeta_n \le b$ be the points of alternation for T_n, that is,

$$T_n(\zeta_i) = \pm(-1)^i\,, \qquad i = 0, 1, \dots , n\,.$$

Note that $T_n - p$ has at least n zeros in $[\zeta_0, \zeta_n]$, one in each $[\zeta_{i-1}, \zeta_i]$, $i = 1, 2, \dots , n$ (we count each internal zero without sign change twice, as in E.10 of Section 3.1). So if $b \ne \zeta_n$, then $T_n - p$ has $n + 1$ zeros on $[a, b]$ including the zero at b, hence $p = T_n$, and the proof is finished. We may thus assume that $T_n(b) = 1$. Assume that $|p'(b)| > |T_n'(b)|$. Since $T_n'(b) \ge 0$, without loss of generality we may assume that $p'(b) > T_n'(b)$, otherwise we study $-p$. Then $T_n - p$ has two zeros on $[\zeta_n, b]$, and hence has $n+1$ zeros in $[a, b]$ (again, we count each internal zero without sign change twice). Thus by E.10 of Section 3.1 we have $p = T_n$, which contradicts the assumption $p'(b) > T_n'(b)$. □

An extension of the above theorem to interior points is considered in E.3.

Theorem 3.3.2. *Suppose $(f_0, \dots, f_{n-1}, g)$ and $(f_0, \dots, f_{n-1}, h)$ are both Chebyshev systems on $[a,b]$ with associated Chebyshev polynomials*

$$T_n := T_n\{f_0, f_1, \dots, f_{n-1}, g; [a,b]\}$$

and

$$S_n := T_n\{f_0, f_1, \dots, f_{n-1}, h; [a,b]\},$$

respectively. Suppose $(f_0, f_1, \dots, f_{n-1}, g, h)$ is also a Chebyshev system. Then the zeros of T_n and S_n interlace (there is exactly one zero of S_n between any two consecutive zeros of T_n).

Proof. Since $(f_0, \dots, f_{n-1}, g, h)$ is a Chebyshev system on $[a,b]$, $T_n \pm S_n$ has at most $n+1$ zeros. However, between any two consecutive alternation points of T_n, of which there are $n+1$, there is a zero of $T_n \pm S_n$ (which may be at an internal alternation point of T_n only if it is a zero without sign change, which is then counted twice). Likewise, there is a zero of $T_n \pm S_n$ between any two consecutive alternation points of S_n. Thus between any three successive alternation points of say T_n there can be at most three zeros of $T_n \pm S_n$. However, if S_n had two zeros between two consecutive zeros of T_n, then there would be three consecutive alternation points of either T_n or S_n with at least four zeros of either $T_n + S_n$ or $T_n - S_n$ between them, which is impossible. □

Theorem 3.3.3. *Suppose $(f_0, f_1, \dots)$ is a Markov system on $[a,b]$ with associated Chebyshev polynomials*

$$T_n := T_n\{f_0, f_1, \dots, f_n; [a,b]\}.$$

Then the zeros of T_n and T_{n-1} strictly interlace (there is exactly one zero of T_{n-1} strictly between any two consecutive zeros of T_n).

Proof. The proof is analogous to that of Theorem 3.3.2. □

Theorem 3.3.4 (Lexicographic Property). *Let $(f_0, f_1, \dots)$ be a Descartes system on $[a,b]$. Suppose $\lambda_0 < \lambda_1 < \dots < \lambda_n$ and $\gamma_0 < \gamma_1 < \dots < \gamma_n$ are nonnegative integers satisfying*

$$\lambda_i \le \gamma_i, \qquad i = 0, 1, \dots, n.$$

Let

$$T_n := T_n\{f_{\lambda_0}, f_{\lambda_1}, \dots, f_{\lambda_n}; [a,b]\}$$

and

$$S_n := T_n\{f_{\gamma_0}, f_{\gamma_0}, \dots, f_{\gamma_n}; [a,b]\}$$

denote the associated Chebyshev polynomials.

Let

$$\alpha_1 < \alpha_2 < \cdots < \alpha_n \quad \textit{and} \quad \beta_1 < \beta_2 < \cdots < \beta_n$$

denote the zeros of T_n *and* S_n*, respectively. Then*

$$\alpha_i \le \beta_i\,, \qquad i = 1, 2, \ldots, n$$

with strict inequality if $\lambda_i \neq \gamma_i$ *for at least one index* i*. (In other words, the zeros of* T_n *lie to the left of the zeros of* S_n*.)*

Proof. It is clearly sufficient to prove the theorem in the case that $\lambda_i = \gamma_i$ for each $i \neq m$, and $\lambda_m < \gamma_m < \lambda_{m+1}$ for a fixed index m, and then to proceed by a sequence of pairwise comparisons (if $m = n$, then λ_{m+1} is meant to be replaced by ∞). So suppose

$$T_n := T_n\{f_{\lambda_0}, \ldots, f_{\lambda_m}, \ldots, f_{\lambda_n}; [a,b]\} = \sum_{i=0}^{n} c_i f_{\lambda_i}$$

and

$$S_n := T_n\{f_{\lambda_0}, \ldots, f_{\gamma_m}, \ldots, f_{\lambda_n}; [a,b]\} = d_m f_{\gamma_m} + \sum_{\substack{i=0 \\ i \neq m}}^{n} d_i f_{\lambda_i}$$

with $\lambda_m < \gamma_m < \lambda_{m+1}$. Then by Theorem 3.3.2 the zeros of S_n and T_n interlace and all that remains to prove is that the largest zero of S_n is larger than the largest zero of T_n. That is, we must show that $\alpha_n < \beta_n$. For this we argue as follows. It follows from Theorem 3.2.5 that the lead coefficient of T_n is less than the lead coefficient of S_n ($c_n < d_n$ provided $m < n$). Since both T_n and S_n have an alternation sequence of length $n+1$ on $[a,b]$, and since

$$\|T_n\|_{[a,b]} = \|S_n\|_{[a,b]} = 1\,, \qquad T_n(b) > 0\,, \quad S_n(b) > 0\,,$$

it follows from E.1 b] that

$$S_n - T_n \in \operatorname{span}\{f_{\lambda_0}, f_{\lambda_1} \ldots, f_{\lambda_n}, f_{\gamma_m}\}$$

has $n+1$ zeros $x_1 \le x_2 \le \cdots \le x_{n+1}$ on $[a,b]$. Therefore, it follows from E.4 c] of Section 3.2 that $(S_n - T_n)(x) > 0$ on (x_{n+1}, b) and $(S_n - T_n)(x) < 0$ on (x_n, x_{n+1}). Hence the assumption $\beta_n \le \alpha_n$ would imply that $S_n - T_n$ has at least $n+2$ zeros on $[a,b]$ (counting each internal zero without sign change twice), which is a contradiction. (Draw a picture and use the alternation characterization of the Chebyshev polynomials T_n and S_n to make the proof of the above statement transparent.) So $\beta_n > \alpha_n$, indeed, and the proof is finished. □

Comments, Exercises, and Examples.

If $H_n := \text{span}(f_0, \dots, f_n)$ is a Chebyshev space on $[a, b]$, A is a compact subset of $[a, b]$, and $p \in (0, \infty)$, then

$$T_n = c\left(f_n - \sum_{k=0}^{n-1} a_k f_k\right)$$

with $a_0, a_1, \dots, a_{n-1} \in \mathbb{R}$ minimizing

$$\int_A \left|f_n - \sum_{k=0}^{n-1} a_k f_k\right|^p$$

is called an L_p Chebyshev polynomial for H_n on A. When $A = [a, b]$ and $p \in (1, \infty]$, the properties of the zeros of these L_p Chebyshev polynomials are explored in Pinkus and Ziegler [79], where much of the material of this section may be found. For example, an L_p analog of Theorem 3.3.2 still holds.

E.1 Existence and Uniqueness of Chebyshev Polynomials. Let $A \subset [a, b]$ be a compact set containing at least $n + 1$ points. Let $(f_0, \dots, f_n)$ be a Chebyshev system on $[a, b]$.

a] Existence of Chebyshev Polynomials. Show that there exists a T_n satisfying properties (3.3.1) to (3.3.3).

Hint: If A contains exactly $n + 1$ points, then the existence of T_n is just the interpolation property of a Chebyshev space formulated in Proposition 3.1.2 b]. So assume that A contains at least $n + 2$ points. Then there is a $\delta > 0$ so that $A \cap [a, c]$ contains at least $n+1$ points for every $c \in (b - \delta, b)$. Show that for every $c \in (b - \delta, b)$, there is a $g_c \in \text{span}\{f_0, \dots, f_n\}$ for which

$$\sup\{f(b) : f \in \text{span}\{f_0, f_1, \dots, f_n\}, \ \|f\|_{[a,c]} = 1\}$$

is attained. Use a variational method to show that g_c satisfies properties (3.3.1) to (3.3.3) with A replaced by $A \cap [a, c]$.

Now let $(c_k)_{k=1}^{\infty}$ be a sequence of numbers from $(b - \delta, b)$ that converges to b. Let $g_{c_k} \in \text{span}\{f_0, \dots, f_n\}$ satisfy properties (3.3.1) to (3.3.3) with A replaced by $A \cap [a, c_k]$. Show that there is a subsequence of $(g_{c_k})_{k=1}^{\infty}$ that converges to a $g \in \text{span}\{f_0, \dots, f_n\}$ uniformly on $[a, b]$. Show that $T_n := g$ satisfies properties (3.3.1) to (3.3.3). □

b] A Lemma for Part c]. Suppose $f, g \in C[a, b]$ with $\|f\|_A = \|g\|_A > 0$ and there are alternation sequences

$$(x_1 < x_2 < \cdots < x_{n+1}) \quad \text{and} \quad (y_1 < y_2 < \cdots < y_{n+1})$$

for f and g, respectively, on A.

Suppose also that

$$\operatorname{sign}(f(x_1)) = \operatorname{sign}(g(y_1)).$$

Show that $f - g$ has at least $n + 1$ zeros on $[a, b]$.

c] **Uniqueness of Chebyshev Polynomials.** Show that the Chebyshev polynomials

$$T_n\{f_0, f_1, \ldots, f_n; A\}$$

satisfying properties (3.3.1) to (3.3.3) are unique.

Hint: Use part a] and E.10 of Section 3.1. □

E.2 More on Chebyshev Polynomials. Let $H_n := \operatorname{span}\{f_0, \ldots, f_n\}$ be a Chebyshev space on $[a, b]$ with associated Chebyshev polynomial denoted by $T_n := T_n\{f_0, \ldots, f_n; [a, b]\}$. Show the following statements.

a] If $1 \in H_n$, then $|T_n(a)| = |T_n(b)| = 1$.

b] If $1 \in H_n$, then T_n is monotone between two successive points of its alternation sequence.

Note that the conclusions of parts a] and b] do not necessarily hold in general.

c] If $T_n =: \sum_{i=0}^{n} a_i f_i$, $a_i \in \mathbb{R}$, then the coefficient sequence of T_n / a_m solves

$$\min_{b_i \in \mathbb{R}} \left\| f_m + \sum_{\substack{i=0 \\ i \neq m}}^{n} b_i f_i \right\|_{[a,b]}$$

uniquely, provided that $\{f_0, \ldots f_{m-1}, f_{m+1}, \ldots f_n\}$ is also a Chebyshev system on $[a, b]$. (So this applies to ordinary polynomials on $[0, 1]$ but not on $[-1, 1]$.)

d] Suppose $(f_0, \ldots, f_n)$ is a Descartes system on $[a, b]$ with associated Chebyshev polynomial $T_n := T_n\{f_0, \ldots, f_n; [a, b]\} =: \sum_{i=0}^{n} a_i f_i$, $a_i \in \mathbb{R}$. Show that $a_n > 0$ and $a_i a_{i+1} < 0$ for each $i = 0, 1, \ldots, n - 1$.

Hint: Use E.4 a] of Section 3.2. □

E.3 Extension of Theorem 3.3.1. *Let $H_n := \operatorname{span}\{f_0, \ldots, f_n\}$ be a Chebyshev space on $[a, b]$ with associated Chebyshev polynomial*

$$T_n := T_n\{f_0, f_1, \ldots, f_n; [a, b]\}$$

and suppose each f_i is differentiable at $x_0 \in [a, b]$.

a] *If $T_n'(x_0) > 0$, then*

$$\max\{p'(x_0) : p \in H_n\,,\ \|p\|_{[a,b]} \leq 1\,,\ p(x_0) = T_n(x_0)\}$$

is attained only by T_n.

b] *If $T_n'(x_0) < 0$, then*

$$\min\{p'(x_0) : p \in H_n\,,\ \ \|p\|_{[a,b]} \leq 1\,,\ \ p(x_0) = T_n(x_0)\}$$

is attained and only by T_n.

Hint: Consider the number of zeros of $T_n - p$. □

E.4 More Lexicographic Properties of Müntz Spaces. Let $[a,b] \subset [0,\infty)$. Suppose

$$\lambda_0 < \lambda_1 < \cdots < \lambda_n \quad \text{and} \quad \gamma_0 < \gamma_1 < \cdots < \gamma_n$$

are arbitrary real numbers if $a > 0$, or arbitrary nonnegative numbers if $a = 0$. Suppose $\lambda_i \leq \gamma_i$ for each i with strict inequality for at least one index i. Let

$$H_n := \operatorname{span}\{x^{\lambda_0}, x^{\lambda_1}, \dots, x^{\lambda_n}\} \quad \text{and} \quad G_n := \operatorname{span}\{x^{\gamma_0}, x^{\gamma_1}, \dots, x^{\gamma_n}\}\,.$$

Denote the associated Chebyshev polynomials for H_n and G_n on $[a,b]$ by

$$T_{n,\lambda} := T_n\{x^{\lambda_0}, x^{\lambda_1}, \dots, x^{\lambda_n}; [a,b]\}$$

and

$$T_{n,\gamma} := T_n\{x^{\gamma_0}, x^{\gamma_1}, \dots, x^{\gamma_n}; [a,b]\}\,,$$

respectively.

a] Show that $\lambda_n \geq 0$ implies $T_{n,\lambda}(b) = 1$.

Hint: $T_{n,\lambda}(b) \neq 1$ would imply that $T_{n,\lambda}'$ has at least $n+1$ distinct zeros in $[a,\infty)$ if $\lambda_0 > 0$ and at least n distinct zeros if $\lambda_0 = 0$. □

b] Let $x_0 = a$ or $x_0 = b$. If $x_0 = a = 0$, then assume that $\lambda_0 = 0$ and $\lambda_1 = 1$. Show that

$$\max\{|p'(x_0)| : p \in H_n\,,\ \ \|p\|_{[a,b]} \leq 1\}$$

is attained uniquely by $\pm T_{n,\lambda}$.

c] Let $x_0 \in [0,\infty) \setminus [a,b]$. If $x_0 = 0$, then assume that $\lambda_0 = 0$. Show that

$$\max\{|p(x_0)| : p \in H_n\,,\ \ \|p\|_{[a,b]} \leq 1\}$$

is attained uniquely by $\pm T_{n,\lambda}$.

Hint for b] and c]: First prove that an extremal $p^* \in H_n$ exists. Then show, by a variational method, that p^* equioscillates $n+1$ times between ± 1 on $[a,b]$. □

d] Let $\lambda_n \geq 0$, $\gamma_n \geq 0$, and $a > 0$. Show that

$$|T_{n,\lambda}'(b)| < |T_{n,\gamma}'(b)|\,.$$

Show also that if $a > 0$ and there exists an index k, $0 \le k \le n$, such that $\lambda_k = \gamma_k = 0$, then

$$|T'_{n,\lambda}(a)| > |T'_{n,\gamma}(a)|\,.$$

e] Let $\lambda_n \ge 0$, $\gamma_n \ge 0$, and $a > 0$. Show that

$$|T_{n,\lambda}(x_0)| < |T_{n,\gamma}(x_0)|\,, \qquad x_0 \in (b, \infty)\,.$$

Show also that if $\lambda_0 \ge 0, \gamma_0 \ge 0$, and $a > 0$, then

$$|T_{n,\lambda}(x_0)| > |T_{n,\gamma}(x_0)|\,, \qquad x_0 \in [0, a)$$

(when $x_0 = 0$, we need the assumption $\lambda_0 = \gamma_0 = 0$).

Hint for d] and e]: Suppose to the contrary that one of the inequalities of parts d] and e] fails. Assume, without loss of generality, that there is an index m such that $\lambda_i = \gamma_i$ whenever $i \ne m$, and $\lambda_m < \gamma_m$. Note that by a], $\lambda_n \ge 0$ and $\gamma_n \ge 0$ imply $T_{n,\lambda}(b) = T_{n,\gamma}(b) = 1$. Also, by E.1 a], $\lambda_k = \gamma_k = 0$ implies $T_{n,\lambda}(a) = T_{n,\gamma}(a) = (-1)^n$. Now use Theorem 3.3.4 to show that

$$T_{n,\lambda} - T_{n,\gamma} \in \operatorname{span}\{x^{\lambda_0}, x^{\lambda_1}, \dots, x^{\lambda_n}, x^{\gamma_m}\}$$

has at least $n+2$ zeros in $(0, \infty)$ (in the cases when $\lambda_n \ge 0$ and $\gamma_n \ge 0$ are assumed) or in $[0, \infty)$ (in the cases when $\lambda_k = \gamma_k = 0$ is assumed). This contradiction finishes the proof. □

f] Let $0 < a < b$. Show that if $\lambda_n \ge 0$ and $\gamma_n \ge 0$, then

$$\max_{p \in H_n} \frac{|p'(b)|}{\|p\|_{[a,b]}} < \max_{q \in G_n} \frac{|q'(b)|}{\|q\|_{[a,b]}}\,.$$

Show also that if there exists an index k, $0 \le k \le n$, such that $\lambda_k = \gamma_k = 0$, then

$$\max_{p \in H_n} \frac{|p'(a)|}{\|p\|_{[a,b]}} > \max_{q \in G_n} \frac{|q'(a)|}{\|q\|_{[a,b]}}\,.$$

Hint: Combine parts b] and d]. □

g] Let $0 < a < b$. Let $\lambda_n \ge 0$ and $\gamma_n \ge 0$. Show that

$$\max_{p \in H_n} \frac{|p(x_0)|}{\|p\|_{[a,b]}} < \max_{q \in G_n} \frac{|q(x_0)|}{\|q\|_{[a,b]}}\,, \qquad x_0 \in (b, \infty)\,.$$

Show also that

$$\max_{p \in H_n} \frac{|p(x_0)|}{\|p\|_{[a,b]}} > \max_{q \in G_n} \frac{|q(x_0)|}{\|q\|_{[a,b]}}\,, \qquad x_0 \in [0, a)$$

(when $x_0 = 0$, we need the assumption $\lambda_0 = \gamma_0 = 0$).

Hint: Combine parts c] and e]. □

h] Extend the validity, with $<$ and $>$ replaced by $\leq$ and $\geq$, respectively, of the inequalities of parts f] and g] to the case when the interval $[a, b] \subset (0, \infty)$ is replaced by $[0, b]$.

i] Suppose there exists an index k, $0 \leq k \leq n$, such that $\lambda_k = \gamma_k = 0$. Let $x_0 \in (0, a)$. Show that the second inequalities of parts e] and g] hold true.

Hint: Modify the arguments given in the hints to parts d], e], f], and g]. Note that $1 \in H_n \cap G_n$ ensures, as in E.1 a], that

$$T_{n,\lambda}(b) = T_{n,\gamma}(b) = 1 \qquad \text{and} \qquad T_{n,\lambda}(a) = T_{n,\gamma}(a) = (-1)^n .$$

□

E.5 Lexicographic Properties of $(\sinh \lambda_0 t, \ldots, \sinh \lambda_n t)$**.** Let

$$0 < \lambda_0 < \lambda_1 < \cdots < \lambda_n \quad \text{and} \quad 0 < \gamma_0 < \gamma_1 < \cdots < \gamma_n .$$

Suppose $\lambda_i \leq \gamma_i$ for each i. Let

$$H_n := \operatorname{span}\{\sinh \lambda_0 t, \sinh \lambda_1 t, \ldots, \sinh \lambda_n t\}$$

and

$$G_n := \operatorname{span}\{\sinh \gamma_0 t, \sinh \gamma_1 t, \ldots, \sinh \gamma_n t\} .$$

Denote the associated Chebyshev polynomials for H_n and G_n on $[0, 1]$ by

$$T_{n,\lambda} := T_n\{\sinh \lambda_0 t, \sinh \lambda_1 t, \ldots, \sinh \lambda_n t; [0, 1]\}$$

and

$$T_{n,\gamma} := T_n\{\sinh \gamma_0 t, \sinh \gamma_1 t, \ldots, \sinh \gamma_n t; [0, 1]\} ,$$

respectively.

a] Let

$$\alpha_1 < \alpha_2 < \ldots < \alpha_n \quad \text{and} \quad \beta_1 < \beta_2 < \cdots < \beta_n$$

denote the zeros of $T_{n,\lambda}$ and $T_{n,\gamma}$, respectively. Show that

$$\alpha_i \leq \beta_i, \qquad i = 1, 2, \ldots, n$$

(in other words, the zeros of $T_{n,\lambda}$ lie to the left of the zeros of $T_{n,\gamma}$).

Outline. By E.2 d] of Section 3.2, $(\sinh \lambda_0 t, \ldots, \sinh \lambda_n t)$ is a Descartes system on $(0, \infty)$. Hence, by Theorem 3.3.3, the zeros of

$$T_{n,\lambda,\delta} := T_n\{\sinh \lambda_0 t, \sinh \lambda_1 t, \ldots, \sinh \lambda_n t; [\delta, 1]\}$$

on $[\delta, 1]$ lie to the left of the zeros of

$$T_{n,\gamma,\delta} := T_n\{\sinh \gamma_0 t,\, \sinh \gamma_1 t,\, \dots,\, \sinh \gamma_n t; [\delta, 1]\}$$

on $[\delta, 1]$ for every $\delta \in (0,1)$. Show that

$$\lim_{\delta \to 0} \|T_{n,\lambda} - T_{n,\lambda,\delta}\| = \lim_{\delta \to 0} \|T_{n,\gamma} - T_{n,\gamma,\delta}\| = 0\,;$$

hence the desired result follows by a continuity argument. □

b] Show that

$$\max\{|p'(0)| : p \in H_n\,,\ \ \|p\|_{[0,1]} \le 1\}$$

is attained uniquely by $\pm T_{n,\lambda}$.

c] Show that

$$T_{n,\lambda}(1) = T_{n,\gamma}(1) = 1\,.$$

Hint: $T_{n,\lambda}(1) \neq 1$ would imply that

$$T'_{n,\lambda} \in \operatorname{span}\{\cosh \lambda_0 t,\, \cosh \lambda_1 t,\, \dots,\, \cosh \lambda_n t\}$$

has at least $n+1$ distinct zeros in $(0,\infty)$. This is impossible, since by E.2 e] of Section 3.2, $(\cosh \lambda_0 t, \dots, \cosh \lambda_n t)$ is a Descartes (hence Chebyshev) system on $(0, \infty)$. □

d] Show that

$$|T'_{n,\lambda}(0)| \ge |T'_{n,\gamma}(0)|\,.$$

Hint: Suppose to the contrary that the above inequality fails to hold. Assume, without loss of generality, that there is an index m such that $\lambda_i = \gamma_i$ whenever $i \neq m$, and $\lambda_m < \gamma_m$. Obviously

$$T_{n,\lambda}(0) = T_{n,\gamma}(0) = 0\,.$$

Part c] implies that $T_{n,\lambda}(1) = T_{n,\gamma}(1) = 1$. Now use part a] and the above observation to show that

$$T_{n,\lambda} - T_{n,\gamma} \in \operatorname{span}\{\sinh \lambda_0 t,\, \sinh \lambda_1 t,\, \dots,\, \sinh \lambda_n t,\, \sinh \gamma_m t\}$$

has at least $n+2$ zeros in $(0, \infty)$. This contradicts E.2 d] of Section 3.2. □

e] Show that

$$\max_{0 \neq p \in H_n} \frac{|p'(0)|}{\|p\|_{[0,1]}} \ge \max_{0 \neq q \in G_n} \frac{|q'(0)|}{\|q\|_{[0,1]}}\,.$$

Hint: Combine parts b] and d]. □

The result of the following exercise has been observed independently by Lubinsky and Ziegler [90] and Kroó and Szabados [94]. Various coefficient estimates for polynomials are discussed in Milovanović, Mitrinović, and Rassias [94]. An estimate for the coefficients of polynomials having a given number of terms is obtained in Baishanski and Bojanic [80]. Approximation by such polynomials is studied in Baishanski [83].

E.6 Coefficient Bounds for Polynomials in a Special Basis. Show that

$$|c_m| \le 2^{-n}\binom{2n}{2m}\|p\|_{[-1,1]}\,, \qquad m = 0, 1, \dots, n$$

for every polynomial p of the form

$$p(x) = \sum_{m=0}^{n} c_m(1-x)^m(x+1)^{n-m}\,, \qquad c_m \in \mathbb{R}\,.$$

Outline. By E.5 a] and b] of Section 2.3

$$T_n(x) = \sum_{m=0}^{n} d_{m,n}(1-x)^m(x+1)^{n-m}\,,$$

where

$$d_{m,n} = (-1)^m 2^{-n}\frac{\binom{n-1/2}{m}\binom{n-1/2}{n-m}}{\binom{n-1/2}{n}} = 2^{-n}\binom{2n}{2m}\,.$$

If $|c_m| > d_{m,n}\|p\|_{[-1,1]}$ for some index m, then the polynomial

$$q(x) = \frac{T_n(x)}{d_{m,n}} - \frac{p(x)}{c_m}$$

has at least n distinct zeros in $(-1,1)$. However,

$$q(x) = \sum_{\substack{j=0\\ j\ne m}}^{n} a_j(1-x)^j(x+1)^{n-j} = (1-x)^n\sum_{\substack{j=0\\ j\ne m}}^{n} a_j\left(\frac{1+x}{1-x}\right)^{n-j}$$

can have at most $n-1$ distinct zeros in $(-1,1)$ since

$$(u^0, u^1, \dots u^{n-m-1}, u^{n-m+1}, u^{n-m+2}, \dots, u^n)$$

is a Chebyshev system on $(0,\infty)$ by E.1 a] of Section 3.1. □

E.7 On the Zeros of the Chebyshev Polynomials for Müntz Spaces. Let $0 =: \lambda_0 < \lambda_1 < \cdots < \lambda_n$, and let

$$H_n := \operatorname{span}\{x^{\lambda_0}, x^{\lambda_1}, \dots, x^{\lambda_n}\}\,.$$

Denote the associated Chebyshev polynomials for H_n on $[0,1]$ by

$$T_n := T_n\{x^{\lambda_0}, x^{\lambda_1}, \dots, x^{\lambda_n}; [0,1]\}\,.$$

a] Let $\epsilon \in \left(0, \frac{1}{2}\right)$. Suppose $\alpha < \beta$ are two consecutive zeros of T_n lying in $[\epsilon, 1-\epsilon]$. Show that

$$\frac{\epsilon^2}{n} \le \beta - \alpha\,.$$

Hint: It is clear that $Q_n(x) := x(1-x)T_n(x)$ has two consecutive zeros $\gamma < \delta$ in $[\alpha, 1]$ such that

$$\delta - \gamma \ge \frac{\epsilon}{n}\,.$$

Show that if

$$\beta - \alpha < \frac{\epsilon^2}{n}\,,$$

then

$$R_n(x) := T_n(x) - T_n\left(\tfrac{\gamma x}{\alpha}\right) \in H_n$$

has at least $n+1$ zeros on $[0, \alpha/\gamma]$, which is a contradiction. □

b] Denote the zeros of T_n in $(0,1)$ by $x_1 < x_2 < \cdots < x_n$. Show that

$$\log x_{k+1} - \log x_k \le \log x_k - \log x_{k-1}\,, \qquad k = 2, 3, \ldots, n-1\,.$$

Hint: Use a zero counting argument, as in the hint to part a]. □

3.4 Müntz-Legendre Polynomials

We examine in some detail the system

$$(x^{\lambda_0}, x^{\lambda_1}, \ldots)$$

on $[0,1]$ which we call a *Müntz system.* In particular, we explicitly construct orthogonal "polynomials" for this system. This allows us to derive various extremal properties of these systems and leads to a very simple proof of the classical Müntz-Szász theorem in Section 4.2.

We adopt the following definition for x^λ:

$$x^\lambda = e^{\lambda \log x}\,, \quad x \in (0, \infty)\,, \quad \lambda \in \mathbb{C} \tag{3.4.1}$$

with value at 0 defined to be the limit of x^λ as $x \to 0$ from $(0,\infty)$ whenever the limit exists.

Given a sequence $\Lambda := (\lambda_i)_{i=0}^\infty$ of complex numbers, an element of $\text{span}\{x^{\lambda_0}, x^{\lambda_1}, \ldots, x^{\lambda_n}\}$ is called a *Müntz polynomial* or a *Λ-polynomial.* We denote the set of all such polynomials by $M_n(\Lambda)$, that is,

$$M_n(\Lambda) := \text{span}\{x^{\lambda_0}, x^{\lambda_1}, \ldots, x^{\lambda_n}\}\,, \tag{3.4.2}$$

where the linear span is over $\mathbb{R}$ or $\mathbb{C}$ according to context. Let

$$M(\Lambda) := \bigcup_{n=0}^{\infty} M_n(\Lambda) = \operatorname{span}\{x^{\lambda_0}, x^{\lambda_1}, \dots\}. \tag{3.4.3}$$

For the $L_2[0,1]$ theory of Müntz systems, we consider

$$\Lambda = (\lambda_i)_{i=0}^{\infty}, \quad \operatorname{Re}(\lambda_i) > -1/2, \quad \text{and} \quad \lambda_i \neq \lambda_j,\ i \neq j, \tag{3.4.4}$$

where $\operatorname{Re}(\lambda)$ denotes the real part of λ. This ensures that the Λ-polynomials

$$p(x) = \sum_{k=0}^{n} a_k x^{\lambda_k}, \qquad a_k \in \mathbb{C}$$

are in $L_2[0,1]$. We can then define the orthogonal Λ-polynomials with respect to Lebesgue measure. We call these *Müntz-Legendre polynomials.* Although we often assume (3.4.4), the following definition requires neither the distinctness of the numbers λ_i nor the assumption $\operatorname{Re}(\lambda_i) > -1/2$.

Definition 3.4.1 (Müntz-Legendre Polynomials). Let $\Lambda := (\lambda_i)_{i=0}^{\infty}$ be a sequence of complex numbers. We define the nth *Müntz-Legendre polynomial* on $(0,\infty)$ by

$$\begin{aligned} L_n(x) &:= L_n\{\lambda_0, \dots, \lambda_n\}(x) \\ &:= \frac{1}{2\pi i} \int_{\Gamma} \prod_{k=0}^{n-1} \frac{t + \overline{\lambda}_k + 1}{t - \lambda_k} \frac{x^t\, dt}{t - \lambda_n}, \qquad n = 0, 1, \dots, \end{aligned} \tag{3.4.5}$$

where the positively oriented, simple closed contour Γ surrounds the zeros of the denominator in the integrand, and $\overline{\lambda}_k$ denotes the conjugate of λ_k.

The orthogonality of the above functions with respect to the Lebesgue measure is proved in Theorem 3.4.3. However, first we give a simple explicit representation of L_n in the case that the numbers λ_i are distinct. This is deduced immediately from evaluating the above integral by the residue theorem.

Proposition 3.4.2. *Let $\Lambda := (\lambda_i)_{i=0}^{\infty}$ be a sequence of distinct complex numbers. Then*

$$L_n\{\lambda_0, \dots, \lambda_n\}(x) = \sum_{k=0}^{n} c_{k,n} x^{\lambda_k}, \qquad x \in (0, \infty) \tag{3.4.6}$$

with

$$c_{k,n} := \frac{\prod_{j=0}^{n-1} (\lambda_k + \overline{\lambda}_j + 1)}{\prod_{j=0, j\neq k}^{n} (\lambda_k - \lambda_j)},$$

where $L_n\{\lambda_0, \dots, \lambda_n\}(x)$ is defined by (3.4.5).

So $L_n\{\lambda_0, \dots, \lambda_n\}$ is indeed a Λ polynomial provided the numbers λ_i are distinct. Its value at $x = 0$ is defined if for all i either $\mathrm{Re}(\lambda_i) > 0$ or $\lambda_i = 0$.

From either Definition 3.4.1 or the above proposition it is obvious that the order of $\lambda_0, \dots, \lambda_{n-1}$ in $L_n\{\lambda_0, \dots, \lambda_n\}$ does not make any difference, as long as λ_n is kept last. For example,

$$L_2\{\lambda_0, \lambda_1, \lambda_2\} = L_2\{\lambda_1, \lambda_0, \lambda_2\}$$

while both, in general, are different from $L_2\{\lambda_0, \lambda_2, \lambda_1\}$. For a fixed sequence Λ, we let $L_n(\Lambda)$, or simply L_n, denote the nth Müntz-Legendre polynomial $L_n\{\lambda_0, \dots, \lambda_n\}$, whenever there is no ambiguity.

An analog of Proposition 3.4.2 can be established even if the numbers λ_i are not distinct, however, in the nondistinct case, $L_n(\Lambda)$ does not belong to the space $M_n(\Lambda)$; see E.7 b]. In the very special case that all the indices are the same we recover the Laguerre polynomials; see E.1.

The orthogonality of $\{L_n\}_{n=0}^{\infty}$ is the content of the main theorem of this section.

Theorem 3.4.3 (Orthogonality). *Let $\Lambda = (\lambda_i)_{i=0}^{\infty}$ be a sequence of complex numbers with $\mathrm{Re}(\lambda_i) > -1/2$ for $i = 0, 1, \dots$. The functions L_n defined by (3.4.5) satisfy*

$$\int_0^1 L_n(x)\overline{L_m(x)}\,dx = \frac{\delta_{n,m}}{1 + \lambda_n + \overline{\lambda}_n} \tag{3.4.7}$$

for all nonnegative integers n and m. (Here $\delta_{n,m}$ is the Kronecker symbol.)

Proof. We may assume that the numbers λ_i are distinct. Note that

$$L_n\{\lambda_0, \lambda_1, \dots, \lambda_n\}(x)$$

is uniformly continuous in $\lambda_0, \dots, \lambda_n$ for x in closed subintervals of $(0, 1]$, and the nondistinct λ_i case can be handled by a limiting argument. We may further suppose that $m \le n$. Since $\mathrm{Re}(\lambda_i) > -1/2$, we can pick a simple closed contour Γ such that Γ lies completely to the right of the vertical line $\mathrm{Re}(t) = -1/2$ and Γ surrounds all zeros of the denominator of the integrand in (3.4.5). When $t \in \Gamma$, we have $\mathrm{Re}(t + \overline{\lambda}_m) > -1$, and

$$\int_0^1 x^{t+\overline{\lambda}_m}\,dx = \frac{1}{t + \overline{\lambda}_m + 1}$$

for every $m = 0, 1, \dots$. Hence Fubini's theorem yields

$$\int_0^1 L_n(x)x^{\overline{\lambda}_m}\,dx = \frac{1}{2\pi i}\int_\Gamma \prod_{k=0}^{n-1} \frac{t + \overline{\lambda}_k + 1}{t - \lambda_k}\,\frac{dt}{(t - \lambda_n)(t + \overline{\lambda}_m + 1)}\,.$$

Notice that for $m < n$, the new factor, $t + \overline{\lambda}_m + 1$, in the denominator can be cancelled, and for $m = n$ the new pole $-(\overline{\lambda}_n + 1)$ is outside Γ since $\mathrm{Re}(-\overline{\lambda}_n - 1) < -1/2$. Changing the contour from Γ to $|t| = R$ with $R > \max\{\lambda_0| + 1, \dots, |\lambda_n| + 1\}$, gives

$$\int_0^1 L_n(x) x^{\overline{\lambda}_m}\, dx = \frac{1}{2\pi i}\int_{|t|=R} \prod_{k=0}^{n-1} \frac{t + \overline{\lambda}_k + 1}{t - \lambda_k} \frac{dt}{(t - \lambda_n)(t + \overline{\lambda}_m + 1)}$$
$$- \frac{\delta_{m,n}}{-\overline{\lambda}_n - 1 - \lambda_n} \prod_{k=0}^{n-1} \frac{-\overline{\lambda}_n + \overline{\lambda}_k}{-\overline{\lambda}_n - 1 - \lambda_k}.$$

On letting $R \to \infty$, we see that the integral on the right-hand side is actually 0, which gives

$$\int_0^1 L_n(x) x^{\overline{\lambda}_m}\, dx = \frac{\delta_{n,m}}{\overline{\lambda}_n + \lambda_n + 1} \prod_{k=0}^{n-1} \frac{\overline{\lambda}_n - \overline{\lambda}_k}{\overline{\lambda}_n + \lambda_k + 1}.$$

Therefore Proposition 3.4.2 and $m \le n$ yield

$$\int_0^1 L_n(x)\overline{L_m(x)}\, dx = \int_0^1 L_n(x) \sum_{k=0}^m \overline{c}_{k,m} x^{\lambda_k}\, dx$$
$$= \overline{c}_{m,m} \int_0^1 L_n(x) x^{\overline{\lambda}_m}\, dx = \frac{\delta_{m,n}}{\lambda_n + \overline{\lambda}_n + 1},$$

and the proof is finished. □

An alternative proof of orthogonality is suggested in E.3. If we let

$$L_n^* := (1 + \lambda_n + \overline{\lambda}_n)^{1/2} L_n\,, \tag{3.4.8}$$

then we get an orthonormal system, that is,

$$\int_0^1 L_n^*(x)\overline{L_m^*(x)}\, dx = \delta_{m,n}\,, \qquad m, n = 0, 1, \dots.$$

We call these L_n^* the *orthonormal Müntz-Legendre polynomials.*

There is also a Rodrigues-type formula for the Müntz-Legendre polynomials (see E.2). Let

$$p_n(x) = \sum_{k=0}^n \frac{x^{\lambda_k}}{\prod_{j=0, j\ne k}^n (\lambda_k - \lambda_j)}.$$

Then

$$L_n(x) = (D_{\lambda_0} D_{\lambda_1} \cdots D_{\lambda_{n-1}})(p_n)(x)\,,$$

where the differential operators D_λ are defined by

$$D_\lambda(f)(x) := x^{-\overline{\lambda}} \frac{d}{dx}(x^{1+\overline{\lambda}} f(x))\,.$$

The following is a differential recurrence formula for $(L_n)_{n=0}^\infty$:

Theorem 3.4.4. *For a fixed sequence* $\Lambda := (\lambda_i)_{i=0}^{\infty}$ *of complex numbers, let* L_n *be defined by (3.4.5). The identity*

$$xL_n'(x) - xL_{n-1}'(x) = \lambda_n L_n(x) + (1 + \overline{\lambda}_{n-1})L_{n-1}(x) \tag{3.4.9}$$

holds for every $x \in (0, \infty)$ *and* $n = 1, 2, \dots$.

Proof. From (3.4.5) we get

$$(x^{-\lambda_n} L_n(x))' = \frac{1}{2\pi i} \int_\Gamma \frac{\prod_{k=0}^{n-2}(t + \overline{\lambda}_k + 1)}{\prod_{k=0}^{n-1}(t - \lambda_k)} (t + \overline{\lambda}_{n-1} + 1) x^{t-\lambda_n - 1}\, dt\,.$$

On multiplying both sides by $x^{\lambda_n + \overline{\lambda}_{n-1} + 1}$, we obtain

$$x^{\lambda_n + \overline{\lambda}_{n-1}+1}(x^{-\lambda_n} L_n(x))' = \frac{1}{2\pi i} \int_\Gamma \frac{\prod_{k=0}^{n-2}(t + \overline{\lambda}_k + 1)}{\prod_{k=0}^{n-1}(t - \lambda_k)} (t + \overline{\lambda}_{n-1} + 1) x^{t+\overline{\lambda}_{n-1}}\, dt\,,$$

and again by the definition of L_{n-1},

$$x^{\lambda_n + \overline{\lambda}_{n-1}+1}(x^{-\lambda_n} L_n(x))' = (x^{\overline{\lambda}_{n-1}+1} L_{n-1}(x))'\,.$$

We finish the proof by simplifying by the product rule and dividing both sides by $x^{\overline{\lambda}_{n-1}}$. □

Corollary 3.4.5. *For a fixed sequence* $\Lambda := (\lambda_i)_{i=0}^{\infty}$ *of complex numbers, let* L_n *and* L_n^* *be defined by (3.4.5) and (3.4.8), respectively. Then for every* $x \in (0, \infty)$ *and for every* $n = 0, 1, \dots,$

a] $\displaystyle xL_n'(x) = \lambda_n L_n(x) + \sum_{k=0}^{n-1} (\lambda_k + \overline{\lambda}_k + 1) L_k(x)\,,$

b] $\displaystyle xL_n^{*\prime}(x) = \lambda_n L_n^*(x) + \sqrt{\lambda_n + \overline{\lambda}_n + 1} \sum_{k=0}^{n-1} \sqrt{\lambda_k + \overline{\lambda}_k + 1}\, L_k^*(x)\,,$

and

c] $\displaystyle xL_n''(x) = (\lambda_n - 1)L_n'(x) + \sum_{k=0}^{n-1} (\lambda_k + \overline{\lambda}_k + 1) L_k'(x)\,.$

Proof. The first identity follows from Theorem 3.4.4 on expressing

$$xL_n'(x) - xL_0'(x)$$

as a telescoping sum. From this and the relation

$$L_k^* = (\lambda_k + \overline{\lambda}_k + 1)^{1/2} L_k$$

we get part b]. Differentiating the identity of part a] gives part c]. □

The values and derivative values of the Müntz-Legendre polynomials at 1 can now all be calculated.

Corollary 3.4.6. *For a fixed sequence $\Lambda := (\lambda_i)_{i=0}^{\infty}$ of complex numbers, let L_n be defined by (3.4.5). Then*

a] $L_n(1) = 1$,

b] $$L'_n(1) = \lambda_n + \sum_{k=0}^{n-1} (\lambda_k + \overline{\lambda}_k + 1),$$

and

c] $$L''_n(1) = (\lambda_n - 1)L'_n(1) + \sum_{k=0}^{n-1} (\lambda_k + \overline{\lambda}_k + 1)L'_k(1).$$

Proof. It suffices to show that $L_n(1) = 1$; the rest follows from Corollary 3.4.5. Notice that

$$L_n(1) = \frac{1}{2\pi i} \int_{\Gamma} \prod_{k=0}^{n-1} \frac{t + \overline{\lambda}_k + 1}{t - \lambda_k} \frac{dt}{t - \lambda_n}.$$

Now, since Γ surrounds all zeros of the denominator, and the degree of the denominator is one higher than that of the numerator, we can evaluate the integral on circles of radius $R \to \infty$ to get the result. □

Comments, Exercises, and Examples.

Müntz polynomials are just exponential polynomials $\sum a_k e^{-\lambda_k t}$ under the change of variables $x = e^{-t}$ and have received considerable scrutiny (Schwartz [59] is a monograph on this topic). The orthogonalizations of Müntz systems exist in the Russian literature (see, for example, Badalyan [55] and Taslakyan [84]; it has been further explored in McCarthy, Sayre, and Shawyer [93]). Borwein, Erdélyi, and Zhang [94b] contains most of the content of this section.

Various properties of the Müntz-Legendre polynomials are examined in the exercises. Note that if $\lambda_0 < \lambda_1 < \lambda_2 < \cdots$, then the Müntz system $(x^{\lambda_0}, x^{\lambda_1}, \dots)$ is a Descartes system on $[a, b]$, $a > 0$, and so we can apply Theorem 3.3.4 to the associated Chebyshev polynomials on $[a, b]$ to deduce how the zeros shift when the exponents are varied lexicographically. Similar results are given for the Müntz-Legendre polynomials in E.7.

E.1 Laguerre Polynomials.

a] Let $L_n\{\lambda_0, \dots, \lambda_n\}(x)$ be defined by (3.4.5). If $\lambda_0 = \cdots = \lambda_n = \lambda$, then

$$L_n\{\lambda_0, \lambda_1, \dots, \lambda_n\}(x) = x^{\lambda} \mathcal{L}_n(-(1 + \lambda + \overline{\lambda}) \log x),$$

where $\mathcal{L}_n$ is the nth Laguerre polynomial orthonormal with respect to the weight e^{-x} on $[0, \infty)$ with $\mathcal{L}_n(0) = 1$ (see E.7 of Section 2.3, where $\mathcal{L}_n$ is denoted by L_n as is standard).

Proof. Since $\lambda_k = \lambda$, (3.4.5) yields

$$L_n\{\lambda_0, \lambda_1, \dots, \lambda_n\}(x) = \frac{1}{2\pi i} \int_\Gamma \frac{x^t (t + \overline{\lambda} + 1)^n}{(t - \lambda)^{n+1}} \, dt \,,$$

where the contour Γ can be taken to be any circle centered at λ. By the residue theorem,

$$\begin{aligned} L_n\{\lambda_0, \lambda_1, \dots, \lambda_n\}(x) &= \frac{1}{n!} \left[\frac{d^n}{dt^n} (x^t (t + \overline{\lambda} + 1)^n \right]_{t=\lambda} \\ &= \frac{1}{n!} \sum_{k=0}^n \binom{n}{k} x^\lambda (\log x)^k [n(n-1)\cdots(k+1)](\lambda + \overline{\lambda} + 1)^k \\ &= x^\lambda \sum_{k=0}^n \frac{1}{k!} \binom{n}{k} (1 + \lambda + \overline{\lambda})^k \log^k x \,. \end{aligned}$$

See also part b]. □

b] Let

$$\mathcal{L}_n(x) := \sum_{k=0}^n \frac{1}{k!} \binom{n}{k} (-x)^k .$$

Then $(\mathcal{L}_n)_{n=0}^\infty$ is an orthonormal sequence of polynomials on $[0, \infty)$ with respect to the inner product

$$\langle f, g \rangle = \int_0^\infty f(x) g(x) e^{-x} \, dx \,.$$

Deduce the orthonormality from a] and Theorem 3.4.3 by substituting

$$y = -(1 + \lambda + \overline{\lambda}) \log x \,.$$

E.2 Rodrigues-Type Formula. Let $\Lambda = (\lambda_i)_{i=0}^\infty$ be a sequence of distinct complex numbers. Let L_n be defined by (3.4.5).

a] Let

$$p_n(x) := \sum_{k=0}^n \frac{x^{\lambda_k}}{\prod_{j=0, j \neq k}^n (\lambda_k - \lambda_j)} \,.$$

Show that

$$p_n(x) = \frac{1}{2\pi i} \int_\Gamma \frac{x^t \, dt}{\prod_{j=0}^n (t - \lambda_j)} \,,$$

where Γ is any contour surrounding $\lambda_0, \lambda_1, \dots, \lambda_n$. Use this to show that

$$p_n^{(k)}(1) = 0 \,, \qquad k = 0, 1, \dots, n-1 \,.$$

b] Show that

$$L_n(x) = (D_{\lambda_0} D_{\lambda_1} \cdots D_{\lambda_{n-1}})(p_n)(x),$$

where

$$D_\lambda(f)(x) := x^{-\overline{\lambda}}(x^{1+\overline{\lambda}} f(x))'.$$

c] If $0 = \lambda_0 < \lambda_1 < \cdots$, then

$$p_n(0) = (-1)^n \prod_{j=1}^{n} \lambda_j^{-1}$$

and $(-1)^n p_n$ is strictly decreasing on $[0, 1]$.

E.3 Another Proof of Orthogonality. Deduce the orthogonality of the sequence $(L_n)_{n=0}^\infty$ on $[0, 1]$ from Theorem 3.4.4 by using integration by parts and induction.

E.4 Integral Recursion. For a given sequence $\Lambda := (\lambda_i)_{i=0}^\infty$ of complex numbers satisfying (3.4.4), let L_n be defined by (3.4.5). Show that

$$L_n(x) = L_{n-1}(x) - (\lambda_n + \overline{\lambda}_{n-1} + 1)x^{\lambda_n} \int_x^1 t^{-\lambda_n - 1} L_{n-1}(t)\, dt, \quad x \in (0, 1].$$

Hint: Use Theorem 3.4.4. □

E.5 On the Maximum of L_n on $[0, 1]$. If $\Lambda = (\lambda_i)_{i=0}^\infty$ is a sequence of nonnegative numbers satisfying

$$\lambda_n \geq \sum_{k=0}^{n-1} (1 + 2\lambda_k), \qquad n = 1, 2, \ldots \tag{3.4.10}$$

and L_n is defined by (3.4.5), then

$$|L_n(x)| < L_n(1) = 1, \qquad x \in [0, 1), \quad n = 2, 3, \ldots.$$

Hint: Use Theorem 3.4.4. □

If $\lambda_k = \rho^k$, then (3.4.10) holds if and only if $\rho \geq 2 + \sqrt{3}$.

E.6 The Reproducing Kernel. Let $\Lambda = (\lambda_i)_{i=0}^\infty$ be as in (3.4.4), and let L_n and L_n^* be defined by (3.4.5) and (3.4.8), respectively. Then for every $p \in M_n(\Lambda)$, we have

$$p(x) = \int_0^1 K_n(x, t) p(t)\, dt,$$

where

$$K_n(x, t) := \sum_{k=0}^{n} L_k^*(x) \overline{L_k^*(t)}$$

is the nth reproducing kernel (see also E.5 of Section 2.2).

E.7 On the Zeros of Müntz-Legendre Polynomials. Assume that

$$(\lambda_0, \lambda_1, \dots, \lambda_n) \subset \left(-\tfrac{1}{2}, \infty\right).$$

a] For a function $f \in C(0,1)$ let $S^-(f)$ and $Z(f)$ denote the number of sign changes and the number of zeros, respectively, of f in $(0,1)$ (we count each zero without sign change twice). Let Φ and $\Psi \in C(0,1)$. Show that if

$$n \le S^-(\alpha\Phi + \beta\Psi) \le Z(\alpha\Phi + \beta\Phi) \le n+1$$

for every real α and β with $\alpha^2 + \beta^2 > 0$, then the zeros of Φ and Ψ strictly interlace.

Proof: This result is due to Pinkus and Ziegler [79]. □

b] Assume that

$$\{\lambda_0, \lambda_1, \dots, \lambda_n\} = \{\widetilde{\lambda}_0, \widetilde{\lambda}_1, \dots, \widetilde{\lambda}_m\},$$

where the numbers $\widetilde{\lambda}_0, \widetilde{\lambda}_1, \dots, \widetilde{\lambda}_m$ are distinct, and let m_j, $j = 0, 1, \dots, m$, be the number of indices $i = 0, 1, \dots, n$ for which $\lambda_i = \widetilde{\lambda}_j$. Show that $L_n\{\lambda_0, \dots, \lambda_n\}$ is in the space

$$H_n := \operatorname{span}\{x^{\lambda_j} (\log x)^i : j = 0, 1, \dots, m, \; i = 0, 1, \dots, m_j - 1\},$$

which is a Chebyshev space on $(0, \infty)$.

Hint: Use the definition and the residue theorem. □

c] Show that $\{L_k\{\lambda_0, \dots, \lambda_k\}\}_{k=0}^n$ is a basis for the Chebyshev space H_n defined in part b].

Hint: Use Theorem 3.4.3 (orthogonality). □

d] Show that $L_n := L_n\{\lambda_0, \dots, \lambda_n\}$ has exactly n distinct zeros in $(0,1)$ and L_n changes sign at each of these zeros.

Hint: Assume to the contrary that the number of sign changes of L_n is less than n. Use part c] to find a function $p \in \operatorname{span}\{L_k\}_{k=0}^{n-1}$ that changes sign exactly at those points in $(0,1)$ where L_n changes sign. Then $\int_0^1 L_n p \neq 0$, which contradicts Theorem 3.4.3. □

e] Suppose $\lambda_n < \lambda_n^*$. Show that the zeros of

$$\Phi := L_n\{\lambda_0, \lambda_1, \dots, \lambda_{n-1}, \lambda_n\}$$

and

$$\Psi := L_n\{\lambda_0, \lambda_1, \dots, \lambda_{n-1}, \lambda_n^*\}$$

in $(0,1)$ strictly interlace.

Hint: Note that Theorem 3.4.3 (orthogonality) implies

$$\int_0^1 (\alpha\Phi + \beta\Psi)p = 0$$

for every $p \in H_{n-1}$, where H_{n-1} is defined by part b] with respect to the sequence $(\lambda_0, \lambda_1, \dots, \lambda_{n-1})$. Use the hint given to part d] to show that $\alpha\Phi + \beta\Psi$ has at least n sign changes in $(0,1)$ whenever α and β are real with $\alpha^2 + \beta^2 > 0$. Use part b] to obtain that $\alpha\Phi + \beta\Psi$ cannot have more than $n+1$ zeros in $(0,1)$ whenever α and β are real with $\alpha^2 + \beta^2 > 0$. Finish the proof by part a]. □

f] Let $\lambda_0, \dots, \lambda_{k-1}, \lambda_{k+1}, \dots, \lambda_n$ be fixed distinct numbers. Suppose

$$(\lambda_{k,i})_{i=1}^{\infty} \subset (-1/2, \infty)$$

is a sequence with $\lim_{i\to\infty} \lambda_{k,i} = \infty$. Show that the largest zero of

$$L_{n,k,i} := L_n\{\lambda_0, \dots, \lambda_{k-1}, \lambda_{k,i}, \lambda_{k+1}, \dots, \lambda_n\}$$

in $(0,1)$ tends to 1.

Outline. Assume, without loss, that $\lambda_{k,i}$ is greater than each of the numbers λ_j, $j = 0, 1, \dots, n$, $j \neq k$. Let

$$g_i(x) := \lambda_{k,i}(L_{n,k,i}(x) - c_{k,n}^{(i)} x^{\lambda_{k,i}}),$$

where

$$c_{k,n}^{(i)} = \frac{\prod_{j=0}^{n-1}(\lambda_{k,i} + \lambda_j + 1)}{\prod_{j=0, j\neq k}^{n}(\lambda_{k,i} - \lambda_j)}$$

is the coefficients of $x^{\lambda_{k,i}}$ in $L_{n,k,i}$. Use (3.4.6) to show that the functions g_i converge uniformly on $[\delta, 1]$, $\delta \in (0,1)$, to a function

$$0 \neq g \in H_{n-1} := \operatorname{span}\{x^{\lambda_0}, \dots, x^{\lambda_{k-1}}, x^{\lambda_{k+1}}, \dots, x^{\lambda_n}\}.$$

Use $L_{n,k,i}(1) = 1$ (see Corollary 3.4.6) and the explicit formula for $c_{k,n}^{(i)}$ to show that $g(1) \leq 0$ and that the functions

$$L_{n,k,i}(x) = (L_{n,n,i}(x) - c_{k,n}^{(i)} x^{\lambda_{k,i}}) + c_{k,n}^{(i)} x^{\lambda_{k,i}}$$

converge to $g(x)$, as $i \to \infty$, for every $x \in (0,1)$.

Now assume that the statement of part f] is false. Then there is an $\epsilon \in (0,1)$ and a subsequence $(\lambda_{k,i_j})_{j=1}^{\infty}$ of $(\lambda_{k,i})_{i=1}^{\infty}$ such that the Müntz-Legendre polynomials L_{n,k,i_j} have no zeros in $[1-\epsilon, 1]$. Deduce from this and

$L'_{n,k,i_j}(1) > 0$, $\lambda_{n,i_j} > 0$ (see Corollary 3.4.6 a]), that g_{i_j} is nondecreasing on $[1-\epsilon, 1]$ whenever $\lambda_{n,i_j} > 0$.

Therefore g is nondecreasing on $[1-\epsilon, 1]$, which, together with $0 \neq g \in H_{n-1}$ and $g(1) \leq 0$, implies that $g(1-\epsilon) < 0$. Hence $L_{n,k,i}(1-\epsilon) < 0$ if i is large enough. Since $L_{n,k,i}(1) = 1$ (see Corollary 3.4.6), each $L_{n,k,i}$ has a zero in $(1-\epsilon, 1)$, provided i is large enough, which contradicts our assumption. □

g] Let Φ and Ψ be as in part e]. Let

$$x_1 < x_2 < \cdots < x_n \quad \text{and} \quad x_1^* < x_2^* < \cdots < x_n^*$$

be the zeros of Φ and Ψ, respectively, in $(0,1)$. Show that $\lambda_n < \lambda_n^*$ implies that

$$x_j < x_j^*, \qquad j = 1, 2, \ldots, n.$$

Hint: Combine parts e] and f]. □

h] Let $\lambda_k \neq \lambda_n$. Show that the zeros of

$$\Phi := L_n\{\lambda_0, \ldots, \lambda_{k-1}, \lambda_k, \lambda_{k+1} \ldots, \lambda_{n-1}, \lambda_n\}$$

and

$$\Psi := L_n\{\lambda_0, \ldots, \lambda_{k-1}, \lambda_n, \lambda_{k+1} \ldots, \lambda_{n-1}, \lambda_k\}$$

in $(0,1)$ strictly interlace.

Hint: Use part a] and arguments similar to those given in the hints to part e]. Note that

$$\Phi(1) = \Psi(1) = 1 \qquad \text{and} \qquad \Psi'(1) - \Phi'(1) = \lambda_n - \lambda_k \neq 0$$

(see Corollary 3.4.6 a]) imply that $\alpha\Phi + \beta\Psi$ is not identically 0 whenever α and β are real with $\alpha^2 + \beta^2 > 0$. □

i] Let Φ and Ψ be as in part h]. Let

$$x_1 < x_2 < \cdots < x_n \quad \text{and} \quad x_1^* < x_2^* < \cdots < x_n^*$$

be the zeros of Φ and Ψ, respectively, in $(0,1)$. Show that $\lambda_k < \lambda_n$ implies that

$$x_j < x_j^*, \qquad j = 1, 2, \ldots, n.$$

Hint: By part h] it is sufficient to prove that $x_n \leq x_n^*$. Let H_n be the Chebyshev space defined in part b]. Corollary 3.4.6 implies that

$$\Phi(1) = \Psi(1) = 1 \qquad \text{and} \qquad \Psi'(1) - \Psi'(1) = \lambda_n - \lambda_k > 0.$$

Deduce from this and part i] that $x_n^* < x_n$ would imply that $0 \neq \Psi - \Phi \in H_n$ has at least $n+1$ distinct zeros in $(0,1]$, which is a contradiction. □

j] Lexicographic Property. Suppose

$$\max_{0 \le i \le n} \lambda_i \le \min_{0 \le j \le n} \mu_j$$

and $\lambda_i < \mu_j$ for some indices i and j. Let

$$x_1 < x_2 < \cdots < x_n \quad \text{and} \quad x_1^* < x_2^* < \cdots < x_n^*$$

be the zeros of

$$L_n\{\lambda_0, \lambda_1, \ldots, \lambda_n\} \quad \text{and} \quad L_n\{\mu_0, \mu_1, \ldots, \mu_n\},$$

respectively, in $(0,1)$. Show that

$$x_j < x_j^*, \qquad j = 1, 2, \ldots, n.$$

Hint: Repeated applications of parts g] and i] give the desired result. □

k] Let $\lambda_0 < \lambda_n$. Let

$$x_1 < x_2 < \cdots < x_n \quad \text{and} \quad x_1^* < x_2^* < \cdots < x_n^*$$

be the zeros of

$$L_n\{\lambda_0, \lambda_1, \ldots, \lambda_n\} \quad \text{and} \quad L_n\{\lambda_n, \lambda_{n-1}, \ldots, \lambda_0\},$$

respectively, in $(0,1)$. Show that $(x_j)_{j=1}^n$ and $(x_j^*)_{j=1}^n$ strictly interlace and

$$x_j < x_j^*, \qquad j = 1, 2, \ldots, n.$$

Hint: Use parts h] and i] and the comment given after Proposition 3.4.2. □

l] Show that the zeros of

$$\Phi := L_{n-1}\{\lambda_0, \ldots, \lambda_{n-1}\} \quad \text{and} \quad \Psi := L_n\{\lambda_0, \ldots, \lambda_n\}$$

in $(0,1)$ strictly interlace.

Hint: Use part a] and arguments similar to those given in the hints for part e]. □

E.8 A Global Estimate for the Zeros. Let $(\lambda_i)_{i=1}^n \subset (-1/2, \infty)$. Assume that $x_1 < x_2 < \cdots < x_n$ are the zeros of $L_n\{\lambda_0, \ldots, \lambda_n\}$ in $(0,1)$. Then

$$\exp\left(-\frac{4n+2}{1+2\lambda_*}\right) < x_1 < x_2 < \cdots < x_n < \exp\left(\frac{-j_1^2}{(1+2\lambda^*)(4n+2)}\right),$$

where $\lambda_* := \min\{\lambda_0, \dots, \lambda_n\}$, $\lambda^* := \max\{\lambda_0, \dots, \lambda_n\}$ and $j_1 > 3\pi/4$ is the smallest positive zero of the Bessel function

$$J_0(z) := \sum_{k=0}^{\infty} \frac{(-z^2)^k}{(k!\,2^k)^2}\,.$$

Proof. Let $\mathcal{L}_n$ be the nth Laguerre polynomial with respect to the weight e^{-x} on $[0, \infty)$, and let the zeros of $\mathcal{L}_n$ be $z_1 < z_2 < \cdots < z_n$. Then by [Szegő [75], pp. 127–131])

$$\frac{j_1^2}{4n+2} < z_1 < z_2 < \cdots < z_n < 4n+2\,,$$

where the upper estimate is asymptotically sharp, and the lower estimate is sharp up to a multiplicative constant (not exceeding $4^4/(9\pi^2)$). Now use E.1 and E.7 j]. □

E.9 The Order of the Zero at 1 of Certain Polynomials. This exercise, due in part to Kós, gives precise estimates on the maximum order of the zero at 1 of a polynomial whose coefficients are bounded in modulus by the leading coefficient.

a] Suppose $a_0, a_1, \dots, a_{n-1}$ are complex numbers with modulus at most 1, and suppose $a_n = 1$. Then the multiplicity of the zero of

$$p(x) = a_0 + a_1 x + a_2 x^2 + \cdots + a_n x^n$$

at 1 is at most $5\sqrt{n}$.

Proof. If p has a zero at 1 of multiplicity m, then for every polynomial f of degree less than m, we have

$$a_0 f(0) + a_1 f(1) + \cdots + a_n f(n) = 0\,. \tag{3.4.11}$$

We construct a polynomial f of degree at most $5\sqrt{n}$, for which

$$f(n) > |f(0)| + |f(1)| + \cdots + |f(n-1)|\,.$$

Equality (3.4.11) cannot hold with this f, so the multiplicity of the zero of p at 1 is at most the degree of f.

Let T_ν be the νth Chebyshev polynomial defined by (2.1.1). Let $k \in \mathbb{N}$, and let

$$g := T_0 + T_1 + \cdots + T_k \in \mathcal{P}_k\,.$$

Note that $g(1) = k+1$. Also, for $0 < y \le \pi$,

$$g(\cos y) = 1 + \cos y + \cos 2y + \cdots + \cos ky = \frac{\sin(k+\frac{1}{2})y + \sin\frac{1}{2}y}{2\sin\frac{1}{2}y}.$$

Hence, for $-1 \le x < 1$,

$$|g(x)| \le \frac{\sqrt{2}}{\sqrt{1-x}}.$$

Let $f(x) := g^4(\frac{2x}{n} - 1)$. Then $f(n) = g^4(1) = (k+1)^4$ and

$$|f(0)| + |f(1)| + \cdots + |f(n-1)| \le \sum_{j=1}^{n} \frac{4}{\left(\frac{2j}{n}\right)^2} < n^2 \sum_{j=1}^{\infty} \frac{1}{j^2} = \frac{\pi^2 n^2}{6}.$$

If $k := \lfloor (\pi^2/6)^{1/4}\sqrt{n} \rfloor$, then

$$f(n) > |f(0)| + |f(1)| + \cdots + |f(n-1)|.$$

In this case the degree of f is $4k \le 5\sqrt{n}$. □

The result of part a] is essentially sharp.

b] For every $n \in \mathbb{N}$, there exists a polynomial

$$p_n(x) = a_0 + a_1 x + \cdots + a_{2n^2} x^{2n^2}$$

such that $a_{2n^2} = 1$, $|a_0|, |a_1|, \ldots, |a_{2n^2-1}|$ are real numbers with modulus at most 1, and p_n has a zero at 1 with multiplicity at least n.

Proof. Define

$$L_n(x) := \frac{(n!)^2}{2\pi i} \int_\Gamma \frac{x^t \, dt}{\prod_{k=0}^{n}(t - k^2)}, \qquad n = 0, 1, \ldots,$$

where the simple closed contour Γ surrounds the zeros of the denominator of the integrand. Then L_n is a polynomial of degree n^2 with a zero of multiplicity at least n at 1. (This can easily be seen by repeated differentiation and then evaluation of the above contour integral by expanding the contour to infinity.)

Also, by the residue theorem,

$$L_n(x) = 1 + \sum_{k=1}^{n} c_{k,n} x^{k^2}$$

where

$$c_{k,n} = \frac{(-1)^n (n!)^2}{\prod_{j=0, j\ne k}^{n} (k^2 - j^2)} = \frac{(-1)^k 2(n!)^2}{(n-k)!(n+k)!}.$$

It follows that

$$|c_{k,n}| \le 2, \qquad k = 1, 2, \ldots, n.$$

Hence,

$$P_n(x) := \frac{L_n(x) + L_n(x^2)}{2}$$

is a polynomial of degree $2n^2$ with a zero of order n at 1. Also P_n has constant coefficient 1 and each of its other coefficients is a real number of modulus less than 1. Now let $p_n(x) := x^{2n^2} P_n(1/x)$. □

c] For every $n \in \mathbb{N}$, there exists a polynomial

$$p_n(x) = a_0 + a_1 x + \cdots + a_n x^n$$

such that $a_n = 1$, $a_0, a_1, \ldots, a_{n-1}$ are real numbers of modulus less than 1, and p has a zero at 1 with multiplicity at least $\lfloor \sqrt{n/2} \rfloor$.

3.5 Chebyshev Polynomials in Rational Spaces

There are very few situations where Chebyshev polynomials can be explicitly computed. Indeed, only the classical case of Section 2.1 is well known.

However, the explicit formulas for the Chebyshev polynomials for the trigonometric rational system

$$(3.5.1) \qquad \left(1, \frac{1 \pm \sin\theta}{\cos\theta - a_1}, \frac{1 \pm \sin\theta}{\cos\theta - a_2}, \ldots, \frac{1 \pm \sin\theta}{\cos\theta - a_n}\right), \quad \theta \in [0, 2\pi)$$

and therefore also for the rational system

$$(3.5.2) \qquad \left(1, \frac{1}{x - a_1}, \frac{1}{x - a_2}, \ldots, \frac{1}{x - a_n}\right), \quad x \in [-1, 1]$$

with distinct real poles outside $[-1, 1]$ are implicitly contained in Achiezer [56].

The case (3.5.1) does not perfectly fit our discussion of Section 3.3 because of the periodicity or because $[0, 2\pi)$ is not a compact subset of $\mathbb{R}$. This leads to nonuniqueness of the Chebyshev polynomials. Note that ordinary polynomials arise as a limiting case of the span of system (3.5.2) on letting all the poles tend to $\pm\infty$.

We are primarily interested in the linear span of (3.5.2) and its trigonometric counterpart obtained with the substitution $x = \cos\theta$. Let

$$(3.5.3) \qquad \mathcal{P}_n(a_1, a_2, \ldots, a_n) := \left\{ \frac{p(x)}{\prod_{k=1}^n |x - a_k|} : p \in \mathcal{P}_n \right\}$$

and

$$(3.5.4) \qquad \mathcal{T}_n(a_1, a_2, \ldots, a_n) := \left\{ \frac{t(\theta)}{\prod_{k=1}^n |\cos\theta - a_k|} : t \in \mathcal{T}_n \right\},$$

where $(a_k)_{k=1}^n \subset \mathbb{C} \setminus [-1, 1]$ is a fixed sequence of poles.

When the poles $a_1, a_2, \ldots, a_n$ are distinct and real, (3.5.3) and (3.5.4) are simply the real spans of the systems

$$(3.5.5) \qquad \left(1, \frac{1}{x-a_1}, \frac{1}{x-a_2}, \ldots, \frac{1}{x-a_n}\right) \quad \text{on } [-1,1]$$

and

$$(3.5.6) \qquad \left(1, \frac{1 \pm \sin\theta}{\cos\theta - a_1}, \frac{1 \pm \sin\theta}{\cos\theta - a_2}, \ldots, \frac{1 \pm \sin\theta}{\cos\theta - a_n}\right) \quad \text{on } [0, 2\pi),$$

respectively.

We can construct Chebyshev polynomials of the first and second kinds, which are analogous to T_n and U_n of Section 2.1, for the spaces $\mathcal{P}_n(a_1, a_2, \ldots, a_n)$ and $\mathcal{T}_n(a_1, a_2, \ldots, a_n)$ as follows. Given a sequence $(a_k)_{k=1}^n \subset \mathbb{C}\backslash[-1,1]$, we define the sequence $(c_k)_{k=1}^n$ by

$$(3.5.7) \qquad a_k = \tfrac{1}{2}(c_k + c_k^{-1}), \qquad |c_k| < 1,$$

that is,

$$(3.5.8) \qquad c_k := a_k - \sqrt{a_k^2 - 1}, \qquad |c_k| < 1.$$

Note that

$$\left(a_k + \sqrt{a_k^2 - 1}\right)\left(a_k - \sqrt{a_k^2 - 1}\right) = 1.$$

In what follows, $\sqrt{a_k^2 - 1}$ is always defined by (3.5.8) (this specifies the choice of root). Let $D := \{z \in \mathbb{C} : |z| < 1\}$, let

$$(3.5.9) \qquad M_n(z) := \left(\prod_{k=1}^n (z - c_k)(z - \overline{c}_k)\right)^{1/2},$$

where the square root is defined so that $M_n^*(z) := z^n M_n(z^{-1})$ is an analytic function in a neighborhood of the closed unit disk $\overline{D}$, and let

$$(3.5.10) \qquad f_n(z) := \frac{M_n(z)}{z^n M_n(z^{-1})}.$$

Note that f_n^2 is actually a finite Blaschke product (see E.12 of Section 4.2). Also, $f_n(z^{-1}) = f_n(z)^{-1}$ whenever $|z| = 1$.

The *Chebyshev polynomials of the first kind* for the spaces

$$\mathcal{P}_n(a_1, a_2, \dots, a_n) \quad \text{and} \quad \mathcal{T}_n(a_1, a_2, \dots, a_n)$$

are now defined by

$$(3.5.11) \quad T_n(x) := \tfrac{1}{2}\left(f_n(z) + f_n(z)^{-1}\right), \quad \text{where } x := \tfrac{1}{2}(z + z^{-1}), \quad |z| = 1$$

and

$$(3.5.12) \qquad \widetilde{T}_n(\theta) := T_n(\cos\theta), \qquad \theta \in \mathbb{R},$$

respectively.

The *Chebyshev polynomials of the second kind* for these two spaces are defined by

$$(3.5.13) \quad U_n(x) := \frac{f_n(z) - f_n(z)^{-1}}{z - z^{-1}}, \quad \text{where } x := \tfrac{1}{2}(z + z^{-1}), \quad |z| = 1$$

and

$$(3.5.14) \qquad \widetilde{U}_n(\theta) := U_n(\cos\theta)\sin\theta, \qquad \theta \in \mathbb{R},$$

respectively.

As we will see, these Chebyshev polynomials preserve many of the elementary properties of the classical trigonometric and algebraic Chebyshev polynomials. This is the content of the next three results.

Theorem 3.5.1 (Chebyshev Polynomials of the First and Second Kinds in Trigonometric Rational Spaces). *Given $(a_k)_{k=1}^n \subset \mathbb{C}\backslash[-1, 1]$, let $\widetilde{T}_n$ and $\widetilde{U}_n$ be defined by (3.5.12) and (3.5.14), respectively. Then the following statements hold:*

a] $\widetilde{T}_n \in \mathcal{T}_n(a_1, a_2, \dots, a_n)$ *and* $\widetilde{U}_n \in \mathcal{T}_n(a_1, a_2, \dots, a_n)$.

b] $\|\widetilde{T}_n\|_{\mathbb{R}} = 1$ *and* $\|\widetilde{U}_n\|_{\mathbb{R}} = 1$.

c] *There exist* $0 = \theta_0 < \theta_1 < \cdots < \theta_n = \pi$ *such that*

$$\widetilde{T}_n(\theta_j) = \widetilde{T}_n(-\theta_j) = (-1)^j, \quad j = 0, 1, \dots, n.$$

d] *There exist* $0 < \tau_1 < \tau_2 < \cdots < \tau_n < \pi$ *such that*

$$\widetilde{U}_n(\tau_j) = -\widetilde{U}_n(-\tau_j) = (-1)^{j-1}, \quad j = 1, 2, \dots, n.$$

e] *For every* $\theta \in \mathbb{R}$,

$$\widetilde{T}_n^2(\theta) + \widetilde{U}_n^2(\theta) = 1.$$

Proof. Observe that there are polynomials $p_1 \in \mathcal{P}_n$ and $p_2 \in \mathcal{P}_{n-1}$ such that

$$(3.5.15) \qquad \widetilde{T}_n(\theta) = T_n(\cos\theta) = \frac{e^{-in\theta}M_n^2(e^{i\theta}) + e^{in\theta}M_n^2(e^{-i\theta})}{2M_n(e^{i\theta})M_n(e^{-i\theta})} = \frac{p_1(\cos\theta)}{\prod_{k=1}^n |\cos\theta - a_k|}$$

and

$$(3.5.16) \qquad \widetilde{U}_n(\theta) = U_n(\cos\theta)\sin\theta = \frac{e^{-in\theta}M_n^2(e^{i\theta}) - e^{in\theta}M_n^2(e^{-i\theta})}{2iM_n(e^{i\theta})M_n(e^{-i\theta})} = \frac{p_2(\cos\theta)\sin\theta}{\prod_{k=1}^n |\cos\theta - a_k|}.$$

Thus a] is proved.

Since $|c_k| < 1$ and f_n^2 is a finite Blaschke product, we have

$$(3.5.17) \qquad |f_n(z)| = 1 \quad \text{whenever} \quad |z| = 1.$$

Now b] follows immediately from (3.5.10) to (3.5.14).

Note that $\widetilde{T}_n(\theta)$ is the real part, and $\widetilde{U}_n(\theta)$ is the imaginary part of $f_n(e^{i\theta})$, that is,

$$f_n(e^{i\theta}) = \widetilde{T}_n(\theta) + i\widetilde{U}_n(\theta), \qquad \theta \in \mathbb{R},$$

which together with (3.5.17) implies e].

To prove parts c] and d], we note that $\widetilde{T}_n(\theta) = \pm 1$ if and only if $f_n(e^{i\theta}) = \pm 1$, and $\widetilde{U}_n(\theta) = \pm 1$ if and only if $f_n(e^{i\theta}) = \pm i$. Since $|c_k| < 1$ for $k = 1, 2, \ldots, n$, f_n^2 has exactly $2n$ zeros in the open unit disk D. Since f_n^2 is analytic in a region containing the closed unit disk $\overline{D}$, c] and d] follow by the argument principle (see, for example, Ash [71]). □

With the transformation $x = \cos\theta = \frac{1}{2}(z + z^{-1})$ and $z = e^{i\theta}$, Theorem 3.5.1 can be reformulated as follows:

Theorem 3.5.2 (Chebyshev Polynomials in Algebraic Rational Spaces). *Given $(a_k)_{k=1}^n \subset \mathbb{C} \setminus [-1, 1]$, let T_n and U_n be defined by (3.5.11) and (3.5.13), respectively. Then*

a] $T_n \in \mathcal{P}_n(a_1, a_2, \ldots, a_n)$ *and* $U_n \in \mathcal{P}_n(a_1, a_2, \ldots, a_n)$.

b] $\|T_n\|_{[-1,1]} = 1$ *and* $\|\sqrt{1-x^2}\, U_n(x)\|_{[-1,1]} = 1$.

c] *There exist* $1 = x_0 > x_1 > \cdots > x_n = -1$ *such that*

$$T_n(x_j) = (-1)^j, \quad j = 0, 1, 2, \ldots, n.$$

d] *There exist* $1 > y_1 > y_2 > \cdots > y_n > -1$ *such that*

$$\sqrt{1 - y_j^2}\, U_n(y_j) = (-1)^{j-1}, \quad j = 1, 2, \ldots, n.$$

e] *For every* $x \in [-1, 1]$,

$$(T_n(x))^2 + (\sqrt{1-x^2}\, U_n(x))^2 = 1.$$

Parts c] and d] of Theorems 3.5.1 and 3.5.2 establish the equioscillation property of the Chebyshev polynomials, which also extends to certain linear combinations of Chebyshev polynomials. In the trigonometric polynomial case this is the fact that $\cos\alpha\cos n\theta + \sin\alpha\sin n\theta = \cos(n\theta - \alpha)$ equioscillates $2n$ times on the unit circle $[0, 2\pi]$. Our next theorem characterizes the Chebyshev polynomials for $\mathcal{T}_n(a_1, a_2, \dots, a_n)$ and records a monotonicity property that we require later.

Theorem 3.5.3 (Chebyshev Polynomials in Trigonometric Rational Spaces). *Let $(a_k)_{k=1}^n \subset \mathbb{C} \setminus [-1, 1]$. Then (i) and (ii) below are equivalent:*

(i) There is an $\alpha \in \mathbb{R}$ such that

$$V = (\cos\alpha)\,\widetilde{T}_n + (\sin\alpha)\,\widetilde{U}_n\,,$$

where $\widetilde{T}_n$ and $\widetilde{U}_n$ are defined by (3.5.12) and (3.5.14).

(ii) $V \in \mathcal{T}_n(a_1, a_2, \dots, a_n)$ has uniform norm 1 on $\mathbb{R}$, and it equioscillates $2n$ times on $\mathbb{R}$ (mod 2π). That is, there exist

$$0 \le \theta_0 < \theta_1 < \cdots < \theta_{2n-1} < 2\pi$$

so that

$$V(\theta_j) = \pm(-1)^j\,, \quad j = 0, 1, \dots, 2n-1\,.$$

Furthermore, if V is of the form in (i) (or characterized by (ii)), then

$$V' = (\cos\alpha)\,\widetilde{T}_n' + (\sin\alpha)\,\widetilde{U}_n'$$

does not vanish between any two consecutive alternation points of V (that is, between θ_{j-1} and θ_j for $j = 1, 2, \dots, 2n-1$ and between θ_{2n-1} and $2\pi + \theta_0$).

Proof. *(i)* $\Rightarrow$ *(ii)*. By Theorem 3.5.1 e] and Cauchy's inequality, we have

$$|(\cos\alpha)\,\widetilde{T}_n + (\sin\alpha)\,\widetilde{U}_n|^2 \le (\cos^2\alpha + \sin^2\alpha)(\widetilde{T}_n^2 + \widetilde{U}_n^2) = 1 \tag{3.5.18}$$

on the real line. From Theorem 3.5.1 c], d], and e], we obtain that $\widetilde{T}_n/\widetilde{U}_n$ oscillates between $+\infty$ and $-\infty$ exactly $2n$ times on $\mathbb{R}$ (mod 2π), and hence it takes the value $\cot\alpha$ exactly $2n$ times. At each such point, (3.5.18) becomes an equality, namely, $(\cos\alpha)\,\widetilde{T}_n + (\sin\alpha)\,\widetilde{U}_n = \pm 1$ with different signs for every two consecutive such points.

(ii) $\Rightarrow$ *(i)*. Let V be as specified in part *(ii)* of the theorem. Let θ_0 be a point where V achieves its maximum on $\mathbb{R}$, so $V(\theta_0) = 1$. We want to show that V is equal to $p := \widetilde{T}_n(\theta_0)\widetilde{T}_n + \widetilde{U}_n(\theta_0)\widetilde{U}_n$. Since $V(\theta_0) = p(\theta_0) = 1$ and $V'(\theta_0) = p'(\theta_0) = 0$, $V - p$ has a zero at θ_0 with multiplicity at least 2. There are at least $2n - 1$ more zeros (we count multiplicities) of $V - p$

in $\mathbb{R}$ (mod 2π), with one between any two consecutive alternation points of p if the first zero of p to the right of θ_0 is greater than the first zero of V to the right of θ_0. If the first zero of V to the right of θ_0 is greater than or equal to the first zero of p to the right of θ_0, then there is one zero of $p - V$ between any two consecutive alternation points of V. In any case $V - p$ has at least $2n + 1$ zeros in $\mathbb{R}$ (mod 2π). Hence $V - p$ is identically 0.

To prove the final part of the theorem let $V \in \mathcal{T}_n(a_1, a_2, \dots, a_n)$ be such that $\|V\|_{\mathbb{R}} = 1$ and V equioscillates $2n$ times on $\mathbb{R}$ (mod 2π) between ± 1. Assume there is a $\theta_0 \in [0, 2\pi)$ such that $|V(\theta_0)| < 1$ and $V'(\theta_0) = 0$. Then $V(\theta_0) \neq 0$; otherwise the numerator of V would have at least $2n + 1$ zeros in $\mathbb{R}$ (mod 2π), which is a contradiction. Observe that there is a trigonometric polynomial $t \in \mathcal{T}_{2n}$ such that

$$V^2(\theta) - V^2(\theta_0) = \frac{t(\theta)}{\prod_{k=1}^n (\cos\theta - a_k)(\cos\theta - \overline{a}_k)}.$$

This t has at least $4n + 1$ zeros in $\mathbb{R}$ (mod 2π), which is a contradiction again. Therefore $V'(\theta) \neq 0$ if $|V(\theta)| < 1$, which means that V is strictly monotone between any two of its consecutive alternation points. □

Under some assumptions on $(a_k)_{k=1}^n$ it is easy to write down the explicit partial fraction decompositions for T_n and U_n.

Theorem 3.5.4. *Let $(a_k)_{k=1}^n \subset \mathbb{C} \setminus [-1, 1]$ be a sequence of distinct numbers such that its nonreal elements are paired by complex conjugation. Let T_n and U_n be the Chebyshev polynomials of the first and second kinds defined by (3.5.11) and (3.5.13), respectively. Then*

$$T_n(x) = A_{0,n} + \frac{A_{1,n}}{x - a_1} + \cdots + \frac{A_{n,n}}{x - a_n} \tag{3.5.19}$$

and

$$U_n(x) = \frac{B_{1,n}}{x - a_1} + \frac{B_{2,n}}{x - a_2} + \cdots + \frac{B_{n,n}}{x - a_n}, \tag{3.5.20}$$

where

$$A_{0,n} = \frac{(-1)^n}{2}\left(c_1^{-1} c_2^{-1} \cdots c_n^{-1} + c_1 c_2 \cdots c_n\right),$$

$$A_{k,n} = \left(\frac{c_k - c_k^{-1}}{2}\right)^2 \prod_{\substack{j=1 \\ j \neq k}}^n \frac{1 - c_k c_j}{c_k - c_j}, \qquad k = 1, 2, \dots, n,$$

and

$$B_{k,n} = \frac{c_k - c_k^{-1}}{2} \prod_{\substack{j=1 \\ j \neq k}}^n \frac{1 - c_k c_j}{c_k - c_j}, \qquad k = 1, 2, \dots, n.$$

Proof. It follows from Theorems 3.5.1 a] and 3.5.2 a] that T_n and U_n can be written as the partial fraction forms above. Now it is quite easy to calculate the coefficients $A_{k,n}$ and $B_{k,n}$. For example,

$$\begin{aligned} A_{0,n} &= \lim_{x\to\infty} T_n(x) = \lim_{z\to 0} \frac{1}{2}\left(\frac{M_n(z)}{z^n M_n(z^{-1})} + \frac{z^n M_n(z^{-1})}{M_n(z)}\right) \\ &= \frac{(-1)^n}{2}(c_1^{-1}c_2^{-1}\cdots c_n^{-1} + c_1c_2\cdots c_n) \end{aligned}$$

and for $k = 1, 2, \ldots, n$,

$$\begin{aligned} A_{k,n} &= \lim_{x\to a_k}(x-a_k)T_n(x) \\ &= \lim_{z\to c_k}\frac{1}{4}(z-c_k)(1-c_k^{-1}z^{-1})\left(\frac{M_n(z)}{z^n M_n(z^{-1})} + \frac{z^n M_n(z^{-1})}{M_n(z)}\right) \\ &= \left(\frac{c_k - c_k^{-1}}{2}\right)^2 \prod_{\substack{j=1\\ j\neq k}}^{n} \frac{1-c_kc_j}{c_k-c_j}\,, \qquad k = 1, 2, \ldots, n\,. \end{aligned}$$

The coefficients $B_{k,n}$ can be calculated in the same fashion. □

Comments, Exercises, and Examples.

The explicit formulas of this section are tremendously useful. They allow, for example, derivation of sharp Bernstein-type inequalities for rational functions; see Section 7.1. Various further properties of these Chebyshev polynomials for rational function spaces are explored in the exercises, which follow, Borwein, Erdélyi, and Zhang [94b]. In particular, the orthogonalization of such rational systems on $[-1,1]$ with respect to the weight $w(x) = (1-x^2)^{-1/2}$ can be made explicit in terms of the Chebyshev polynomials. Various other aspects of these orthogonalizations may be found in Achiezer [56], Bultheel et al. [91], and Van Assche and Vanherwegen [92].

E.1 Further Properties of $\widetilde{T}_n$ and T_n. Given $(a_k)_{k=1}^n \subset \mathbb{C} \setminus [-1,1]$, let $(c_k)_{k=1}^n$ be defined by

$$c_k := a_k - \sqrt{a_k^2 - 1}\,, \qquad |c_k| < 1\,,$$

as before. We introduce the *Bernstein factors*

$$B_n(x) := \sum_{k=1}^n \operatorname{Re}\left(\frac{\sqrt{a_k^2-1}}{a_k - x}\right)$$

and

$$\widetilde{B}_n(\theta) := B_n(\cos\theta) == \sum_{k=1}^n \operatorname{Re}\left(\frac{\sqrt{a_k^2-1}}{a_k - \cos\theta}\right),$$

where the choice of $\sqrt{a_k^2 - 1}$ is determined by the restriction $|c_k| < 1$. Note that for $x \in [-1, 1]$, we have

$$\mathrm{Re}\left(\frac{\sqrt{a_k^2-1}}{a_k - x}\right) = \mathrm{Re}\left(\frac{c_k^{-1} - c_k}{\frac{1}{2}c_k^{-1} - x}\right) \geq \frac{(1-|c_k|^2)(1-|c_k|)^2}{|1-2c_k x|^2} > 0\,.$$

The following result generalizes the trigonometric identities

$$(\cos nt)' = -n \sin nt\,, \qquad (\sin nt)' = n\cos nt\,,$$

and

$$((\cos nt)')^2 + ((\sin nt)')^2 = n^2\,,$$

which are limiting cases (if $n \in \mathbb{N}$ and $t \in \mathbb{R}$ are fixed, then $\lim \widetilde{B}_n(t) = n$ as all $a_k \to \pm\infty$).

a] Show that, on the real line,

$$\widetilde{T}_n' = -\widetilde{B}_n \widetilde{U}_n\,, \qquad \widetilde{U}_n' = \widetilde{B}_n \widetilde{T}_n\,,$$

and

$$(\widetilde{T}_n')^2 + (\widetilde{U}_n')^2 = \widetilde{B}_n^2\,.$$

Hint: For example,

$$\begin{aligned}\widetilde{T}_n'(\theta) &= \frac{1}{2}\left(f_n'(e^{i\theta}) - \frac{f_n'(e^{i\theta})}{f_n^2(e^{i\theta})}\right) ie^{i\theta}\\ &= \frac{-e^{i\theta} f_n'(e^{i\theta})}{f_n(e^{i\theta})}\,\frac{f_n(e^{i\theta}) - f_n(e^{i\theta})^{-1}}{2i} = -\widetilde{B}_n(\theta)\widetilde{U}_n(\theta)\,.\end{aligned}$$

□

b] If $V := (\cos\alpha)\widetilde{T}_n + (\sin\alpha)\widetilde{U}_n$ for some $\alpha \in \mathbb{R}$, then

$$(V')^2 + \widetilde{B}_n^2 V^2 = \widetilde{B}_n^2$$

holds on the real line.

c] **The Derivative of T_n at ± 1.** Let T_n be defined by (3.5.11). Then

$$T_n'(1) = \left(\sum_{k=1}^n \mathrm{Re}\left(\frac{1+c_k}{1-c_k}\right)\right)^2$$

and

$$T_n'(-1) = (-1)^n \left(\sum_{k=1}^n \mathrm{Re}\left(\frac{1-c_k}{1+c_k}\right)\right)^2.$$

d] **Contour Integral for T_n.** Show that

$$T_n(x) = \frac{1}{2\pi i} \int_\gamma \left(\prod_{j=1}^{n} \frac{(t-c_j)(t-\overline{c}_j)}{(1-c_jt)(1-\overline{c}_jt)} \right)^{1/2} \frac{t-x}{t^2-2tx+1}\, dt$$

for every $x \in [-1,1]$, where γ is a circle centered at the origin with radius $1 < r < \min\{|c_j^{-1}| : 1 \le j \le n\}$, and the square root in the integrand is an analytic function of t in a neighborhood of γ.

Hint: Cauchy's integral formula and the map $x = \frac{1}{2}(z + z^{-1})$ give

$$\begin{aligned} T_n(x) &= \frac{1}{2}\left(\frac{M_n(z)}{z^n M_n(z^{-1})} + \frac{z^n M_n(z^{-1})}{M_n(z)} \right) \\ &= \frac{1}{2\pi i} \int_\gamma \frac{1}{2} \frac{M_n(t)}{t^n M_n(t^{-1})} \left(\frac{1}{t-z} + \frac{1}{t-z^{-1}} \right) dt \\ &= \frac{1}{2\pi i} \int_\gamma \frac{M_n(t)}{t^n M_n(t^{-1})} \frac{t-x}{t^2-2tx+1}\, dt\,, \end{aligned}$$

where M_n is defined by (3.5.9). □

E.2 Orthogonality. Given $(a_k)_{k=1}^\infty \subset \mathbb{R}\setminus[-1,1]$, let $(c_k)_{k=1}^\infty$ be defined by

$$a_k = \tfrac{1}{2}(c_k + c_k^{-1}), \qquad c_k = a_k - \sqrt{a_k^2 - 1}\,, \quad c_k \in (-1,1)$$

and let $(T_n)_{n=0}^\infty$ be defined by (3.5.11).

a] Show that

$$\int_{-1}^{1} T_n(x) T_m(x) \frac{dx}{\sqrt{1-x^2}} = \frac{\pi}{2}(-1)^{n+m}(1 + c_1^2 \cdots c_m^2) c_{m+1} \cdots c_n$$

for all integers $0 \le m \le n$. (The empty product is understood to be 1.)

b] Given $a \in \mathbb{R} \setminus [-1,1]$, let $c \in (-1,1)$ be defined by

$$a = \tfrac{1}{2}(c + c^{-1})\,, \qquad c = a - \sqrt{a^2-1}\,, \quad c \in (-1,1)\,.$$

Show that

$$\int_{-1}^{1} T_n(x) \frac{1}{x-a} \frac{dx}{\sqrt{1-x^2}} = \frac{2\pi}{c - c^{-1}} \prod_{j=1}^{n} \frac{c - c_j}{1 - cc_j}\,.$$

c] Show that

$$\int_{-1}^{1} T_n(x) \frac{dx}{\sqrt{1-x^2}} = (-1)^n \pi c_1 c_2 \cdots c_n$$

and

$$\int_{-1}^{1} T_n(x) \frac{1}{x-a_k} \frac{dx}{\sqrt{1-x^2}} = 0\,, \qquad k = 1, 2, \dots, n\,.$$

Given a sequence $(a_k)_{k=1}^{\infty} \subset \mathbb{R} \backslash [-1, 1]$, we define

$$R_0 := 1\,, \qquad R_n := T_n + c_n T_{n-1}$$

and

$$R_0^* := \frac{1}{\sqrt{\pi}}\,, \qquad R_n^* := \sqrt{\frac{2}{\pi(1-c_n^2)}}\,(T_n + c_n T_{n-1})\,.$$

The following part of this exercise indicates that these simple linear combinations of T_n and T_{n-1} give the orthogonalization of the rational system

$$\left(1, \frac{1}{x-a_1}, \frac{1}{x-a_2}, \dots\right)$$

whenever $(a_k)_{k=1}^{\infty} \subset \mathbb{R} \setminus [-1, 1]$ is a sequence of distinct real numbers.

d] Show that, for all nonnegative integers n and m,

$$\int_{-1}^{1} R_n^*(x) R_m^*(x) \frac{dx}{\sqrt{1-x^2}} = \delta_{m,n}\,,$$

where $\delta_{m,n}$ is the Kronecker symbol.

Proof. Let $m \le n$. By part c],

$$\int_{-1}^{1} R_n(x) \frac{1}{x-a_k} \frac{dx}{\sqrt{1-x^2}} = 0$$

holds for $k = 1, 2, \dots, n-1$. Also

$$\begin{aligned}\int_{-1}^{1} R_n(x) \frac{dx}{\sqrt{1-x^2}} &= \int_{-1}^{1} (T_n(x) + c_n T_{n-1}(x)) \frac{dx}{\sqrt{1-x^2}} \\ &= (-1)^n (c_1 c_2 \cdots c_n) + c_n (-1)^{n-1} (c_1 c_2 \cdots c_{n-1}) = 0\,.\end{aligned}$$

This implies that

$$\int_{-1}^{1} R_n(x) R_m(x) \frac{dx}{\sqrt{1-x^2}} = 0\,, \qquad m = 0, 1, \dots, n-1\,.$$

Finally, it follows from part a] that

$$\int_{-1}^{1} R_n^*(x)^2 \frac{dx}{\sqrt{1-x^2}} = 1\,.$$

□

e] Assume $(a_k)_{k=1}^{\infty} \subset \mathbb{R} \setminus [-1, 1]$. Then T_n and R_n have exactly n zeros in $[-1, 1]$, and the zeros of T_{n-1} and T_n strictly interlace.

E.3 Extension of Theorems 3.5.1 and 3.5.3. Given $(a_k)_{k=1}^{2n} \subset \mathbb{C} \setminus \mathbb{R}$, let

$$\mathcal{T}_n(a_1, a_2, \ldots, a_{2n}) := \left\{ \frac{t(\theta)}{\prod_{k=1}^{2n} |\sin((\theta - a_k)/2)|} : t \in \mathcal{T}_n \right\}.$$

Without loss of generality we may assume that

$$\operatorname{Im}(a_k) > 0\,, \qquad k = 1, 2, \ldots, 2n\,.$$

a] Show that there is a polynomial $q_{2n} \in \mathcal{P}_{2n}^c$ of the form

$$q_{2n}(z) = \gamma \prod_{k=1}^{2n} (z - c_k)\,, \qquad |c_k| < 1\,, \quad \gamma \in \mathbb{C}$$

such that

$$|q_{2n}(e^{i\theta})| = \prod_{k=1}^{2n} |\sin((\theta - a_k)/2)|\,, \qquad \theta \in \mathbb{R}\,.$$

Hint: Use the fact that $|z - c| = |1 - \overline{c}z|$ whenever $|z| = 1$ and $c \in \mathbb{C}$. □

Associated with $q_{2n} \in \mathcal{P}_{2n}^c$ defined in part a], let

$$M_n(z) := \sqrt{q_{2n}(z)}$$

and

$$M_n^*(z) := \left(\overline{\gamma} \prod_{k=1}^{2n} (1 - \overline{c}z) \right)^{1/2},$$

where the square roots are defined so that M_n^* is analytic in a neighborhood of the closed unit disk, and M_n is analytic in a neighborhood of the complement of the open unit disk. Let

$$f_n(z) := \frac{M_n(z)}{M_n^*(z)}\,.$$

For $\theta \in \mathbb{R}$, we define

$$\widetilde{T}_n(\theta) := \operatorname{Re}(f_n(e^{i\theta})) = \frac{1}{2} \left(\frac{M_n(e^{i\theta})}{M_n^*(e^{i\theta})} + \frac{M_n^*(e^{i\theta})}{M_n(e^{i\theta})} \right)$$

and

$$\widetilde{U}_n(\theta) := \operatorname{Im}(f_n(e^{i\theta})) = \frac{1}{2i} \left(\frac{M_n(e^{i\theta})}{M_n^*(e^{i\theta})} - \frac{M_n^*(e^{i\theta})}{M_n(e^{i\theta})} \right).$$

Using the new (extended) definitions, show the following:

b] $\widetilde{T}_n \in \mathcal{T}_n(a_1, a_2, \dots, a_{2n})$ and $\widetilde{U}_n \in \mathcal{T}_n(a_1, a_2, \dots, a_{2n})$.

c] $\|\widetilde{T}_n\|_{\mathbb{R}} = 1$ and $\|\widetilde{U}_n\|_{\mathbb{R}} = 1$.

d] There are numbers $\theta_1 < \theta_2 < \cdots < \theta_{2n}$ in $[-\pi, \pi)$ such that

$$\widetilde{T}(\theta_j) = \pm(-1)^j, \qquad j = 1, 2, \dots, 2n.$$

e] There are numbers $\tau_1 < \tau_2 < \cdots < \tau_{2n}$ in $[-\pi, \pi)$ such that

$$\widetilde{U}(\tau_j) = \pm(-1)^j, \qquad j = 1, 2, \dots, 2n.$$

f] $\widetilde{T}(\theta)^2 + \widetilde{U}(\theta)^2 = 1$ for every $\theta \in \mathbb{R}$.

g] Both $\widetilde{T}_n$ and $\widetilde{U}_n$ have exactly $2n$ simple zeros in the period $[-\pi, \pi)$, and the zeros of $\widetilde{T}_n$ and $\widetilde{U}_n$ strictly interlace.

h] The statements of Theorem 3.5.3 remain valid.

E.4 Extension of the Bernstein Factor $\widetilde{B}_n$. Let

$$(a_k)_{k=1}^{2n} \subset \mathbb{C} \setminus \mathbb{R}, \qquad \operatorname{Im}(a_k) > 0.$$

With the notation of the previous exercise we define

$$\widetilde{B}_n(\theta) := \frac{e^{i\theta} f_n'(e^{i\theta})}{f_n(e^{i\theta})}, \qquad \theta \in \mathbb{R}.$$

a] Show that for every $\theta \in \mathbb{R}$,

$$\widetilde{B}_n(\theta) = \sum_{k=1}^{2n} \frac{1 - |c_k|^2}{|c_k - e^{i\theta}|^2} = \sum_{k=1}^{2n} \frac{1 - |e^{ia_k}|^2}{|e^{ia_k} - e^{i\theta}|^2}.$$

b] Show that, on the real line,

$$\widetilde{T}_n' = -\widetilde{B}_n \widetilde{U}_n, \qquad \widetilde{U}_n' = \widetilde{B}_n \widetilde{T}_n,$$

and

$$(\widetilde{T}_n')^2 + (\widetilde{U}_n')^2 = \widetilde{B}_n^2.$$

c] Show that

$$(V')^2 + \widetilde{B}_n^2 V^2 = \widetilde{B}_n^2$$

holds on the real line for every V of the form

$$V = (\cos\alpha)\,\widetilde{T}_n + (\sin\alpha)\,\widetilde{U}_n, \qquad \alpha \in \mathbb{R}.$$

E.5 Chebyshev Polynomials for $\mathcal{P}_n(a_1, a_2, \dots, a_n)$ on $\mathbb{R}$. Let

$$(a_k)_{k=1}^n \subset \mathbb{C} \setminus \mathbb{R} \qquad \text{with} \qquad \operatorname{Im}(a_k) > 0\,, \qquad k = 1, 2, \dots, n\,.$$

Let

$$M_n(z) := \left(\prod_{k=1}^n (z - a_k) \right)^{1/2}$$

and

$$M_n^*(z) := \left(\prod_{k=1}^n (z - \overline{a}_k) \right)^{1/2},$$

where the square roots are defined so that M_n^* is analytic in a neighborhood of the closed upper half-plane, and M_n is analytic in a neighborhood of the closed lower half-plane. Let

$$f_n(z) := \frac{M_n(z)}{M_n^*(z)}\,.$$

For $x \in \mathbb{R}$, we define

$$T_n(x) := \operatorname{Re}(f_n(x)) = \frac{1}{2} \left(\frac{M_n(x)}{M_n^*(x)} + \frac{M_n^*(x)}{M_n(x)} \right)$$

and

$$U_n(x) := \operatorname{Im}(f_n(x)) = \frac{1}{2i} \left(\frac{M_n(x)}{M_n^*(x)} - \frac{M_n^*(x)}{M_n(x)} \right).$$

Show the following:

a] $T_n \in \mathcal{P}_n(a_1, a_2, \dots, a_n)$ and $U_n \in \mathcal{P}_n(a_1, a_2, \dots, a_n)$.

b] $\|T_n\|_{\mathbb{R}} = 1$ and $\|U_n\|_{\mathbb{R}} = 1$.

c] There are real numbers $x_1 > x_2 > \cdots > x_{n-1}$ such that

$$T_n(x_j) = (-1)^j\,, \quad \lim_{x \to \infty} T_n(x) = 1\,, \quad \text{and} \quad \lim_{x \to -\infty} T_n(x) = (-1)^n\,.$$

d] There are real numbers $y_1 > y_2 > \cdots > y_{n-1}$ such that

$$U_n(y_j) = (-1)^{j+1} \quad \text{and} \quad \lim_{x \to \pm\infty} U_n(x) = 0\,.$$

e] $T_n(x)^2 + U_n(x)^2 = 1$ for every $x \in \mathbb{R}$.

f] Both T_n and U_n have exactly n simple zeros on $\mathbb{R}$, and the zeros of T_n and U_n strictly interlace.

g] The following statements are equivalent:

(i) There exists an $\alpha \in \mathbb{R}$ such that

$$V = (\cos \alpha) T_n + (\sin \alpha) U_n .$$

(ii) $V \in \mathcal{P}_n(a_1, a_2, \cdots, a_n)$ has uniform norm 1 on $\mathbb{R}$, and it equioscillates n times on the extended real line. That is, there are extended real numbers $\infty \geq z_1 > z_2 > \cdots > z_n > -\infty$ such that

$$V(z_j) = \pm(-1)^j , \qquad j = 1, 2, \ldots, n ,$$

where

$$V(\infty) := \lim_{x \to \infty} V(x) .$$

h] With the notation of part g], V is strictly monotone on each of the intervals

$$(z_1, \infty), \ (z_2, z_1), \ \ldots, \ (z_n, z_{n-1}), \ (-\infty, z_n) .$$

E.6 Bernstein Factor on $\mathbb{R}$. Let $(a_k)_{k=1}^n \subset \mathbb{C} \setminus \mathbb{R}$ with

$$\operatorname{Im}(a_k) > 0 , \qquad k = 1, 2, \ldots, n .$$

With the notation of E.5 let

$$B_n(x) := \frac{f_n'(x)}{f_n(x)} , \qquad x \in \mathbb{R} .$$

a] Show that

$$B_n(x) = \sum_{k=1}^{n} \frac{2\operatorname{Im}(a_k)}{|x - a_k|^2} , \qquad x \in \mathbb{R} .$$

b] Show that, on the real line,

$$T_n' = -B_n U_n , \qquad U_n' = B_n T_n ,$$

and

$$(T_n')^2 + (U_n')^2 = B_n^2 .$$

c] Show that

$$(V')^2 + B_n^2 V^2 = B_n^2$$

holds on the real line for every V of the form

$$V = (\cos \alpha) T_n + (\sin \alpha) U_n .$$

E.7 Coefficient Bounds in Nondense Rational Function Spaces. Suppose $(a_k)_{k=1}^{\infty} \subset \mathbb{R} \setminus [-1,1]$ is a sequence of distinct numbers satisfying

$$\sum_{j=1}^{\infty} \sqrt{1 - |a_j|^{-2}} < \infty\,.$$

Show that there are numbers $K_j > 0$ such that

$$|D_{j,n}| \leq K_j \|p\|_{[-1,1]}\,, \qquad j = 0, 1, \dots, n\,, \quad n \in \mathbb{N}$$

for every $p \in \mathcal{P}_n(a_1, a_2, \dots, a_n)$ of the form

$$p(x) = D_{0,n} + \frac{D_{1,n}}{x - a_1} + \cdots + \frac{D_{n,n}}{x - a_n}\,, \qquad D_{j,n} \in \mathbb{R}\,.$$

Hint: Use E.2 c] of Section 3.3 and Theorem 3.5.4. □

4

Denseness Questions

Overview

We give an extended treatment of when various Markov spaces are dense. In particular, we show that denseness, in many situations, is equivalent to denseness of the zeros of the associated Chebyshev polynomials. This is the principal theorem of the first section. Various versions of Weierstrass' classical approximation theorem are then considered. The most important is in Section 4.2 where Müntz's theorem concerning the denseness of $\text{span}\{1, x^{\lambda_1}, x^{\lambda_2}, \dots\}$ is analyzed in detail. The third section concerns the equivalence of denseness of Markov spaces and the existence of unbounded Bernstein inequalities. In the final section we consider when rational functions derived from Markov systems are dense. Included is the surprising result that rational functions from a fixed infinite Müntz system are always dense.

4.1 Variations on the Weierstrass Theorem

Much of the utility of polynomials stems from the fact that all continuous functions on a finite closed interval are uniform limits of them. This is the well-known Weierstrass approximation theorem. There are numerous proofs of this; several are presented in the exercises. Another proof follows from the main theorem of this section.

Associated with a Markov system $\mathcal{M} := (f_0, f_1, \dots)$ on $[a,b]$ we define, as in Section 3.3, the Chebyshev polynomials

$$T_n := T_n\{f_0, f_1, \dots, f_n; [a,b]\}.$$

Denote the zeros of T_n by $(a \leq)x_1 < x_2 < \cdots < x_n(\leq b)$. Let $x_0 := a$ and $x_{n+1} := b$. The *mesh* of T_n is defined by

$$M_n := M_n(T_n : [a,b]) := \max_{1 \leq i \leq n+1} |x_i - x_{i-1}|. \tag{4.1.1}$$

This is a measure of the maximal gap between two consecutive zeros of T_n with respect to the interval $[a,b]$.

For a sequence $(T_n)_{n=0}^{\infty}$ of Chebyshev polynomials associated with a fixed Markov system on $[a,b]$, we have

$$\lim_{n\to\infty} M_n = 0 \qquad \text{if and only if} \qquad \liminf_{n\to\infty} M_n = 0.$$

This follows from the fact that if $m < n$, then T_m cannot have more than one zero between any two consecutive zeros of T_n.

Our main result shows the strong connection between the denseness of the real span of an infinite Markov system $\mathcal{M}$ of C^1 functions on $[a,b]$ in $C[a,b]$ and the density of the zeros of the associated Chebyshev polynomials.

Theorem 4.1.1. *Suppose $\mathcal{M} := (1, f_1, f_2, \dots)$ is an infinite Markov system on $[a,b]$ with each $f_i \in C^1[a,b]$. Then* span $\mathcal{M}$ *is dense in $C[a,b]$ if and only if*

$$\lim_{n\to\infty} M_n = 0,$$

where M_n is the mesh of the associated Chebyshev polynomials.

Proof. The *only if* part of this result is the easier part and we offer the following proof. Suppose span $\mathcal{M}$ is dense in $C[a,b]$, while $\liminf_{n\to\infty} M_n > 0$. Then there exists an interval $[c,d] \subset [a,b]$ that contains no zero of T_n for infinitely many n, say, for $n_1 < n_2 < \cdots$. Consider the piecewise linear function F defined as follows. Let $c < y_1 < y_2 < y_3 < y_4 < d$, and let

$$F(x) := \begin{cases} 0, & x \in \{a, c, d, b\} \\ 2, & x \in \{y_1, y_3\} \\ -2, & x \in \{y_2, y_4\} \end{cases}$$

and be linear elsewhere. Since span $\mathcal{M}$ is dense in $C[a,b]$, there exists a $k \in \mathbb{N}$ and a $p \in \text{span}\{1, f_1, \dots, f_{n_k}\}$ with

$$\|p - F\|_{[a,b]} < 1. \tag{4.1.2}$$

Now $p - T_{n_k}$ has at least $n_k - 2$ zeros on $[a, c] \cup [d, b]$ because T_{n_k} has at least n_k extrema on these intervals. The four extrema of F on (c, d) together with (4.1.2) guarantee at least three more zeros of $p - T_{n_k}$ on (c, d). Hence $p - T_{n_k}$ has at least $n_k + 1$ zeros and vanishes identically. This contradicts (4.1.2).

The *if* part of the theorem follows from the next theorem and E.8 a] of Section 3.2. This exercise shows that $(f_1', f_2', \dots)$ is a weak Markov system on $[a, b]$. □

The phenomenon formulated in Theorem 4.1.1 is quite general, and we prove a rather more general result than is needed for the preceding theorem. The *modulus of continuity* ω_f of a function $f : [a, b] \mapsto \mathbb{R}$ is defined by

$$\omega_f(\delta) := \sup_{\substack{|x-y|<\delta \\ x,y\in[a,b]}} |f(x) - f(y)| . \tag{4.1.3}$$

Theorem 4.1.2. *Suppose that*

$$H_n := \operatorname{span}\{1, g_1, g_2, \dots , g_n\}$$

is a Chebyshev space on $[a, b]$ *with associated Chebyshev polynomial* T_n. *Suppose each* $g_i \in C^1[a, b]$ *and* $(g_1', \dots , g_n')$ *is a weak Chebyshev system on* $[a, b]$ *(weak Chebyshev systems are defined in E.8 of Section 3.2). Let* $H_n' := \operatorname{span}\{g_1', \dots , g_n'\}$. *If* $f \in C[a, b]$, *then there exists an* $h_n \in H_n$ *such that*

$$\|h_n - f\|_{[a,b]} \le C\omega_f\left(\sqrt{\delta_n}\right),$$

where

$$\delta_n := M_n(T_n : [a, b]) .$$

Here C *is a constant depending only on* a *and* b.

Proof. Suppose $a < c < d < b$ and $S_n \in H_n$ is the best uniform approximation from H_n to F on $[a, c] \cup [d, b]$, where

$$F(x) := \begin{cases} 0, & x \in [a, c] \\ 1, & x \in [d, b] . \end{cases}$$

We claim the following:

$$S_n \text{ is monotone on } [c, d] \tag{4.1.4}$$

and

$$\|S_n - F\|_{[a,c]\cup[d,b]} \le \frac{5\delta_n}{(d - c)} . \tag{4.1.5}$$

Let $\eta := n+1$ be the dimension of the Chebyshev space H_n. Since S_n is a best approximation to F on $[a,c] \cup [d,b]$, there exist $\eta+1$ points in this set where the maximum error

$$\epsilon_n := \|F - S_n\|_{[a,c]\cup[d,b]} \tag{4.1.6}$$

occurs with alternating sign (see Theorem 3.1.6). Suppose $m+1$ of these points $y_0 < \cdots < y_m$ lie in $[a,c]$, and $\eta - m$ of these points $y_{m+1} < \cdots < y_\eta$ lie in $[d,b]$. Then S_n' has at least $m-1$ sign changes in (a,c) (one at each alternation point in $[a,c]$ except possibly at the endpoints a and c). Likewise, S_n' has at least $\eta - m - 2$ sign changes in (d,b). So S_n' has at least $\eta - 3$ sign changes in $(a,c) \cup (d,b)$. Note that this count excludes y_m and y_{m+1}. Thus S_n' has at most one more sign change in (a,b) unless S_n' vanishes identically (which is not possible for $\eta \geq 2$). Now suppose S_n' has a sign change on (c,d). Then, since there is at most one sign change of S_n' in (c,d), it cannot be the case that both $y_m = c$ and $y_{m+1} = d$ and S_n' changes sign at neither c nor d, otherwise

$$\operatorname{sign}(S_n(c) - f(c)) = \operatorname{sign}(S_n(d) - f(d))$$

as a consideration of the two cases shows. But if $y_m \neq c$ or $y_{m+1} \neq d$ or S_n' changes sign at either c or d, then we have accounted for all the sign changes of S_n' by accounting for the (possible) one additional sign change (either S_n' vanishes with sign change at c or d or one of y_m or y_{m+1} is an interior alternation point of S_n where S_n' vanishes). Thus S_n' has no zeros with sign change in (c,d) and (4.1.4) is proved.

To prove (4.1.5) we proceed as follows. With ϵ_n defined by (4.1.6),

$$D_n := \epsilon_n T_n - S_n$$

has at least m zeros on $[a,c]$ and

$$D_n^* := D_n + 1 = 1 + \epsilon_n T_n - S_n$$

has at least $\eta - m - 1$ zeros on $[d,b]$ (counting each internal zero without sign change twice). Thus D_n' has at least $\eta - 3$ sign changes on $[a,c] \cup [d,b]$. Suppose T_n has at least four alternation points on an interval $[\gamma, \delta] \subset (c,d)$, and suppose that

$$S_n(\delta) - S_n(\gamma) < 2\epsilon_n\,.$$

Then, because of (4.1.4) and the oscillation of T_n on $[\gamma,\delta]$,

$$D_n + \frac{S_n(\gamma) + S_n(\delta)}{2} = \epsilon_n T_n - \left[S_n - \frac{S_n(\gamma) + S_n(\delta)}{2}\right]$$

has at least three zeros on $[\gamma, \delta]$ and hence

$$D_n' = \left(D_n + \frac{S_n(\gamma) + S_n(\delta)}{2} \right)'$$

has at least two sign changes on $[\gamma, \delta]$. This, however, gives that $D_n' \in H_n'$ has a total of at least $\eta - 1 = n$ sign changes, which is impossible. In particular,

$$S_n(\delta) - S_n(\gamma) \geq 2\epsilon_n$$

on any interval $[\gamma, \delta] \subset (c, d)$ where T_n has at least 4 alternation points. Thus,

$$S_n(d) - S_n(c) \geq \frac{(d-c)}{5\delta_n} 2\epsilon_n \, .$$

However, since S_n is a best approximation to F on $[a, c] \cup [d, b]$,

$$S_n(d) - S_n(c) \leq 1 + 2\epsilon_n$$

and we can deduce (4.1.5) on comparing these last two inequalities and noting that $\epsilon_n \leq \frac{1}{2}$.

The proof is now a routine argument, which for simplicity, is presented on the interval $[a, b] := [0, 1]$. Let

$$V(x) := f(0) + \sum_{i=0}^{m-1} \left(f\left(\tfrac{i+1}{m}\right) - f\left(\tfrac{i}{m}\right) \right) S_{n,i}(x) \, ,$$

where, for $i = 0, 1, \ldots, m-1$, $S_{n,i} \in H_n$ is the best uniform approximation to

$$F_{n,i}(x) := \begin{cases} 0 \, , & x \in \left[0, \tfrac{i+1}{m}\right) \\ 1 \, , & x \in \left[\tfrac{i+1}{m}, 1\right] \end{cases}$$

on $\left[0, \frac{i}{m}\right] \cup \left[\frac{i+1}{m}, 1\right]$. Let

$$\widetilde{f}(x) := f(0) + \sum_{i=0}^{m-1} \left(f\left(\tfrac{i+1}{m}\right) - f\left(\tfrac{i}{m}\right) \right) F_{n,i}(x) \, .$$

Then repeated applications of (4.1.5) with the intervals $[a, c] := \left[0, \frac{i}{m}\right]$ and $[d, b] := \left[\frac{i+1}{m}, 1\right]$ yield for every $x \in [0, 1]$ that

$$\begin{aligned} |V(x) - f(x)| &\leq |V(x) - \widetilde{f}(x)| + |\widetilde{f}(x) - f(x)| \\ &\leq \sum_{i=0}^{m-1} \left(f\left(\tfrac{i+1}{m}\right) - f\left(\tfrac{i}{m}\right) \right) \left(S_{n,i}(x) - F_{n,i}(x) \right) + \omega_f\left(\tfrac{1}{m}\right) \\ &\leq (m-1)(5\delta_n m)\, \omega_f\left(\tfrac{1}{m}\right) + 2\,\omega_f\left(\tfrac{1}{m}\right) + \omega_f\left(\tfrac{1}{m}\right) \, . \end{aligned}$$

Hence, with $m := \lfloor \delta_n^{-1/2} \rfloor$,

$$\|V - f\|_{[0,1]} \leq C\, \omega_f\left(\sqrt{\delta_n}\right) .$$

□

An immediate corollary to Theorem 4.1.1 is the Weierstrass theorem.

Corollary 4.1.3. *The polynomials are dense in* $C[-1,1]$.

Proof. $\mathcal{M} = (1, x, x^2, \dots)$ is an infinite Markov system of C^1 functions on $[-1,1]$. The associated Chebyshev polynomials are just the usual Chebyshev polynomials T_n (see Section 2.1) and

$$M_n \le \frac{\pi}{n}, \qquad n = 1, 2, \dots$$

is obvious from E.1 of Section 2.1. □

Also from the last part of the proof of Theorem 4.1.2 we have the following corollary.

Corollary 4.1.4. *Suppose* $\mathcal{M} := (1, f_1, f_2, \dots)$ *is an infinite Markov system on* $[a,b]$ *with each* $f_i \in C^1[a,b]$. *Then for each* $n \in \mathbb{N}$, *there exists a*

$$p_n \in \operatorname{span}\{1, f_1, f_2, \dots, f_n\}$$

such that

$$\|p_n - f\|_{[a,b]} \le C(1 + m^2 M_n)\,\omega_f\left(\tfrac{1}{m}\right)$$

for every $m \in \mathbb{N}$, *where* C *is a constant depending only on* a *and* b.

Comments, Exercises, and Examples.

The Weierstrass approximation theorem of 1885 (see Weierstrass [15]) is one of the very basic theorems of approximation theory. It, of course, requires that clear distinctions be made about the nature of convergence (pointwise versus uniform) and the region of convergence (intervals versus complex domains). Weierstrass, the preeminent analyst of the last third of the nineteenth century, was principal in insisting that such distinctions be clearly made. His famous and profoundly surprising example of a nowhere differentiable continuous function dates from 1872. A number of proofs of his approximation theorem and its many generalizations are explored in the exercises. Theorem 4.1.1 was proved by Borwein [90]. The *only if* part of this theorem can be found in Kroó and Peherstorfer [92].

Applications of the methods and results of this section can be found in Borwein [91b], Borwein and Saff [92], and Lorentz, Golitschek, and Makovoz [92]. The last two papers give an application to weighted incomplete polynomials, where the zeros of the Chebyshev polynomials are often dense in a subinterval (see also Mhaskar and Saff [85]).

E.1 The Weierstrass Approximation Theorem. *Every real-valued continuous function on a finite closed interval* $[a,b]$ *can be uniformly approximated by polynomials with real coefficients.*

Every complex-valued continuous function on a finite closed interval $[a,b]$ *can be uniformly approximated by polynomials with complex coefficients.*

More precisely, in the real case, let

$$E_n := E_n(f : [a,b]) := \inf_{p \in \mathcal{P}_n} \|f - p\|_{[a,b]} .$$

The Weierstrass approximation theorem asserts that

$$\lim_{n \to \infty} E_n(f : [a,b]) = 0 , \qquad f \in C[a,b] .$$

The following steps outline an elementary proof basically due to Lebesgue [1898]. Parts a] to d] deal with the real version (first statement) of the theorem. The complex version (second statement) of the theorem can easily be reduced to the real version; see part e].

a] Every continuous function on $[a,b]$ can be uniformly approximated by piecewise linear functions.

Hint: Consider the piecewise linear function that interpolates f at n equally spaced points and use the uniform continuity of f. □

b] It suffices to prove that $|x|$ can be uniformly approximated by polynomials on $[-1,1]$.

Hint: Use part a]. □

c] Approximation to $|x|$. Show that

$$\lim_{n \to \infty} E_n(|x| : [-1,1]) = 0 .$$

Hint: The Taylor series expansion of $f(z) := \sqrt{1-z}$ yields

$$\sqrt{1-z} = 1 - \frac{1}{2} z + \frac{1}{2 \cdot 4} z^2 - \frac{1 \cdot 3}{2 \cdot 4 \cdot 6} z^3 + \cdots$$

and the convergence is uniform for $0 \le z \le 1$. (By Abel's theorem, a power series converges uniformly on every closed subinterval of the set of points in $\mathbb{R}$ where it converges; see, for example, Stromberg [81]). Thus,

$$\begin{aligned} |x| &= \sqrt{x^2} = \sqrt{1-(1-x^2)} \\ &= 1 - \frac{1}{2}(1-x^2) + \frac{1}{2 \cdot 4}(1-x^2)^2 - \frac{1 \cdot 3}{2 \cdot 4 \cdot 6}(1-x^2)^3 + \cdots \end{aligned}$$

and the convergence is uniform for $-1 \le x \le 1$. □

d] An Alternative to c]. Let

$$Q_0(x) := 1 \qquad \text{and} \qquad Q_{n+1}(x) := \frac{1}{2}(1 - x^2 + Q_n^2(x)) .$$

Show that

$$0 \le Q_{n+1}(x) \le Q_n(x) \le 1, \qquad n = 0, 1, \ldots , \quad x \in [-1,1]$$

and $Q_n(x) \to 1 - |x|$ uniformly on $[-1,1]$ as $n \to \infty$.

Hint: First show the pointwise convergence and then use Dini's theorem (see, for example, Royden [88]). □

e] Complex Version of the Weierstrass Approximation Theorem. *Every complex-valued continuous function on a finite closed interval* $[a, b]$ *can be uniformly approximated by polynomials with complex coefficients.*

It can be shown that

$$E_n(|x| : [-1, 1]) \sim \frac{c}{n},$$

where $0.280168 < c < 0.280174$. Bernstein [13] established the above asymptotic with weaker bounds on c, namely, $0.278 < c < 0.286$, and observed that $\frac{1}{2}\pi^{-1/2} = 0.282$ is roughly the average of these bounds. The stronger bounds on c, due to Varga and Carpenter, show that $c \neq \frac{1}{2}\pi^{-1/2}$, but it is open whether or not c is some familiar constant; see Varga [90].

E.2 The Stone-Weierstrass Theorem. *If* X *is a compact Hausdorff space, then a subalgebra* $\mathcal{A}$ *of* $C(X)$, *which contains* $f = 1$ *and separates points, is dense in* $C(X)$.

A subalgebra $\mathcal{A}$ of $C(X)$ is a vector space of functions that is closed under multiplication (here, addition and multiplication are pointwise). Separating points means that for any two distinct $x, y \in X$, there exists an $f \in \mathcal{A}$ such that $f(x) \neq f(y)$.

a] Observe that the set $\mathcal{P} := \cup_{n=0}^{\infty} \mathcal{P}_n$ of all polynomials with real coefficients is a subalgebra of $C[a, b]$ that separates points, and hence the Stone-Weierstrass theorem implies the Weierstrass approximation theorem.

b] Observe that the real polynomials in x^2 form a subalgebra of $C[-1, 1]$ that does not separate points.

We outline a standard proof of the Stone-Weierstrass theorem. Let $\overline{\mathcal{A}}$ denote the closure of a subalgebra $\mathcal{A} \subset C(X)$ in the uniform norm.

c] If $f \in \overline{\mathcal{A}}$, then $|f| \in \overline{\mathcal{A}}$.

Proof. If $f \in \mathcal{A}$, then $p(f) \in \mathcal{A}$ for any polynomial p. Now choose p_n such that $p_n(x) \to |x|$ on the interval $[-\|f\|, \|f\|]$. □

d] Let

$$(f \vee g)(x) := \max\{f(x), g(x)\} \quad \text{and} \quad (f \wedge g)(x) := \min\{f(x), g(x)\}.$$

Show that if $f, g \in \overline{\mathcal{A}}$, then so are $f \vee g$ and $f \wedge g$.

Hint:

$$f \vee g = \frac{1}{2}(f + g + |f - g|)\,, \quad \text{and} \quad f \wedge g = \frac{1}{2}(f + g - |f - g|)\,.$$

□

e] If $p, q \in X$ are distinct and $\lambda, \mu \in \mathbb{R}$, then there exists $f \in \mathcal{A}$ with

$$f(p) = \lambda \qquad \text{and} \qquad f(q) = \mu\,.$$

Hint: Let $g \in \mathcal{A}$ be such that $g(p) \neq g(q)$ and consider

$$f := \frac{\lambda - \mu}{g(p) - g(q)} \cdot g + \frac{\mu g(p) - \lambda g(q)}{g(p) - g(q)} \cdot 1\,.$$

□

f] Completion of Proof. Let $f \in C(X)$. For each $p, q \in X$, let f_{pq} be an element of $\mathcal{A}$ with $f_{pq}(p) = f(p)$ and $f_{pq}(q) = f(q)$. Fix $\epsilon > 0$ and define open sets

$$V_{pq} := \{x \in X : f_{pq}(x) < f(x) + \epsilon\}\,.$$

Now $\{V_{pq} : p \in X\}$ is an open cover of the compact Hausdorff space X, so for each $q \in X$ we can pick a finite subcover

$$\{V_{p_1 q}, V_{p_2 q}, \dots, V_{p_n q}\}$$

of X. We let

$$f_q := \min\{f_{p_1 q}, f_{p_2 q}, \dots, f_{p_n q}\}\,.$$

Observe that $f_q \in \overline{\mathcal{A}}$ by part e], and

$$f_q(x) < f(x) + \epsilon\,, \qquad x \in X\,.$$

g] Continued. Let

$$V_q := \{x \in X : f_q(x) > f(x) - \epsilon\}\,,$$

where f_q is defined in part f] for every $q \in X$. Then $\{V_q : q \in X\}$ is an open cover of the compact Hausdorff space X, so we can extract a finite subcover

$$\{V_{q_1}, V_{q_2}, \dots, V_{q_m}\}$$

of X. Now let

$$g := \max\{f_{q_1}, f_{q_2}, \dots, f_{q_m}\}\,.$$

Note that $g \in \overline{\mathcal{A}}$ by part e], and

$$f(x) = \epsilon < g(x) < f(x) + \epsilon\,, \qquad x \in X\,,$$

which finishes the proof. □

The next exercise presents pretty theorems due to Bohman [52] and Korovkin [53] on the convergence of sequences of positive linear operators. The exercise after that gives some applications that include different proofs of the Weierstrass theorem via convergence of special polynomials, such as the Bernstein polynomials.

An operator L on $C(X)$ is called *monotone* if

$$f \le g \quad \text{implies} \quad L(f) \le L(g)$$

(here $f \le g$ means $f(x) \le g(x)$ for all $x \in X$).

E.3 Monotone Operator Theorems.

Korovkin's First Theorem. *Let $(L_n)_{n=1}^{\infty}$ be a sequence of monotone linear operators on $C(K)$ (the set of continuous, 2π periodic, real-valued functions on $\mathbb{R}$). Let*

$$f_0(x) := 1\,, \quad f_1(x) := \sin x\,, \quad f_2(x) := \cos x\,.$$

Then

$$\lim_{n\to\infty} \|L_n(f) - f\|_K = 0$$

for all $f \in C(K)$ if and only if

$$\lim_{n\to\infty} \|L_n(f_i) - f_i\|_K = 0\,, \qquad i = 0, 1, 2\,.$$

Korovkin's Second Theorem. *Let $(L_n)_{n=1}^{\infty}$ be a sequence of monotone linear operators on $C[a,b]$. Let*

$$f_0(x) := 1\,, \quad f_1(x) := x\,, \quad f_2(x) := x^2\,.$$

Then

$$\lim_{n\to\infty} \|L_n(f) - f\|_{[a,b]} = 0$$

for all $f \in C[a,b]$ if and only if

$$\lim_{n\to\infty} \|L_n(f_i) - f_i\|_{[a,b]} = 0\,, \qquad i = 0, 1, 2\,.$$

Korovkin's theorem in a more general setting can be found in Lorentz [86a].

a] **Proof of Korovkin's Second Theorem.** The *only if* part of the theorem is trivial. For the *if* part, observe that the pointwise convergence of $(L_n)_{n=1}^{\infty}$ can be easily proved since, for any preassigned $\epsilon > 0$ at any fixed x_0, one can find parabolas $y = p_1(x) := a_1x^2 + b_1x + c_1$ and $y = p_2(x) := a_2x^2 + b_2x + c_2$ such that

$$p_1(x) < f(x) < p_2(x)\,, \qquad x \in [a, b]$$

with

$$|f(x_0) - p_1(x_0)| < \epsilon \qquad \text{and} \qquad |f(x_0) - p_2(x_0)| < \epsilon\,.$$

Now use the continuity of f and the compactness of $[a, b]$ to make the above argument uniform on the interval $[a, b]$. □

b] Proof of Korovkin's First Theorem.

Hint: Modify the proof of Korovkin's second theorem. □

E.4 Bernstein Polynomials. The nth *Bernstein polynomial* for a function $f \in C[0, 1]$ is defined by

$$B_n(f)(x) := \sum_{k=0}^{n} f\left(\frac{k}{n}\right)\binom{n}{k} x^k (1-x)^{n-k}\,, \qquad n = 1, 2, \ldots\,.$$

a] Let

$$f_0(x) := 1\,, \quad f_1(x) := x\,, \quad f_2(x) := x^2\,.$$

Show that

$$B_n(f_0) = f_0\,, \quad B_n(f_1) = f_1\,, \quad B_n(f_2) = \tfrac{n-1}{n} f_2 + \tfrac{1}{n} f_1$$

for every $n = 0, 1, 2, \ldots$.

b] Use Korovkin's second theorem and part a] to show that

$$\lim_{n\to\infty} \|B_n(f) - f\|_{[0,1]} = 0$$

for every $f \in C[0, 1]$.

For more on Bernstein polynomials, see Lorentz [86b].

E.5 The Fourier and Fejér Operators. For $f \in C(K)$, let

$$S_n(f)(x) := \frac{1}{2\pi}\int_{-\pi}^{\pi} f(t+x)\left(\frac{\sin\left(n+\frac{1}{2}\right)t}{2\sin\frac{1}{2}t}\right) dt\,, \qquad n = 0, 1, \ldots$$

and

$$F_n(f)(x) := \frac{1}{2\pi n}\int_{-\pi}^{\pi} f(t+x)\left(\frac{\sin\frac{1}{2}nt}{\sin\frac{1}{2}t}\right)^2 dt\,, \qquad n = 0, 1, \ldots\,.$$

The operator S_n is called the *Fourier operator*, while the operator F_n is called the *Fejér operator*.

a] Show that $S_n(f)$ is the nth partial sum of the Fourier series of f, that is,

$$S_n(f)(x) = \frac{a_0}{2} + \sum_{k=1}^{n} (a_k \cos kx + b_k \sin kx),$$

where

$$a_k = \frac{1}{\pi} \int_{-\pi}^{\pi} f(t) \cos kt \, dt$$

and

$$b_k = \frac{1}{\pi} \int_{-\pi}^{\pi} f(t) \sin kt \, dt.$$

Hint:

$$\frac{\sin\left(n + \frac{1}{2}\right) t}{2 \sin \frac{1}{2} t} = \frac{1}{2} + \sum_{k=1}^{n} \cos kt$$

and

$$S_n(f)(x) = \frac{1}{\pi} \int_{-\pi}^{\pi} f(t+x) \left(\frac{1}{2} + \sum_{k=1}^{n} \cos kt \right) dt.$$

□

b] $F_n(f)$ is the Cesàro mean of $S_0, S_1, \dots S_{n-1}$, that is,

$$F_n(f) = \frac{S_0(f) + S_1(f) + \cdots + S_{n-1}(f)}{n}.$$

Hint:

$$\sum_{k=0}^{n-1} \frac{\sin\left(k + \frac{1}{2}\right) t}{\sin \frac{1}{2} t} = \left(\frac{\sin \frac{1}{2} nt}{\sin \frac{1}{2} t} \right)^2.$$

□

c] **Fejér's Theorem.** *For every $f \in C(K)$, $F_n(f) \to f$ uniformly on $\mathbb{R}$.*
Hint: Each F_n is obviously a monotone operator on $C(K)$, so it suffices to prove the uniform convergence of $(F_n)_{n=1}^{\infty}$ on $\mathbb{R}$ only for f_i, $i = 0, 1, 2$, as defined in Korovkin's first theorem. However, this is obvious, since

$$F_n(f_0) = f_0, \quad F_n(f_1) = \tfrac{n-1}{n} f_1, \quad F_n(f_2) = \tfrac{n-1}{n} f_2$$

for every $n = 1, 2, \dots$. □

d] The set $\mathcal{T} := \cup_{n=0}^{\infty} \mathcal{T}_n$ of all real trigonometric polynomials is dense in $C(K)$, the set of all continuous, 2π periodic, real-valued functions. The set $\mathcal{T}^c := \cup_{n=0}^{\infty} \mathcal{T}_n^c$ of all complex trigonometric polynomials is dense in $C(K)$, the set of all continuous, 2π periodic, complex-valued functions.

Hint: This follows from Fejér's theorem. This is also a corollary of the Stone-Weierstrass theorem (see E.2). □

The remaining parts of the exercise follow Lorentz [86a]. Suppose that $L_n : C(K) \mapsto \mathcal{T}_n$ is a linear operator. We say that L_n preserves the elements of $\mathcal{T}_n$ if $L_n(t) = t$ for every $t \in \mathcal{T}_n$. A canonical example for such a linear operator L_n is the Fourier operator S_n. The purpose of the remaining part of the exercise is to show that the Fourier operator S_n is extremal among linear operators preserving the elements of $\mathcal{T}_n$ in the sense that it has the smallest norm. This leads to the result of Faber, Nikolaev, and Lozinskii (see part g]) that for arbitrary linear operators L_n preserving the elements of $\mathcal{T}_n$, $n = 1, 2, \dots$, the sequence $(L_n(f))_{n=1}^{\infty}$ cannot converge for every $f \in C(K)$.

e] **Berman's Generalization of a Formula of Faber and Marcinkiewicz.** Let f_a denote the *a-translation* of a function $f \in C(K)$, that is, $f_a(x) := f(x + a)$. Suppose L_n is a linear operator preserving $\mathcal{T}_n$. Show that

$$\frac{1}{2\pi} \int_{-\pi}^{\pi} L_n(f_t)(x - t)\, dt = S_n(f)(x)$$

for every $f \in C(K)$ and $x \in K$.

Hint: Let

$$A_n(x) := \frac{1}{2\pi} \int_{-\pi}^{\pi} L_n(f_t)(x - t)\, dt\,.$$

Show that $A_n(f) = S_n(f)$ for every $f \in \mathcal{T}_n$. Prove that $A_n(f) = S_n(f)$ for every f of the form $f(x) = \cos mx$ or $f(x) = \sin mx$, where m is an integer greater than n. Conclude that $A_n(f) = S_n(f)$ for every $f \in \mathcal{T} := \cup_{n=0}^{\infty} \mathcal{T}_n$. Note that $\mathcal{T}$ is dense in $C(K)$. This means that to complete the proof, it is sufficient to show that $A_n : C(K) \to \mathcal{T}_n$ and $S_n : C(K) \to \mathcal{T}_n$ are continuous. Observe that $\|A_n\| \le \|L_n\|$ and $\|S_n\| \le c \log n$ for some $c > 0$; see also part f]. □

f] **The Norm of the Fourier Operator S_n.** Show that

$$\|S_n\| := \sup\left\{ \frac{\|S_n(f)\|_K}{\|f\|_K} : \ f \in C(K) \right\} = \frac{1}{2\pi} \int_{-\pi}^{\pi} \left| \frac{\sin\left(n + \frac{1}{2}\right) t}{2 \sin \frac{1}{2} t} \right| dt\,.$$

Use this to prove that there exist two constants $c_1 > 0$ and $c_2 > 0$ independent of n such that

$$c_1 \log n \le \|S_n\| \le c_2 \log n\,, \qquad n = 2, 3, \dots\,.$$

Actually, it can be proved that

$$\|S_n\| = \frac{4}{\pi^2} \log n + O(1)\,, \qquad n = 2, 3, \dots\,.$$

See, for example, Lorentz [86a].

g] The Norm of Operators that Preserve Trigonometric Polynomials. Let $L_n : C(K) \to \mathcal{T}_n$ be a linear operator preserving the elements of $\mathcal{T}_n$. Show that

$$\|L_n\| \geq \|S_n\| \geq c_1 \log n\,,$$

where $c_1 > 0$ is a constant independent of n.

Hint: Use parts e] and f]. □

E.6 Polynomials in x^{λ_n}. Given $n \in \mathbb{N}$ and $\lambda_n \in \mathbb{R}$, let

$$\mathcal{P}_n(\lambda_n) := \{p_n(x^{\lambda_n}) : p_n \in \mathcal{P}_n\}.$$

Suppose $\delta \in (0,1)$ and $\lambda_n \geq 1$ for all $n \in \mathbb{N}$. Then $\cup_{n=1}^{\infty} \mathcal{P}_n(\lambda_n)$ is dense in $C[\delta, 1]$ if and only if

$$\limsup_{n\to\infty} \frac{\log n}{\lambda_n} \geq \frac{1}{2} \log \frac{1}{\delta}\,.$$

To prove the above statement, proceed as follows (see also Borwein [91b]). Denote the Chebyshev polynomial for $\mathcal{P}_n(\lambda_n)$ on $[\delta, 1]$ by $T_{n,\delta}$. Denote the zeros of $T_{n,\delta}$ in $[\delta, 1]$ by

$$x_{1,n}^{(\delta)} < x_{2,n}^{(\delta)} < \cdots < x_{n,n}^{(\delta)}\,.$$

Let $x_{0,n}^{(\delta)} := \delta$ and $x_{n+1,n}^{(\delta)} := 1$. Let

$$M_n(\delta) := \max_{1\leq i\leq n+1} \left(x_{i,n}^{(\delta)} - x_{i-1,n}^{(\delta)}\right)\,.$$

a] Show that

$$T_{n,\delta}(x) = T_n\left(\frac{2}{1-\delta^{\lambda_n}}\, x^{\lambda_n} - \frac{1+\delta^{\lambda_n}}{1-\delta^{\lambda_n}}\right), \qquad x \in [\delta, 1]\,,$$

where T_n is the Chebyshev polynomial of degree n as defined by (2.1.1).

b] Let $\delta := \liminf_{n\to\infty} x_{1,n}^{(0)}$. Show that if

$$\liminf_{n\to\infty} \left(\max_{2\leq i\leq n+1} \left(x_{i,n}^{(0)} - x_{i-1,n}^{(0)}\right)\right) = 0\,,$$

then $\liminf_{n\to\infty} M_n(\delta) = 0$.

Hint: Count the zeros of $T_{n,\delta} - T_{n,0} \in \mathcal{P}_n(\lambda_n)$ in $[\delta, 1]$. □

c] Let $\delta := \liminf_{n\to\infty} x_{1,n}^{(0)}$, as in part b]. Show that

$$\frac{1}{2}\log\frac{1}{\delta} = \limsup_{n\to\infty}\frac{\log n}{\lambda_n}$$

whenever the right-hand side is finite.

Hint: Use the explicit formula for $T_{n,0}$ given in part a]. □

d] Let δ be defined by

$$\frac{1}{2}\log\frac{1}{\delta} = \limsup_{n\to\infty}\frac{\log n}{\lambda_n}.$$

Suppose $\delta > 0$. Show that

$$\liminf_{n\to\infty}(x_{2,n}^{(0)} - x_{1,n}^{(0)}) = 0.$$

Hint: Use parts a] and c]. □

e] Let $\delta := \liminf_{n\to\infty} x_{1,n}^{(0)}$, as in parts b] and c]. Suppose $\delta > 0$. Show that

$$x_{i,n}^{(0)} - x_{i-1,n}^{(0)} \leq \frac{2}{\delta}\left(x_{2,n}^{(0)} - x_{1,n}^{(0)}\right)$$

for every sufficiently large $n \in \mathbb{N}$ and for every $i = 2, 3, \dots, n+1$.

Hint: Count the zeros of

$$T_{n,0}(x) - T_{n,0}(\alpha_{i,n}x) \in \mathcal{P}_n(\lambda_n),$$

where

$$\alpha_{i,n} := \frac{x_{i-1,n}^{(0)}}{x_{1,n}^{(0)}} < \frac{2}{\delta}$$

for every sufficiently large $n \in \mathbb{N}$ and for every $i = 2, 3, \dots, n+1$. □

f] Let δ be defined by

$$\frac{1}{2}\log\frac{1}{\delta} = \limsup_{n}\frac{\log n}{\lambda_n}.$$

Suppose $\delta > 0$. Show that $\cup_{n=1}^{\infty}\mathcal{P}_n(\lambda_n)$ is dense in $C[\delta, 1]$.

Hint: By parts a] to e], $\liminf_{n\to\infty} M_n(\delta) = 0$. Now apply Theorem 4.1.2. □

g] Let $0 \leq y < \delta$, where δ is as in part f]. Show that $\cup_{n=1}^{\infty}\mathcal{P}_n(\lambda_n)$ is not dense in $C[y, 1]$.

Hint: Show that there exists a constant c depending only on δ (and not on n or y) such that

$$|p(y)| \leq c\|p\|_{[\delta,1]}$$

for every $p \in \cup_{n=1}^{\infty}\mathcal{P}_n(\lambda_n)$ and $y \in [0, \delta]$. Now use E.4 c] of Section 3.3 and part a]. □

E.10 of Section 6.2 extends part g] of the above exercise. Namely, if $0 \leq \widetilde{\delta} < \delta$, where δ is the same as in part f] and $A \subset [0, 1]$ is a set of Lebesgue measure at least $1 - \widetilde{\delta}$, then $\cup_{n=1}^{\infty}\mathcal{P}_n(\lambda_n)$ is not dense in $C(A)$.

E.7 The Weierstrass Theorem in L_p. *Let $[a,b]$ be a finite interval and $p \in (0,\infty)$. Show that both $C[a,b] \cap L_p[a,b]$ and the set $\mathcal{P} := \cup_{n=0}^{\infty} \mathcal{P}_n$ of all real algebraic polynomials are dense in $L_p[a,b]$.*

Hint: The proof of the first statement is a routine measure theoretic argument; see Rudin [87]. The second statement follows from the first and the Weierstrass approximation theorem; see E.1. □

E.8 Density of Polynomials with Integer Coefficients.

a] Suppose $f \in C[0,1]$ and $f(0)$ and $f(1)$ are integers. Show that for every $\epsilon > 0$ there is a polynomial p with integer coefficients such that

$$\|f - p\|_{[0,1]} < \epsilon\,.$$

Outline. By E.4, there is an integer $n > 2/\epsilon$ so that

$$\|f - B_n(f)\|_{[0,1]} < \frac{\epsilon}{2}\,.$$

Let

$$\widetilde{B}_n(f) := \sum_{k=0}^{n} \left\lfloor f\left(\frac{k}{n}\right)\binom{n}{k}\right\rfloor x^k(1-x)^{n-k}\,.$$

Show that if $x \in [0,1]$, then

$$\begin{aligned}0 \le B_n(f)(x) - \widetilde{B}_n(f)(x) &\le \sum_{k=1}^{n-1} x^k(1-x)^{n-k}\\ &\le \frac{1}{n}\sum_{k=0}^{n}\binom{n}{k}x^k(1-x)^{n-k} = \frac{1}{n} < \frac{\epsilon}{2}\,.\end{aligned}$$

Note that $\widetilde{B}_n(f)$ is a polynomial with integer coefficients, and

$$\begin{aligned}\|f - \widetilde{B}_n(f)\|_{[0,1]} &\le \|f - B_n(f)\|_{[0,1]} + \|B_n(f) - \widetilde{B}(f)\|_{[0,1]}\\ &< \frac{\epsilon}{2} + \frac{\epsilon}{2} = \epsilon\,.\end{aligned}$$

□

b] Suppose the interval $[a,b]$ does not contain an integer. Show that polynomials with integer coefficients form a dense set in $C[a,b]$.

Proof 1. This is an immediate consequence of part a]. □

Proof 2. Assume, without loss of generality, that $[a,b] \subset (0,1)$. By the Weierstrass approximation theorem, it is sufficient to prove that for every $\epsilon > 0$ there is a polynomial p with integer coefficients such that

$$\left\| \tfrac{1}{2} - p \right\|_{[a,b]} < \epsilon\,,$$

since then all real numbers, and hence all $p \in \mathcal{P}_n$, can be approximated by polynomials with integer coefficients.

The existence of such a polynomial p follows from the identity

$$\frac{1}{2} = \frac{1-x}{1-(1-2(1-x))} = \sum_{k=0}^{\infty} (1-x)(1-2(1-x))^k\,,$$

where the infinite sum converges uniformly on $[a,b] \subset (0,1)$. □

E.9 Weierstrass Theorem on Arcs. Let ∂D denote the unit circle of the complex plane.

a] Show that the set $\mathcal{P}^c := \cup_{n=0}^{\infty} \mathcal{P}_n^c$ of all polynomials with complex coefficients is not dense in $C(\partial D)$.

Hint: Use the orthonormality of the system $((2\pi)^{-1/2} e^{in\theta})_{n=-\infty}^{\infty}$ on $[-\pi,\pi]$ to show that if k is a positive integer and $p \in \mathcal{P}^c$, then

$$2\pi \|z^{-k} - p(z)\|_{C(\partial D)} \geq \|e^{-ik\theta} - p(e^{i\theta})\|_{L_2[-\pi,\pi]} \geq 2\pi\,.$$

So none of the functions $z^{-1}, z^{-2}, \ldots$ is in the uniform closure of $\mathcal{P}^c$ on ∂D. □

b] Let $A \subset \partial D$ be an arc of length less than 2π. Then the set $\mathcal{P}^c$ of all polynomials with complex coefficients is dense in $C(A)$.

This is a special case of Mergelyan's theorem (see, for example, Rudin [87]).

Proof. Without loss of generality, we may assume that A is symmetric with respect to the real line. By E.5 d], it is sufficient to prove that $f(z) := z^{-1}$ is in the uniform closure $\overline{\mathcal{P}^c}$ of $\mathcal{P}^c$ on ∂D (this already implies that each z^k, $k \in \mathbb{Z}$, is in $\overline{\mathcal{P}^c}$). By E.11 j] of Section 2.1, $\mathrm{cap}(A) < 1$. By E.11 g] of Section 2.1, $\mu(A) = \mathrm{cap}(A)$, where $\mu(A)$ denotes the Chebyshev constant of A. Hence

$$0 \leq \mu(A) < \alpha < 1$$

with some α. Recalling the definition of $\mu(A)$, we can deduce that there are monic polynomials $p_n \in \mathcal{P}_n^c$ such that

$$\|p_n\|_A \leq \alpha^n\,, \qquad n = 1, 2, \ldots\,.$$

For $n = 1, 2, \ldots$, let

$$q_n(z) := z^{n-1} p_n(1/z) = z^{-1} + r_{n-1}(z)\,,$$

where $r_{n-1} \in \mathcal{P}_{n-1}^c$. Since A is symmetric with respect to the real line,

$$\|z^{-1} + r_{n-1}(z)\|_A = \|q_n\|_A = \|p_n\|_A \leq \alpha^n \underset{n\to\infty}{\longrightarrow} 0\,.$$

Hence $f(z) = z^{-1}$ is in $\overline{\mathcal{P}^c}$, which finishes the proof. □

4.2 Müntz's Theorem

A very attractive variant of the Weierstrass theorem characterizes exactly when the linear span of a system of monomials

$$\mathcal{M} := (x^{\lambda_0}, x^{\lambda_1}, \dots)$$

is dense in $C[0,1]$ or $L_2[0,1]$.

Theorem 4.2.1 (Full Müntz Theorem in $C[0,1]$). *Suppose $(\lambda_i)_{i=1}^\infty$ is a sequence of distinct positive numbers. Then*

$$\operatorname{span}\{1, x^{\lambda_1}, x^{\lambda_2}, \dots\}$$

is dense in $C[0,1]$ if and only if

$$\sum_{i=1}^{\infty} \frac{\lambda_i}{\lambda_i^2 + 1} = \infty .$$

Note that when $\inf_i \lambda_i > 0$,

$$\sum_{i=1}^{\infty} \frac{\lambda_i}{\lambda_i^2 + 1} = \infty \quad \text{if and only if} \quad \sum_{i=1}^{\infty} \frac{1}{\lambda_i} = \infty .$$

Müntz studied only this case, and his theorem is usually given in terms of the second condition.

When $\lambda_i \geq 1$ for each $i = 1, 2, \dots$, the above theorem follows by a simple trick from the L_2 version of Müntz's theorem. The proof of the general case is left as a guided exercise. The difficult case to deal with is the one where 0 and ∞ are both cluster points of the sequence $(\lambda_i)_{i=0}^\infty$; see E.18.

Theorem 4.2.2 (Full Müntz Theorem in $L_2[0,1]$). *Suppose $(\lambda_i)_{i=0}^\infty$ is a sequence of distinct real numbers greater than $-1/2$. Then*

$$\operatorname{span}\{x^{\lambda_0}, x^{\lambda_1}, \dots\}$$

is dense in $L_2[0,1]$ if and only if

$$\sum_{i=0}^{\infty} \frac{2\lambda_i + 1}{(2\lambda_i + 1)^2 + 1} = \infty .$$

The proof of the following full L_1 version of Müntz's theorem is presented as E.19.

Theorem 4.2.3 **(Full Müntz Theorem in $L_1[0,1]$).** *Suppose $(\lambda_i)_{i=0}^\infty$ is a sequence of distinct real numbers greater than -1. Then*

$$\operatorname{span}\{x^{\lambda_0}, x^{\lambda_1}, \dots\}$$

is dense in $L_1[0,1]$ if and only if

$$\sum_{i=0}^{\infty} \frac{\lambda_i + 1}{(\lambda_i + 1)^2 + 1} = \infty \, .$$

Now we formulate a general Müntz-type theorem in $L_p[0,1]$, that contains the above $C[0,1]$, $L_2[0,1]$, and $L_1[0,1]$ results as special cases. The proof of this theorem is outlined in E.20.

Theorem 4.2.4 **(Full Müntz Theorem in $L_p[0,1]$).** *Let $p \in [1,\infty)$. Suppose $(\lambda_i)_{i=0}^\infty$ is a sequence of distinct real numbers greater than $-1/p$. Then*

$$\operatorname{span}\{x^{\lambda_0}, x^{\lambda_1}, \dots\}$$

is dense in $L_p[0,1]$ if and only if

$$\sum_{i=0}^{\infty} \frac{\lambda_i + \frac{1}{p}}{\left(\lambda_i + \frac{1}{p}\right)^2 + 1} = \infty \, .$$

The full version of Müntz's theorem for arbitrary distinct real exponents on an interval $[a,b]$, $0 < a < b$, is given in E.7 and E.9.

Proof of Theorem 4.2.1 assuming Theorem 4.2.2 and each $\lambda_i \geq 1$. We need the following two inequalities:

$$\begin{aligned}
(4.2.1) \quad \left| x^m - \sum_{i=0}^{n} a_i x^{\lambda_i} \right| &= \left| \int_0^x \left(m t^{m-1} - \sum_{i=0}^{n} a_i \lambda_i t^{\lambda_i - 1} \right) dt \right| \\
&\leq \int_0^1 \left| m t^{m-1} - \sum_{i=0}^{n} a_i \lambda_i t^{\lambda_i - 1} \right| dt \\
&\leq \left(\int_0^1 \left| m t^{m-1} - \sum_{i=0}^{n} a_i \lambda_i t^{\lambda_i - 1} \right|^2 dt \right)^{1/2}
\end{aligned}$$

for every $x \in [0,1]$ and $m = 1, 2, \dots$, and

$$(4.2.2) \qquad \left(\int_0^1 \left| t^m - \sum_{i=0}^{n} a_i t^{\lambda_i} \right|^2 dt \right)^{1/2} \leq \left\| x^m - \sum_{i=0}^{n} a_i x^{\lambda_i} \right\|_{[0,1]}$$

for every $m = 0, 1, 2, \ldots$. The assumption that $\lambda_i \geq 1$ for each i implies that

$$\sum_{i=0}^{\infty} \frac{\lambda_i}{\lambda_i^2 + 1} = \infty \quad \text{if and only if} \quad \sum_{i=0}^{\infty} \frac{2(\lambda_i - 1) + 1}{(2(\lambda_i - 1) + 1)^2 + 1} = \infty$$

and

$$\sum_{i=0}^{\infty} \frac{\lambda_i}{\lambda_i^2 + 1} = \infty \quad \text{if and only if} \quad \sum_{i=0}^{\infty} \frac{2\lambda_i + 1}{(2\lambda_i + 1)^2 + 1} = \infty\,.$$

If $\sum_{i=0}^{\infty} \lambda_i/(\lambda_i^2 + 1) = \infty$, then (4.2.1), together with Theorem 4.2.2 and the Weierstrass approximation theorem (see E.1 of Section 4.1), shows that

$$\operatorname{span}\{1, x^{\lambda_1}, x^{\lambda_2}, \ldots\}$$

is dense in $C[0,1]$.

If the above span is dense in $C[0,1]$, then (4.2.2), together with E.7 of Section 4.1, shows that it is also dense in $L_2[0,1]$. Hence Theorem 4.2.2 implies $\sum_{i=1}^{\infty} \lambda_i/(\lambda_i^2 + 1) = \infty$. □

Proof of Theorem 4.2.2. We consider the approximation to x^m by elements of $\operatorname{span}\{x^{\lambda_0}, \ldots, x^{\lambda_{n-1}}\}$ in $L_2[0,1]$, and we assume $m > -\frac{1}{2}$ and $m \neq \lambda_i$ for any i. In the notation of Section 3.4 we define

$$\Lambda := (\lambda_0, \lambda_1, \ldots, \lambda_{n-1}, m)$$

and study L_n^*, the nth orthonormal Müntz-Legendre polynomial associated with Λ. By (3.4.8) and (3.4.6) we have (with $\lambda_n := m$)

$$L_n^*(x) = a_n x^m + \sum_{i=0}^{n-1} a_i x^{\lambda_i}\,,$$

where

$$|a_n| = \sqrt{1 + 2m} \prod_{i=0}^{n-1} \left| \frac{m + \lambda_i + 1}{m - \lambda_i} \right| .$$

It follows from $\|L_n^*\|_{L^2[0,1]} = 1$ and orthogonality that L_n^*/a_n is the error term in the best $L_2[0,1]$ approximation to x^m from $\operatorname{span}\{x^{\lambda_0}, \ldots, x^{\lambda_{n-1}}\}$ (why?). Therefore

$$\min_{b_i \in \mathbb{C}} \left\| x^m - \sum_{i=0}^{n-1} b_i x^{\lambda_i} \right\|_{L^2[0,1]} = \frac{1}{|a_n|} = \frac{1}{\sqrt{1 + 2m}} \prod_{i=0}^{n-1} \left| \frac{m - \lambda_i}{m + \lambda_i + 1} \right| .$$

So, for a nonnegative integer m different from any of the exponents λ_i,

$$x^m \in \overline{\mathrm{span}}\{x^{\lambda_0}, x^{\lambda_1}, \dots\} \tag{4.2.3}$$

(where $\overline{\mathrm{span}}$ denotes the $L_2[0,1]$ closure of the span) if and only if

$$\lim_{n\to\infty} \prod_{i=0}^{n-1} \left| \frac{m-\lambda_i}{m+\lambda_i+1} \right| = 0\,.$$

That is, (4.2.3) holds if and only if

$$\lim_{n\to\infty} \prod_{\substack{i=0\\ \lambda_i > m}}^{n-1} \left| 1 - \frac{2m+1}{m+\lambda_i+1} \right| \prod_{\substack{i=0\\ -1/2<\lambda_i\le m}}^{n-1} \left| 1 - \frac{2\lambda_i+1}{m+\lambda_i+1} \right| = 0\,.$$

Hence (4.2.3) holds if and only if either

$$\sum_{\substack{i=0\\ \lambda_i>m}}^{\infty} \frac{1}{2\lambda_i+1} = \infty \quad \text{or} \quad \sum_{\substack{i=0\\ -1/2<\lambda_i\le m}}^{\infty} (2\lambda_i+1) = \infty\,,$$

which is the case if and only if

$$\sum_{i=0}^{\infty} \frac{2\lambda_i+1}{(2\lambda_i+1)^2+1} = \infty\,,$$

and the proof can be finished by the Weierstrass approximation theorem (see E.1 of Section 4.1). □

Comments, Exercises, and Examples.

Theorem 4.2.1 (in the case when $\inf\{\lambda_i : i \in \mathbb{N}\} > 0$) and Theorem 4.2.2 were proved independently by Müntz [14] and Szász [16]. Szász [16] proved the full version of Theorem 4.2.2. Theorem 4.2.1 is to be found in Borwein and Erdélyi [to appear 5]. Much of Theorem 4.2.4 is stated in Schwartz [59] without proof and may be deduced by his methods. Indeed, Schwartz's method appears to give Theorem 4.2.4 for $p \in [1,2]$. Johnson (private communication) and Operstein [to appear] show how to derive the full Theorem 4.2.4 from Theorem 4.2.1 as does E.20; see also E.7 of the next section.

Less complete versions of the results presented in this section are often called the Müntz-Szász Theorems. A 1912 version due to Bernstein can be found in his collected works.

A variant on our proof of Müntz's theorem is presented in E.2. A distinct proof based on possible zero sets of analytic functions may be found in Feinerman and Newman [76]; see also E.10, where this method is explored for denseness questions for $\{\cos \lambda_k\theta\}$.

Extensions of Müntz's theorem abound. For example, generalizations to complex exponents are considered in Luxemburg and Korevaar [71], to angular regions in Anderson [72] and with an exponential weight on $[0, \infty)$ in Fuchs [46]. It is a nontrivial problem to establish a Müntz-type theorem on an interval $[a, b]$, $a > 0$, in which case the elements of the sequence $\Lambda = (\lambda_i)_{i=0}^{\infty}$ are allowed to be *arbitrary distinct real* numbers. This is the content of E.7; it is due to Clarkson and Erdős [43] (in the case when each λ_i is a nonnegative integer) and Schwartz [59] (in the general case). It is shown in Section 6.2 that if $\Lambda = (\lambda_i)_{i=0}^{\infty}$ is an increasing sequence of nonnegative real numbers, then the interval $[0, 1]$ in Müntz's theorem (Theorem 4.2.1) can be replaced by an arbitrary compact set $A \subset [0, \infty)$ of positive Lebesgue measure.

The exercises also explore in detail the closure of Müntz spaces in the nondense cases. This study was initiated by Clarkson and Erdős [43], who treated the case when the exponents are nonnegative integers. The considerably harder general case is due to Schwartz [59].

Denseness questions about quotients and products of Müntz polynomials from a given Müntz space are discussed in Sections 4.4 and 6.2, respectively.

Some of the literature on the multivariate versions of Müntz's theorem can be found in Ogawa and Kitahara [87], Bloom [90], and Kroó and Szabados [94].

E.1 Another Proof of Some Cases of Müntz's Theorem.

a] Golitsckek's Proof of Müntz's Theorem when $\sum_{i=1}^{\infty} 1/\lambda_i = \infty$. Suppose that $(\lambda_i)_{i=1}^{\infty}$ is a sequence of distinct, positive real numbers satisfying $\sum_{i=1}^{\infty} 1/\lambda_i = \infty$. Golitschek [83] gives the following simple argument to show that $\operatorname{span}\{1, x^{\lambda_1}, x^{\lambda_2}, \dots\}$ is dense in $C[0, 1]$.

Proof. Assume that $m \neq \lambda_k$, $k = 0, 1, \dots$, and define the functions Q_n inductively: $Q_0(x) := x^m$ and

$$Q_n(x) := (\lambda_n - m)x^{\lambda_n} \int_x^1 Q_{n-1}(t)t^{-1-\lambda_n}\, dt\,, \qquad n = 1, 2, \dots .$$

Show, by induction on n, that each Q_n is of the form

$$Q_n(x) = x^m - \sum_{i=0}^{n} a_{n,i} x^{\lambda_i}\,, \qquad a_{n,i} \in \mathbb{R}\,.$$

Show also that

$$\|Q_0\|_{[0,1]} = 1 \qquad \text{and} \qquad \|Q_n\|_{[0,1]} \leq \left|1 - \frac{m}{\lambda_n}\right| \cdot \|Q_{n-1}\|_{[0,1]}\,,$$

so

$$\|Q_n\|_{[0,1]} \leq \prod_{i=0}^{n} \left|1 - \frac{m}{\lambda_i}\right| \to 0 \qquad \text{as } n \to \infty\,.$$

□

b] Another Proof of Müntz's Theorem when $\lambda_i \to c > 0$. Suppose that $(\lambda_i)_{i=1}^{\infty}$ is a sequence of distinct positive real numbers that converges to $c > 0$. Show, without using the arguments given in the proof of Müntz's theorem, that $\operatorname{span}\{1, x^{\lambda_1}, x^{\lambda_2}, \dots\}$ is dense in $C[a,b]$, $a > 0$.

Hint: Let k be a nonnegative integer. Use divided differences to approximate $x^c \log^k x$ uniformly on $[a,b]$. Finish the proof by using the Weierstrass approximation theorem (see E.1 of Section 4.1). □

E.2 Another Proof of Müntz's Theorem in $L_2[0,1]$.

a] Gram's Lemma. *Let $(V, \langle \cdot, \cdot \rangle)$ be an inner product space, and let $g \in V$. Suppose $\{f_1, \dots, f_n\}$ is a basis for an n-dimensional subspace P of V. Then the distance d_n from g to P is given by*

$$d_n := \inf\{\langle g-p, g-p\rangle^{1/2} : p \in P\} = \left(\frac{G(f_1, f_2, \dots, f_n, g)}{G(f_1, f_2, \dots, f_n)}\right)^{1/2},$$

where G is the Gram determinant

$$G(f_1, f_2, \dots, f_m) := \begin{vmatrix} \langle f_1, f_1\rangle & \dots & \langle f_1, f_m\rangle \\ \vdots & \ddots & \vdots \\ \langle f_m, f_1\rangle & \dots & \langle f_m, f_m\rangle \end{vmatrix}.$$

Proof. As in Theorem 2.2.3, the best approximation to g from P is given by

$$f^* = \sum_{i=1}^{n} c_i f_i,$$

where the c_i are uniquely determined by the orthogonality conditions

$$\langle f^* - g, f_k\rangle = 0, \qquad k = 1, 2, \dots, n.$$

Since

$$d_n^2 = \langle g - f^*, g - f^*\rangle,$$

we are led to a system of $n+1$ equations

$$\sum_{i=1}^{n} c_i \langle f_i, f_k\rangle = \langle g, f_k\rangle, \qquad k = 1, 2, \dots, n$$

and

$$\sum_{i=1}^{n} c_i \langle f_i, g\rangle + d_n^2 = \langle g, g\rangle.$$

Solving this system by using Cramer's rule, we get the desired result. □

b] As in E.3 of Section 3.2,

$$\begin{vmatrix} \frac{1}{\alpha_1+\beta_1} & \cdots & \frac{1}{\alpha_1+\beta_n} \\ \vdots & \ddots & \vdots \\ \frac{1}{\alpha_n+\beta_1} & \cdots & \frac{1}{\alpha_n+\beta_n} \end{vmatrix} = \frac{\prod_{1\le i<j\le m} (\alpha_j - \alpha_i)(\beta_j - \beta_i)}{\prod_{1\le i,j\le n} (\alpha_i + \beta_j)}$$

for arbitrary complex numbers α_i and β_j with $\alpha_i + \beta_j \neq 0$.

c] Let $\gamma, \lambda_0, \dots, \lambda_n$ be distinct real numbers greater than $-1/2$. Then the $L_2[0,1]$ distance d_n from x^γ to $\text{span}\{x^{\lambda_0}, \dots, x^{\lambda_n}\}$ is given by

$$d_n = \frac{1}{\sqrt{2\gamma+1}} \prod_{i=0}^{n} \left| \frac{\gamma - \lambda_i}{\gamma + \lambda_i + 1} \right| .$$

Hint: In $L_2[0,1]$,

$$\langle x^a, x^b \rangle = \int_0^1 x^a x^b \, dx = \frac{1}{a+b+1}, \qquad a, b \in (-1/2, \infty) .$$

Now apply parts a] and b]. □

d] Complete the proof of Müntz's theorem in $L_2[0,1]$.

E.3 More on Müntz's Theorem in the Nondense Case. We assume throughout this exercise that $(\lambda_i)_{i=0}^\infty$ is a sequence of nonnegative real numbers satisfying

$$\sum_{i=1}^{\infty} \frac{1}{\lambda_i} < \infty$$

and the gap condition

$$\inf\{\lambda_i - \lambda_{i-1} : i \in \mathbb{N}\} > 0$$

holds. Some of the results of this exercise hold even if the above gap condition is removed (see the later exercises).

a] Show that

$$0 < \prod_{\substack{i=0\\ i\neq m}}^{\infty} \left| \frac{\lambda_i + \lambda_m}{\lambda_i - \lambda_m} \right| = \exp(\gamma_m \lambda_m) ,$$

where $\gamma_m \to 0$ as $m \to \infty$.

Hint: First show that the above infinite product exists. Write the above product as

$$\prod_{\substack{i=0\\ \lambda_i<\lambda_m}}^{\infty} \left| \frac{\lambda_i + \lambda_m}{\lambda_i - \lambda_m} \right| \prod_{\substack{i=0\\ \lambda_i\in(\lambda_m, 2\lambda_m)}}^{\infty} \left| \frac{\lambda_i + \lambda_m}{\lambda_i - \lambda_m} \right| \prod_{\substack{i=0\\ \lambda_i\ge 2\lambda_m}}^{\infty} \left| \frac{\lambda_i + \lambda_m}{\lambda_i - \lambda_m} \right|$$

and estimate the three factors above separately. □

b] Deduce that if $\lambda_i \neq \lambda_m$ for each i, then

$$\left\|x^{\lambda_m} - p(x)\right\|_{L_2[0,1]} \geq \frac{1}{\sqrt{2\lambda_m+1}} \prod_{i=0}^{\infty} \left| \frac{(\lambda_i + \frac{1}{2}) - (\lambda_m + \frac{1}{2})}{(\lambda_i + \frac{1}{2}) + (\lambda_m + \frac{1}{2})} \right|$$
$$= \exp(-\gamma_m(\lambda_m))$$

for every $p \in \operatorname{span}\{x^{\lambda_0}, \dots, x^{\lambda_n}\}$, where $\gamma_m \to 0$ as $m \to \infty$.

c] Show that for every $\epsilon > 0$ there is a constant c_ϵ depending only on ϵ and $(\lambda_i)_{i=0}^{\infty}$ (but not on the number of terms in p) such that

$$|a_i| \leq c_\epsilon (1+\epsilon)^{\lambda_i} \|p\|_{L_2[0,1]}$$

for every $p \in \operatorname{span}\{x^{\lambda_0}, x^{\lambda_1}, \dots\}$ of the form $p(x) = \sum_{i=0}^{n} a_i x^{\lambda_i}$.

Hint: Use part b]. □

d] Bounded Bernstein-Type Inequality. Let $\lambda_0 = 0$ and $\lambda_1 \geq 1$. Show that for every $\epsilon \in (0,1)$ there is a constant c_ϵ depending only on ϵ and $(\lambda_i)_{i=0}^{\infty}$ (but not on the number of terms in p) such that

$$\|p'\|_{[0,1-\epsilon]} \leq c_\epsilon \|p\|_{L_2[0,1]}$$

and hence

$$\|p'\|_{[0,1-\epsilon]} \leq c_\epsilon \|p\|_{[0,1]}$$

for every $p \in \operatorname{span}\{x^{\lambda_0}, x^{\lambda_1}, \dots\}$.

Hint: Use part c]. □

The result of the next part is due to Clarkson and Erdős [43].

e] The Closure of a Nondense Müntz Space. Suppose $f \in C[0,1]$ and there exist $p_n \in \operatorname{span}\{x^{\lambda_0}, x^{\lambda_1}, \dots\}$ of the form

$$p_n(x) = \sum_{i=0}^{k_n} a_{i,n} x^{\lambda_i}\,, \qquad a_{i,n} \in \mathbb{R}\,, \quad n = 1, 2, \dots$$

such that $\lim_{n\to\infty} \|p_n - f\|_{[0,1]} = 0$. Show that f is of the form

$$f(x) = \sum_{i=0}^{\infty} a_i x^{\lambda_i}\,, \qquad a_{i,n} \in \mathbb{R}\,, \quad x \in [0,1)\,.$$

Show also that f can be extended analytically throughout the region

$$\{z \in \mathbb{C} \backslash (-\infty, 0] : |z| < 1\}$$

and

$$\lim_{n\to\infty} a_{i,n} = a_i\,, \qquad i = 0, 1, \dots\,.$$

If $(\lambda_i)_{i=0}^{\infty}$ is a sequence of distinct nonnegative integers, then f can be extended analytically throughout the open unit disk.

Hint: Use part c]. □

If $(\lambda_i)_{i=0}^\infty$ is *lacunary* (that is, $\inf\{\lambda_{i+1}/\lambda_i : i \in \mathbb{N}\} > 1$), then the uniform closure of $\text{span}\{x^{\lambda_0}, x^{\lambda_1}, \dots\}$ on $[0,1]$ is exactly

$$\left\{ f \in C[0,1] : f(x) = \sum_{i=0}^\infty a_i x^{\lambda_i}, \quad x \in [0,1] \right\}.$$

If $(\lambda_i)_{i=0}^\infty$ is not lacunary, then this fails, namely, there exists a function f of the form

$$f(x) = \sum_{i=0}^\infty a_i x^{\lambda_i}, \qquad x \in [0,1)$$

in the uniform closure of $\text{span}\{x^{\lambda_0}, x^{\lambda_1}, \dots\}$ on $[0,1]$ such that the right-hand side does not converge at the endpoint 1; see Clarkson and Erdős [43].

f] Bounded Chebyshev-Type Inequality. Show that for every $\epsilon \in (0,1)$ there exists a constant c_ϵ depending only on ϵ and $(\lambda_i)_{i=0}^\infty$ (but not on the number of terms in p) such that

$$\|p\|_{[0,1]} \le c_\epsilon \|p\|_{[1-\epsilon,1]}$$

for every $p \in \text{span}\{x^{\lambda_0}, x^{\lambda_i}, \dots\}$.

Outline. Using the scaling $x \to x^{1/\lambda_1}$, without loss of generality we may assume that $\lambda_1 = 1$. Suppose there exists a sequence

$$(p_m)_{m=1}^\infty \subset \text{span}\{x^{\lambda_0}, x^{\lambda_1}, \dots,\}, \qquad m = 1, 2, \dots$$

such that

$$0 < A_m := \|p_m\|_{[0,1]} \to \infty$$

while

$$\|p_m\|_{[1-\epsilon,1]} = 1, \qquad m = 1, 2, \dots.$$

Let $q_m := p_m/A_m$. Note that $\|q_m\|_{[0,1]} = 1$ for each m, and $\|q_m\|_{[1-\epsilon,1]} \to 0$ as $m \to \infty$. Then, by part d],

$$\|q_m'\|_{[0,1-\delta]} \le c_\delta$$

for every $\delta \in (0,1)$. Hence $(q_m)_{m=0}^\infty$ is a sequence of uniformly bounded and equicontinuous functions on closed subintervals of $[0,1)$, and by the Arzela-Ascoli theorem (see, for example, Rudin [87]) we may extract a uniformly convergent subsequence on $[0, 1-\epsilon/2]$. This subsequence, by part e], converges uniformly to a function F analytic on $(0, 1-\epsilon/2)$, but since $\|q_m\|_{[1-\epsilon,1]} \to 0$, F must be identically zero. This is a contradiction since $\|q_m\|_{[0,1]} = 1$ and $\|q_m\|_{[0,1-\epsilon]} = \|q_m\|_{[0,1]}$ for every sufficiently large m. □

g] Suppose $(q_m)_{m=1}^\infty \subset \text{span}\{x^{\lambda_0}, x^{\lambda_1}, \dots\}$ and $\|q_m\|_{[a,b]} \le 1$ for each m, where $0 \le a < b$. Show that there is a subsequence of $(q_m)_{m=1}^\infty$ that converges uniformly on every closed subinterval of $[0,b)$.

Hint: Use parts f] and d] and the Arzela-Ascoli theorem. □

E.4 Müntz's Theorem with Real Exponents on $[a, b]$, $a > 0$. Suppose $(\lambda_i)_{i=-\infty}^{\infty}$ is a set of distinct real numbers satisfying

$$\sum_{\substack{i=-\infty \\ \lambda_i \neq 0}}^{\infty} \frac{1}{|\lambda_i|} < \infty$$

with $\lambda_i < 0$ for $i < 0$ and $\lambda_i \geq 0$ for $i \geq 0$. Suppose that the gap condition

$$\inf\{\lambda_i - \lambda_{i-1} : i \in \mathbb{Z}\} > 0$$

holds. Associated with

$$p(x) := \sum_{i=-n}^{n} a_i x^{\lambda_i}, \qquad n = 0, 1, \ldots$$

let

$$p^-(x) := \sum_{i=-n}^{-1} a_i x^{\lambda_i} \qquad \text{and} \qquad p^+(x) := \sum_{i=0}^{n} a_i x^{\lambda_i}.$$

Let $0 < a < b$.

a] Show that there exists a constant c depending only on a, b, and $(\lambda_i)_{i=-\infty}^{\infty}$ (but not on the number of terms in p) such that

$$\|p^+\|_{[a,b]} \leq c\|p\|_{[a,b]} \qquad \text{and} \qquad \|p^-\|_{[a,b]} \leq c\|p\|_{[a,b]}$$

for every $p \in \operatorname{span}\{x^{\lambda_i}\}_{i=-\infty}^{\infty}$.

Outline. It is sufficient to prove only the first inequality; the second inequality follows from the first by the substitution $y = x^{-1}$. If the first inequality fails to hold, then there exists a sequence $(p_n)_{n=1}^{\infty} \subset \operatorname{span}\{x^{\lambda_i}\}_{i=-\infty}^{\infty}$ such that

$$\|p_n^+\|_{[a,b]} = 1, \qquad n = 1, 2, \ldots, \qquad \text{and} \qquad \lim_{n \to \infty} \|p_n\|_{[a,b]} = 0.$$

Since $p = p^+ + p^-$, the above relations imply that

$$\|p_n^-\|_{[a,b]} \leq K < \infty, \qquad n = 1, 2, \ldots.$$

By E.3 g] and E.3 e], there exists a subsequence $(n_i)_{i=1}^{\infty}$ such that $(p_{n_i}^+)_{i=1}^{\infty}$ converges uniformly on every closed subinterval of $[0, b)$ to a function f analytic on

$$D_b := \{z \in \mathbb{C} \backslash (-\infty, 0] : |z| < b\}$$

of the form

$$f(z) = \sum_{i=0}^{\infty} a_i z^{\lambda_i}, \qquad z \in D_b,$$

while $(p_{n_i}^-)_{i=1}^\infty$ converges uniformly on every closed subinterval of (a, ∞) to a function g analytic on

$$E_a := \{z \in \mathbb{C} \setminus (-\infty, 0] : |z| > a\}$$

of the form

$$g(z) = \sum_{i=-\infty}^{-1} a_i z^{\lambda_i}\,, \qquad z \in E_a\,, \qquad \lim_{\substack{x\to\infty \\ x\in\mathbb{R}}} g(x) = 0\,.$$

Now $\lim_{i\to\infty} \|p_{n_i}\|_{[a,b]} = 0$ and $p_{n_i} = p_{n_i}^+ + p_{n_i}^-$ imply that $f + g = 0$ on (a, b). Show that

$$h(z) := \begin{cases} f(e^z)\,, & \text{Re}z < \log b \\ -g(e^z)\,, & \text{Re}z > \log a \end{cases}$$

is a well-defined bounded entire function, and hence $h = 0$ on $\mathbb{C}$ by Liouville's theorem. From this, deduce that

$$f = 0 \;\text{ on }\; [0, b) \qquad \text{and} \qquad g = 0 \;\text{ on }\; (a, \infty)\,.$$

Hence, for every $y \in (a, b)$,

$$\lim_{i\to\infty} \|p_{n_i}^+\|_{[a,y]} = 0$$

and

$$\lim_{i\to\infty} \|p_{n_i}^+\|_{[y,b]} = \lim_{i\to\infty} \|p_{n_i} - p_{n_i}^-\|_{[y,b]} = 0\,.$$

Therefore

$$\lim_{i\to\infty} \|p_{n_i}^+\|_{[a,b]} = 0\,,$$

which contradicts $\|p_n^+\|_{[a,b]} = 1, \;\; n = 1, 2, \ldots$. □

b] The Closure of Müntz Polynomials. Let $f \in C[0,1]$, and suppose there exist Müntz polynomials $p_n \in \text{span}\{x^{\lambda_i}\}_{i=-\infty}^\infty$ of the form

$$p_n(x) = \sum_{i=-k_n}^{k_n} a_{i,n} x^{\lambda_i}\,, \qquad n = 1, 2, \ldots$$

such that $\lim_{n\to\infty} \|p_n - f\|_{[a,b]} = 0$. Show that f is of the form

$$f(x) = \sum_{i=-\infty}^{\infty} a_i x^{\lambda_i}\,, \qquad x \in (a, b)\,,$$

where

$$f^+(x) := \sum_{i=1}^{\infty} a_i x^{\lambda_i}\,, \qquad x \in [0,b)\,,$$

$$f^-(x) := \sum_{i=-\infty}^{-1} a_i x^{\lambda_i}\,, \qquad x \in (a,\infty)\,, \qquad \lim_{x\to\infty} f^-(x) = 0\,,$$

f can be extended analytically throughout the region

$$\{z \in \mathbb{C} \setminus (-\infty,0] : a < |z| < b\}\,,$$

and

$$\lim_{n\to\infty} a_{i,n} = a_i\,, \qquad i \in \mathbb{Z}\,.$$

Hint: Use part a] and E.3 e]. □

E.5 Removing the Gap Conditions. Assume throughout this exercise that $0 \le \lambda_0 < \lambda_1 < \cdots$ and $\sum_{i=1}^{\infty} 1/\lambda_i < \infty$.

a] **Bounded Chebyshev-Type Inequality.** Show that for every $\epsilon \in (0,1)$ there is a constant c_ϵ depending only on ϵ and $(\lambda_i)_{i=0}^{\infty}$ (but not on the number of terms in p) such that

$$\|p\|_{[0,1]} \le c_\epsilon \|p\|_{[1-\epsilon,1]}$$

for every $p \in \operatorname{span}\{x^{\lambda_0}, x^{\lambda_1}, \dots\}$. (This is the inequality of E.3 f] without the gap condition $\inf\{\lambda_i - \lambda_{i-1} : i \in \mathbb{N}\} > 0$.)

Hint: Assume, without loss of generality, that $\lambda_0 = 0$. Observe that $\lim_{i\to\infty} \lambda_i/i = \infty$. Choose $m \in \mathbb{N}$ such that $\lambda_i > 2i$ whenever $i > m$. Define $\Gamma := (\gamma_i)_{i=1}^{\infty}$ by

$$\gamma_i := \begin{cases} \min\{\lambda_i, i\}\,, & i = 0,1,\dots,m \\ \frac{1}{2}\lambda_i + i\,, & i = m+1, m+2, \dots\,. \end{cases}$$

Then

$$0 = \gamma_0 < \gamma_1 < \cdots\,, \qquad \sum_{i=1}^{\infty} \frac{1}{\gamma_i} < \infty\,, \qquad \gamma_i \le \lambda_i\,, \qquad i = 0,1,\dots$$

and $\inf\{\gamma_i - \gamma_{i-1} : i \in \mathbb{N}\} > 0$. Now use E.3 g] of Section 3.3 with $[a,b] = [1-\epsilon,1]$ and E.3 f] of this section. □

b] **Bounded Bernstein-Type Inequality.** Suppose $\lambda_0 = 0$ and $\lambda_1 \ge 1$. Prove that for every $\epsilon \in (0,1)$ there is a constant c_ϵ depending only on ϵ and $(\lambda_i)_{i=0}^{\infty}$ (but not on the number of terms in p) such that

$$\|p'\|_{[0,1-\epsilon]} \le c_\epsilon \|p\|_{[0,1]}$$

for every $p \in \operatorname{span}\{x^{\lambda_0}, x^{\lambda_1}, \dots\}$. (This is the second inequality of E.3 d] without the gap condition $\inf\{\lambda_i - \lambda_{i-1} : i \in \mathbb{N}\} > 0$.)

Hint: Define the sequence Γ as in the hint given to part a]. Now use E.3 f] of Section 3.3 with $[a,b]$, $a \in (0, 1-\epsilon]$, E.3 g] of this section, and part a] of this exercise. □

c] Let $0 \le a < b$. Show that $\text{span}\{x^{\lambda_0}, x^{\lambda_1}, \dots\}$ is not dense in $C[a,b]$.

Hint: Use part a] and Theorem 4.2.1 (full Müntz's theorem in $C[0,1]$). □

d] Let

$$\lambda_{2k-1} := k^2\,, \qquad \lambda_{2k} := k^2 + 2^{-k^2}\,, \qquad k = 1, 2, \dots\,,$$
$$a_{2k-1} := 2^{k^2}\,, \qquad a_{2k} := -2^{k^2}\,, \qquad k = 1, 2, \dots\,.$$

Show that the function $f(x) := \sum_{i=1}^{\infty} \left(a_{2i-1} x^{\lambda_{2i-1}} + a_{2i} x^{\lambda_{2i}}\right)$ is a well-defined continuous function on $[0,1]$. Show also that $\sum_{i=1}^{\infty} a_i x^{\lambda_i}$ does not converge for any $x \in (0, \infty)$ (hence the conclusions of E.3 e] are not valid without a gap condition).

E.6 A Comparison Theorem. *Let $0 \le k \le n$ be fixed integers. Assume*

$$\lambda_0 < \lambda_1 < \cdots < \lambda_k < 0 < \lambda_{k+1} < \lambda_{k+2} < \cdots < \lambda_n\,,$$
$$\gamma_0 < \gamma_1 < \cdots < \gamma_k < 0 < \gamma_{k+1} < \gamma_{k+2} < \cdots < \gamma_n\,,$$

and

$$|\gamma_i| \le |\lambda_i|\,, \qquad i = 0, 1, \dots, n$$

with strict inequality for at least one index i. Let

$$H_n := \text{span}\{x^{\lambda_0}, x^{\lambda_1}, \dots, x^{\lambda_n}\} \quad \text{and} \quad G_n := \text{span}\{x^{\gamma_0}, x^{\gamma_1}, \dots, x^{\gamma_n}\}$$

and let $0 < a < b$. Then

$$\min_{p \in G_n} \|1 - p\|_{[a,b]} < \min_{p \in H_n} \|1 - p\|_{[a,b]}\,.$$

Hint: Let $q^* \in H_n$ be the best approximation to $x^0 \equiv 1$ on $[a,b]$. Let

$$r(x) = (-1)x^0 + \sum_{i=0}^{n} x^{\gamma_i} \in \text{span}\{x^{\gamma_0}, \dots, x^{\gamma_k}, x^0, x^{\gamma_{k+1}}, \dots, x^{\gamma_n}\}$$

interpolate

$$q^* - 1 \in \text{span}\{x^{\lambda_0}, \dots, x^{\lambda_k}, x^0, x^{\lambda_{k+1}}, \dots, x^{\lambda_n}\}$$

at the $n+1$ distinct zeros, $x_1, x_2, \dots, x_{n+1}$, of $q^* - 1$ on $[a,b]$ (see Theorem 3.1.6). Use Theorem 3.2.5 to show that

$$|r(x)| \le |q(x)|\,, \qquad x \in [a,b]$$

with strict inequality for $x \ne x_i$. Finally show that if $p^* := r + 1$, then

$$\min_{p \in G_n} \|1 - p\|_{[a,b]} \le \|1 - p^*\|_{[a,b]} < \|1 - q^*\|_{[a,b]} = \min_{p \in H_n} \|1 - p\|_{[a,b]}\,.$$

□

E.7 Full Müntz Theorem on $[a, b]$, $a > 0$. Let $(\lambda_i)_{i=0}^{\infty}$ be a sequence of distinct real numbers, and let $0 < a < b$. Show that $\operatorname{span}\{x^{\lambda_0}, x^{\lambda_1}, \dots\}$ is dense in $C[a, b]$ if and only if

$$\sum_{\substack{i=0 \\ \lambda_i \neq 0}}^{\infty} \frac{1}{|\lambda_i|} = \infty .$$

Hint: Distinguish the following cases.

Case 1: The sequence $(\lambda_i)_{i=0}^{\infty}$ has a cluster point $0 \neq \lambda \in \mathbb{R}$. Use Theorem 4.2.1 to show that $\operatorname{span}\{x^{\lambda_0}, x^{\lambda_1}, \dots\}$ is dense in $C[a, b]$.

Case 2: The point 0 is a cluster point of $(\lambda_i)_{i=1}^{\infty}$. Use Case 1 to show first that $\operatorname{span}\{x^{\lambda_0+1}, x^{\lambda_1+1}, \dots\}$ is dense in $C[a, b]$, and recall that $a > 0$.

Case 3: The sequence $(\lambda_i)_{i=0}^{\infty}$ does not have any (finite) cluster points, and either

$$\sum_{\substack{i=0 \\ \lambda_i > 0}}^{\infty} \frac{1}{\lambda_i} = \infty \qquad \text{or} \qquad \sum_{\substack{i=0 \\ \lambda_i < 0}}^{\infty} \frac{1}{|\lambda_i|} = \infty .$$

Use Theorem 4.2.1 to show that $\operatorname{span}\{x^{\lambda_0}, x^{\lambda_1}, \dots\}$ is dense in $C[a, b]$.

Case 4:

$$\sum_{\substack{i=0 \\ \lambda_i \neq 0}}^{\infty} \frac{1}{|\lambda_i|} < \infty .$$

Without loss of generality we may assume that $0 \notin \{\lambda_i\}_{i=0}^{\infty}$ (why?). By a change of scaling, we may also assume that $[a, b] = [1 - \epsilon, 1]$. Let

$$\{\widetilde{\lambda}_i\}_{i=-\infty}^{\infty} = \{\lambda_i\}_{i=0}^{\infty}, \quad \text{where} \quad \cdots < \widetilde{\lambda}_{-2} < \widetilde{\lambda}_{-1} < 0 < \widetilde{\lambda}_0 < \widetilde{\lambda}_1 < \cdots .$$

Show that there is a sequence $(\gamma_i)_{i=-\infty}^{\infty}$ satisfying

$$\cdots < \gamma_{-2} < \gamma_{-1} < 0 < \gamma_0 < \gamma_1 < \cdots ,$$

$$|\gamma_i| < |\widetilde{\lambda}_i| , \qquad i \in \mathbb{Z} , \qquad \sum_{i=-\infty}^{\infty} \frac{1}{|\gamma_i|} < \infty ,$$

and the gap condition

$$\inf\{\gamma_i - \gamma_{i-1} : i \in \mathbb{Z}\} > 0 .$$

Use E.4 a], E.5 a], and E.2 c] to show that

$$1 \notin \overline{\operatorname{span}}\{x^{\gamma_i}\}_{i=-\infty}^{\infty} ,$$

where $\overline{\operatorname{span}}\{x^{\gamma_i}\}_{i=-\infty}^{\infty}$ denotes the uniform closure of the span on $[a, b]$. Finally use E.6 to show that

$$1 \notin \overline{\operatorname{span}}\{x^{\widetilde{\lambda}_i}\}_{i=-\infty}^{\infty} = \overline{\operatorname{span}}\{x^{\lambda_i}\}_{i=0}^{\infty} .$$

□

E.8 Further Results for Nonnegative Sequences with No Gap Condition. Assume throughout this exercise that $0 \le \lambda_0 < \lambda_1 < \cdots$, $\sum_{i=1}^{\infty} 1/\lambda_i < \infty$, and $0 \le a < b$.

a] Show that for every $\epsilon \in (0, b)$ there is a constant c_ϵ depending only on ϵ, a, b, and $(\lambda_i)_{i=0}^{\infty}$ (but not on the number of terms in p) such that

$$\|p\|_{[0,b-\epsilon]} \le c_\epsilon \int_a^b |p(x)|\, dx$$

for every $p \in \operatorname{span}\{x^{\lambda_0}, x^{\lambda_1}, \dots\}$.

Hint: Assume that $b = 1$; the general case can be reduced to this by scaling. Use parts a] and b] of E.5 with

$$\widetilde{p}(x) := \int_0^x p(t)\, dt \in \operatorname{span}\{x^{\lambda_0+1}, x^{\lambda_1+1}, \dots\}.$$

□

b] Assume

$$(p_n)_{n=1}^{\infty} \subset \operatorname{span}\{x^{\lambda_0}, x^{\lambda_1}, \dots\}$$

converges to an $f \in C[a, b]$ uniformly on $[a, b]$. Show that f can be extended analytically throughout the region

$$D_b := \{z \in \mathbb{C} \setminus (-\infty, 0] : 0 < |z| < b\}$$

and the convergence is uniform on every closed subset of D_b.

Hint: This part of the exercise is difficult. A proof of a more general statement can be found in Schwartz [59, pp. 38–48]. □

c] Suppose $(p_n)_{n=1}^{\infty} \subset \operatorname{span}\{x^{\lambda_0}, x^{\lambda_1}, \dots\}$ and $\|p_n\|_{[a,b]} \le 1$ for each n. Show that there is a subsequence of $(p_n)_{n=1}^{\infty}$ that converges uniformly on every closed subinterval of $[0, b)$. (So the conclusion of E.3 g] holds without the gap condition $\inf\{\lambda_i - \lambda_{i-1} : i \in \mathbb{N}\} > 0$.)

Hint: Use parts a] and b] of E.5 and the Arzela-Ascoli theorem. □

d] Let K be a closed subset of D_b defined in part b]. Show that there is a constant c_K depending only on K, a, b, and $(\lambda_i)_{i=0}^{\infty}$ (but not on the number of terms in p) such that

$$\|p\|_K \le c_K \|p\|_{[a,b]}$$

for every $p \in \operatorname{span}\{x^{\lambda_0}, x^{\lambda_1}, \dots\}$.

Hint: Use parts c] and b]. (If the gap condition $\inf\{\lambda_i - \lambda_{i-1} : i \in \mathbb{N}\} > 0$ holds, then the simpler result of E.3 e] can be used instead of part b] of this exercise.) □

E.9 Full Müntz Theorem on $[a,b]$, $a>0$, in L_q Norm. Schwartz [59] gives the following results: Suppose $(\lambda_i)_{i=-\infty}^{\infty}$ is a sequence of distinct real numbers. For a finite set Γ of integers and

$$p(x) = \sum_{i\in\Gamma} a_i x^{\lambda_i}\,, \qquad a_i \in \mathbb{R}\,,$$

let

$$p^-(x) := \sum_{\substack{i\in\Gamma\\ \lambda_i<0}} a_i x^{\lambda_i} \qquad \text{and} \qquad p^+(x) := \sum_{\substack{i\in\Gamma\\ \lambda_i\ge 0}} a_i x^{\lambda_i}\,.$$

Let $0<a<b$ and $1 \le q \le \infty$.

Theorem 4.2.5. *Suppose*

$$\sum_{\substack{i=-\infty\\ \lambda_i\neq 0}}^{\infty} \frac{1}{|\lambda_i|} < \infty\,.$$

Then there exists a constant c depending only on a, b, q, and $(\lambda_i)_{i=-\infty}^{\infty}$ (but not on the number of terms in p) such that

$$\|p^+\|_{L_q[a,b]} \le c\|p\|_{L_q[a,b]} \qquad \textit{and} \qquad \|p^-\|_{L_q[a,b]} \le c\|p\|_{L_q[a,b]}$$

for every $p \in \operatorname{span}\{x^{\lambda_i}\}_{i=-\infty}^{\infty}$.

Theorem 4.2.6. *Suppose that $0<a<b$. Then $\operatorname{span}\{x^{\lambda_0}, x^{\lambda_1}, \dots\}$ is dense in $L_q[a,b]$ if and only if*

$$\sum_{\substack{i=0\\ \lambda_i\neq 0}}^{\infty} \frac{1}{|\lambda_i|} = \infty\,. \tag{4.2.4}$$

a] Prove the two above results under the gap condition

$$\inf\{\lambda_i - \lambda_{i-1} : i \in \mathbb{Z}\} > 0\,.$$

Hint to Theorem 4.2.5: When $q=\infty$ see E.4 a]. If $1 \le q < \infty$, then modify the proof suggested in the hints to E.4 a], by using E.8 a]. □

Hint to Theorem 4.2.6: If (4.2.4) holds, then the fact that $\operatorname{span}\{x^{\lambda_0}, x^{\lambda_1}, \dots\}$ is dense in $L_q[a,b]$ follows from E.7 and the obvious inequality

$$\|p\|_{L_q[a,b]} \le (b-a)^{1/q}\|p\|_{[a,b]}\,.$$

Now suppose (4.2.4) does not hold. Use Theorem 4.2.5 and E.8 a] to show that for every $\epsilon \in (0, \frac{1}{2}(b-a))$ there exists a constant c_ϵ depending only on a, b, q, and $(\lambda_i)_{i=-\infty}^{\infty}$ (but not on the number of terms in p) such that

$$\|p\|_{[a+\epsilon,b-\epsilon]} \le c_\epsilon \|p\|_{L_q[a,b]}$$

for every $p \in \operatorname{span}\{x^{\lambda_i}\}_{i=-\infty}^{\infty}$. Now show that the above inequality implies that $\operatorname{span}\{x^{\lambda_0}, x^{\lambda_1}, \dots\}$ is not dense in $L_q[a,b]$. □

E.10 Denseness of $\mathrm{span}\{\cos\lambda_k\theta\}$ **and** $\mathrm{span}\{z^{\lambda_k}\}$**.** Throughout this exercise the span is assumed to be over $\mathbb{C}$. Let

$$D_R := \{z \in \mathbb{C} : |z| < R\} \qquad \text{and} \qquad C_R := \{z \in \mathbb{C} : |z| = R\}\,.$$

a] Show that $(\cos n\theta)_{n=0}^\infty$ is a complete orthogonal system in $L_2[0,\pi]$. So no term, $\cos n\theta$, can be removed if we wish to preserve denseness of the span in $L_2[0,\pi]$.

b] Let $A := \{e^{i\theta} : \theta \in [0,\delta]\}$. Suppose $(\lambda_k)_{k=0}^\infty$ is a sequence of distinct complex numbers satisfying

$$|\lambda_k| \le k e^{1-\delta}\,, \qquad k = 1, 2, \ldots\,.$$

If $\delta \in [1, 2\pi]$, then $\mathrm{span}\{z^{\lambda_0}, z^{\lambda_1}, \ldots\}$ is dense in $L_2(A)$.

The proof of part b] is outlined in parts c], d], and e].

c] **Jensen's Formula.** Suppose h is a nonnegative integer and

$$f(z) = \sum_{k=h}^\infty c_k z^k\,, \qquad c_h \neq 0$$

is analytic on a disk of radius greater than R, and suppose that the zeros of f in $\overline{D}_R \setminus \{0\}$ are $a_1, a_2, \ldots, a_n$, where each zero is listed as many times as its multiplicity. Then

$$\log|c_h| + h\log R + \sum_{k=1}^n \log\frac{R}{|a_k|} = \frac{1}{2\pi}\int_0^{2\pi} \log|f(Re^{i\theta})|\,d\theta\,.$$

Proof. This is a simple consequence of Poisson's formula (see, for example, Ahlfors [53]), which states that

$$\log|F(0)| = \frac{1}{2\pi}\int_0^{2\pi} \log|F(Re^{i\theta})|\,d\theta$$

whenever the function F is analytic and zero-free in an open region containing the closed disk $\overline{D}_R$. Now, in the above notation, if we let

$$F(z) := f(z)\left(\frac{R}{z}\right)^h \prod_{k=1}^n \frac{R^2 - \overline{a}_k z}{R(z - a_k)}$$

and apply Poisson's formula to F, we get the required result by noting that $|F(z)| = |f(z)|$ whenever $|z| = R$. (The case where f has zeros on the boundary of D requires an additional limiting argument.) □

d] Let $-\infty \le a < b \le \infty$. Suppose $(f_k)_{k=0}^{\infty}$ is a sequence in $L_2(a,b)$ and $\text{span}\{f_0, f_1, \dots\}$ is not dense in $L_2(a,b)$. Then there exists a nonzero $g \in L_2(a,b)$ such that

$$\int_a^b f_k(x)g(x)\,dx = 0\,, \qquad k = 0, 1, \dots\,.$$

This is an immediate consequence of the Riesz representation theorem (see E.7 g] of Section 2.2) and the Hahn-Banach theorem. The second theorem says that if $\text{span}\{f_0, f_1, \dots\}$ is not dense in a Banach space, then there exists a nonzero continuous linear functional vanishing on $\{f_0, f_1, \dots\}$. The first theorem gives the form of the functional; see Rudin [73].

e] Prove b] as follows: Suppose $\text{span}\{z^{\lambda_0}, z^{\lambda_1}, \dots\}$ is not dense in $L_2(A)$. Then by d] there exists a $g \in L_2[0,\delta]$ such that

$$f(z) := \int_0^\delta \exp(i(z+1)\theta)g(\theta)\,d\theta$$

vanishes at $z = \lambda_k$, $k = 0, 1, \dots$. Also observe that f is an entire function, and there is an absolute constant $\alpha > 0$ such that

$$|f(z)| \le \alpha \exp(\delta |z|)\,.$$

Use E.8 a] of Section 2.2 to show that $f \ne 0$. Let $R > |\lambda_0|$ be an integer. Applying c] on D_R and exponentiating, we obtain

$$|c_h| \exp((\delta - 1)R)\frac{R^{R+1}}{R!} \le |c_h| R^h \prod_{\substack{k=0\\ \lambda_k \ne 0}}^{R} \frac{R}{|\lambda_k|} \le \alpha \exp(\delta R)\,,$$

where c_h is the first nonvanishing coefficient of the Taylor series expansion of f around 0. However,

$$\lim_{R\to\infty} \frac{R^{R+1}}{R!\,e^R} = \infty\,,$$

which is a contradiction and finishes the proof. □

f] Let $A := [0,\delta]$. Suppose $(\lambda_k)_{k=0}^{\infty}$ is a sequence of distinct complex numbers satisfying

$$|\lambda_k| \le k e^{1-\delta}\,, \qquad k = 1, 2, \dots\,.$$

Show that if $\delta \in [1,\pi]$, then $\text{span}\{\cos \lambda_0\theta, \cos \lambda_1\theta, \dots\}$ is dense in $L_2(A)$.

Hint: Proceed as in the proof of part b]. □

g] Suppose $(\lambda_k)_{k=0}^\infty$ is a sequence of distinct real numbers satisfying

$$0 \le \lambda_k \le k\,, \qquad k = 0, 1, \dots\,.$$

Then $\operatorname{span}\{\cos\lambda_0\theta,\ \cos\lambda_1\theta,\ \dots\}$ is dense in $L_2[0, \pi - \epsilon]$ for any $\epsilon > 0$.

Proof. This is harder; see Boas [54, p. 235]. □

h] Suppose $(\lambda_k)_{k=0}^\infty$ is a sequence of distinct complex numbers satisfying $0 \le |\lambda_k| \le k$. Suppose f is an entire function such that $\|f\|_{D_R} \le \alpha e^R$ for all $R > 0$ with an absolute constant $\alpha > 0$, and $\operatorname{span}\{f(\beta z) : \beta \in \mathbb{C}\}$ is dense in $L_2(C_1)$. Then $\operatorname{span}\{f(\lambda_0 z), f(\lambda_1 z), \dots\}$ is dense in $L_2(C_1)$.

E.11 On the Hardy Space $\mathcal{H}_\infty$. We denote by $\mathcal{H}_\infty$ the class of functions that are analytic and bounded on $D := \{z \in \mathbb{C} : |z| < 1\}$. We let

$$\|f\|_{\mathcal{H}_\infty} := \|f\|_D = \sup_{u \in D} |f(u)|\,.$$

a] If $f \in \mathcal{H}_\infty$, then

$$f(z) = \sum_{n=0}^{\infty} a_n z^n\,, \qquad z \in D\,,$$

where

$$|a_n| \le \|f\|_{\mathcal{H}_\infty}\,.$$

Hint: By Cauchy's integral formula

$$|a_n| = \frac{1}{n!}|f^{(n)}(0)| \le \frac{1}{2\pi}\int_{|t|=R}\left|\frac{f(t)}{t^{n+1}}\right||dt| \le R^{-n}\|f\|_{\mathcal{H}_\infty}$$

holds for every $R \in (0,1)$. □

b] If $f \in \mathcal{H}_\infty$, then

$$|f'(z)| \le \left(\frac{1}{1-|z|}\right)^2 \|f\|_{\mathcal{H}_\infty}\,, \qquad |z| < 1$$

and

$$|f^{(n)}(z)| \le n!\left(\frac{1}{1-|z|}\right)^{n+1} \|f\|_{\mathcal{H}_\infty}\,, \qquad |z| < 1\,.$$

c] $\mathcal{H}_\infty$ is a Banach algebra.

Hint: See Rudin [73]. □

E.12 Blaschke Products. A product of the form

$$B(z) := z^k \prod_{i=1}^{\infty} \left(\frac{\alpha_i - z}{1 - \overline{\alpha}_i z} \right) \frac{|\alpha_i|}{\alpha_i}, \qquad \alpha_i \in \mathbb{C} \setminus \{0\}$$

with $k \in \mathbb{Z}$ is called a *Blaschke product.* Let $D := \{z \in \mathbb{C} : |z| < 1\}$.

a] Let

$$\varphi_\alpha(z) := \frac{\alpha - z}{1 - \overline{\alpha} z}, \qquad \alpha \in \mathbb{C}.$$

Show that $|\varphi_\alpha(z)| = 1$ whenever $|z| = 1$ and

$$\varphi'_\alpha(z) = \frac{|\alpha|^2 - 1}{(1 - \overline{\alpha} z)^2}.$$

b] Show that if $|\alpha| < 1$, then $\varphi_\alpha(z)$ maps the closed unit disk $\overline{D}$ one-to-one onto itself.

c] A Minimization Property. Let $\beta_1, \beta_2, \dots, \beta_n$ be fixed complex numbers with $|\beta_i| > 1, \ i = 1, \dots n$. Show that

$$\min_{a_i \in \mathbb{C}} \left\| 1 - \sum_{i=1}^{n} \frac{a_i}{z - \beta_i} \right\|_D = \prod_{i=1}^{n} |\beta_i|^{-1}$$

and that the minimum is attained by the normalized finite Blaschke product

$$1 - \sum_{i=1}^{n} \frac{a_i^*}{z - \beta_i} = \left(\prod_{i=1}^{n} \beta_i \right)^{-1} \prod_{i=1}^{n} \left(\frac{\overline{\beta}_i^{-1} - z}{1 - \beta_i^{-1} z} \right).$$

Hint: Suppose that the statement is false. Then there are some $a_i \in \mathbb{C}$ such that

$$\left| 1 - \sum_{i=1}^{n} \frac{a_i}{z - \beta_i} \right| < \left(\prod_{i=1}^{n} |\beta_i| \right)^{-1} = \left| 1 - \sum_{i=1}^{n} \frac{a_i^*}{z - \beta_i} \right|$$

for all $z \in \mathbb{C}$ with $|z| = 1$. Now Rouché's theorem implies that

$$\sum_{i=1}^{n} \frac{a_i - a_i^*}{z - \beta_i}$$

has n zeros in the disk D, which is a contradiction. □

d] Suppose $(\beta_n)_{n=1}^{\infty}$ is a sequence in D satisfying

$$\beta_1 = \beta_2 = \dots = \beta_k = 0, \qquad \beta_n \neq 0, \qquad n = k+1, k+2, \dots$$

and

$$\sum_{n=1}^{\infty}(1-|\beta_n|)<\infty\,.$$

Then

$$B(z):=z^k\prod_{n=k+1}^{\infty}\left(\frac{\beta_n-z}{1-\overline{\beta}_n z}\right)\frac{|\beta_n|}{\beta_n}$$

defines a bounded analytic function on D (that is, $B\in\mathcal{H}_\infty$), which vanishes at z if and only if $z=\beta_j$ for some $j=1,2,\dots$, in which case the multiplicity of z_0 in $B(z)$ is the same as the multiplicity of β_j in $(\beta_n)_{n=1}^{\infty}$.

e] Suppose $(\beta_n)_{n=1}^{\infty}$ is a sequence in D satisfying

$$\sum_{n=1}^{\infty}(1-|\beta_n|)=\infty\,.$$

Denote the multiplicity of β_j in $(\beta_n)_{n=1}^{\infty}$ by m_j. Suppose $f\in\mathcal{H}_\infty$ has a zero at each β_j with multiplicity m_j. Then $f=0$.

Hint: Suppose $\|f\|_D>0$. Without loss of generality we may assume that $f(0)\neq 0$. By Jensen's formula (see E.10 c]),

$$\sum_{\substack{n=1\\|\beta_n|<R}}^{\infty}\log\frac{R}{|\beta_n|}+\log|f(0)|=\frac{1}{2\pi}\int_0^{2\pi}\log|f(Re^{i\theta}|\,d\theta\leq\log\|f\|_D$$

for every $R\in(0,1)$. Letting R tend to 1, we obtain

$$\sum_{n=1}^{\infty}\log\frac{1}{|\beta_n|}<\log\|f\|_D-\log|f(0)|<\infty\,.$$

Hence

$$\sum_{n=1}^{\infty}(1-|\beta_n|)<\infty\,,$$

which contradicts the assumption. □

Note that the conclusion of part e] holds for the larger *Nevanlinna class N*, which is defined as the set of those analytic functions f on D for which

$$\sup_{R\in(0,1)}\frac{1}{2\pi}\int_0^{2\pi}\log^+|f(Re^{i\theta})|d\theta<\infty\,,$$

where $\log^+ x:=\max\{\log x,0\}$.

f] Let

$$B(z) := z^k \prod_{i=1}^{n} \left(\frac{\alpha_i - z}{1 - \overline{\alpha}_i z} \right) \frac{|\alpha_i|}{\alpha_i}, \qquad \alpha_i \in \mathbb{C} \setminus \{0\}$$

be a finite Blaschke product. Show that

$$|B'(z)| = k + \sum_{i=1}^{n} \frac{1 - |\alpha_i|^2}{|z - \alpha_i|^2}, \qquad |z| = 1.$$

Hint: Consider B'/B, where $|B(z)| = 1$ whenever $|z| = 1$. □

g] Suppose

$$\sum_{i=1}^{\infty} (1 - |\alpha_i|) < \infty, \qquad \alpha_i \in (0, 1)$$

and

$$B(z) := \prod_{i=1}^{\infty} \left(\frac{\alpha_i - z}{1 - \overline{\alpha}_i z} \right) \frac{|\alpha_i|}{\alpha_i}.$$

Show that

$$\|((1 - z)^2 B(z))'\|_D \leq \left(4 + 2 \sum_{i=1}^{\infty} (1 - \alpha_i^2) \right) \|B\|_D.$$

Hint: Use f]. □

E.13 Yet Another Proof of Müntz's Theorem when $\inf\{\lambda_i : i \in \mathbb{N}\} > 0$. As in E.10, this proof requires a consequence of the Hahn-Banach theorem and the Riesz representation theorem which we state in a]. For details, the reader is referred to Feinerman and Newman [76] and Rudin [87]. We assume throughout the exercise that $\lambda_0 := 0$ and that $(\lambda_k)_{k=1}^{\infty}$ is a sequence of distinct positive numbers satisfying $\inf\{\lambda_k : k \in \mathbb{N}\} > 0$.

a] $\text{span}\{1, x^{\lambda_1}, x^{\lambda_2}, \dots\}$ is not dense if and only if there exists a nonzero finite Borel measure μ on $[0, 1]$ with

$$\int_0^1 t^{\lambda_k} \, d\mu(t) = 0, \qquad k = 0, 1, 2 \dots.$$

b] Show that $\sum_{k=1}^{\infty} 1/\lambda_k = \infty$ implies that $\text{span}\{1, x^{\lambda_1}, x^{\lambda_2}, \dots\}$ is dense in $C[0, 1]$.

Outline. Suppose there is a nonzero finite Borel measure μ on $[0, 1]$ such that

$$\int_0^1 t^{\lambda_k} \, d\mu(t) = 0, \qquad k = 0, 1, \dots.$$

Let

$$f(z) = \int_0^1 t^z \, d\mu(t) , \qquad \mathrm{Re}(z) > 0 .$$

Show that

$$g(z) := f\left(\frac{1+z}{1-z}\right) \in \mathcal{H}_\infty$$

and

$$g\left(\frac{\lambda_k - 1}{\lambda_k + 1}\right) = 0 \qquad \text{with} \qquad \left|\frac{\lambda_k - 1}{\lambda_k + 1}\right| < 1 , \qquad k = 1, 2, \ldots .$$

Note that $\sum_{k=1}^{\infty} 1/\lambda_k = \infty$ and $\inf\{\lambda_k : k \in \mathbb{N}\} > 0$ imply

$$\sum_{k=1}^{\infty} \left(1 - \left|\frac{\lambda_k - 1}{\lambda_k + 1}\right|\right) = \infty .$$

Hence E.12 e] yields that $g = 0$ on the open unit disk. Therefore $f(z) = 0$ whenever $\mathrm{Re}(z) > 0$, so

$$f(n) = \int_0^1 t^n \, d\mu(t) = 0 , \qquad n = 1, 2, \ldots .$$

Note that

$$\int_0^1 t^0 \, d\mu(t) = 0$$

also holds because of the choice of μ. Now the Weierstrass approximation theorem yields that

$$\int_0^1 f(t) \, d\mu(t) = 0$$

for every $f \in C[0,1]$, which contradicts the fact that the Borel measure μ is nonzero. So part a] implies that $\mathrm{span}\{1, x^{\lambda_1}, x^{\lambda_2}, \ldots\}$ is dense in $C[0,1]$. □

c] Show that $\sum_{k=1}^{\infty} 1/\lambda_k < \infty$ implies that $\mathrm{span}\{1, x^{\lambda_1}, x^{\lambda_2} \ldots\}$ is not dense in $C[0,1]$.

Outline. Show under the above assumption that

$$f(z) = \int_0^1 t^z \left\{ \frac{1}{2\pi} \int_{-\infty}^{\infty} f(-1+is) e^{-is \log t} \, ds \right\} dt , \qquad \mathrm{Re}(z) > -1$$

if f is defined by

$$f(z) := \frac{z}{(2+z)^3} \prod_{n=1}^{\infty} \frac{\lambda_n - z}{2 + \lambda_n + z} .$$

Show that

$$d\mu(t) = \left\{ \frac{1}{2\pi} \int_{-\infty}^{\infty} f(-1+is) e^{-is\log t} \, ds \right\} dt$$

defines a nonzero finite Borel measure, μ, on $[0,1]$ such that

$$\int_0^1 t^{\lambda_k} \, d\mu(t) = 0 \, , \qquad k = 0, 1, 2, \dots$$

as is required by part a]. For the above, show that

$$f(z) = -\frac{1}{2\pi} \int_{-\infty}^{\infty} \frac{f(is-1)}{is-1-z} \, ds \, , \qquad \mathrm{Re}(z) > -1$$

and use that

$$\frac{1}{1+z-is} = \int_0^1 t^{z-is} \, dt \, .$$

□

E.14 Another Proof of Denseness of Müntz Spaces when $\lambda_i \to 0$. Suppose $\Lambda := (\lambda_i)_{i=1}^\infty$ is a sequence of distinct positive numbers with $\lim\limits_{i\to\infty} \lambda_i = 0$. Show that

$$M(\Lambda) := \mathrm{span}\{1, x^{\lambda_1}, x^{\lambda_2}, \dots\}$$

is dense in $C[0,1]$ if and only if $\sum_{i=1}^\infty \lambda_i = \infty$.

Hint: If $\sum_{i=1}^\infty \lambda_i = \infty$, then $\lim\limits_{i\to\infty} \lambda_i = 0$ implies that

$$\sum_{i=1}^{\infty} \left(1 - \left| \frac{\lambda_i - 1}{\lambda_i + 1} \right| \right) = \infty \, .$$

So the outline of the proof of E.13 b] yields that $M(\Lambda)$ is dense in $C[0,1]$.

If $\eta := \sum_{i=1}^\infty \lambda_i < \infty$, then, by Theorem 6.1.1, the inequality

$$\|xp'(x)\|_{[0,1]} \le 9\eta \, \|p\|_{[0,1]}$$

holds for every $p \in M(\Lambda)$. Use this inequality to show that $M(\Lambda)$ fails to be dense in $C[0,1]$. □

E.15 Denseness of Müntz Spaces with Complex Exponents. Suppose $\Lambda := (\lambda_i)_{i=1}^\infty$ is a sequence of complex numbers satisfying

$$\mathrm{Re}(\lambda_i) > 0 \, , \qquad i = 1, 2, \dots \, .$$

Show that if

$$\sum_{n=1}^{\infty} \left(1 - \left| \frac{\lambda_n - 1}{\lambda_n + 1} \right| \right) = \infty \, ,$$

then $\mathrm{span}\{1, x^{\lambda_1}, x^{\lambda_2}, \dots\}$ is dense in $C[0,1]$. (In this exercise the span is taken over $\mathbb{C}$, and $C[0,1]$ denotes the set of all complex-valued continuous functions on $[0,1]$.)

E.16 Christoffel Functions for Nondense Müntz Spaces. Let $\Lambda = (\lambda_i)_{i=0}^{\infty}$ be a sequence of distinct complex numbers with $\mathrm{Re}(\lambda_i) > -\frac{1}{2}$ for each i. As in Section 3.4, let $L_k^* := L_k^*\{\lambda_0, \dots, \lambda_k\}$ denote the associated orthonormal Müntz-Legendre polynomials on $[0, 1]$.

a] Let K_n be defined by

$$\frac{1}{K_n(y)} := \inf \left\{ \int_0^1 |p(t)|^2 \, dt : p \in \operatorname{span}\{x^{\lambda_0}, x^{\lambda_1}, \dots, x^{\lambda_n}\}, \ p(y) = 1 \right\}.$$

Show that

$$K_n(y) = \sum_{k=0}^{n} |L_k^*(y)|^2 .$$

The function $1/K_n$ is called the nth *Christoffel function associated with* Λ.

Hint: Proceed as in the hint to E.13 of Section 2.3. □

In the rest of the exercise we assume that $(\lambda_i)_{i=0}^{\infty}$ is a sequence of non-negative integers. We use this assumption for treating (higher) derivatives, although some weaker assumptions would lead to the same conclusions.

b] Suppose $\sum_{i=1}^{\infty} 1/\lambda_i < \infty$. Show that for every $\epsilon \in (0, 1)$ and $m \in \mathbb{N}$, there exists a constant $c_{\epsilon,m}$ depending only on Λ, ϵ, and m such that

$$\|p^{(m)}\|_{[0,1-\epsilon]} \le c_{\epsilon,m} \|p\|_{L_2[0,1]}$$

for every $p \in \operatorname{span}\{x^{\lambda_0}, x^{\lambda_1}, \dots\}$.

Hint: Use E.3 c]. □

c] Show that the following statements are equivalent:

(1) $\operatorname{span}\{x^{\lambda_0}, x^{\lambda_1}, \dots\}$ is not dense in $C[0, 1]$.

(2) $\sum_{i=1}^{\infty} 1/\lambda_i < \infty$.

(3) $\sum_{k=0}^{\infty} (L_k^*)^2$ converges uniformly on $[0, 1-\epsilon]$ for all $\epsilon \in (0, 1)$.

(4) There exists an $x \in [0, 1)$ so that $\sum_{k=0}^{\infty} (L_k^*(x))^2 < \infty$.

Outline. The equivalence of (1) and (2) is the content of Müntz's theorem (Theorem 4.2.1). To see that (2) implies (3), first observe that

$$\sum_{k=0}^{\infty} \left| (L_k^*)^{(m)}(y) \right|^2 = \sup \left\{ \left| p^{(m)}(y) \right|^2 : \ p \in \operatorname{span}\{x^{\lambda_0}, x^{\lambda_1}, \dots\}, \quad \|p\|_{L_2[0,1]} = 1 \right\},$$

which can be proved similarly to part a]. Hence by part b], for every $\epsilon \in (0, 1)$, there exists a constant c_ϵ depending only on ϵ such that

$$\sum_{k=0}^{\infty} \left(L_k^*(x)\right)^2 \le c_\epsilon \qquad \text{and} \qquad \sum_{k=0}^{\infty} \left(L_k^{*\prime}(x)\right)^2 \le c_\epsilon\,, \qquad x \in [0, 1-\epsilon]\,.$$

Since $\left(\sum_{k=0}^{n} (L_k^*)^2\right)' = 2\sum_{k=0}^{n} L_k^* L_k^{*\prime}$ on $[0,\infty)$, applying the Cauchy-Schwarz inequality, we obtain that

$$\left|\left(\sum_{k=0}^{n} (L_k^*)^2\right)'(x)\right| \le 2c_\epsilon\,, \qquad x \in [0, 1-\epsilon]\,.$$

Therefore the functions $\sum_{k=0}^{n} \left(L_k^*\right)^2$, $n = 1, 2, \ldots$, are uniformly bounded and equicontinuous on $[0, 1-\epsilon]$, which implies the uniform convergence of the functions K_n on $[0, 1-\epsilon]$ by the Arzela-Ascoli theorem. Since (3) obviously implies (4), what remains to be proven is that (4) implies (1). This can be easily done by part a]. □

d] Let $\epsilon \in (0,1)$ and $m \in \mathbb{N}$ be fixed. Show that if $\sum_{i=0}^{\infty} 1/\lambda_i < \infty$, then $\sum_{k=0}^{\infty} ((L_k^*)^{(m)})^2$ converges uniformly on $[0, 1-\epsilon]$.

Hint: Modify the argument given in the hints to part c]. □

e] Show that if $\sum_{i=0}^{\infty} 1/\lambda_i < \infty$, then

$$\lim_{k\to\infty} \|(L_k^*)^{(m)}\|_{[0,1-\epsilon]} = 0$$

for every $\epsilon \in (0,1)$ and $m \in \mathbb{N}$.

E.17 Chebyshev-Type Inequality with Explicit Bound via the Paley-Wiener Theorem. The method outlined in this exercise was suggested by Halász. A function f is called *entire* if it is analytic on the complex plane. An entire function f is called a function of *exponential type* δ if there exists a constant c depending only on f such that

$$|f(z)| \le c\exp(\delta|z|)\,, \qquad z \in \mathbb{C}\,.$$

The collection of all such entire functions of exponential type δ is denoted be E^δ. The Paley-Wiener theorem characterizes the functions F that can be written as the Fourier transform of some function $f \in L_2[-\delta, \delta]$.

Theorem (Paley-Wiener). *Let $\delta \in (0,\infty)$. Then $F \in E^\delta \cap L_2(\mathbb{R})$ if and only if there exists an $F \in L_2[-\delta,\delta]$ such that*

$$F(z) = \int_{-\delta}^{\delta} f(t)e^{itz}\,dt\,.$$

For a proof see, for example, Rudin [87].

In the rest of the exercise let $\Lambda = (\lambda_k)_{k=0}^{\infty}$ be an increasing sequence with $\lambda_0 = 0$ and $\sum_{k=1}^{\infty} 1/\lambda_k < \infty$.

E.5 says that

$$C(\epsilon, \Lambda) := \sup \left\{ \frac{\|p\|_{[0,1]}}{\|p\|_{[1-\epsilon,1]}} : \; p \in \operatorname{span}\{x^{\lambda_0}, x^{\lambda_1}, \dots\} \right\} < \infty$$

holds for every $\epsilon \in (0,1)$. In this exercise we establish an explicit bound for $C(\epsilon, \Lambda)$.

a] Show that

$$C(\epsilon, \Lambda) = \sup \left\{ \frac{|p(0)|}{\|p\|_{[1-\epsilon,1]}} : \; p \in \operatorname{span}\{x^{\lambda_0}, x^{\lambda_1}, \dots\} \right\}$$

for every $\epsilon \in (0,1)$.

Hint: Use $\lambda_0 = 0$, E.4 c] of Section 3.3, and the monotonicity of the Chebyshev polynomial

$$T_n\{x^{\lambda_0}, x^{\lambda_1}, \dots, x^{\lambda_n}; [1-\epsilon, 1]\}$$

on $[0, 1-\epsilon]$. □

b] Assume that

(1) $F \in E^{\delta} \cap L_2(\mathbb{R})$;

(2) $F(i\lambda_k) = 0, \;\; k = 1, 2, \dots$ (i is the imaginary unit); and

(3) $F(0) = 1$.

Show that

$$|P(\infty)| \le \|F\|_{L_2(\mathbb{R})} \, \|P\|_{L_2[-\delta,\delta]}$$

for every $P \in \operatorname{span}\{e^{-\lambda_0 t}, e^{-\lambda_1 t}, \dots\}$.

Outline. By the Paley-Wiener theorem

$$F(z) = \int_{-\delta}^{\delta} f(t) e^{itz} \, dt$$

for some $f \in L_2[-\delta, \delta]$. Now if

$$P(t) = a_0 + \sum_{k=1}^{n} a_k e^{-\lambda_k t},$$

then

$$\begin{aligned} \int_{-\delta}^{\delta} f(t) P(t) \, dt &= a_0 \int_{-\delta}^{\delta} f(t) \, dt + \sum_{k=1}^{n} a_k \int_{-\delta}^{\delta} f(t) e^{-\lambda_k t} \, dt \\ &= a_0 F(0) + \sum_{k=1}^{n} a_k F(i\lambda_k) = a_0 = P(\infty). \end{aligned}$$

Hence by the Cauchy-Schwarz inequality and the L_2 inversion theorem of Fourier transforms, we obtain that

$$|P(\infty)| \leq \|f\|_{L_2[-\delta,\delta]} \, \|P\|_{L_2[-\delta,\delta]} \leq \|F\|_{L_2(\mathbb{R})} \, \|P\|_{L_2[-\delta,\delta]} \, .$$

□

Given $\delta \in (0,1)$, let $N \in \mathbb{N}$ be chosen so that

$$\sum_{k=N+1}^{\infty} \frac{1}{\lambda_k} \leq \frac{\delta}{3} \, .$$

Let

$$\sigma_k := A\lambda_k \qquad \text{with} \qquad A := \frac{\delta}{3N} \, .$$

Let

$$F(z) := \frac{\sin(\delta z/3)}{\delta z/3} \times \prod_{k=1}^{N} \left(\left(1 - \frac{z}{i\lambda_k} \right) \frac{\sin(\sigma_k z/\lambda_k)}{\sigma_k z/\lambda_k} \right) \prod_{k=N+1}^{\infty} \left(1 - \left(\frac{\sin(z/\lambda_k)}{\sin i} \right)^4 \right),$$

where i is the imaginary unit.

c] Show that $F \in E^\delta$.

d] Observe that $F(0) = 1, \;\; F(i\lambda_k) = 0, \;\; k = 1, 2, \dots$, and

$$|F(t)| \leq \frac{\sin(\delta t/3)}{\delta t/3} \prod_{k=1}^{N} \left(2 + \frac{1}{\sigma_k} \right), \qquad t \in \mathbb{R} \, .$$

e] Show that

$$|P(\infty)| \leq \frac{3c}{\delta} \prod_{k=1}^{N} \left(2 + \frac{1}{\sigma_k} \right) \|P\|_{[-\delta,\delta]}$$

for every $P \in \operatorname{span}\{e^{-\lambda_0 t}, e^{-\lambda_1 t}, \dots\}$ with $c := \|t^{-1} \sin t\|_{L_2(\mathbb{R})}$.

Hint: Use parts b], c], and d]. □

f] Let $\lambda_k := k^\alpha, \;\; \alpha > 1$. Show that there exists a constant c_α depending only on α such that

$$\|p\|_{[0,1]} \leq \exp\left(c_\alpha \epsilon^{1/(1-\alpha)}\right) \|p\|_{[1-\epsilon,1]}$$

for every $p \in \operatorname{span}\{x^{\lambda_0}, x^{\lambda_1}, \dots\}$ and for every $\epsilon \in (0, 1/2]$.

Proof. Let

$$\delta := -\frac{1}{2}\log(1-\epsilon)\,. \tag{4.2.5}$$

Observe that N in part e] can be chosen so that

$$N := \left\lfloor \left(\frac{\delta(\alpha-1)}{3}\right)^{1/(1-\alpha)} \right\rfloor + 1\,. \tag{4.2.6}$$

Also, σ_k in part d] is of the form $\sigma_k = \delta k^\alpha (3N)^{-1}$. Let $M+1$ be the smallest value of $k \in \mathbb{N}$ for which

$$\frac{1}{\sigma_k} < 1\,, \qquad \text{that is}\,, \qquad \frac{3N}{k^\alpha \delta} \le 1\,.$$

Note that

$$M := \left\lfloor \left(\frac{3N}{\delta}\right)^{1/\alpha} \right\rfloor.$$

If $0 < M < N$, then

$$\begin{aligned}
\prod_{k=1}^N \left(2 + \frac{1}{\sigma_k}\right) &= \prod_{k=1}^N \left(2 + \frac{3N}{\delta k^\alpha}\right) \\
&\le \left(\prod_{k=1}^M \frac{9N}{\delta k^\alpha}\right)\left(\prod_{k=M+1}^N 3\right) \le \left(\frac{9N}{\delta}\right)^M \left(\frac{M}{e}\right)^{-\alpha M} 3^{N-M} \\
&= \left(\frac{9e^\alpha N}{\delta}\right)^M M^{-\alpha M} 3^{N-M} \\
&\le \left(\frac{9e^\alpha N}{\delta}\right)^M \left(\frac{1}{2}\left(\frac{3N}{\delta}\right)^{1/\alpha}\right)^{-\alpha M} 3^{N-M} \\
&\le (3(2e)^\alpha)^M 3^{N-M} \le (3(2e)^\alpha)^N\,,
\end{aligned}$$

and the theorem follows by (4.2.5), (4.2.6), and part e].

If $N \le M$, then

$$\begin{aligned}
\prod_{k=1}^N \left(2 + \frac{1}{\sigma_k}\right) &= \prod_{k=1}^N \left(2 + \frac{3N}{\delta k^\alpha}\right) \\
&\le \left(\prod_{k=1}^N \frac{9N}{\delta k^\alpha}\right) \le \left(\frac{9N}{\delta}\right)^N \left(\frac{N}{e}\right)^{-\alpha N} = \left(\frac{9e^\alpha N^{(1-\alpha)}}{\delta}\right)^N \\
&\le \left(\frac{9e^\alpha}{\delta}\right)^N \left(\left(\frac{\delta(\alpha-1)}{3}\right)^{1/(1-\alpha)}\right)^{(1-\alpha)N} \\
&\le \left(\frac{9e^\alpha}{\delta}\right)^N \left(\frac{\delta(\alpha-1)}{3}\right)^N \le (3e^\alpha(\alpha-1))^N\,,
\end{aligned}$$

and the theorem follows by (4.2.5), (4.2.6), and part e].

If $M = 0$, then

$$\prod_{k=1}^{N}\left(2+\frac{1}{\sigma_k}\right) \leq \prod_{k=1}^{N} 3 = 3^N\,,$$

and the theorem follows by (4.2.5), (4.2.6), and part e]. □

The next part of the exercise shows that the result of part f] is close to sharp.

g] Let $\lambda_k := k^\alpha,\ \ \alpha > 1$. Let $\epsilon \in (0, 1/2]$. Show that there exists a constant c_α depending only on $\alpha > 0$ so that

$$\sup\left\{\frac{|p(0)|}{\|p\|_{[1-\epsilon,1]}} : p \in \operatorname{span}\{x^{\lambda_0}, x^{\lambda_1}, \dots\}\right\} \geq \exp\left(c_\alpha \epsilon^{1/(1-\alpha)}\right).$$

Proof. Let $n \in \mathbb{N}$ be a fixed. We define $\gamma_k := kn^{\alpha-1},\ \ k = 0, 1, \dots$. Let $T_n(x) := ((x-1)/2)^n$ and

$$Q_n(x) := T_n\left(\frac{2x^{n^{\alpha-1}}}{1-(1-\epsilon)^{n^{\alpha-1}}} - \frac{1+(1-\epsilon)^{n^{\alpha-1}}}{1-(1-\epsilon)^{n^{\alpha-1}}}\right).$$

Then $Q_n \in \operatorname{span}\{x^{\gamma_0}, \dots x^{\gamma_n}\}$, and by E.3 g] of Section 3.3 we obtain that

$$\begin{aligned}\sup\left\{\frac{|p(0)|}{\|p\|_{[1-\epsilon,1]}} : p \in \operatorname{span}\{x^{\lambda_0}, x^{\lambda_1}, \dots\}\right\} &\geq \frac{|Q_n(0)|}{\|Q_n\|_{[1-\epsilon,1]}} = |Q_n(0)| \\ &= \left(\frac{1}{1-(1-\epsilon)^{n^{\alpha-1}}}\right)^n.\end{aligned}$$

Now let n be the smallest integer satisfying $n^{\alpha-1} \geq \epsilon^{-1}$. Since $(1-\epsilon)^{1/\epsilon}$ is bounded away from 0 on $(0, 1/2]$, the result follows. □

E.18 Completion of the Proof of Theorem 4.2.1. The case when $\lambda_i \geq 1$ for each i has already been proved. The only real remaining difficulty is part d].

a] Prove Theorem 4.2.1 in the case when $\inf\{\lambda_i : i \in \mathbb{N}\} > 0$.

Hint: Use the scaling $x \to x^{1/\delta}$ and the already proved case. □

b] Show that if $(\lambda_i)_{i=1}^\infty \subset (0, \infty)$ has a cluster point $\lambda \in (0, \infty)$, then $\operatorname{span}\{1, x^{\lambda_1}, x^{\lambda_2}, \dots\}$ is dense in $C[0, 1]$.

Hint: Use part a]. □

c] Suppose $(\lambda_i)_{i=1}^\infty \subset (0, \infty)$ and $\lambda_i \to 0$. Then $\operatorname{span}\{1, x^{\lambda_1}, x^{\lambda_2}, \dots\}$ is dense in $C[0, 1]$ if and only if $\sum_{i=1}^\infty \lambda_i = \infty$.

Hint: This is the content of E.14. □

d] Suppose

$$\{\lambda_i : i \in \mathbb{N}\} = \{\alpha_i : i \in \mathbb{N}\} \cup \{\beta_i : i \in \mathbb{N}\}$$

with

$$\lim_{i\to\infty} \alpha_i = 0 \qquad \text{and} \qquad \lim_{i\to\infty} \beta_i = \infty\,.$$

Show that $\operatorname{span}\{1, x^{\lambda_1}, x^{\lambda_2}, \dots\}$ is dense in $C[0,1]$ if and only if

$$\sum_{i=1}^{\infty} \alpha_i + \sum_{i=1}^{\infty} \frac{1}{\beta_i} = \infty\,. \tag{4.2.7}$$

Outline. If (4.2.7) holds, then the denseness of $\operatorname{span}\{1, x^{\lambda_1}, x^{\lambda_2}, \dots\}$ in $C[0,1]$ follows from parts a] and c]. Now assume that (4.2.7) does not hold, so

$$\sum_{i=1}^{\infty} \alpha_i < \infty \qquad \text{and} \qquad \sum_{i=1}^{\infty} \frac{1}{\beta_i} < \infty\,.$$

For notational convenience, let

$$\begin{aligned} T_{n,\alpha} &:= T_n\{1, x^{\alpha_1}, \dots, x^{\alpha_n} : [0,1]\}\,, \\ T_{n,\beta} &:= T_n\{1, x^{\beta_1}, \dots, x^{\beta_n} : [0,1]\}\,, \\ T_{2n,\alpha,\beta} &:= T_{2n}\{1, x^{\alpha_1}, \dots, x^{\alpha_n}, x^{\beta_1}, \dots, x^{\beta_n} : [0,1]\} \end{aligned}$$

(we use the notation introduced in Section 3.3).

It follows from Theorem 6.1.1 (Newman's inequality) and the Mean Value Theorem that for every $\epsilon > 0$ there exists a $k_1(\epsilon) \in \mathbb{N}$ depending only on $(\alpha_i)_{i=1}^{\infty}$ and ϵ (and not on n) such that $T_{n,\alpha}$ has at most $k_1(\epsilon)$ zeros in $[\epsilon, 1)$ and at least $n - k_1(\epsilon)$ zeros in $(0, \epsilon)$.

Similarly, E.5 b] and the Mean Value Theorem imply that for every $\epsilon > 0$ there exists a $k_2(\epsilon) \in \mathbb{N}$ depending only on $(\beta_i)_{i=1}^{\infty}$ and ϵ (and not on n) so that $T_{n,\beta}$ has at most $k_2(\epsilon)$ zeros in $(0, 1-\epsilon]$ and at least $n - k_2(\epsilon)$ zeros $(1-\epsilon, 1)$.

Now, on counting the zeros of $T_{n,\alpha} - T_{2n,\alpha,\beta}$ and $T_{n,\beta} - T_{2n,\alpha,\beta}$, we can deduce that for every $\epsilon > 0$ there exists a $k(\epsilon) \in \mathbb{N}$ depending only on $(\lambda_i)_{i=1}^{\infty}$ and ϵ (and not on n) so that $T_{2n,\alpha,\beta}$ has at most $k(\epsilon)$ zeros in $[\epsilon, 1-\epsilon]$.

Let $\epsilon := \frac{1}{4}$ and $k := k\left(\frac{1}{4}\right)$. Pick $k+4$ points

$$\tfrac{1}{4} < \eta_0 < \eta_1 < \cdots < \eta_{k+3} < \tfrac{3}{4}$$

and a function $f \in C[0,1]$ such that $f(x) = 0$ for all $x \in \left[0, \frac{1}{4}\right] \cup \left[\frac{3}{4}, 1\right]$, while

$$f(\eta_i) := 2 \cdot (-1)^i\,, \qquad i = 0, 1, \dots\,.$$

Assume that there exists a $p \in \text{span}\{1, x^{\lambda_1}, x^{\lambda_2}, \dots\}$ such that

$$\|f - p\|_{[0,1]} < 1 .$$

Then $p - T_{2n,\alpha,\beta}$ has at least $2n + 1$ zeros in $(0, 1)$. However, for sufficiently large n,

$$p - T_{2n,\alpha,\beta} \in \text{span}\{1, x^{\lambda_1}, \dots, x^{\lambda_{2n}}\}$$

so it can have at most $2n$ zeros in $[0, \infty)$. This contradiction shows that $\text{span}\{1, x^{\lambda_1}, x^{\lambda_2}, \dots\}$ is not dense in $C[0, 1]$. □

e] Prove Theorem 4.2.1 in full generality.

Outline. Combine parts a] to d]. □

E.19 Proof of Theorem 4.2.3. Prove Theorem 4.2.3.

Proof. Assume that

$$\text{span}\{x^{\lambda_0}, x^{\lambda_1}, \dots\}$$

is dense in $L_1[0, 1]$. Let m be a fixed nonnegative integer. Let $\epsilon > 0$. Choose a

$$p \in \text{span}\{x^{\lambda_0}, x^{\lambda_1}, \dots\}$$

such that

$$\|x^m - p(x)\|_{L_1[0,1]} < \epsilon .$$

Now let

$$q(x) := \int_0^x p(t)\, dt \in \text{span}\{x^{\lambda_0 + 1}, x^{\lambda_1 + 1}, \dots\} .$$

Then

$$\left\| \frac{x^{m+1}}{m+1} - q(x) \right\|_{[0,1]} < \epsilon .$$

So the Weierstrass approximation theorem yields that

$$\text{span}\{1, x^{\lambda_0 + 1}, x^{\lambda_1 + 1}, \dots\}$$

is dense in $C[0, 1]$, and Theorem 2.1 implies that

$$\sum_{i=0}^{\infty} \frac{\lambda_i + 1}{(\lambda_i + 1)^2 + 1} = \infty .$$

Now assume that

$$\sum_{i=0}^{\infty} \frac{\lambda_i + 1}{(\lambda_i + 1)^2 + 1} = \infty. \tag{4.2.8}$$

By the Hahn-Banach theorem and the Riesz representation theorem

$$\operatorname{span}\{x^{\lambda_0}, x^{\lambda_1}, \dots\}$$

is not dense in $L_1[0,1]$ if and only if there exists a $0 \neq h \in L_\infty[0,1]$ satisfying

$$\int_0^1 t^{\lambda_i} h(t)\, dt = 0, \qquad i = 0, 1, \dots .$$

Suppose there exists a $0 \neq h \in L_\infty[0,1]$ such that

$$\int_0^1 t^{\lambda_i} h(t)\, dt = 0\,, \qquad i = 0, 1, \dots .$$

Let

$$f(z) := \int_0^1 t^z h(t)\, dt\,, \qquad \operatorname{Re}(z) > -1\,.$$

Then

$$g(z) := f\left(\frac{1+z}{1-z} - 1\right)$$

is a bounded analytic function on the open unit disk that satisfies

$$g\left(\frac{\lambda_i}{\lambda_i + 2}\right) = 0 \qquad \text{with} \qquad \left|\frac{\lambda_i}{\lambda_i + 2}\right| < 1\,, \qquad i = 0, 1, \dots .$$

Note that (4.2.8) implies

$$\sum_{i=1}^{\infty} \left(1 - \left|\frac{\lambda_i}{\lambda_i + 2}\right|\right) = \infty\,.$$

Hence Blaschke's theorem (E.12 e]) yields that $g = 0$ on the open unit disk. Therefore $f(z) = 0$ whenever $\operatorname{Re}(z) > -1$, so

$$f(n) = \int_0^1 t^n h(t)\, dt = 0\,, \qquad n = 0, 1, \dots .$$

Now the Weierstrass approximation theorem yields

$$\int_0^1 u(t) h(t)\, dt = 0$$

for every $u \in C[0,1]$, which contradicts the fact that $0 \neq h$. So

$$\operatorname{span}\{x^{\lambda_0}, x^{\lambda_1}, \dots\}$$

is dense in $L_1[0,1]$. □

E.20 Proof of Theorem 4.2.4.

a] Show that if

$$\sum_{i=0}^{\infty} \frac{\lambda_i + \frac{1}{p}}{\left(\lambda_i + \frac{1}{p}\right)^2 + 1} = \infty \,, \tag{4.2.9}$$

then $\mathrm{span}\{x^{\lambda_0}, x^{\lambda_1}, \dots\}$ is dense in $L_p[0,1]$.

Outline. By the Hahn-Banach theorem and the Riesz representation theorem

$$\mathrm{span}\{x^{\lambda_0}, x^{\lambda_1}, \dots\}$$

is not dense in $L_p[0,1]$ if and only if there exists a $0 \neq h \in L_q[0,1]$ satisfying

$$\int_0^1 t^{\lambda_i} h(t)\, dt = 0 \,, \qquad i = 0, 1, \dots \,,$$

where q is the conjugate exponent of p defined by $p^{-1} + q^{-1} = 1$.

Suppose there exists a $0 \neq h \in L_q[0,1]$ such that

$$\int_0^1 t^{\lambda_i} h(t)\, dt = 0 \,, \qquad i = 0, 1, \dots \,.$$

Let

$$f(z) := \int_0^1 t^z h(t)\, dt \,, \qquad \mathrm{Re}(z) > -\tfrac{1}{p} \,.$$

Use Hölder's inequality to show that

$$g(z) := f\left(\frac{1+z}{1-z} - \frac{1}{p}\right)$$

is a bounded analytic function on the open unit disk that satisfies

$$g\left(\frac{\lambda_i + \frac{1}{p} - 1}{\lambda_i + \frac{1}{p} + 1}\right) = 0 \qquad \text{with} \qquad \left|\frac{\lambda_i + \frac{1}{p} - 1}{\lambda_i + \frac{1}{p} + 1}\right| < 1 \,, \qquad i = 0, 1, \dots \,.$$

Note that (4.2.9) implies

$$\sum_{i=1}^{\infty} \left(1 - \left|\frac{\lambda_i + \frac{1}{p} - 1}{\lambda_i + \frac{1}{p} + 1}\right|\right) = \infty \,.$$

Hence Blaschke's theorem (E.12 e]) yields that $g = 0$ on the open unit disk. Therefore $f(z) = 0$ whenever $\mathrm{Re}(z) > -\frac{1}{p}$, so

$$f(n) = \int_0^1 t^n h(t)\,dt = 0\,, \qquad n = 0, 1, \ldots\,.$$

Now the Weierstrass approximation theorem yields

$$\int_0^1 u(t)h(t)\,dt = 0$$

for every $u \in C[0,1]$, which contradicts the fact that $0 \neq h$. So

$$\mathrm{span}\{x^{\lambda_0}, x^{\lambda_1}, \ldots\}$$

is dense in $L_p[0,1]$. □

b] Show that if

$$\sum_{i=0}^{\infty} \frac{\lambda_i + \frac{1}{p}}{\left(\lambda_i + \frac{1}{p}\right)^2 + 1} < \infty\,,$$

then $\mathrm{span}\{x^{\lambda_0}, x^{\lambda_1}, \ldots\}$ is not dense in $L_p[0,1]$.

Outline. This follows from E.7 of Section 4.3. □

c] Suppose $(\lambda_i)_{i=0}^{\infty}$ is a sequence of distinct positive numbers. Let $p \in [1,\infty)$. Show that $\mathrm{span}\{e^{-\lambda_0 t}, e^{-\lambda_t}, \ldots\}$ is dense in $L_p[0,\infty)$ if and only if

$$\sum_{i=0}^{\infty} \frac{\lambda_i}{\lambda_i^2 + 1} = \infty\,.$$

Outline. Use parts a] and b] and the substitution $x = e^{-t}$. □

E.21 Müntz Theorem on $[a,b]$ with $a < 0 < b$. *Suppose $\Lambda := (\lambda_i)_{i=1}^{\infty}$ is a sequence of distinct nonnegative integers, and suppose $a < 0 < b$. Then* $\mathrm{span}\{1, x^{\lambda_1}, x^{\lambda_2}, \ldots\}$ *is dense in $C[a,b]$ if and only if*

$$\sum_{\substack{i=1 \\ \lambda_i \text{ is even}}}^{\infty} \frac{1}{\lambda_i} = \infty \qquad \textit{and} \qquad \sum_{\substack{i=1 \\ \lambda_i \text{ is odd}}}^{\infty} \frac{1}{\lambda_i} = \infty\,.$$

E.22 The Zeros of the Chebyshev Polynomials in Nondense Müntz Spaces. Let $(\lambda_i)_{i=0}^{\infty}$ be a sequence of distinct nonnegative real numbers with $\lambda_0 := 0$ and $\sum_{i=1}^{\infty} 1/\lambda_i < \infty$. Let

$$T_n := T_n\{\lambda_0, \lambda_1, \ldots, \lambda_n; [0,1]\}$$

be the Chebyshev polynomials for $\mathrm{span}\{x^{\lambda_0}, \ldots, x^{\lambda_n}\}$ on $[0,1]$. Let

$$Z := \{x \in [0,1] : T_n(x) = 0 \text{ for some } n \in \mathbb{N}\}\,.$$

Let Z' be the set of all limit points of Z and Z'' be the set of all limit points of Z'. Show that $Z'' = \{1\}$.

Hint: Use the bounded Bernstein-type inequality of E.5 b] and the interlacing property of the zeros of the Chebyshev polynomials T_n. □

4.3 Unbounded Bernstein Inequalities

In Section 4.1 we characterized the denseness of C^1 Markov spaces by the behavior of the zeros of their associated Chebyshev polynomials. The principal result of this section is a characterization of denseness of Markov spaces by whether or not they have an unbounded Bernstein inequality.

Definition 4.3.1 (Unbounded Bernstein Inequality). Let $\mathcal{A}$ be a subset of $C^1[a,b]$. We say that $\mathcal{A}$ has an *everywhere unbounded Bernstein inequality* if

$$\sup\left\{\frac{\|p'\|_{[\alpha,\beta]}}{\|p\|_{[a,b]}} : \ 0 \neq p \in \mathcal{A}\right\} = \infty$$

for every $[\alpha,\beta] \subset [a,b], \ \ \alpha \neq \beta$.

The subset

$$\mathcal{A} := \{x^2 p(x) : p \in \mathcal{P}_n, \ \ n = 0, 1, \dots\}$$

has an everywhere unbounded Bernstein inequality despite the fact that $f'(0) = 0$ for every $f \in \mathcal{A}$.

The next result shows that in most instances the Chebyshev polynomial is close to extremal for Bernstein-type inequalities. This is a theme that will be explored further in later chapters.

Theorem 4.3.2 (A Bernstein-Type Inequality for Chebyshev Spaces). *Let* $(1, f_1, \dots, f_n)$ *be a Chebyshev system on* $[a,b]$ *such that each* f_i *is differentiable at* $x_0 \in [a,b]$. *Let*

$$T_n := T_n\{1, f_1, \dots, f_n; [a,b]\}$$

be the associated Chebyshev polynomial. Then

$$\frac{|p'(x_0)|}{\|p\|_{[a,b]}} \leq \frac{2}{1 - |T_n(x_0)|}\,|T_n'(x_0)|$$

for every $0 \neq p \in \operatorname{span}\{1, f_1, \dots, f_n\}$, *provided* $|T_n(x_0)| \neq 1$.

Proof. Let $a = y_0 < y_1 < \cdots < y_n = b$ denote the extreme points of T_n, that is,

$$T_n(y_i) = (-1)^{n-i}, \qquad i = 0, 1, \dots, n$$

(see the definition and E.1 a] in Section 3.3). Let $y_k \leq x_0 \leq y_{k+1}$ and

$$0 \neq p \in H_n := \operatorname{span}\{1, f_1, \dots, f_n\}.$$

If $p'(x_0) = 0$, then there is nothing to prove. Assume that $p'(x_0) \neq 0$. Then we may normalize p so that

$$\|p\|_{[a,b]} = 1 \qquad \text{and} \qquad \text{sign}(p'(x_0)) = \text{sign}(T_n(y_{k+1}) - T_n(y_k))\,.$$

Let $\delta := |T_n(x_0)|$. Let $\epsilon \in (0, 1)$ be fixed. Then there exists a constant η with $|\eta| \leq \delta + \frac{1}{2}(1 - \delta)$ such that

$$\eta + \tfrac{1}{2}(1 - \delta)(1 - \epsilon)p(x_0) = T_n(x_0)\,.$$

Now let

$$q(x) := \eta + \tfrac{1}{2}(1 - \delta)(1 - \epsilon)p(x)\,.$$

Then

$$\|q\|_{[a,b]} < 1\,, \qquad q(x_0) = T_n(x_0)\,,$$

and

$$\text{sign}(q'(x_0)) = \text{sign}(T_n(y_{k+1}) - T_n(y_k))\,.$$

If the desired inequality did not hold for p, then for a sufficiently small $\epsilon > 0$

$$|q'(x_0)| > |T_n'(x_0)|\,,$$

so

$$h(x) := q(x) - T_n(x)$$

would have at least three zeros in (y_k, y_{k+1}). But h has at least one zero in each of (y_i, y_{i+1}). Hence $h \in H_n$ has at least $n + 2$ zeros in $[a, b]$, which is a contradiction. □

We now state the main result.

Theorem 4.3.3 (Characterization of Denseness by Unbounded Bernstein Inequality). *Suppose $\mathcal{M} := (f_0, f_1, \dots)$ is an infinite Markov system on $[a, b]$ with each $f_i \in C^2[a, b]$, and suppose that $(f_1/f_0)'$ does not vanish on (a, b). Then* span $\mathcal{M}$ *is dense in $C[a, b]$ if and only if* span $\mathcal{M}$ *has an everywhere unbounded Bernstein inequality.*

Proof. The *only if* part of this Theorem is obvious. A good uniform approximation on $[a, b]$ to a function with uniformly large derivative on a subinterval $[\alpha, \beta] \subset [a, b]$ must have large derivative at some points in $[\alpha, \beta]$.

In the other direction we use Theorems 4.3.2 and 4.1.1 in the following way. Without loss of generality we may assume that $f_0 = 1$ (why?). If span $\mathcal{M}$ is not dense in $C[a, b]$, then, by Theorem 4.1.1, there exists a subinterval $[\alpha, \beta] \subset [a, b]$, where all elements of a subsequence of the sequence of associated Chebyshev polynomials,

$$(T_n\{1, f_1, \dots, f_n; [a,b]\}),$$

have no zeros. It remains to show that from this subsequence we can pick another subsequence (T_{n_i}) and a subinterval $[c,d] \subset [\alpha, \beta]$ with

$$\|T_{n_i}\|_{[c,d]} < 1 - \delta \tag{4.3.1}$$

and

$$\|T'_{n_i}\|_{[c,d]} < \gamma \tag{4.3.2}$$

for some absolute constants $\delta > 0$ and $\gamma > 0$. The result will now follow from Theorem 4.3.2. A proof that the above choice of (T_{n_i}) is possible is outlined in E.1. □

Theorem 4.3.3 has the following interesting corollary.

Corollary 4.3.4. *Suppose $(\alpha_k)_{k=1}^{\infty} \subset \mathbb{R} \setminus [-1,1]$ is a sequence of distinct real numbers. Then*

$$\text{span}\left\{1, \frac{1}{x-\alpha_1}, \frac{1}{x-\alpha_2}, \dots\right\}$$

is dense in $C[-1,1]$ if and only if

$$\sum_{k=1}^{\infty} \sqrt{\alpha_k^2 - 1} = \infty .$$

(Here, unlike in Section 3.5, $\sqrt{\alpha_k^2 - 1}$ denotes the principal square root of $\alpha_k^2 - 1$.)

Proof. A combination of Theorem 4.3.3 and Corollary 7.1.3 yields the *only if* part of the corollary.

The Chebyshev polynomials T_n (of the first kind) and U_n (of the second kind) for the Chebyshev space

$$\text{span}\left\{1, \frac{1}{x-\alpha_1}, \dots, \frac{1}{x-\alpha_n}\right\}$$

on $[-1,1]$ were introduced in Section 3.5. The properties of

$$\widetilde{T}_n(\theta) := T_n(\cos\theta) \qquad \text{and} \qquad \widetilde{U}_n(\theta) := U_n(\cos\theta)\sin\theta ,$$

established in Section 3.5, include

$$\|\widetilde{T}_n\|_{\mathbb{R}} = 1 \quad \text{and} \quad \|\widetilde{U}_n\|_{\mathbb{R}} = 1 , \tag{4.3.3}$$

(4.3.4) $$\widetilde{T}_n^2 + \widetilde{U}_n^2 = 1\,,$$

(4.3.5) $$(\widetilde{T}_n')^2 + (\widetilde{U}_n')^2 = \widetilde{B}_n^2\,,$$

(4.3.6) $$\widetilde{T}_n' = -\widetilde{B}_n \widetilde{U}_n\,,$$

and

(4.3.7) $$\widetilde{U}_n' = \widetilde{B}_n \widetilde{T}_n\,,$$

where

$$\widetilde{B}_n(\theta) := \sum_{k=1}^{n} \frac{\sqrt{\alpha_k^2 - 1}}{|\alpha_k - \cos\theta|}\,, \qquad \theta \in \mathbb{R}$$

($\sqrt{\alpha_k^2 - 1}$ denotes the principal square root of $\alpha_k^2 - 1$) and the identities hold on the real line. Suppose

$$\sum_{k=1}^{\infty} \sqrt{\alpha_k^2 - 1} = \infty\,.$$

Then

(4.3.8) $$\lim_{n\to\infty} \min_{\theta\in[\alpha,\beta]} B_n(\theta) = \infty\,, \qquad 0 < \alpha < \beta < \pi\,.$$

Assume that there is an interval $[a, b] \subset (-1, 1)$ such that

$$\sup_{n\in\mathbb{N}} \|T_n'\|_{[a,b]} < \infty\,.$$

Let $\alpha := \arccos b$ and $\beta := \arccos a$. Then

$$\sup_{n\in\mathbb{N}} \|\widetilde{T}_n'\|_{[\alpha,\beta]} < \infty\,.$$

It follows from properties (4.3.6) and (4.3.8) that

$$\lim_{n\to\infty} \|\widetilde{U}_n\|_{[\alpha,\beta]} = 0\,,$$

and hence, by property (4.3.4),

$$\lim_{n\to\infty} \|\widetilde{T}_n^2 - 1\|_{[\alpha,\beta]} = 0\,.$$

Thus, by properties (4.3.7) and (4.3.8),

$$\lim_{n\to\infty} \min_{\theta\in[\alpha,\beta]} |\widetilde{U}_n'(\theta)| = \infty\,,$$

that is,

$$\lim_{n\to\infty} |\widetilde{U}_n(\beta) - \widetilde{U}_n(\alpha)| = \infty\,,$$

which contradicts property (4.3.3). Hence

$$\sup_{n\in\mathbb{N}} \frac{\|T_n'\|_{[a,b]}}{\|T_n\|_{[-1,1]}} = \sup_{n\in\mathbb{N}} \|T_n'\|_{[a,b]} = \infty$$

for every $[a, b] \subset (-1, 1)$, which, together with Theorem 4.3.3 shows the *if* part of the corollary. □

Comments, Exercises, and Examples.

We followed Borwein and Erdélyi [95a] in this section. Corollary 4.3.4, to be found in Achiezer [56], is proved by using entirely different methods.

E.1 The Crucial Detail in the Proof of Theorem 4.3.3. Suppose that $\mathcal{M} := (1, f_1, f_2, \dots)$ is an infinite Markov system of C^2 functions on $[a, b]$ and f_1' does not vanish on (a, b). Suppose that the sequence of associated Chebyshev polynomials (T_n) has a subsequence (T_{n_i}) with no zeros on some subinterval $[\alpha, \beta]$ of $[a, b]$. Show that there exists another subinterval $[c, d]$ and another infinite subsequence (T_{n_i}) such that for some $\delta > 0$ and $\gamma > 0$, and for each n_i,

$$\|T_{n_i}\|_{[c,d]} < 1 - \delta \qquad \text{and} \qquad \|T'_{n_i}\|_{[c,d]} < \gamma\,.$$

Outline. For both inequalities first choose a subinterval $[c_1, d_1] \subset [\alpha, \beta]$ and a subsequence $(n_{i,1})$ of (n_i) such that each alternation point of each $T_{n_{i,1}}$ is outside $[c_1, d_1]$. Then choose a subsequence $(n_{i,2})$ of $(n_{i,1})$ so that either each $T_{n_{i,2}}$ is increasing or each $T_{n_{i,2}}$ is decreasing on $[c_1, d_1]$. Study the first case; the second is analogous. Let $[c_2, d_2]$ be the middle third of $[c_1, d_1]$. If the first inequality fails to hold with $[c_2, d_2]$ and $(n_{i,2})$, then there is a subsequence $(n_{i,3})$ of $(n_{i,2})$ such that $\|T_{n_{i,3}}\|_{[c_2,d_2]} \to 1$ as $n_{i,3} \to \infty$. Hence, there is a subsequence $(n_{i,4})$ of $(n_{i,3})$ such that either

$$\max_{c_2 \le x \le d_2} T_{n_{i,4}}(x) \to 1 \qquad \text{or} \qquad \min_{c_2 \le x \le d_2} T_{n_{i,4}}(x) \to -1\,.$$

Once again, study the first case; the second is analogous. Since each $T_{n_{i,3}}$ is increasing on $[c_1, d_1]$,

$$\lim_{n_{i,4} \to \infty} \|1 - T_{n_{i,4}}\|_{[d_2,d_1]} = 0\,.$$

Now choose $g := a_0 + a_1 f_1 + a_2 f_2$ so that g has two distinct zeros α_1 and α_2 in $[d_2, d_1]$, $\|g\|_{[\alpha_1,\alpha_2]} < 1$, and g is positive on (α_1, α_2). Let $\beta := \max_{\alpha_1 \le x \le \alpha_2} g(x)$ and $\widetilde{g} := g + 1 - \beta$. Show that $T_{n_{i,4}} - \widetilde{g}$ has at least $n+1$ distinct zeros in $[a, b]$ if $n_{i,4}$ is large enough, which is a contradiction.

For the second inequality, note that E.4 of Section 3.2 implies that $(f_1', \dots, f_n')$ is a weak Chebyshev system on $[a, b]$, and so is

$$\left(\left(\frac{T_2'}{T_1'}\right)', \left(\frac{T_3'}{T_1'}\right)', \dots, \left(\frac{T_n'}{T_1'}\right)' \right), \qquad n = 2, 3, \dots\,.$$

From this deduce that each $(T'_{n_{i,2}}/T_1')'$ has at most one sign change in $[c_2, d_2]$. Choose a subinterval $[c_3, d_3] \subset [c_2, d_2]$ and a subsequence $(n_{i,5})$ of

$(n_{i,2})$ so that none of $(T'_{n_{i,5}}/T'_1)'$ changes sign in $[c_3, d_3]$. Choose a subsequence $(n_{i,6})$ of $(n_{i,5})$ so that either each $T'_{n_{i,6}}/T'_1$ is increasing or each $T'_{n_{i,6}}/T'_1$ is decreasing on $[c_3, d_3]$. Again, consider the first case; the second is analogous. Let $[c_4, d_4]$ be the middle third of $[c_3, d_3]$. If the second inequality fails to hold with $[c_4, d_4]$ and $(n_{i,6})$, then there is a subsequence $(n_{i,7})$ of $(n_{i,6})$ such that either

$$\lim_{n_{i,7}\to\infty} \max_{c_4 \le x \le d_4} \frac{T'_{n_{i,7}}(x)}{T'_1(x)} = \infty$$

or

$$\lim_{n_{i,7}\to\infty} \min_{c_4 \le x \le d_4} \frac{T'_{n_{i,7}}(x)}{T'_1(x)} = -\infty\,.$$

Once again we just treat the first case; the second is analogous. In this case, for every $K > 0$ there is an $N \in \mathbb{N}$ such that for every $n_{i,7} \ge N$ we have

$$T'_{n_{i,7}}(x) > K\,, \qquad x \in [d_4, d_3]\,.$$

Hence

$$K(d_3 - d_4) \le \int_{d_4}^{d_3} T'_{n_{i,7}}(x)\,dx = T_{n_{i,7}}(d_3) - T_{n_{i,7}}(d_4) \le 2\,,$$

which is a contradiction. □

E.2 On the Uniform Closure of Nondense Markov Spaces. Suppose that $\mathcal{M} = (f_0, f_1, \dots)$ is an infinite Markov system on $[a,b]$ with each $f_i \in C^2[a,b]$, and suppose that $(f_1/f_0)'$ does not vanish on (a,b). Suppose that $\operatorname{span}\mathcal{M}$ fails to be dense in $C[a,b]$.

Show that there exists a subinterval $[\alpha,\beta] \subset [a,b]$, $\alpha < \beta$, such that every $g \in C[a,b]$ in the uniform closure of $\operatorname{span}\mathcal{M}$ on $[a,b]$ is differentiable on $[\alpha,\beta]$.

Hint: By Theorem 4.3.3 there exists an interval $[\alpha,\beta] \subset [a,b]$ and a constant $\eta \in \mathbb{R}$ so that

$$\|h'\|_{[\alpha,\beta]} \le \eta \|h\|_{[a,b]}$$

for every $h \in \operatorname{span}\mathcal{M}$. Suppose $g \in C[a,b]$ and

$$\lim_{n\to\infty} \|h_n - g\|_{[a,b]} = 0.$$

Choose $n_i \in \mathbb{N}$ such that

$$\|g - h_{n_i}\|_{[a,b]} \le 2^{-i}\,, \qquad i = 0, 1, \dots\,.$$

Then

$$g = h_{n_0} + \sum_{i=1}^{\infty} (h_{n_i} - h_{n_{i-1}})\,.$$

Since

$$\|(h_{n_i} - h_{n_{i-1}})'\|_{[\alpha,\beta]} \le \eta 2^{1-i}\,,$$

it follows that g is differentiable on $[\alpha,\beta]$. □

E.3 An Analog of Theorem 4.3.3 with Applications. Suppose that $\mathcal{M} = (f_0, f_1, \dots)$ is an extended complete Chebyshev (ECT) system of C^∞ functions on $[a, b]$, as in E.3 of Section 3.1. Note that E.3 a] of Section 3.1 implies that f_0 does not vanish on $[a, b]$.

a] Show that the differential operator D defined by

$$D(f) := \left(\frac{f}{f_0} \right)', \qquad f \in C^1[a, b]$$

maps $\mathcal{M}$ to $\mathcal{M}_D$, where

$$\mathcal{M}_D := \left(\left(\frac{f_1}{f_0} \right)', \left(\frac{f_2}{f_0} \right)', \dots \right)$$

and $\mathcal{M}_D$ is once again an ECT system of C^∞ functions on $[a, b]$.

Hint: Use E.8 b] of Section 3.2. □

b] The differential operators $D^{(n)}(f)$ are defined for every $f \in C^n[a, b]$ as follows. Let

$$F_{i,0} := f_i\,, \qquad i = 0, 1, 2, \dots\,,$$

$$F_{i,n} := \left(\frac{F_{i+1,n-1}}{F_{0,n-1}} \right)', \qquad i = 0, 1, 2, \dots\,, \qquad n = 1, 2, \dots\,,$$

$$D^{(0)}(f) := f, \quad D^{(n)}(f) := \left(\frac{D^{(n-1)}(f)}{F_{0,n-1}} \right)', \qquad n = 1, 2, \dots\,.$$

Let

$$\mathcal{M}_{D^{(1)}} := \mathcal{M}_D \qquad \text{and} \qquad \mathcal{M}_{D^{(n)}} := \left(\mathcal{M}_{D^{(n-1)}} \right)_D, \qquad n = 2, 3, \dots\,.$$

Show that if span $\mathcal{M}_{D^{(n)}}$ is dense in $C[a, b]$, then so is span $\mathcal{M}$.

c] Suppose that span $\mathcal{M}$ fails to be dense in $C[c, d]$ for every subinterval $[c, d] \subset [a, b]$, $c < d$. Show that for each $n \in \mathbb{N}$, there exists an interval $[\alpha_n, \beta_n] \subset [a, b]$, $\alpha_n < \beta_n$, such that

$$\sup \left\{ \frac{\|D^{(n)}(f)\|_{[\alpha_n, \beta_n]}}{\|f\|_{[a,b]}} : 0 \neq f \in \operatorname{span} \mathcal{M} \right\} < \infty\,.$$

Hint: Use Theorem 4.3.1 and induction on n. □

d] Suppose that span $\mathcal{M}$ fails to be dense in $C[c, d]$ for every subinterval $[c, d] \subset [a, b]$ $c < d$. Show that for each $n \in \mathbb{N}$, there exists an interval $[\alpha_n, \beta_n] \subset [a, b]$, $\alpha_n < \beta_n$, where every $g \in C[a, b]$ in the uniform closure of span $\mathcal{M}$ on $[a, b]$ is n times continuously differentiable.

Hint: Use part c]. The argument is similar to the one given in the hint to E.2. □

e] Suppose that span $\mathcal{M}$ fails to be dense in $C[c,d]$ for every subinterval $[c,d] \subset [a,b]$, $c < d$.

Show that every function in the uniform closure of span $\mathcal{M}$ on $[a,b]$ is C^∞ on a dense subset of $[a,b]$.

(This is the case for Müntz systems

$$\mathcal{M} := (x^{\lambda_0}, x^{\lambda_1}, \dots), \qquad \lambda_i \in \mathbb{R}, \qquad \sum_{\substack{i=1 \\ \lambda_i \neq 0}}^{\infty} \frac{1}{|\lambda_i|} < \infty$$

on $[a,b]$, $0 \le a < b$; see E.7 of Section 4.2.)

E.4 Bounded Bernstein-Type Inequality for Nondense Müntz Spaces. Suppose $(\lambda_i)_{i=1}^\infty$ is a sequence of distinct positive numbers satisfying

$$\sum_{i=1}^{\infty} \frac{\lambda_i}{\lambda_i^2 + 1} < \infty.$$

Then for every $\epsilon > 0$, there is a constant c_ϵ such that

$$|p'(x)| \le \frac{c_\epsilon}{x} \|p\|_{[0,1]}, \qquad x \in (0, 1-\epsilon]$$

for every

$$p \in \operatorname{span}\{1, x^{\lambda_1}, x^{\lambda_2}, \dots\}.$$

To prove this proceed as follows. Let $\lambda_0 := 0$, and let

$$T_n := T_n\{\lambda_0, \lambda_1, \dots, \lambda_n; [0,1]\}$$

be the Chebyshev polynomial for $\operatorname{span}\{1, x^{\lambda_1}, \dots, x^{\lambda_n}\}$ on $[0,1]$. Let

$$M(\Lambda) := \operatorname{span}\{1, x^{\lambda_1}, x^{\lambda_2}, \dots\}.$$

a] Observe that for every $\epsilon > 0$ there exists a $k_\epsilon \in \mathbb{N}$ depending only on $(\lambda_i)_{i=1}^\infty$ and ϵ (and not on n) such that T_n has at most k_ϵ zeros in $[\epsilon, 1-\epsilon]$.

Hint: This is proved in the outline of the proof of E.18 d] of Section 4.2. □

b] Show that every nonempty $(a,b) \subset (0,1)$ contains a nonempty subinterval (α, β) for which there are integers $0 < n_1 < n_2 < \cdots$ such that none of the Chebyshev polynomials T_{n_i} vanishes on (α, β).

Hint: Use part a]. □

c] Show that every nonempty $(a, b) \subset (0, 1)$ contains a nonempty subinterval (α, β) such that

$$\sup \left\{ \frac{\|p'\|_{[\alpha,\beta]}}{\|p\|_{[0,1]}} : \ 0 \neq p \in M(\Lambda) \right\} < \infty .$$

Hint: Modify the proof of Theorem 4.3.3. □

d] Finish the proof of the initial statement of the exercise.

Hint: Use part c] and a linear scaling. □

E.5 The Closure of Nondense Müntz Spaces in $C[0,1]$. Suppose $(\lambda_i)_{i=1}^{\infty}$ is a sequence of distinct positive numbers satisfying

$$\sum_{i=1}^{\infty} \frac{\lambda_i}{\lambda_i^2 + 1} < \infty .$$

Show that

$$\overline{\text{span}}\{1, x^{\lambda_1}, x^{\lambda_2}, \dots\} \subset C^{\infty}(0, 1) ,$$

that is, if f is the uniform limit of a sequence from $\text{span}\{1, x^{\lambda_1}, x^{\lambda_2}, \dots\}$, then f is infinitely many times differentiable on $(0, 1)$.

Hint: Use E.4 with the substitution $x = e^{-t}$. □

E.6 A Nondense Markov Space with Unbounded Bernstein Inequality on a Subinterval. One may incorrectly suspect that nondense Markov spaces on $[a, b]$ can be characterized by an *everywhere bounded* Bernstein inequality on (a, b), at least under the assumptions of Theorem 4.3.3. The purpose of this exercise is to show that this is far from true, and in a sense, Theorem 4.3.3 is the best possible result.

The same construction can be used to give a nondense Markov space on $[a, b]$ such that the set

$$Z := \{x \in [a, b] : T_n(x) = 0 \ \text{ for some } \ n \in \mathbb{N}\}$$

is neither dense nor nowhere dense in $[a, b]$. Here $(T_n)_{n=0}^{\infty}$ is the sequence of associated Chebyshev polynomials on $[a, b]$.

We construct an infinite Markov system on $(-\infty, \infty)$ as follows. Suppose $\Lambda := (\lambda_i)_{i=0}^{\infty}$ is a sequence of even integers satisfying

$$0 = \lambda_0 < \lambda_1 < \lambda_2 < \cdots , \qquad \sum_{i=1}^{\infty} \frac{1}{\lambda_i} < \infty .$$

Suppose $m > 0$. Let $\varphi_k \in C(-\infty, \infty)$, $k = 0, 1, \dots$, be defined by

$$\varphi_{2k}(x) := \begin{cases} x^{2k+m}, & x \geq 0 \\ c_{2k} x^{\lambda_{2k}}, & x \leq 0 \end{cases}$$

and

$$\varphi_{2k+1}(x) := \begin{cases} x^{2k+m+1}, & x \geq 0 \\ -c_{2k+1} x^{\lambda_{2k+1}}, & x \leq 0, \end{cases}$$

where $(c_i)_{i=0}^\infty$ is a sequence of positive numbers associated with a fixed sequence of integers $0 := n_0 < n_1 < n_2 < \cdots$ and a constant $\delta > 0$, and it is chosen as follows. Let

$$m_j := n_{j+1} - n_j - 1, \qquad j = 0, 1, \ldots.$$

Let

$$T_n(x) := \cos(n \arccos(2x-1)) =: \sum_{i=0}^n \alpha_{i,n} x^i$$

be the nth Chebyshev polynomial on $[0,2]$. Now choose the constants $c_i > 0$ such that

$$\sum_{j=0}^\infty \sum_{i=n_j}^{n_{j+1}-1} c_i \delta^{\lambda_i} |\alpha_{i-n_j, m_j}| \leq 1.$$

a] Show that $(\varphi_0, \varphi_1, \ldots)$ is a Markov system on $(-\infty, \infty)$.

Hint: If

$$p(x) = \sum_{i=0}^n a_i \varphi_i(x), \qquad a_i \in \mathbb{R},$$

then

$$p(x) = \sum_{i=0}^n a_i x^{\lambda_i}, \qquad x \in [0, \infty),$$

while

$$p(x) = \sum_{i=0}^n (-1)^i a_i x^{\lambda_i}, \qquad x \in (-\infty, 0].$$

Now apply Theorem 3.2.4 (Descartes' rule of signs). □

b] Show that $\operatorname{span}\{\varphi_0, \varphi_1, \ldots\}$ is not dense in $C[-\delta, 2]$.

c] Show that there is a sequence of integers $0 := n_0 < n_1 < n_2 < \cdots$ such that

$$\sup\left\{ \frac{\|p'\|_{[\alpha,\beta]}}{\|p\|_{[-\delta,2]}} : \quad 0 \neq p \in \operatorname{span}\{\varphi_0, \varphi_1, \ldots\} \right\} = \infty$$

for every nonempty interval $[\alpha, \beta] \subset [0, 2]$, while

$$\sup\left\{ \frac{|p'(x)|}{\|p\|_{[-\delta,2]}} : \quad 0 \neq p \in \operatorname{span}\{\varphi_0, \varphi_1, \ldots\} \right\} < \infty$$

for every $x \in (-\delta, 0)$.

d] Suppose the sequence $(\varphi_i)_{i=0}^\infty$ is defined associated with a fixed sequence of integers $0 := n_0 < n_1 < n_2 < \cdots$ and $\delta := -2$. Let

$$T_n := T_n\{\varphi_0, \varphi_1, \dots, \varphi_n; [-2,2]\}$$

be the Chebyshev polynomial for $\operatorname{span}\{\varphi_0, \varphi_1, \dots, \varphi_n\}$ on $[-2,2]$. Let

$$Z := \{x \in [-2,2] : T_n(x) = 0 \text{ for some } n \in \mathbb{N}\}$$

and let Z' be the set of all limit points of Z, and let Z'' be the set of all limit points of Z'. Show that the sequence of integers $0 := n_0 < n_1 < n_2 < \cdots$ in the definition of $(\varphi_i)_{i=0}^\infty$ can be chosen so that

$$Z'' \cap (-2,0) = \emptyset \qquad \text{and} \qquad [0,2] \cup \{-2\} \subset Z''.$$

E.7 Nikolskii-Type Inequalities for Nondense Müntz Spaces. Suppose that $p \in [1,\infty]$. Suppose $(\lambda_i)_{i=0}^\infty$ is a sequence of distinct real numbers greater than $-1/p$ satisfying

$$\sum_{i=0}^\infty \frac{\lambda_i + \frac{1}{p}}{\left(\lambda_i + \frac{1}{p}\right)^2 + 1} < \infty .$$

Show that for every $\epsilon > 0$, there exists a constant $c_\epsilon > 0$ depending only on ϵ and p so that

$$|q(x)| \le c_\epsilon x^{-1/p} \|q\|_{L_p[0,1]}$$

for every $q \in \operatorname{span}\{x^{\lambda_0}, x^{\lambda_1}, \dots\}$ and for every $x \in [0, 1-\epsilon]$.

In particular, for every $\epsilon > 0$, there exists a constant $c_\epsilon > 0$ depending only on ϵ so that

$$\|q\|_{[\epsilon,1-\epsilon]} \le c_\epsilon \|q\|_{L_p[0,1]}$$

for every $q \in \operatorname{span}\{x^{\lambda_0}, x^{\lambda_1}, \dots\}$.

Thus, $\operatorname{span}\{x^{\lambda_0}, x^{\lambda_1}, \dots\}$ is not dense in $L_p[0,1]$.

Hint: Use Hölder's inequality to show, as in Operstein [to appear], that the operator

$$J : L_p[0,1] \mapsto L_\infty[0,1]$$

defined by

$$J(\varphi)(0) := 0, \qquad J(\varphi)(x) := x^{1/p-1} \int_0^x \varphi(t)\,dt, \quad x > 0$$

is a bounded linear operator. That is, there is a constant $c \ge 0$ such that

$$\|J(\varphi)\|_{L_\infty[0,1]} \le c\,\|\varphi\|_{L_p[0,1]}$$

for every $\varphi \in L_p[0,1]$. Now observe that $q \in \operatorname{span}\{x^{\lambda_0}, x^{\lambda_1}, \dots\}$ implies that $J(q) \in \operatorname{span}\{x^{\mu_0}, x^{\mu_1}, \dots\}$, where $\mu_i := \lambda_i + \frac{1}{p}$, and so $(\mu_i)_{i=0}^\infty$ is a sequence of positive numbers satisfying

$$\sum_{i=0}^{\infty} \frac{\mu_i}{\mu_i^2 + 1} < \infty \,.$$

Therefore, by E.4, for every $\epsilon > 0$, there exists a constant $c_\epsilon > 0$ such that

$$|(J(q))'(x)| \leq \frac{c_\epsilon}{x} \, \|J(q)\|_{[0,1]} \leq \frac{c_\epsilon}{x} \, c \, \|q\|_{L_p[0,1]}$$

for every $q \in \operatorname{span}\{x^{\lambda_0}, x^{\lambda_1}, \dots\}$ and for every $x \in (0, 1-\epsilon]$. Note that for $x \in (0,1)$,

$$(J(q))'(x) = x^{1/p-1} q(x) - \left(1 - \tfrac{1}{p}\right) x^{1/p-2} \int_0^x q(t)\, dt \,,$$

where

$$\left| \int_0^x q(t)\, dt \right| = x^{1-1/p} |J(q)(x)| \leq c x^{1-1/p} \|q\|_{L_p[0,1]} \,.$$

Therefore

$$x^{1/p} |q(x)| - \left(1 - \tfrac{1}{p}\right) \|q\|_{L_p[0,1]} \leq c_\epsilon c \, \|q\|_{L_p[0,1]}$$

for every $q \in \operatorname{span}\{x^{\lambda_0}, x^{\lambda_1}, \dots\}$ and for every $x \in (0, 1-\epsilon]$. □

E.8 The Closure of Nondense Müntz Spaces in $L_p[0,1]$. Let $p \in [1, \infty]$. Suppose $(\lambda_i)_{i=0}^\infty$ is a sequence of distinct real numbers greater than $-1/p$ satisfying

$$\sum_{i=0}^{\infty} \frac{\lambda_i + \frac{1}{p}}{\left(\lambda_i + \frac{1}{p}\right)^2 + 1} < \infty \,.$$

Show that if f is a function in the $L_p[0,1]$ closure of

$$\operatorname{span}\{x^{\lambda_0}, x^{\lambda_1}, \dots\} \,,$$

then $f \in C^\infty(0,1)$, that is, f is infinitely many times differentiable on $(0,1)$.

Hint: Combine E.5 and E.7. □

4.4 Müntz Rationals

A surprising and beautiful theorem, conjectured by Newman and proved by Somorjai [76], states that rational functions derived from any infinite Müntz system are always dense in $C[a,b]$, $a \geq 0$. More specifically, we have the following result.

Theorem 4.4.1 (Denseness of Müntz Rationals). *Let $(\lambda_i)_{i=0}^{\infty}$ be any sequence of distinct real numbers. Suppose $a > 0$. Then*

$$\left\{ \frac{\sum_{i=0}^{n} a_i x^{\lambda_i}}{\sum_{i=0}^{n} b_i x^{\lambda_i}} : \; a_i, b_i \in \mathbb{R}, \;\; n \in \mathbb{N} \right\}$$

is dense in $C[a,b]$.

The same result holds when $a = 0$, however, the proof in this case requires a few more technical details; see E.1 b].

The proof of this theorem, primarily due to Somorjai, rests on the next theorem. We introduce the following notation. A function Z defined on (a,b) is called an *ϵ-zoomer* ($\epsilon > 0$) at $\zeta \in (a,b)$ if

$$(4.4.1) \qquad \begin{aligned} Z(x) &> 0\,, & x &\in [a,b]\,, \\ Z(x) &\leq \epsilon\,, & x &< \zeta - \epsilon\,, \\ Z(x) &\geq \epsilon^{-1}, & x &> \zeta + \epsilon\,. \end{aligned}$$

While (approximate) δ-functions are the building blocks for polynomial approximations, the existence of ϵ-zoomers is all that is needed for rational approximations. More precisely, we have the following result.

Theorem 4.4.2 (Existence of Zoomers and Denseness). *Let S be a linear subspace of $C[a,b]$. Suppose that S contains an ϵ-zoomer for every $\epsilon > 0$ at every $\zeta \in (a,b)$. Then*

$$R(S) := \left\{ \frac{p}{q} : p, q \in S \right\}$$

is dense in $C[a,b]$.

Proof. It suffices to consider the case $[a,b] = [0,1]$. Let $n \in \mathbb{N}$ and $\epsilon > 1/n$ be fixed. We construct a partition of unity inductively as follows. Let Z_0 be any positive function in S and choose functions $Z_1, Z_2, \ldots, Z_n \in S$ positive on $[0,1]$ so that

$$Z_k(x) < \epsilon Z_{k-1}(x)\,, \qquad x < \frac{k}{n} - \epsilon$$

and

$$Z_k(x) > \epsilon^{-1} Z_{k-1}(x)\,, \qquad x > \frac{k-1}{n} + \epsilon$$

(which the existence of ϵ-zoomers allows for). Let

$$\Delta_k(x) := \frac{Z_k(x)}{\sum_{i=0}^n Z_i(x)}, \qquad k = 0, 1, \ldots, n. \tag{4.4.2}$$

So

$$\Delta_k(x) > 0, \qquad x \in [0, 1]$$

and

$$\sum_{k=0}^{n} \Delta_k(x) = 1.$$

Since for every $x \in [0, 1]$,

$$\sum_{\substack{k=0 \\ \frac{k}{n} - \epsilon > x}}^{n} Z_k(x) + \sum_{\substack{k=0 \\ \frac{k}{n} + \epsilon < x}}^{n} Z_k(x) < 2\epsilon \sum_{k=0}^{n} Z_k(x),$$

we also know that

$$\sum_{\substack{k=0 \\ \left|\frac{k}{n} - x\right| > \epsilon}}^{n} \Delta_k(x) \le 2\epsilon, \qquad x \in [0, 1].$$

Now let $f \in C[0, 1]$, and consider the approximation

$$\sum_{k=0}^{n} f\left(\tfrac{k}{n}\right) \Delta_k(x) \in R(S).$$

Then

$$\begin{aligned}
&\left| f(x) - \sum_{k=0}^{n} f\left(\tfrac{k}{n}\right) \Delta_k(x) \right| \\
&\le \left| \sum_{\substack{k=0 \\ \left|\frac{k}{n} - x\right| \le \epsilon}}^{n} \left(f(x) - f\left(\tfrac{k}{n}\right)\right) \Delta_k(x) \right| + \left| \sum_{\substack{k=0 \\ \left|\frac{k}{n} - x\right| > \epsilon}}^{n} \left(f(x) - f\left(\tfrac{k}{n}\right)\right) \Delta_k(x) \right| \\
&\le \omega_f(\epsilon) + 2\epsilon\, \omega_f(1),
\end{aligned}$$

which finishes the proof. □

We can now finish the proof of Theorem 4.4.1:

Proof of Theorem 4.4.1. We may suppose, on passing to a subsequence if necessary, that $(\lambda_i)_{i=0}^{\infty}$ is a convergent sequence (possibly converging to infinity). We may also assume that $a := 1/b$ with $b > 1$. Since $C[1/b, b]$ is invariant under $x \to 1/x$, we may assume that $(\lambda_i)_{i=0}^\infty$ has a nonnegative (possibly infinite) limit.

Case 1: The sequence $(\lambda_i)_{i=0}^\infty$ has a finite nonnegative limit. Then Müntz's theorem on $[a,b]$, $a>0$ (see E.7 of Section 4.2), yields that the Müntz polynomials themselves associated with $(\lambda_i)_{i=0}^\infty$ are already dense in $C[1/b,b]$.

Case 2: The sequence $(\lambda_i)_{i=0}^\infty$ tends to infinity. In this case $(x/\zeta)^{\lambda_i}$ is an ϵ-zoomer at $\zeta \in (a,b)$ for sufficiently large λ_i, and the result follows from Theorem 4.4.2. □

Comments, Exercises, and Examples.

A comparison between Müntz's theorem and the main result of this section shows the power of a single division in approximation. In what other contexts does allowing a division create a spectacularly different result? Newman [78, p. 12] conjectures that if $\mathcal{M}$ is any infinite Markov system on $[0,1]$, then the set

$$\left\{\frac{p}{q} : p, q \in \operatorname{span} \mathcal{M}\right\}$$

of rational functions is dense in $C[0,1]$.

Newman calls this a "wild conjecture in search of a counterexample." It does, however, hold for both

$$\mathcal{M} = (x^{\lambda_0}, x^{\lambda_1}, \dots), \qquad \lambda_i \geq 0 \text{ are distinct}$$

(see E.1 b]) and

$$\mathcal{M} = \left(\frac{1}{x-\alpha_1}, \frac{1}{x-\alpha_2}, \dots\right), \qquad \alpha_i \in \mathbb{R} \setminus [0,1] \text{ are distinct}$$

(see E.2). A counterexample to the full generality of this conjecture is presented in E.6. However, the characterization of the class of Markov systems for which it holds remains as an interesting question. In particular, it is open if Newman's conjecture holds for Descartes systems.

The reader is referred to Newman [78] for an extensive treatment of these matters; see also Zhou [92a]. In [78, p. 50] Newman asks about the denseness of the products $(\sum a_i x^{i^2})(\sum b_i x^{i^2})$ in $C[0,1]$ (see E.3). He speculates that this "extra" multiplication of Müntz polynomials should not carry the utility of the "extra" division. This is proved in Section 6.2, where it is shown that products pq of Müntz polynomials from nondense Müntz spaces never form a dense set in $C[0,1]$.

E.1 Denseness of Müntz Rationals on $[0,1]$**.** A function C defined on $[a,b]$ is called an ϵ-*crasher*, $\epsilon > 0$, at $\zeta \in (a,b)$ if

$$\begin{aligned} C(x) &> 0\,, & x &\in [a,b]\,,\\ C(x) &\leq \epsilon\,, & x &> \zeta+\epsilon\,,\\ C(x) &\geq \epsilon^{-1}\,, & x &\leq \zeta-\epsilon\,. \end{aligned}$$

a] Let S be a linear subspace of $C[a,b]$. Suppose that S contains an ϵ-crasher for every $\epsilon > 0$ at every $\zeta \in (a,b)$. Show that

$$R(S) := \left\{ \frac{p}{q} : p, q \in S \right\}$$

is dense in $C[a,b]$.

Hint: The argument is a trivial modification of the proof of Theorem 4.4.2. □

Let $(\lambda_i)_{i=0}^\infty$ be a sequence of distinct real numbers. Let $R(\Lambda)$ be the space of functions $f \in C[0,1]$ of the form

$$f(x) = \frac{\sum_{i=0}^n a_i x^{\lambda_i}}{\sum_{i=0}^n b_i x^{\lambda_i}}, \qquad a_i, b_i \in \mathbb{R}, \quad n \in \mathbb{N}, \quad x \in (0,1].$$

b] Show that if $(\lambda_i)_{i=0}^\infty$ is a sequence of distinct nonnegative real numbers with $\lambda_0 := 0$, then $R(\Lambda)$ is dense in $C[0,1]$.

Hint: As in the proof of Theorem 4.4.1, we may suppose, on passing to a subsequence if necessary, that $(\lambda_i)_{i=1}^\infty$ is a convergent sequence (possibly converging to ∞). Distinguish the following two cases.

Case 1: $\lim_{i\to\infty} \lambda_i = \infty$. Given $\epsilon > 0$ and $\zeta \in (0,1)$, show that

$$Z(x) := \frac{\epsilon}{2} + \left(\frac{x}{\zeta}\right)^{\lambda_i}$$

is an ϵ-zoomer at ζ on $[0,1]$ if λ_i is large enough. Use Theorem 4.4.2 to finish the proof.

Case 2: $(\lambda_i)_{i=1}^\infty$ is a sequence with finite limit. Given $\epsilon > 0$ and $\zeta \in (0,1)$, show that

$$C(x) := \frac{\epsilon}{2} + (-1)^n \frac{\epsilon}{4} T_n\{\lambda_0, \lambda_1, \dots, \lambda_n; [\zeta, 1]\}(x)$$

is an ϵ-crasher at ζ on $[0,1]$ if n is large enough. This can be proved by using Müntz's theorem (E.7 of Section 4.2) on $[\zeta - \epsilon, 1]$ with $\epsilon \in (0, \zeta)$, E.3 c] of Section 3.3, and the monotonicity of C on $[0, \zeta]$. Finish the proof by part a]. □

The result of E.1 b] is due to Bak and Newman [78]. Zhou [92b] extends this result to sequences of arbitrary distinct real numbers.

E.2 Another Markov-System with Dense Rationals. Let $(\beta_i)_{i=1}^\infty$ be any sequence of distinct numbers in $\mathbb{R} \setminus [0,1]$. Show that

$$\left\{ \frac{\sum_{i=1}^n \frac{a_i}{x-\beta_i}}{\sum_{i=1}^n \frac{b_i}{x-\beta_i}} : \ a_i, b_i \in \mathbb{R}, \ n \in \mathbb{N} \right\}$$

is dense in $C[0,1]$.

E.3 Products of Müntz Spaces. Associated with a sequence $\Lambda := (\lambda_i)_{i=0}^\infty$ of real numbers, let

$$M^2(\Lambda) := \left\{ \left(\sum_{i=0}^n a_i x^{\lambda_i} \right) \left(\sum_{i=0}^n b_i x^{\lambda_i} \right) : \ a_i, b_i \in \mathbb{R}, \ n \in \mathbb{N} \right\}.$$

a] Show that if $(\lambda_k)_{k=0}^\infty = (k^\alpha)_{k=0}^\infty$, $\alpha > 2$, then $M^2(\Lambda)$ is not dense in $C[0,1]$.

b] Show that with $(\lambda_k)_{k=0}^\infty = (k^2)_{k=0}^\infty$ the nondenseness of $M^2(\Lambda)$ does not follow from Müntz's theorem since $\sum_{k \in \Gamma} 1/k = \infty$, where Γ is the set of natural numbers k of the form $k = n^2 + m^2$ with nonnegative integers n and m.

It is shown in Section 6.2 that $M^2(\Lambda)$ is not dense in $C[0,1]$ whenever Λ is a sequence of nonnegative real numbers satisfying $\sum_{k=1}^\infty 1/\lambda_k < \infty$, so the "extra multiplication" is of no spectacular utility.

It is not always possible to extend Theorem 4.4.2 to the case when the numerators and the denominators are coming from different infinite Müntz spaces. Somorjai [76] shows that

$$\left\{ \frac{\sum_{i=0}^n a_i x^{\lambda_i}}{\sum_{i=0}^n b_i x^{\delta_i}} : \ a_i, b_i \in \mathbb{R}, \ n \in \mathbb{N} \right\}$$

is not dense in $C[0,1]$ when, for example,

$$\eta^2 \lambda_k < \eta \delta_{k+1} < \lambda_{k+1}, \qquad k = 0, 1, \ldots$$

for some $\eta > 1$.

E.4 Nondense Ratios of Müntz Spaces. Suppose $0 \le \lambda_0 < \lambda_1 < \cdots$. Let $a > 0$. Show that

$$\left\{ \frac{\sum_{i=0}^n a_i x^{\lambda_i}}{\sum_{i=0}^n b_i x^{-\lambda_i}} : \ a_i, b_i \in \mathbb{R}, \ n \in \mathbb{N} \right\}$$

is dense in $C[a,b]$ if and only if $\sum_{i=1}^\infty 1/\lambda_i = \infty$.

Hint: For one direction use Müntz's theorem. For the other direction use E.5 a] and b] and E.8 b] and c] of Section 4.2. □

E.5 On the Rate of Approximation by Müntz Rationals. Let $(\lambda_i)_{i=0}^\infty$ be a fixed sequence of nonnegative real numbers. We wish to estimate the error of the best uniform approximation to $f \in C[0,1]$ on $[0,1]$ from $\mathrm{span}\{x^{\lambda_0}, \ldots, x^{\lambda_n}\}$. We let

$$R_n^*(f) := \inf \left\{ \left\| f(x) - \frac{\sum_{i=0}^n a_i x^{\lambda_i}}{\sum_{i=0}^n b_i x^{\lambda_i}} \right\|_{[0,1]} : \ a_i, b_i \in \mathbb{R} \right\},$$

where the infimum is taken for all $a_i, b_i \in \mathbb{R}$, $i = 0, 1, \ldots, n$.

In the case when $\lambda_{n+1} - \lambda_n \geq 1,\ n = 0, 1, \ldots$, Newman [78] claims (without proof) that there exists a constant C independent of n such that

$$R_n^*(f) = C\omega_f\left(\tfrac{1}{n}\right)$$

and that this is the best possible. He conjectures, in Newman [78], that this estimate holds for every sequence $(\lambda_i)_{i=0}^\infty$ of distinct nonnegative real numbers.

a] Observe in the last line of the proof of Theorem 4.4.2, with $\epsilon = \frac{2}{n}$, we have

$$\left| f(x) - \sum_{k=0}^{n} f\left(\tfrac{k}{n}\right) \Delta_k(x) \right| \leq 6\,\omega_f\left(\tfrac{1}{n}\right)$$

since $\omega_f(1) \leq n\omega_f\left(\frac{1}{n}\right)$.

b] What growth conditions on the sequence $(\lambda_i)_{i=0}^\infty$ guarantees that

$$R_n^*(f) = O\left(\omega_f\left(\tfrac{1}{n}\right)\right)?$$

Hint: Estimate the "degree" of the zoomers Z_k defined in Theorem 4.4.2. Use the zoomers defined in Case 1 of the hint for E.1 b]. □

c] Let

$$f(x) := \begin{cases} x\sin(1/x), & x \in \mathbb{R} \setminus \{0\} \\ 0, & x = 0. \end{cases}$$

Show that

$$R_n^*(f) \geq c_1 n^{-1} \geq c_2\,\omega_f\left(\tfrac{1}{n}\right)$$

for every sequence $(\lambda_i)_{i=0}^\infty$ of nonnegative real numbers, where c_1 and c_2 are positive constants independent of $(\lambda_i)_{i=0}^\infty$.

Hint: $\left(\sum_{i=0}^n a_i x^{\lambda_i}\right) / \left(\sum_{i=0}^n b_i x^{\lambda_i}\right)$ has at most n zeros on $[0, 1]$. □

E.6 A Markov System with Nondense Rationals. This example outlines a construction of Markov systems on $[-1, 1]$ whose rationals are not dense in $C[-1, 1]$. It follows and corrects Borwein and Shekhtman [93].

We construct an infinite Markov system on $(-\infty, \infty)$ as follows. Suppose that $\Lambda := (\lambda_i)_{i=0}^\infty$, where $2\lambda_i + 1 = 7^{2i}$. Then Λ is a sequence of even integers satisfying

$$0 = \lambda_0 < \lambda_1 < \lambda_2 < \cdots, \qquad \sum_{i=1}^\infty \frac{1}{\lambda_i} < \infty.$$

Let $\varphi_k \in C[-1, 1],\ k = 0, 1, \ldots$, be defined by

$$\varphi_{2k}(x) := x^{\lambda_{2k}}$$

and

$$\varphi_{2k+1}(x) := \begin{cases} x^{\lambda_{2k+1}}, & x \geq 0 \\ -x^{\lambda_{2k+1}}, & x \leq 0. \end{cases}$$

a] Show that $(\varphi_0, \varphi_1, \dots)$ is a Markov system on $(-\infty, \infty)$.

Hint: If

$$p(x) = \sum_{i=0}^{n} a_i \varphi_i(x), \qquad a_i \in \mathbb{R},$$

then

$$p(x) = \sum_{i=0}^{n} a_i x^{\lambda_i}, \qquad x \in [0, \infty),$$

while

$$p(x) = \sum_{i=0}^{n} (-1)^i a_i x^{\lambda_i}, \qquad x \in (-\infty, 0].$$

Now apply Theorem 3.2.4 (Descartes' rule of signs). □

b] There exists an absolute constant $c > 0$ (independent of n) such that

$$\left\| \sum_{i=0}^{n} a_i x^{\lambda_i} \right\|_{L_2[0,1]} \leq c \left\| \sum_{i=0}^{n} a_i x^{\lambda_i} \right\|_{L_2[1/2,1]}$$

for all choices of $a_i \in \mathbb{R}$.

Hint: Use E.8 a] of Section 4.2 and Hölder's inequality. □

c] The inequalities

$$\frac{1}{3} \sum_{i=0}^{n} |a_i|^2 \leq \int_0^1 \left(\sum_{i=0}^{n} a_i \sqrt{2\lambda_i + 1}\, x^{\lambda_i} \right)^2 dx \leq \frac{5}{3} \sum_{i=0}^{n} |a_i|^2$$

hold for all choices of $a_i \in \mathbb{R}$.

Proof. We have

$$\int_0^1 \left(\sum_{i=0}^{n} a_i \sqrt{2\lambda_i + 1}\, x^{\lambda_i} \right)^2 dx$$

$$= \sum_{i=0}^{n} |a_i|^2 + 2 \sum_{k=1}^{n} \sum_{j=0}^{n-k} a_j a_{j+k} \frac{\sqrt{2\lambda_j + 1}\sqrt{2\lambda_{j+k} + 1}}{\lambda_j + \lambda_{j+k} + 1}.$$

Here

$$\frac{\sqrt{2\lambda_j + 1}\sqrt{2\lambda_{j+k} + 1}}{\lambda_j + \lambda_{j+k} + 1} = 2\, \frac{7^j 7^{j+k}}{7^{2j} + 7^{2j+2k}} < 2\, \frac{7^{2j+k}}{7^{2j+2k}} = 2 \cdot 7^{-k},$$

so on applying the Cauchy-Schwarz inequality n times, we get

$$2\left|\sum_{k=1}^{n}\sum_{j=0}^{n-k} a_j a_{j+k} \frac{\sqrt{2\lambda_j+1}\sqrt{2\lambda_{j+k}+1}}{\lambda_j+\lambda_{j+k}+1}\right|$$
$$\leq 4\left(\sum_{k=1}^{n} 7^{-k}\right)\sum_{i=0}^{n}|a_i|^2 \leq \frac{2}{3}\sum_{i=0}^{n}|a_i|^2\,,$$

and the result follows. □

d] The inequality

$$\left\|\sum_{i=0}^{n} a_i x^{\lambda_i}\right\|_{L_2[0,1]} \leq \sqrt{5}\left\|\sum_{i=0}^{n}(-1)^i a_i x^{\lambda_i}\right\|_{L_2[0,1]}$$

holds for all choices of $a_i \in \mathbb{R}$.

Hint: Use part c]. □

e] The rational functions of the form

$$\frac{\sum_{i=0}^{n} a_i\varphi_i}{\sum_{i=0}^{n} b_i\varphi_i}\,, \quad a_i, b_i \in \mathbb{R}\,, \quad n \in \mathbb{N}$$

are not dense in $C[-1,1]$.

Proof. Consider $f \in C[-1,1]$ defined by

$$f(x) := \begin{cases} 1 & \text{if } \; x \in [-1,-1/2] \\ 0 & \text{if } \; x \in [0,1] \\ -2x & \text{if } \; x \in [-1/2,0]\,. \end{cases}$$

We show that f is not uniformly approximable on $[-1,1]$ by the above rational functions. Suppose that

$$\left\|\frac{\sum_{i=0}^{n} a_i\varphi_i}{\sum_{i=0}^{n} b_i\varphi_i} - f\right\|_{[-1,1]} < \epsilon < 1\,, \qquad a_i, b_i \in \mathbb{R}\,.$$

This implies

$$\left\|\frac{\sum_{i=0}^{n} a_i x^{\lambda_i}}{\sum_{i=0}^{n} b_i x^{\lambda_i}}\right\|_{[0,1]} < \epsilon \tag{4.4.3}$$

and

$$\left\|\frac{\sum_{i=0}^{n}(-1)^i a_i x^{\lambda_i}}{\sum_{i=0}^{n}(-1)^i b_i x^{\lambda_i}} - 1\right\|_{[1/2,1]} < \epsilon\,. \tag{4.4.4}$$

Without loss of generality we may assume that

$$\left\| \sum_{i=0}^{n} b_i x^{\lambda_i} \right\|_{L_2[0,1]} = 1\,. \tag{4.4.5}$$

From (4.4.3) and (4.4.5) it follows that

$$\left\| \sum_{i=0}^{n} a_i x^{\lambda_i} \right\|_{L_2[0,1]} < \epsilon\,. \tag{4.4.6}$$

Part d], together with (4.4.3) and (4.4.5), implies that

$$\left\| \sum_{i=0}^{n} (-1)^i a_i x^{\lambda_i} \right\|_{L_2[0,1]} < \sqrt{5}\,\epsilon\,. \tag{4.4.7}$$

Part d], together with (4.4.5), also implies that

$$\left\| \sum_{i=0}^{n} (-1)^i b_i x^{\lambda_i} \right\|_{L_2[0,1]} \geq \frac{1}{\sqrt{5}}\,. \tag{4.4.8}$$

Combining part b] and (4.4.8), we obtain that

$$\left\| \sum_{i=0}^{n} (-1)^i b_i x^{\lambda_i} \right\|_{L_2[1/2,1]} \geq \frac{1}{\sqrt{5}\,c}\,. \tag{4.4.9}$$

Now (4.4.7) and (4.4.9) yield the right-hand side of

$$1 - \epsilon \leq \frac{\|\sum_{i=0}^{n} (-1)^i a_i x^{\lambda_i}\|_{L_2[1/2,1]}}{\|\sum_{i=0}^{n} (-1)^i b_i x^{\lambda_i}\|_{L_2[1/2,1]}} \leq 5c\epsilon$$

while (4.4.4) yields the left-hand side of it. This shows that $\epsilon > 0$ cannot be arbitrarily small. □

5

Basic Inequalities

Overview

The classical inequalities for algebraic and trigonometric polynomials are treated in the first section. These include the inequalities of Remez, Bernstein, Markov, and Schur. The second section deals with Markov's and Bernstein's inequalities for higher derivatives. The final section is concerned with the size of factors of polynomials.

5.1 Classical Polynomial Inequalities

We start with the classical inequalities of Remez, Bernstein, Markov, and Schur. The most basic and general of these is probably due to Remez. How large can $\|p\|_{[-1,1]}$ be if $p \in \mathcal{P}_n$ and

$$m(\{x \in [-1,1] : |p(x)| \leq 1\}) \geq 2 - s$$

holds? The inequality of Remez [36] answers this question. His inequality and its trigonometric analog can be extended to generalized nonnegative polynomials (discussed in Appendix 4) by a simple density argument. These extensions also play a central role in the proof of various other Bernstein, Markov, Nikolskii, and Schur type inequalities for generalized nonnegative polynomials, where simple density arguments do not work.

Theorem 5.1.1 (Remez Inequality). *The inequality*

$$\|p\|_{[-1,1]} \le T_n\left(\frac{2+s}{2-s}\right)$$

holds for every $p \in \mathcal{P}_n$ and $s \in (0,2)$ satisfying

$$m\left(\{x \in [-1,1] : |p(x)| \le 1\}\right) \ge 2 - s\,.$$

Here T_n is the Chebyshev polynomial of degree n defined by (2.1.1). Equality holds if and only if

$$p(x) = \pm T_n\left(\frac{\pm 2x + s}{2-s}\right).$$

Proof. The proof is essentially a perturbation argument that establishes that an extremal polynomial is of the required form. Let $\|\cdot\|$ denote the uniform norm on $[-1,1]$, and let

$$\mathcal{P}_n(s) := \left\{p \in \mathcal{P}_n : m\left(\{x \in [-1,1] : |p(x)| \le 1\}\right) \ge 2 - s\right\}. \tag{5.1.1}$$

The set $\mathcal{P}_n(s)$ is compact, say, in the uniform norm on $[-1,1]$, by E.1, and the function $p \to \|p\|$ is continuous. Hence there is a $p^* \in \mathcal{P}_n(s)$ such that

$$\|p^*\| = \sup_{p \in \mathcal{P}_n(s)} \|p\|\,.$$

First assume that $p^*(1) = \|p^*\|$. We claim that all the zeros of p^* are real and lie in $[-1,1)$. Indeed, if p^* vanishes at a nonreal z, then for sufficiently small $\eta > 0$ and $\epsilon > 0$,

$$q(x) := (1+\eta)p^*(x)\left(1 - \frac{\epsilon(x-1)^2}{(x-z)(x-\overline{z})}\right)$$

is in $\mathcal{P}_n(s)$ and contradicts the maximality of p^*. If p^* has a real zero z outside $[-1,1]$, then, in similar fashion,

$$q(x) := (1+\eta)p^*(x)\left(1 - \epsilon\ \mathrm{sign}(z)\frac{1-x}{z-x}\right)$$

contradicts the maximality of p^*.

Now we show that $|p^*(\zeta)| = \|p^*\|$ cannot occur with $\zeta \in (-1,1)$. To see this, assume, without loss of generality, that $p^*(\zeta) = \|p^*\|$. Then the polynomials

$$q_1(x) := p^*\left(\frac{\zeta-1}{2} + \frac{\zeta+1}{2}x\right) \quad \text{and} \quad q_2(x) := p^*\left(\frac{\zeta+1}{2} + \frac{\zeta-1}{2}x\right)$$

satisfy $q_j(1) = \|q_j\| = \|p^*\|$, $j = 1, 2$. Since $p^* \in \mathcal{P}_n(s)$ is extremal, the Lebesgue measure $m(\Omega_j)$ of

$$\Omega_j := \{x \in [-1,1] : |q_j(x)| > 1\}$$

is at least s, otherwise $q_j + \epsilon$ with sufficiently small $\epsilon > 0$ contradicts the maximality of p^*. On the other hand,

$$\frac{1+\zeta}{2} m(\Omega_1) + \frac{1-\zeta}{2} m(\Omega_2) = m(\{x \in [-1,1] : p^*(x) > 1\}) \leq s\,.$$

Hence $m(\Omega_1) = m(\Omega_2) = s$, which means that $q_j \in \mathcal{P}_n(s)$, $j = 1, 2$, are extremal polynomials attaining their uniform norm on $[-1, 1]$ at 1. It now follows from the first part of the proof that q_1 and q_2 have all their zeros in $[-1, 1]$, which is impossible since the number of zeros of q_j, $j = 1, 2$, in $[-1, 1]$ is equal to the number of zeros of p^* in $[-1, \zeta)$ and in $(\zeta, 1]$, respectively. This contradiction proves that $|p^*(\zeta)| < \|p^*\|$ for every $\zeta \in (-1, 1)$. Hence either $p^*(1) = \pm\|p^*\|$ or $p^*(-1) = \pm\|p^*\|$.

Without loss of generality we may assume that $p^*(1) = \|p^*\|$, otherwise we consider $\pm p^*(-x) \in \mathcal{P}_n(s)$. We now have

$$\text{(5.1.2)} \qquad p^*(1) = \|p^*\| > |p^*(x)|\,, \quad -1 < x < 1\,,$$

and each zero of p^* lies in $[-1, 1)$.

Next we prove that

$$\text{(5.1.3)} \qquad |p^*(x)| \leq 1\,, \qquad -1 \leq x \leq 1 - s\,.$$

Assume to the contrary that for some $-1 \leq \zeta_3 < \zeta_2 < \zeta_1 < 1$,

$$\begin{cases} |p^*(x)| > 1\,, & x \in I_1 := (\zeta_1, 1] \\ |p^*(x)| \leq 1\,, & x \in I_2 := [\zeta_2, \zeta_1] \\ |p^*(x)| > 1\,, & x \in I_3 := (\zeta_3, \zeta_2)\,. \end{cases}$$

Let $x_1, x_2, \dots, x_m$ be the zeros of p^* in (ζ_2, ζ_1). Since all zeros of p^* are in $[-1, 1)$, we have $m \geq 1$, otherwise $p^{*\prime}$ would vanish at an x larger than the largest zero of p^*, which is a contradiction. The remaining $n - m$ zeros of p^* lie in $[-1, \zeta_3)$. We set

$$p_1(x) := \prod_{j=1}^{m} (x - x_j)\,, \qquad p_2 := \frac{p^*}{p_1}\,.$$

The polynomial $q(x) := p_1(x + h)p_2(x)$ with $0 < h < \zeta_2 - \zeta_3$ has the following properties:

(1) If $|p^*(x)| \le 1$ for some $x \in [-1, \zeta_3]$, then $|q(x)| \le 1$.

(2) For each $x \in I_2$ we have $|q(x-h)| \le |p^*(x)| \le 1$.

(3) $q(1) = p_1(1+h)p_2(1) > p_1(1)p_2(1) = p^*(1) = \|p^*\|$.

These properties show that $q \in \mathcal{P}_n(s)$ contradicts the maximality of p^*.

By E.2, among all polynomials $p \in \mathcal{P}_n$ with $\|p\| \le 1$, the Chebyshev polynomial T_n increases fastest for $x > 1$. Hence, by a linear transformation, we see that the four polynomials

$$p^*(x) = \pm T_n\left(\frac{\pm 2x + s}{2-s}\right)$$

are the only extremal polynomials. In particular,

$$\|p^*\| = T_n\left(\frac{2+s}{2-s}\right).$$

□

The next theorem establishes a Remez-type inequality for trigonometric polynomials. Throughout this section, as before, $K := \mathbb{R} \pmod{2\pi}$.

Theorem 5.1.2. *The inequality*

$$\|t\|_K \le \exp(4ns)$$

holds for every $t \in \mathcal{T}_n$ and $s \in (0, \pi/2]$ satisfying

$$m(\{\theta \in [-\pi, \pi) : |t(\theta)| \le 1\}) \ge 2\pi - s\,. \tag{5.1.4}$$

The inequality

$$\|t\|_K \le T_n\left(\frac{2+\sigma}{2-\sigma}\right), \qquad \sigma = 1 - \cos(s/2)\,, \quad 0 < s < 2\pi\,,$$

also holds for every even $t \in \mathcal{T}_n$ and $s \in (0, 2\pi)$ satisfying (5.1.4), and equality holds if and only if

$$t(\theta) = \pm T_n\left(\frac{\pm 2\cos\theta + \sigma}{2-\sigma}\right), \qquad \sigma = 1 - \cos(s/2)\,.$$

Proof. We prove the second part first. Suppose $t \in \mathcal{T}_n$ is even and satisfies $|t(\theta)| \le 1$ on $K \setminus \Omega$ with $m(\Omega) \le s$. Then the polynomial $p \in \mathcal{P}_n$ defined by $t(\theta) := p(\cos\theta)$ satisfies $|p(x)| \le 1$ on $[-1,1] \setminus \Omega'$, where

$$\Omega' := \{x \in [-1,1] : x = \cos\theta\,,\ \theta \in \Omega \cap [0,\pi]\}\,.$$

It is easy to see that $m(\Omega') \le 1 - \cos(s/2) =: \sigma$, where equality holds if and only if $\Omega := [-s/2, s/2]$. Hence, Theorem 5.1.1 implies that

$$\|t\|_K \le T_n\left(\frac{2+\sigma}{2-\sigma}\right),$$

where equality holds if and only if p is of the form given in the theorem.

The first part of the theorem can be easily obtained from the second part as follows. Let $t \in \mathcal{T}_n$ satisfy (5.1.4). Without loss of generality we may assume that $t(0) = \|t\|_K$. The polynomial

$$\widetilde{t}(\theta) := \tfrac{1}{2}(t(\theta) + t(-\theta)) \in \mathcal{T}_n$$

is even and

$$m(\{\theta \in [-\pi,\pi) : |\widetilde{t}(\theta)| \le 1\}) \ge 2\pi - 2s\,.$$

Hence, the second part of the theorem yields that

$$\|\widetilde{t}\|_{L_\infty(K)} \le T_n\left(\frac{2+\sigma}{2-\sigma}\right),$$

where $\sigma := 1 - \cos s = 2\sin^2(s/2) \le 1$ for every $s \in (0, \pi/2]$. Since

$$T_n(x) \le \left(x + \sqrt{x^2-1}\right)^n, \qquad x \ge 1\,,$$

we have

$$\begin{aligned}
T_n\left(\frac{2+\sigma}{2-\sigma}\right) &\le \left(\frac{1+\sqrt{2\sigma}+\sigma/2}{1-\sigma/2}\right)^n \le \left(1+\sqrt{2\sigma}+\tfrac{7}{2}\sigma\right)^n \\
&\le \exp\left(n\left(\sqrt{2\sigma}+\tfrac{7}{4}\sigma\right)\right) \le \exp\left(n\left(s+\tfrac{7}{4}s^2\right)\right) \\
&\le \exp(4ns)
\end{aligned}$$

for every $s \in (0, \pi/2]$. Therefore

$$\|t\|_K = t(0) = \widetilde{t}(0) \le T_n\left(\frac{2+\sigma}{2-\sigma}\right) \le \exp\left(n\left(s+\tfrac{7}{4}s^2\right)\right) \le \exp(4ns)$$

for every $s \in (0, \pi/2]$, and the theorem is proved. Note that we have proved slightly more than we claimed in the statement of the theorem. That is,

$$\|t\|_K \le \exp\left(n\left(s+\tfrac{7}{4}s^2\right)\right)$$

for every $t \in \mathcal{T}_n$ satisfying (5.1.4). $\square$

We now prove the basic inequality that bounds the derivative of a trigonometric polynomial in terms of its maximum modulus on the period K.

Theorem 5.1.3 (Bernstein-Szegő Inequality). *The inequality*

$$t'(\theta)^2 + n^2 t^2(\theta) \le n^2 \|t\|_K^2\,, \qquad \theta \in K$$

holds for every $t \in \mathcal{T}_n$. Equality holds if and only if $|t(\theta)| = \|t\|_K$ or t is of the form $t(\tau) = \beta\cos(n\tau - \alpha)$ with $\alpha, \beta \in \mathbb{R}$.

Proof. Assume that there are $t \in \mathcal{T}_n$ and $\theta \in K$ such that $\|t\|_K < 1$ and

$$t'(\theta)^2 + n^2 t^2(\theta) > n^2\,. \tag{5.1.5}$$

For the sake of brevity let $T_{n,\alpha}(\tau) := \cos(n\tau - \alpha)$. It is easy to see that there exists an $\alpha \in K$ such that

$$T_{n,\alpha}(\theta) = t(\theta) \qquad \text{and} \qquad \text{sign}(T'_{n,\alpha}(\theta)) = \text{sign}(t'(\theta))\,. \tag{5.1.6}$$

Since $T'_{n,\alpha}(\theta)^2 + n^2 T^2_{n,\alpha}(\theta) = n^2$, (5.1.5) and (5.1.6) imply that

$$|t'(\theta)| > |T'_{n,\alpha}(\theta)| \qquad \text{and} \qquad \text{sign}(T'_{n,\alpha}(\theta) = \text{sign}(t'(\theta))\,.$$

Hence E.4 yields that $0 \ne t - T_{n,\alpha} \in \mathcal{T}_n$ has at least $2n+2$ distinct zeros in K, which is a contradiction. To find all the extremal polynomials, see the hint to E.5. □

As a corollary of Theorem 5.1.3 we have $\|t'\|_K \le n\|t\|_K$ for every $t \in \mathcal{T}_n$, and by induction on m we obtain the following theorem:

Theorem 5.1.4 (Bernstein's Inequality). *The inequality*

$$\|t^{(m)}\|_K \le n^m \|t\|_K$$

holds for every $t \in \mathcal{T}_n$.

Corollary 5.1.5. *The inequality of Theorem 5.1.4 remains true for all $t \in \mathcal{T}_n^c$.*

Proof. Choose $\alpha \in \mathbb{R}$ such that $e^{i\alpha}t^{(m)}$ attains the value $\|t^{(m)}\|_K$, say, at $\theta = \tau$. Now $\widetilde{t}(\theta) := \text{Re}(e^{i\alpha}t(\theta)) \in \mathcal{T}_n$ and $\|\widetilde{t}\|_K \le \|t\|_K$. On applying Theorem 5.1.4 to $\widetilde{t} \in \mathcal{T}_n$, we obtain

$$\|t^{(m)}\|_K = e^{i\alpha}t^{(m)}(\tau) = \widetilde{t}^{(m)}(\tau) \le n^m \|\widetilde{t}\|_K \le n^m \|t\|_K\,.$$

□

The above corollary implies the following algebraic polynomial version of Bernstein's inequality on the unit disk.

Corollary 5.1.6. *The inequality*

$$\|p'\|_D \le n\,\|p\|_D$$

holds for every $p \in \mathcal{P}_n^c$*, where* $D := \{z \in \mathbb{C} : |z| < 1\}$.

Proof. If $p \in \mathcal{P}_n^c$ then $t(\theta) := p(e^{i\tau}) \in \mathcal{T}_n^c$ and by Corollary 5.1.5 we have

$$|p'(z)| = |-ie^{-i\tau}t'(\tau)| \le n\|t\|_K = n\|p\|_D\,, \qquad z = e^{i\tau}\,.$$

The maximum principle (see E.1 d] of Section 1.2) finishes the proof. □

From Corollary 5.1.5, by the substitution $x = \cos\tau$, we get the algebraic polynomial case of Bernstein's inequality on $[-1,1]$.

Theorem 5.1.7 (Bernstein's Inequality). *The inequality*

$$|p'(x)| \le \frac{n}{\sqrt{1-x^2}}\,\|p\|_{[-1,1]}\,, \qquad -1 < x < 1$$

holds for every $p \in \mathcal{P}_n^c$.

The next theorem improves the previous result if x is close to ± 1.

Theorem 5.1.8 (Markov's Inequality). *The inequality*

$$\|p'\|_{[-1,1]} \le n^2\,\|p\|_{[-1,1]}$$

holds for every $p \in \mathcal{P}_n$.

A proof can be given as a simple combination of Theorem 5.1.7 and the next theorem.

Theorem 5.1.9 (Schur's Inequality). *The inequality*

$$\|p\|_{[-1,1]} \le n\,\left\|p(x)\sqrt{1-x^2}\right\|_{[-1,1]}$$

holds for every $p \in \mathcal{P}_{n-1}$.

Proof. Let

$$x_k := \cos\tfrac{(2k-1)\pi}{2n}\,, \qquad k = 1, 2, \dots, n\,,$$

so the numbers x_k are the zeros of the Chebyshev polynomial

$$T_n(x) = \cos(n \arccos x)\,.$$

Assume that $p \in \mathcal{P}_{n-1}$ and $\|p(x)\sqrt{1-x^2}\|_{[-1,1]} \le 1$. If $|y| \le x_n$, then

$$\begin{aligned}|p(y)| &\le (1-y^2)^{-1/2} \le (1-x_n^2)^{-1/2} \\ &= \left(\sin\frac{\pi}{2n}\right)^{-1} \le \left(\frac{2}{\pi}\frac{\pi}{2n}\right)^{-1} = n\,.\end{aligned}$$

Now let $|y| \in (x_n, 1]$. Without loss of generality we may assume that $y \in (x_n, 1]$. The Lagrange interpolation polynomial of a $p \in \mathcal{P}_{n-1}$ with nodes $x_1, x_2, \dots, x_n$ is just p itself, hence E.6 of Section 1.1 yields

$$\begin{aligned}|p(y)| &= \left|\sum_{k=1}^n p(x_k)\frac{T_n(y)}{T_n'(x_k)(y-x_k)}\right| \\ &= \frac{1}{n}\left|\sum_{k=1}^n p(x_k)\sqrt{1-x_k^2}\,\frac{T_n(y)}{y-x_k}\right| \\ &\le \frac{1}{n}\sum_{k=1}^n \frac{T_n(y)}{y-x_k} \le \frac{1}{n}T_n'(y) \le \frac{1}{n}T_n'(1) = n\,,\end{aligned}$$

where we use the facts that

$$\frac{T_n(y)}{y-x_k} > 0\,, \qquad k = 1, 2, \dots, n\,,$$

T_n' is increasing on (x_n, ∞), and $T_n'(1) = n^2$. □

Proof of Theorem 5.1.8. Let $p \in \mathcal{P}_n$. Then $p' \in \mathcal{P}_{n-1}$ and Theorems 5.1.4 and 5.1.9 yield

$$\|p'\|_{[-1,1]} \le n\left\|p'(x)\sqrt{1-x^2}\right\|_{[-1,1]} \le n^2\|p\|_{[-1,1]}\,,$$

and the theorem is proved. □

Comments, Exercises, and Examples.

Every result of this section was proved by the person it is named after; see Remez [36], Bernstein [12], Szegő [28] or [82], A. A. Markov [1889], Schur [19], and M. Riesz [14]. However, earlier less complete versions of these basic polynomial inequalities also appear in the literature. Some of the proofs were simplified later. For example, in the proof of Theorem 5.1.1 we followed the method given in Erdélyi [89b], while in the proof of Theorem 5.1.8 a method of Pólya and Szegő [76] is used. Theorem 5.1.3 is also obtained in Corput and Schaake [35], however, it follows from an earlier result of Szegő [28]. Theorem 5.1.2 was established in Erdélyi [92a] in a slightly weaker form. Various extensions of the inequalities of this section are discussed in Sections 5.2 to 7.2 and in the appendices. Rahman and Schmeisser [83] also offers a collection of Markov- and Bernstein-type inequalities. Some further classical inequalities are in E.5 of Appendix 3.

An interesting extension of Markov's inequality is due to Bojanov [82a]. It states that

$$\|p'\|_{L_q[-1,1]} \le \|T_n'\|_{L_q[-1,1]} \|p\|_{[-1,1]}$$

for every $p \in \mathcal{P}_n$ and $q \in [1, \infty]$, where T_n is the Chebyshev polynomial of degree n as in (2.1.1).

The following result is due to Szegő [25]. The inequality

$$|p'(0)| \le cn^{2\alpha} \|p\|_{D_\alpha}$$

holds for every $p \in \mathcal{P}_n^c$, where c is an absolute constant and

$$D_\alpha := \{z \in \mathbb{C} : |z| \le 1, \ |\arg(z)| \le \pi(1-\alpha)\}, \qquad \alpha \in (0,1].$$

Throughout the exercises T_n denotes the Chebyshev polynomial of degree n as defined by (2.1.1).

E.1 A Detail in the Proof of Theorem 5.1.1. Show that the sets $\mathcal{P}_n(s)$ defined by (5.1.1) are compact in the uniform norm on $[-1,1]$ for every fixed $n \in \mathbb{N}$ and $s \in (0,2)$.

E.2 Chebyshev's Inequality. Prove that

$$|p(y)| \le |T_n(y)| \cdot \|p\|_{[-1,1]}, \qquad y \in \mathbb{R} \setminus [-1,1]$$

for every $p \in \mathcal{P}_n$, and equality holds if and only if $p = cT_n$ for some $c \in \mathbb{R}$.

Extend the above inequality to every $p \in \mathcal{P}_n^c$ and find all $p \in \mathcal{P}_n^c$ for which equality holds.

Hint: If $p \in \mathcal{P}_n$, $\|p\|_{[-1,1]} = 1$, and $|p(y)| > |T_n(y)|$ for some $y \in \mathbb{R} \setminus [-1,1]$, then with $\lambda := T_n(y)p(y)^{-1} \in [-1,1]$, the polynomial $\lambda p - T_n \in \mathcal{P}_n$ has at least $n+1$ zeros (counting multiplicities). □

E.3 Trigonometric Chebyshev Polynomials on Subintervals of K.

a] For $n \in \mathbb{N}$ and $\omega \in (0, \pi)$, let

$$Q_{n,\omega}(\theta) := T_{2n}\left(\frac{\sin(\theta/2)}{\sin(\omega/2)}\right).$$

Show that $Q_{n,\omega} \in \mathcal{T}_n$ attains the values $\pm\|Q_{n,\omega}\|_{[-\omega,\omega]} = \pm 1$ with alternating sign $2n+1$ times on $[-\omega, \omega]$.

b] Prove that

$$|t(\theta)| \le Q_{n,\omega}(\theta)\|t\|_{[-\omega,\omega]}, \qquad \theta \in K \setminus [-\omega, \omega]$$

for every $t \in \mathcal{T}_n$, and for every fixed $\theta \in K \setminus [-\omega, \omega]$. Equality holds if and only if $p = cQ_{n,\omega}$ for some $c \in \mathbb{R}$.

c] Show that there exist absolute constants $c_1 > 0$ and $c_2 > 0$ such that

$$\tfrac{1}{2}\exp(c_1 n(\pi - \omega)) \le \|Q_{n,\omega}\|_K = Q_{n,\omega}(\pi) \le \exp(c_2 n(\pi - \omega)).$$

E.4 A Zero Counting Lemma. *Let $\theta \in K$ be fixed. Suppose f and g are continuous functions on $K := \mathbb{R} \pmod{2\pi}$, differentiable at a fixed $\theta \in K$, and suppose f and g have the properties*

(1) $\|f\|_K = 1, \quad \|g\|_K < 1,$

(2) *there are* $-\pi \le x_1 < x_2 < \cdots < x_{2n} < \pi$ *so that* $f(x_j) = (-1)^j$,

(3) $f(\theta)) = g(\theta), \quad |g'(\theta)| > |f'(\theta)|, \quad$ *and* $\quad \operatorname{sign}(g'(\theta) = \operatorname{sign}(f'(\theta)).$

Then $f - g$ has at least $2n + 2$ distinct zeros on K.

E.5 Sharpness of Theorem 5.1.3. Let $\theta \in K$ be fixed. Suppose $t \in \mathcal{T}_n$ and

$$t'(\theta)^2 + n^2 t^2(\theta) = n^2 \|t\|_K^2 \,.$$

Show that either $|t(\theta)| = \|t\|_K$ or t is of the form

$$t(\tau) = \beta \cos(n\tau - \alpha)\,, \qquad \alpha, \beta \in \mathbb{R}\,.$$

Hint: Suppose $\|t\|_K = 1$ and $|t(\theta)| < 1$. Choose $\alpha \in K$ such that (5.1.6) is satisfied and show that $t - T_{n,\alpha} \in \mathcal{T}_n$ has at least $2n + 2$ zeros on K (counting multiplicities). □

E.6 Sharpness of Theorem 5.1.4. Show that Theorem 5.1.4 is sharp and equality holds if and only if t is of the form

$$t(\tau) = \beta \cos(n\tau - \alpha)\,, \qquad \alpha, \beta \in \mathbb{R}\,.$$

E.7 Sharpness of Corollary 5.1.6. Show that Corollary 5.1.5 is sharp and equality holds if and only if p is of the form $p(z) = cz^n, \ \ c \in \mathbb{C}$.

E.8 Sharpness of Theorem 5.1.7. Show that for a fixed integer $n \ge 1$, Theorem 5.1.7 is sharp if and only if x is a zero of the Chebyshev polynomial T_n, that is,

$$x = \cos \frac{(2k-1)\pi}{2n}\,, \qquad k = 1, 2, \ldots, n\,,$$

and $p = cT_n$ for some $c \in \mathbb{R}$.

E.9 Sharpness of Theorem 5.1.9. Show that Theorem 5.1.9 is sharp and equality holds if and only if $p = cU_n$ for some $c \in \mathbb{R}$, where U_n is the Chebyshev polynomial of the second kind defined in E.10 of Section 2.1.

E.10 Sharpness of Theorem 5.1.8. Show that Theorem 5.1.8 is sharp and equality holds if and only if $p = cT_n$ for some $c \in \mathbb{R}$.

E.11 A Property of the Zeros of $t \in \mathcal{T}_n$. Let $\theta \in K$ be fixed. Show that every $t \in \mathcal{T}_n$ has at most

$$M := enr|t(\theta)|^{-1} \|t\|_K$$

zeros (counting multiplicities) in the interval $[\theta - r, \theta + r], \ \ r > 0$.

Proof. Assume that $t \in \mathcal{T}_n$ has $m > M$ zeros in $[\theta - r, \theta + r]$. Interpolate t at these m zeros by a Hermite interpolation polynomial of degree at most $m-1$ (see E.7 of Section 1.1). This gives the identically zero polynomial. The formula for the remainder term of the Hermite interpolation polynomial and Theorem 5.1.4 (Bernstein's inequality) yield that there exists a $\zeta \in (\theta - r, \theta + r)$ such that

$$\begin{aligned} |t(\theta)| &= \frac{1}{m!} r^m |t^{(m)}(\zeta)| < \left(\frac{e}{m}\right)^m r^m n^m \|t\|_K \\ &< \left(\frac{enr}{M}\right)^m \|t\|_K \leq |t(\theta)|^m \|t\|_K^{1-m} \leq |t(\theta)|\,, \end{aligned}$$

which is impossible. □

E.12 A Property of the Zeros of a $p \in \mathcal{P}_n$. Show that every $p \in \mathcal{P}_n$ has at most

$$M := \frac{e}{\sqrt{2}} n \sqrt{r}\, |p(1)|^{-1} \|p\|_{[-1,1]}$$

zeros (counting multiplicities) in $[1-r, 1]$, $r > 0$.

Hint: Use the substitution $x = \cos\tau$, E.11, and the inequality

$$\cos r < 1 - \tfrac{1}{4} r^2\,, \qquad 0 < r \leq 2\,.$$

□

E.13 Riesz's Lemma.

a] *Suppose $t \in \mathcal{T}_n$ and $t(\alpha) = \|t\|_K = 1$ for some $\alpha \in K$. Then*

$$t(\theta) \geq \cos(n(\theta - \alpha))\,, \qquad \theta \in \left[\alpha - \tfrac{\pi}{2n}, \alpha + \tfrac{\pi}{2n}\right],$$

and equality holds for a fixed $\theta \in \left[\alpha - \frac{\pi}{2n}, \alpha + \frac{\pi}{2n}\right]$ if and only if t is of the form $t(\tau) = \cos(n(\tau - \alpha))$. In particular, t does not vanish in $\left(\alpha - \frac{\pi}{2n}, \alpha + \frac{\pi}{2n}\right)$.

Hint: If this were false, then

$$q(\tau) := t(\tau) - \cos(n(\tau - \alpha))$$

would have more than $2n$ zeros on K (counting multiplicities). □

b] *Suppose $p \in \mathcal{P}_n$ and $p(1) = \|p\|_{[-1,1]} = 1$. Then that*

$$p(x) \geq T_n(x)\,, \qquad x \in \left[\cos \tfrac{\pi}{2n}, 1\right],$$

and equality holds for a fixed $x \in \left[\cos\frac{\pi}{2n}, 1\right]$ if and only if $p = T_n$, where T_n is the Chebyshev polynomial of degree n as defined by (2.1.1). In particular, p does not vanish in $\left(\cos\frac{\pi}{2n}, 1\right]$.

The next two exercises follows Erdélyi [88] and Erdélyi and Szabados [89b].

E.14 A Markov-Type Inequality for Trigonometric Polynomials on $[-\omega,\omega]$. Show that there exists a constant $0 < c \leq 16\pi$ such that

$$\|s'\|_{[-\omega,\omega]} \leq \left(n + c\frac{\pi-\omega}{\omega}n^2\right)\|s\|_{[-\omega,\omega]}$$

for every $s \in \mathcal{T}_n$ and $\omega \in (0,\pi]$.

Proof. If $\pi - \omega > (2n)^{-1}$, then

$$(3\tan^2(\omega/2)+1)^{1/2} < 8n\,,$$

and E.19 c] gives the result. If $\pi - \omega \leq (2n)^{-1}$, then Theorem 5.1.4 (Bernstein's inequality) combined with the Mean Value Theorem yields

$$\|s\|_{[-\pi,\pi]} \leq \|s\|_{[-\omega,\omega]} + (\pi-\omega)n\,\|s\|_{[-\pi,\pi]}\,,$$

and hence

$$\|s\|_{[-\omega,\omega]} \geq (1-n(\pi-\omega))\,\|s\|_{[-\pi,\pi]}$$

for every $s \in \mathcal{T}_n$. Therefore, using Theorem 5.1.4 (Bernstein's inequality) and $\pi - \omega \leq (2n)^{-1}$, we get

$$\begin{aligned}\|s'\|_{[-\omega,\omega]} &\leq \|s'\|_{[-\pi,\pi]} \leq n\,\|s\|_{[-\pi,\pi]} \\ &\leq \frac{n}{1-n(\pi-\omega)}\,\|s\|_{[-\omega,\omega]} \leq (n+2(\pi-\omega)n^2)\|s\|_{[-\omega,\omega]}\end{aligned}$$

for every $s \in \mathcal{T}_n$. □

E.15 Schur-Type Inequality for $\mathcal{T}_n$ on $[-\omega,\omega]$. Let $\omega \in (0,2\pi]$. Show that

$$\|s\|_{[-\omega,\omega]} \leq \frac{2n+1}{\sin(\omega/2)}\left\|s(\tau)\left(\tfrac{1}{2}(\cos\tau - \cos\omega)\right)^{1/2}\right\|_{[-\omega,\omega]}$$

for every $s \in \mathcal{T}_n$, and equality holds if and only if s is of the form

$$s(\tau) = c\,\frac{\sin\left[(2n+1)\arccos\frac{\sin(\tau/2)}{\sin(\omega/2)}\right]}{(\cos\tau-\cos\omega)^{1/2}}\,,\qquad c\in\mathbb{R}\,.$$

Note that the right-hand side of the above is an element of $\mathcal{T}_n$.

Hint: Define $2n+1$ distinct points in $(-\omega,\omega)$ by

$$\tau_k := 2\arcsin\left(\sin\frac{\omega}{2}\sin\frac{k\pi}{2n+1}\right)\,,\qquad k=0,\pm1,\ldots,\pm n\,.$$

Distinguish two cases in estimating $|s_n(\theta)|$ for $\theta \in [-\omega,\omega]$.

Case 1: $|\theta| \le |\tau_n|$. Use the inequality

$$\frac{1}{2}(\cos\theta - \cos\omega) > \frac{\sin^2(\omega/2)}{(2n+1)^2}$$

to get the desired result.

Case 2: $\tau_n < |\theta| \le \omega$. Let

$$M_n(s) := \max_{|k|\le n}\{|s(\tau_k)|((\cos\tau_k - \cos\omega)/2)^{1/2}\}.$$

Let $n \in \mathbb{N}$, $\omega \in (0,\pi]$, and $\theta \in [-\omega, -\tau_n) \cup (\tau_n, \omega]$ be fixed. Show that there is an $\widetilde{s}_n \in \mathcal{T}_n$ such that

$$\frac{|\widetilde{s}_n(\theta)|}{M_n(\widetilde{s}_n)} = \max_{s\in\mathcal{T}_n}\frac{|s(\theta)|}{M_n(s)}\,.$$

Show by a variational method that $\widetilde{s}_n$ is of the form

$$\widetilde{s}_n(\tau) = c\,\frac{\sin\left[(2n+1)\arccos\frac{\sin(\tau/2)}{\sin(\omega/2)}\right]}{(\cos\tau - \cos\omega)^{1/2}}\,, \qquad c \in \mathbb{R}\,.$$

□

E.16 Another Proof of Schur's Inequality.

a] Prove Theorem 5.1.9 (Schur's inequality) by using the method given in the hint to E.15.

b] Prove the result of E.15 by using interpolation, as in the proof of Theorem 5.1.9.

Proof. See Erdélyi and Szabados [89b]. □

E.17 Growth of Polynomials in the Complex Plane. Let

$$D := \{z \in \mathbb{C} : |z| < 1\} \qquad \text{and} \qquad D_\varrho := \{z \in \mathbb{C} : |z| < \varrho\}.$$

a] Show that

$$|p(z)| \le |z|^n\,\|p\|_D$$

for every $p \in \mathcal{P}_n^c$ and $z \in \mathbb{C} \setminus \overline{D}$.

Find all $p \in \mathcal{P}_n^c$ for which equality holds.

Hint: Apply the maximum principle (see E.1 d] of Section 1.2) with D and

$$q(z) := z^n p(z^{-1}) \in \mathcal{P}_n^c\,.$$

□

b] Show that the transformation $z = M(w) := \frac{1}{2}(w + w^{-1})$ can be written as

$$x = \tfrac{1}{2}(\varrho + \varrho^{-1})\cos\theta\,, \qquad y = \tfrac{1}{2}(\varrho - \varrho^{-1})\sin\theta\,,$$

where $z = x + iy$, $x, y \in \mathbb{R}$ and $w = \varrho e^{i\theta}$, $\varrho > 0$, $\theta \in K$.

c] For $\varrho > 0$, let E_ϱ be the image of the circle ∂D_ϱ under M. Show that E_ϱ is the ellipse

$$\frac{x^2}{a^2} + \frac{y^2}{b^2} = 1 \quad \text{with semiaxes} \quad a := \tfrac{1}{2}(\varrho + \varrho^{-1}) \quad \text{and} \quad b := \tfrac{1}{2}|\varrho - \varrho^{-1}|\,.$$

Furthermore, $E_\varrho = E_{\varrho^{-1}}$, and E_1 is the interval $[-1, 1]$ covered twice.

d] Show that

$$|p(z)| \le \varrho^n \, \|p\|_{[-1,1]}$$

for every $p \in \mathcal{P}_n^c$, $z \in E_\varrho$, $\varrho > 1$.

Proof. Applying the maximum principle (see E.1 d] of Section 1.2) with $D = D_1$ and

$$Q(w) := w^n p\left(\tfrac{1}{2}(w + w^{-1})\right) \in \mathcal{P}_{2n}^c\,,$$

we obtain

$$|Q(w)| \le \|Q\|_{D_1} = \|p\|_{[-1,1]}\,, \qquad w \in \partial D_{\varrho^{-1}}\,.$$

This, together with part c], yields

$$|p(z)| \le \varrho^n \|p\|_{[-1,1]}\,, \qquad z \in E_\varrho\,.$$

□

e] Show that there exists an absolute constant c such that

$$|p(z)| \le c\,\|p\|_{-1,1]}\,, \qquad p \in \mathcal{P}_n^c$$

whenever

$$z = x + iy\,, \quad x, y \in \mathbb{R}\,, \quad |x| \le 1 + \frac{1}{n^2}\,, \quad |y| \le \frac{\sqrt{|1 - x^2|_+}}{n} + \frac{1}{n^2}$$

($|1 - x^2|_+ := \max\{1 - x^2, 0\}$).

Hint: Use part d]. □

f] Prove the following Markov-Bernstein inequality. There is a constant $c(m)$ depending only on m such that

$$|p^{(m)}(x)| \le c(m)\left(\min\left\{n^2, \frac{n}{\sqrt{1 - x^2}}\right\}\right)^m \|p\|_{[-1,1]}$$

for every $p \in \mathcal{P}_n^c$ and $x \in [-1, 1]$.

Hint: Use part e] and Cauchy's integral formula (see E.1 a] of Section 1.2). □

Bernstein established Theorem 5.1.4 in order to prove inverse theorems of approximation. Bernstein's method is presented in the proof of the next exercise, which is one of the simplest cases. However, several other inverse theorems of approximation can be proved by straightforward modifications of the proof of this exercise. That is why Bernstein- and Markov-type inequalities play a significant role in approximation theory. Direct and inverse theorems of approximation and related matters may be found in many books on approximation theory, including Cheney [66], Lorentz [86a], and DeVore and Lorentz [93].

E.18 An Inverse Theorem of Approximation. Let Lip_α, $\alpha \in (0,1]$, denote the family of all real-valued functions g defined on K satisfying

$$\sup\left\{\frac{|g(x)-g(y)|}{|x-y|^\alpha} : x \neq y \in K\right\} < \infty.$$

For $f \in C(K)$, let

$$E_n(f) := \inf\{\|t-f\|_K : t \in \mathcal{T}_n\}.$$

An example for a direct theorem of approximation is stated in part a]. Part b] deals with its inverse result.

a] *Suppose f is m times differentiable on K and $f^{(m)} \in \mathrm{Lip}_\alpha$ for some $\alpha \in (0,1]$. Then there is a constant C depending only on f so that*

$$E_n(f) \le Cn^{-(m+\alpha)}, \qquad n = 1, 2, \ldots .$$

Proof. See, for example, Lorentz [86a]. □

b] *Suppose m is a nonnegative integer, $\alpha \in (0,1)$, and $f \in C(K)$. Suppose there is a constant $C > 0$ depending only on f such that*

$$E_n(f) \le Cn^{-(m+\alpha)}, \qquad n = 1, 2, \ldots .$$

Then f is m times continuously differentiable on K and $f^{(m)} \in \mathrm{Lip}_\alpha$.

Outline. We show only that f is m times continuously differentiable on K. The rest can be proved similarly, but its proof requires more technical details. See, for example, Lorentz [86a].

For each $k \in \mathbb{N}$, let $Q_{2^k} \in \mathcal{T}_{2^k}$ be chosen so that

$$\|Q_{2^k} - f\|_K \le C\,2^{-k(m+\alpha)}.$$

Then

$$\|Q_{2^{k+1}} - Q_{2^k}\|_K \le 2C\, 2^{-k(m+\alpha)}\,.$$

Now

$$f(\theta) = Q_{2^0}(\theta) + \sum_{k=1}^{\infty}(Q_{2^{k+1}} - Q_{2^k})(\theta)\,, \qquad \theta \in K\,,$$

and by Theorem 5.1.4 (Bernstein's inequality)

$$\begin{aligned}
|Q_{2^0}^{(j)}| + \left|\sum_{k=1}^{\infty}(Q_{2^{k+1}} - Q_{2^k})^{(j)}(\theta)\right| \\
&\le \|Q_1\|_K + \sum_{k=1}^{\infty}\left(2^{k+1}\right)^j \|Q_{2^{k+1}} - Q_{2^k}\|_K \\
&\le \|Q_1\|_K + \sum_{k=1}^{\infty}\left(2^{k+1}\right)^j 2C\, 2^{-k(m+\alpha)} \\
&\le \|Q_1\|_K + 2^{j+1}C\sum_{k=1}^{\infty}(2^{j-m-\alpha})^k < \infty
\end{aligned}$$

for every $\theta \in K$ and $j = 0, 1, \dots, m$, since $\alpha > 0$. Now we can conclude that $f^{(j)}(\theta)$ exists and

$$f^{(j)}(\theta) = Q_1^{(j)}(\theta) + \sum_{k=1}^{\infty}(Q_{2^{k+1}} - Q_{2^k})^{(j)}(\theta)$$

for every $\theta \in K$ and $j = 0, 1, \dots, m$. The fact that $f^{(m)} \in C(K)$ can be seen by the Weierstrass M-test. □

The next exercise follows Videnskii [60].

E.19 Videnskii's Inequalities. The main results of this exercise are the Bernstein- (part b]) and Markov-type (part c]) inequalities for trigonometric polynomials on an interval shorter than the period.

Let $\omega \in (0, \pi)$,

$$t_n(\theta) := Q_{n,\omega}(\theta) = \cos\left(2n \arccos\left(\frac{\sin(\theta/2)}{\sin(\omega/2)}\right)\right),$$

and

$$u_n(\theta) := \sin\left(2n \arccos\left(\frac{\sin(\theta/2)}{\sin(\omega/2)}\right)\right).$$

a] Recall that $t_n \in \mathcal{T}_n$ by E.3 a].

b] Show that

$$|s_n'(\theta)| \le |t_n'(\theta) + iu_n'(\theta)| \, \|s_n\|_{[-\omega,\omega]} = n\left(1 - \frac{\cos^2(\omega/2)}{\cos^2(\theta/2)}\right)^{-1/2} \|s_n\|_{[-\omega,\omega]}$$

for every $s_n \in \mathcal{T}_n$ and $\theta \in (-\omega, \omega)$. Equality holds if and only if $s_n = ct_n$ for some $c \in \mathbb{R}$ and $s_n(\theta) = 0$.

Hint: First show that for every $n \in \mathbb{N}$, $\omega \in (0, \pi]$, and $\theta \in [-\omega, \omega]$, there exists an $\widetilde{s} \in \mathcal{T}_n$ such that

$$\frac{|\widetilde{s}'(\theta)|}{\|\widetilde{s}\|_{[-\omega,\omega]}} = \max_{s_n \in \mathcal{T}_n} \frac{|s_n'(\theta)|}{\|s_n\|_{[-\omega,\omega]}} .$$

Use a variational method to show that either

$$\|\widetilde{s}_n\|_{[-\omega,\omega]} = \|\widetilde{s}_n\|_{[-\pi,\pi]}$$

or there exist $\alpha \in (-\pi, -\omega]$ and $\beta \in [\omega, \pi)$ such that $\widetilde{s}_n = cT_n$ for some $c \in \mathbb{R}$, where, as in Section 3.3,

$$T_n := T_n\{1, \cos\tau, \sin\tau, \ldots, \cos n\tau, \sin n\tau; [\alpha, \beta]\}$$

is the Chebyshev polynomial for $\mathcal{T}_n$ on $[\alpha, \beta]$. In the first case use Theorem 5.1.4 (Bernstein's inequality). In the second case observe that

$$T_n(\tau) = t_n\left(\frac{\sin((\tau - \gamma)/2)}{\sin(\widetilde{\omega}/2)}\right)$$

with

$$\gamma := \tfrac{1}{2}(\alpha + \beta) \qquad \text{and} \qquad \widetilde{\omega} := \tfrac{1}{2}(\beta - \alpha) .$$

□

c] Show that if $2n > (3\tan^2(\omega/2) + 1)^{1/2}$, then

$$\|s_n'\|_{[-\omega,\omega]} \le t_n'(\omega)\|s_n\|_{[-\omega,\omega]} = 2n^2 \cot(\omega/2)\|s_n\|_{[-\omega,\omega]}$$

for every $s_n \in \mathcal{T}_n$, and equality holds if and only if $s_n = ct_n$, $c \in \mathbb{R}$.

Outline. Let $-\omega = \xi_0 < \xi_1 < \cdots < \xi_{2n} = \omega$ be the points where

$$t_n(\xi_j) = (-1)^j , \qquad j = 0, 1, \ldots, 2n$$

and

$$u_n(\xi_j) = 0 , \qquad j = 1, 2, \ldots, 2n - 1 ,$$

and let $\theta_1 < \theta_2 < \cdots < \theta_{2n}$ be the zeros of t_n (which lie in $(-\omega, \omega)$). Note that

$$u_n'(\theta_j) = 0\,, \qquad j = 1, 2, \ldots, 2n\,.$$

Step 1. Show that $2n > (3\tan^2(\omega/2) + 1)^{1/2}$ implies $t_n''(\theta) > 0$ for every $\theta \in [\xi_{2n-1}, \omega]$, so t_n' is increasing on $[\xi_{2n-1}, \omega]$.

Step 2. Use part b] to show that if $\theta \in [\theta_1, \theta_{2n}]$, then

$$|t_n'(\theta)| \leq |t_n'(\theta) + iu_n'(\theta)| \leq |t_n'(\theta_{2n}) + iu_n'(\theta_{2n})| = |t_n'(\theta_{2n})|\,.$$

Step 3. Deduce from Steps 1 and 2 that

$$|t_n'(\theta)| < |t_n'(\pm\omega)|\,, \qquad \theta \in (-\omega, \omega)\,.$$

Step 4. Show that there is an $\widetilde{s} \in \mathcal{T}_n$ such that

$$\frac{\|\widetilde{s}_n'\|_{[-\omega,\omega]}}{\|\widetilde{s}_n\|_{[-\omega,\omega]}} = \max_{s \in \mathcal{T}_n} \frac{\|s'\|_{[-\omega,\omega]}}{\|s\|_{[-\omega,\omega]}}\,.$$

For the rest of the proof let $\widetilde{s}_n$ be normalized by $\|\widetilde{s}\|_{[-\omega,\omega]} = 1$ and let $\theta^* \in [-\omega, \omega]$ be chosen so that

$$|\widetilde{s}_n'(\theta^*)| = \|\widetilde{s}_n'(\theta)\|_{[-\omega,\omega]}\,.$$

Let $(a_1 < a_2 < \cdots < a_m)$ be an alternation sequence of maximal length for $\widetilde{s}_n \in C[-\omega, \omega]$ on $[-\omega, \omega]$. We would like to show that $m = 2n + 1$. Clearly $m < 2n + 2$.

Step 5. Use a variational method to show that $2n \leq m$.

Step 6. Use a variational method to show that $\theta^* = \pm\omega$ implies $m = 2n+1$, so $m = 2n$ implies $\theta^* \in (-\omega, \omega)$.

Step 7. Show by a variational method that

$$|s_n'(\pm\omega)| \leq |t_n'(\pm\omega)|\, \|s_n\|_{[-\omega,\omega]}$$

for every $s_n \in \mathcal{T}_n$, and equality holds if and only if $s_n = ct_n$, $c \in \mathbb{R}$. In particular, if $m = 2n$, then

$$|\widetilde{s}_n'(\pm\omega)| < |t_n'(\pm\omega)|\,.$$

Step 8. Use part b] and Step 3 to show that

$$|\widetilde{s}_n'(\theta)| \leq t_n'(\theta_{2n}) < t_n'(\omega)\,, \qquad \theta \in [\theta_1, \theta_{2n}]\,.$$

Step 9. Use part b] to show that $m = 2n$ implies

$$|\widetilde{s}_n'(\theta_j)| < |t_n'(\theta_j)| = (-1)^j t_n'(\theta_j)\,, \qquad j = 1, 2, \ldots, 2n\,.$$

Step 10. Suppose $m = 2n$. Use Steps 6 and 8 to show that

$$\theta^* \in (-\omega, \theta_1) \cup (\theta_{2n}, \omega)\,.$$

Step 11. Suppose $m = 2n$. Use the defining property of $\widetilde{s}_n$ and Steps 1 and 10 to show that

$$|\widetilde{s}_n'(\theta^*)| = \|\widetilde{s}_n'\|_{[-\omega,\omega]} \geq |t_n'(\pm\omega)| > |t_n'(\theta^*)|\,.$$

Step 12. Suppose $m = 2n$ and $s_n'(\theta^*) \geq 0$. Use Steps 7 to 11 to show that $t_n' - \widetilde{s}'{}_n$ has at least $2n + 1$ distinct zeros in $(-\omega, \omega)$, a contradiction.

Step 13. Show that $m = 2n + 1$ and $\widetilde{s}_n = \pm t_n$. □

E.20 Inequalities for Entire Functions of Exponential Type. Entire functions of (exponential) type τ are defined in E.17 of Section 4.2. Denote by E_τ the set of all entire functions of exponential type at most τ. This exercise collects some of the interesting inequalities known for E_τ. Since a trigonometric polynomial of degree n belongs to E_n, these results can be viewed as extensions of the corresponding inequalities for trigonometric polynomials. More on various inequalities for entire functions of exponential type may be found in Rahman and Schmeisser [83].

a] **Bernstein's Inequality.** The inequality

$$\|f'\|_{\mathbb{R}} \leq \tau\,\|f\|_{\mathbb{R}}\,, \qquad x \in \mathbb{R}$$

holds for every $f \in E_\tau$.

Proof. See Bernstein [23] or Rahman and Schmeisser [83]. □

b] **Extension of the Bernstein-Szegő Inequality.** The inequality

$$(f'(x))^2 + (\tau f(x))^2 \leq \tau^2 \|f\|_{\mathbb{R}}^2\,, \qquad x \in \mathbb{R}$$

holds for every $f \in E_\tau$ taking real values on the real line.

Proof. See Duffin and Schaeffer [37]. □

c] **The Growth of $f \in E_\tau$.** The inequality

$$|f(x + iy)| \leq e^{\tau|y|}\|f\|_{\mathbb{R}}\,, \qquad x, y \in \mathbb{R}$$

holds for every $f \in E_\tau$.

Proof. See Rahman and Schmeisser [83]. □

d] The Growth of $f \in E_\tau$ Taking Real Values on $\mathbb{R}$. The inequality

$$|f(x+iy)| \le (\cosh \tau y)\, \|f\|_{\mathbb{R}}\,, \qquad x, y \in \mathbb{R}$$

holds for every $f \in E_\tau$ taking real values on the real line.

Proof. See Schaeffer and Duffin [38]. □

e] Bernstein-Type Inequality in L_p. Let $p \in (0, \infty)$. The inequality

$$\|f'\|_{L_p(\mathbb{R})} \le \tau\, \|f\|_{L_p(\mathbb{R})}$$

holds for every $f \in E_\tau$.

Proof. See Rahman and Schmeisser [90]. □

E.21 Markov-Type Inequality on Connected Subsets of the Complex Plane. Let E be a connected compact set of the complex plane. Then

$$\|p'\|_E \le \frac{e}{2}\, \frac{n^2}{\operatorname{cap}(E)}\, \|p\|_E$$

for every $p \in \mathcal{P}_n^c$.

Proof. See Pommerenke [59c]. □

Erdős conjectured that the constant $\frac{e}{2}$ in the above inequality can be replaced by $\frac{1}{2}$. This result would contain Theorem 5.1.8 (Markov's inequality) as a special case. However, Rassias, Rassias, and Rassias [77] disproved the conjecture. Erdős still speculates that $\frac{e}{2}$ in Pommerenke's inequality may be replaced by $\frac{1}{2}(1 + o(1))$.

The result of the next exercise is formulated so that its proof is elementary at the expense of precision and generality.

E.22 The Interval where the Sup Norm of a Weighted Polynomial Lives. Suppose $w = \exp(-Q)$, where

(1) $Q : \mathbb{R} \to \mathbb{R}$ is continuous and even,

(2) Q' is continuous and positive in $(0, \infty)$,

(3) $tQ'(t)$ is increasing in $(0, \infty)$, and

(4) $\lim_{t \to 0+} tQ'(t) = 0 \quad$ and $\quad \lim_{t \to \infty} tQ'(t) = \infty$.

Let $a_n > 0$ be chosen so that

$$a_n^n w(a_n) = \max_{x \in \mathbb{R}} |x^n w(x)|\,.$$

a] Show that

$$n = a_n Q'(a_n) \qquad \text{and} \qquad 0 < a_1 < a_2 < a_3 < \cdots\,.$$

b] Show that

$$\|pw\|_{\mathbb{R}} \leq \|pw\|_{[-2a_{2n}, 2a_{2n}]}$$

for every $p \in \mathcal{P}_n^c$.

Proof. Using Chebyshev's inequality (see E.2 of Section 5.1) and the explicit form (2.1.1) of the Chebyshev polynomial T_n, we can deduce that

$$|p(x)| \leq \left| \frac{x}{a} + \sqrt{\left(\frac{x}{a}\right)^2 - 1} \right|^n \|p\|_{[-a,a]} \leq \left(\frac{2|x|}{a}\right)^n \|p\|_{[-a,a]}$$

for every $p \in \mathcal{P}_n^c$, $x \in \mathbb{R} \setminus [-a, a]$, and $a > 0$. Choosing $a := a_n$, and using the fact that w is decreasing on $[0, \infty)$, we obtain

$$|(pw)(x)| \leq \left(\frac{2|x|}{a_n}\right)^n \frac{w(x)}{w(a_n)} \|pw\|_{[-a_n, a_n]}$$

for every $p \in \mathcal{P}_n^c$ and $x \in \mathbb{R} \setminus [-a_n, a_n]$. Now if $x \in \mathbb{R} \setminus [-2a_{2n}, 2a_{2n}]$, then

$$\begin{aligned}
\frac{|x|^n w(x)}{a_n^n w(a_n)} &\leq \exp\left(\int_{a_n}^{|x|} \frac{d}{dt} \log(t^n w(t))\right) dt \\
&= \exp\left(\int_{a_n}^{|x|} \frac{n - tQ'(t)}{t} \, dt\right) \\
&\leq \exp\left(\int_{a_{2n}}^{2a_{2n}} \frac{n - 2n}{t} \, dt\right) = 2^{-n},
\end{aligned}$$

where we used $a_{2n} \geq a_n$, $tQ'(t) \geq 2n$ for $t \geq a_{2n}$, and $tQ'(t) \geq n$ for $t \geq a_n$. Combining the above inequality with the previous one, we obtain that

$$|(pw)(x)| \leq \|pw\|_{[-a_n, a_n]}, \qquad x \in \mathbb{R} \setminus [-2a_{2n}, 2a_{2n}]$$

for every $p \in \mathcal{P}_n^c$, from which the result follows. □

c] Let $Q(x) := |x|^\alpha$, $\alpha > 0$. Show that Q satisfies the assumptions of the exercise, and

$$a_n = \left(\frac{n}{\alpha}\right)^{1/\alpha}.$$

The idea of infinite-finite range inequalities, of which E.22 b] is an example, goes back to Freud (a_n is the *Freud number*); see Nevai's survey paper [86]. The sharp form of these is due to Mhaskar and Saff [85]; see also Lubinsky and Saff [88] and Saff and Totik [to appear]. They are not that difficult to prove, but need the maximum principle for subharmonic functions.

E.23 A Theorem of Markov. Let T_n be the Chebyshev polynomial of degree n defined by (2.1.1). Then

$$T_n(x) = \sum_{k=0}^{n} c_{k,n} x^k .$$

The *Markov numbers* $M_{k,n}$, $0 \le k \le n$, are defined by

$$M_{k,n} := \begin{cases} |c_{k,n}| & \text{if } k \equiv n \pmod 2 \\ |c_{k,n-1}| & \text{if } k \equiv n-1 \pmod 2 . \end{cases}$$

These can be explicitly computed from (2.1.1).

a] The inequalities

$$|a_{k,n}| \le M_{k,n} \, \|p\|_{[-1,1]}$$

hold for every $p \in \mathcal{P}_n$ of the form

$$p(x) = \sum_{k=0}^{n} a_{k,n} x^k, \qquad a_{k,n} \in \mathbb{R} .$$

Proof. See, for example, Natanson [64]. □

b] Show that

$$|p'(0)| \le (2n-1) \, \|p\|_{[-1,1]}$$

for every $p \in \mathcal{P}_{2n}$.

Hint: Use part a]. □

c] Show that

$$|p'(x)| \le \frac{2n-1}{1-|x|} \, \|p\|_{[-1,1]}$$

for every $p \in \mathcal{P}_{2n}$ and $x \in (-1,1)$.

Hint: Use part b] and a linear transformation. □

Of course, part c] gives a better result than Theorem 5.1.7 (Bernstein's Inequality) only if x is very close to 0. This is exactly the case we need in our application of part c] in E.4 c] of Section 6.1.

5.2 Markov's Inequality for Higher Derivatives

From Theorem 5.1.8 (Markov's Inequality), by induction on m it follows that

$$\|p^{(m)}\|_{[-1,1]} \le (n(n-1)\cdots(n-m+1))^2 \|p\|_{[-1,1]}$$

for every $p \in \mathcal{P}_n$.

However, this is not the best possible result. The main result of this section is the following inequality of Duffin and Schaeffer [41], which gives a sharp improvement of the above. E.2 d] extends the following result to polynomials with complex coefficients:

Theorem 5.2.1. *If $p \in \mathcal{P}_n$ satisfies*

$$\left|p\left(\cos \tfrac{j\pi}{n}\right)\right| \le 1\,, \qquad j = 0, 1, \dots, n\,,$$

then for every $m = 1, 2, \dots, n$,

$$|p^{(m)}(x+iy)| \le |T_n^{(m)}(1+iy)|\,, \qquad x \in [-1,1]\,, \quad y \in \mathbb{R}\,,$$

where T_n is the Chebyshev polynomial of degree n defined by (2.1.1). Equality can occur only if $p = \pm T_n$.

To prove this inequality we need three lemmas, which are of some interest in their own right.

Lemma 5.2.2. *Suppose $\lambda_1, \lambda_2, \dots, \lambda_n$ are distinct real numbers,*

$$q(z) := c \prod_{j=1}^{n} (z - \lambda_j)\,, \qquad 0 \ne c \in \mathbb{R}\,,$$

and $p \in \mathcal{P}_n^c$ satisfies

$$|p'(\lambda_j)| \le |q'(\lambda_j)|\,, \qquad j = 1, 2, \dots, n\,.$$

Then, for every $m \in \mathbb{N}$,

$$|p^{(m)}(x)| \le |q^{(m)}(x)| \tag{5.2.1}$$

whenever x is a zero of $q^{(m-1)}$.

Proof. For $m = 1$, inequality (5.2.1) is simply a restatement of the assumption of the lemma, so consider the case $m = 2$. The Lagrange interpolation formula (see E.6 of Section 1.1) gives

$$\frac{p'(z)}{q(z)} = \sum_{j=1}^{n} \frac{p'(\lambda_j)}{q'(\lambda_j)} \frac{1}{z - \lambda_j} = \sum_{j=1}^{n} \frac{\delta_j}{z - \lambda_j}\,, \tag{5.2.2}$$

where, by the hypotheses of the lemma, $|\delta_j| \le 1$. There is a similar expression for $q'(z)/q(z)$ in which each δ_j is equal to 1. On differentiating (5.2.2), we obtain

$$\frac{p''(z)q(z) - p'(z)q'(z)}{q(z)^2} = -\sum_{j=1}^{n} \frac{\delta_j}{(z - \lambda_j)^2}\,.$$

Thus at the points x where $q'(x) = 0$ we have

$$\left|\frac{p''(x)}{q(x)}\right| = \left|\sum_{j=1}^{n} \frac{\delta_j}{(x-\lambda_j)^2}\right| \leq \sum_{j=1}^{n} \frac{1}{(x-\lambda_j)^2} = \left|\frac{q''(x)}{q(x)}\right|,$$

and it follows that the lemma is true for $m = 2$.

The proof for $m > 2$ is by induction. Let $|p^{(m)}(x)| \leq |q^{(m)}(x)|$ at the zeros of $q^{(m-1)}$ (which are real and distinct). Applying the previous argument to $p^{(m)}$ and $q^{(m)}$ instead of p' and q', we obtain

$$|p^{(m+1)}(x)| \leq |q^{(m+1)}(x)|$$

at the zeros of $q^{(m)}$. This completes the induction. □

Lemma 5.2.3. *Let $q \in \mathcal{P}_n$ have n distinct zeros in $(-\infty, b)$, and suppose that in a strip of the complex plane it satisfies the inequality*

$$(5.2.3) \qquad |q(x+iy)| \leq |q(b+iy)|, \qquad x \in [a,b], \quad y \in \mathbb{R}.$$

Suppose also that $p' \in \mathcal{P}_{n-1}$ satisfies

$$(5.2.4) \qquad |p'(x)| \leq |q'(x)|$$

whenever x is a zero of q. Then the derivatives of p and q satisfy

$$(5.2.5) \qquad |p^{(m)}(x+iy)| \leq |q^{(m)}(b+iy)|, \qquad x \in [a,b], \quad y \in \mathbb{R}.$$

Proof. First we show that at every point $x_0 + iy_0$ in the strip

$$|p'(x_0+iy_0)| \leq |q'(b+iy_0)|.$$

Let

$$q(z) := c\prod_{j=1}^{n}(z-\lambda_j), \qquad 0 \neq c \in \mathbb{R}, \quad \lambda_j \in (-\infty, b).$$

Let $h(z)$ be another polynomial with the same leading coefficient as q and whose zeros are obtained by reflecting about x_0 those zeros of q that lie to the right of x_0. Thus

$$h(z) := c\prod_{j=1}^{n}(z-\beta_j),$$

where

$$(5.2.6) \qquad \beta_j := \begin{cases} 2x_0 - \lambda_j & \text{if } \lambda_j > x_0 \\ \lambda_j & \text{if } \lambda_j \leq x_0 . \end{cases}$$

Then on the line $z = x_0 + iy,\ \ y \in \mathbb{R}$, we have $|z - \beta_j| = |z - \lambda_j|$, so

$$|h(x_0 + iy)| = |q(x_0 + iy)|\,. \tag{5.2.7}$$

We now show that

$$|p'(x_0 + iy_0)| \le |h'(x_0 + iy_0)|\,. \tag{5.2.8}$$

Note that

$$\frac{h'(z)}{h(z)} = \sum_{j=1}^{n} \frac{1}{z - \beta_j}\,, \tag{5.2.9}$$

and recalling (5.2.2), we have

$$\frac{p'(z)}{q(z)} = \sum_{j=1}^{n} \frac{\delta_j}{z - \lambda_j} \tag{5.2.10}$$

with $\delta_j \in [-1, 1]$ for each j. Comparing the right-hand sides of (5.2.9) and (5.2.10), respectively, at $z = x_0 + iy_0$, we obtain

$$\begin{aligned}
\left|\sum_{j=1}^{n} \frac{\delta_j}{x_0 + iy_0 - \lambda_j}\right| &= \left|\sum_{j=1}^{n} \frac{\delta_j (x_0 - \lambda_j)}{(x_0 - \lambda_j)^2 + y_0^2} - i \sum_{j=1}^{n} \frac{y_0 \delta_j}{(x_0 - \lambda_j)^2 + y_0^2}\right| \\
&\le \left|\sum_{j=1}^{n} \frac{x_0 - \beta_j}{(x_0 - \beta_j)^2 + y_0^2} - i \sum_{j=1}^{n} \frac{y_0}{(x_0 - \beta_j)^2 + y_0^2}\right| \\
&= \left|\sum_{j=1}^{n} \frac{1}{x_0 - \beta_j + iy_0}\right|
\end{aligned}$$

since by construction $|x_0 - \lambda_j| = x_0 - \beta_j$. Therefore

$$\left|\frac{p'(x_0 + iy_0)}{q(x_0 + iy_0)}\right| \le \left|\frac{h'(x_0 + iy_0)}{h(x_0 + iy_0)}\right|\,,$$

and (5.2.7) yields (5.2.8).

Let $\alpha \in \mathbb{C},\ \ |\alpha| < 1$, be an arbitrary constant and let

$$\varphi(z) := q(z) - \alpha h(z + x_0 - b)\,.$$

Let Γ be the simple closed curve consisting of a segment of the line $z = b + iy,\ \ y \in \mathbb{R}$, and the portion of a circle with center at b and radius ϱ that lies to the right of this line. Relations (5.2.3) and (5.2.7) show that on the line segment of Γ,

$$|q(z)| > |\alpha h(z + x_0 - b)|\,.$$

If ϱ is sufficiently large, the same inequality is true for the circular portion of Γ since q and h have the same leading coefficient. Thus Rouché's theorem gives that φ and h have the same number of zeros inside Γ. We conclude that φ has no zeros on or to the right of the line $z = b + iy,\ \ y \in \mathbb{R}$. The last statement, together with Theorem 1.3.1, implies that φ' has no zeros on the line $z = b + iy,\ \ y \in \mathbb{R}$. Thus for $|\alpha| < 1$,

$$q'(b + iy_0) - \alpha h'(x_0 + iy_0) \neq 0\,,$$

which, together with (5.2.8), yields

$$|p'(x_0 + iy_0)| \le |h'(x_0 + iy_0)| \le |q'(b + iy_0)|\,.$$

This proves the lemma for $m = 1$.

We turn now to the case $m > 1$. Applying the lemma with $m = 1$ when $p = q$, we have

$$|q'(x + iy)| \le |q'(b + iy)|\,, \qquad x \in [a, b]\,, \quad y \in \mathbb{R}\,.$$

Thus q' satisfies all the requirements that are imposed on the interpolating polynomial q (with n replaced by $n - 1$) in the lemma, and by Lemma 5.2.2, $|p''(x)| \le |q''(x)|$ at the $n-1$ zeros of q'. Applying the lemma to $p'(x)$ instead of p in the case $m = 1$, for which it has already been proved, we have

$$|p''(x + iy)| \le |q''(b + iy)|\,, \qquad x \in [a, b]\,, \quad y \in \mathbb{R}\,,$$

which proves that inequality (5.2.5) is true for $m = 2$. Repetition of this argument completes the proof for larger values of m. □

Lemma 5.2.4. *Suppose $p \in \mathcal{P}_n^c$ and $|p(\lambda_j)| \le 1$ for*

$$\lambda_j := \cos \tfrac{j\pi}{n}\,, \qquad j = 0, 1, \ldots, n\,.$$

Then

$$|p'(x_k)| \le \frac{n}{\sqrt{1 - x_k^2}} \qquad \text{for} \quad x_k := \cos \tfrac{(2k-1)\pi}{2n}\,, \qquad k = 0, 1, \ldots, n\,.$$

For every fixed k, equality holds if and only if $p = cT_n$ for some $c \in \mathbb{C}$ with $|c| = 1$.

Proof. Let

$$f(x) = (1 - x^2)T_n'(x) = a \prod_{j=0}^{n} \left(x - \cos \tfrac{j\pi}{n}\right).$$

Then by the differential equation for T_n (see E.3 e] of Section 2.1), we have

$$f'(x) = -xT_n'(x) - n^2 T_n(x)\,.$$

Differentiating the Lagrange interpolation formula for p (see E.6 of Section 1.1) gives

$$p'(x) = \sum_{j=0}^{n} \frac{p(\lambda_j)}{f'(\lambda_j)} \frac{(x-\lambda_j)f'(x) - f(x)}{(x-\lambda_j)^2}\,.$$

For the zeros of T_n, this reduces to

$$p'(x) = -T_n'(x) \sum_{j=0}^{n} \frac{p(\lambda_j)}{f'(\lambda_j)} \frac{(1-x\lambda_j)}{(x-\lambda_j)^2} \tag{5.2.11}$$

since at these points

$$\begin{aligned}(x-\lambda_j)f'(x) - f(x) &= -x(x-\lambda_j)T_n'(x) - (1-x^2)T_n'(x) \\ &= -(1-x\lambda_j)T_n'(x)\,.\end{aligned}$$

In the same way, we obtain for the zeros of T_n,

$$T_n'(x) = -T_n'(x) \sum_{j=0}^{n} \frac{T_n(\lambda_j)}{f'(\lambda_j)} \frac{(1-x\lambda_j)}{(x-\lambda_j)^2}$$

and since $T_n(\lambda_j)$ and $f'(\lambda_j)$ are of opposite sign ($f'(\lambda_j) = -n^2 T_n(\lambda_j)$), this gives

$$T_n'(x) = T_n'(x) \sum_{j=0}^{n} \left| \frac{1}{f'(\lambda_j)} \right| \frac{(1-x\lambda_j)}{(x-\lambda_j)^2}\,. \tag{5.2.12}$$

Since $|p(\lambda_j)| \le 1$ in (5.2.11), on comparing (5.2.11) and (5.2.12), we obtain for every zero x of T_n that

$$|p'(x)| \le |T_n'(x)| = \frac{n}{\sqrt{1-x^2}}\,.$$

For a fixed zero of T_n, the equality occurs if and only if

$$p(\lambda_j) = cT_n(\lambda_j)\,, \qquad j = 0, 1, \ldots, n$$

for some $c \in \mathbb{C}$, $|c| = 1$, that is, if and only if $p = cT_n$ for some $c \in \mathbb{C}$ with $|c| = 1$. □

Proof of Theorem 5.2.1. Suppose $p \ne T_n$ satisfies the assumption of the theorem. Then by Lemma 5.2.4 there exists a constant $\alpha > 1$ such that $|\alpha p'(x)| \le |T_n'(x)|$ at the zeros of T_n. Applying Lemma 5.2.3 with p and q replaced by αp and T_n (assumption (5.2.3) is satisfied with $[a,b] := [-1,1]$ by E.1 b]), we obtain

$$|p^{(m)}(x+iy)| \le \alpha^{-1} \left| T_n^{(m)}(1+iy) \right|, \qquad x \in [-1,1]\,, \quad y \in \mathbb{R}$$

for every $m = 1, 2, \ldots, n$. If $p = T_n$, then we have the same inequality with α^{-1} replaced by 1. □

Comments, Exercises, and Examples.

The inequality

$$\|p^{(m)}\|_{[-1,1]} \le T_n^{(m)}(1) \cdot \|p\|_{[-1,1]}, \quad p \in \mathcal{P}_n$$

was first proved by V. A. Markov [16]. He was the brother of the more famous A. A. Markov who proved the above inequality for $m = 1$ in [1889] (see Theorem 5.1.8). However, their ingenious proofs are rather complicated. Bernstein presented a shorter variational proof of V. A. Markov's inequality in 1938 (see Bernstein [58], which includes a complete list of Bernstein's publications). Our discussion in this section follows Duffin and Schaeffer [41].

E.1 A Property of Chebyshev Polynomials.

a] Let $(\alpha_i)_{i=1}^{2n}$ be a sequence of $2n$ nonnegative numbers, and let (α_i') be a rearrangement of this sequence according to magnitude,

$$\alpha_1' \ge \alpha_2' \ge \cdots \ge \alpha_{2n}' \ge 0.$$

Show that for every $y \ge 0$,

$$(\alpha_1\alpha_2 + y)(\alpha_3\alpha_4 + y)\cdots(\alpha_{2n-1}\alpha_{2n} + y) \tag{5.2.13}$$

is not greater than

$$(\alpha_1'\alpha_2' + y)(\alpha_3'\alpha_4' + y)\cdots(\alpha_{2n-1}'\alpha_{2n}' + y).$$

Proof. If α_1 and α_3 are at least as large as any of the remaining numbers α_i, then

$$(\alpha_1\alpha_3 + y)(\alpha_2\alpha_4 + y) - (\alpha_1\alpha_2 + y)(\alpha_3\alpha_4 + y) = y(\alpha_1 - \alpha_4)(\alpha_3 - \alpha_2) \ge 0.$$

This shows that the numbers α_i in (5.2.13) can be rearranged so that the two largest occur in the same factor without decreasing (5.2.13). Then the two largest of the remaining numbers α_i may be brought into the same factor without decreasing (5.2.13), and so on. □

b] Show that the Chebyshev polynomials T_n defined by (2.1.1) satisfy the inequality

$$|T_n(x+iy)| \le |T_n(1+iy)|, \qquad x \in [-1,1], \quad y \in \mathbb{R}.$$

Proof. We have

$$|T_n(x+iy)|^2 = c^2 \prod_{j=1}^{n} ((x - \cos\theta_j)^2 + y^2),$$

where $\theta_j = \frac{(2j-1)\pi}{2n}$ and $c = 2^{n-1}$. With $x = \cos\theta$ we write

$$|T_n(x+iy)|^2 = c^2 \prod_{j=1}^{n} \left(\tfrac{1}{4}\left|e^{i\theta} - e^{-i\theta_j}\right|^2 \left|e^{i\theta} - e^{i\theta_j}\right|^2 + y^2\right).$$

Geometrically, $e^{\pm i\theta_j}$, $j = 1, 2, \dots, n$, represent $2n$ points equally distributed on the unit circle. Connect these points by chords to the point $e^{i\theta}$. Then the lengths of these $2n$ chords are given by $|e^{i\theta} - e^{\pm i\theta_j}|$. If θ is increased or decreased by any multiple of $\frac{\pi}{n}$, we obtain a new set of chords, but the aggregate of their lengths is unchanged. Choose φ such that

$$\varphi \equiv \theta \pmod{\tfrac{\pi}{n}}, \qquad -\tfrac{\pi}{2n} \le \varphi \le \tfrac{\pi}{2n}.$$

If $x^* = \cos\varphi$, then

$$T_n(x^*+iy) = c^2 \prod_{j=1}^{n} \left(\tfrac{1}{4}\left|e^{i\varphi} - e^{-i\theta_j}\right|^2 \left|e^{i\varphi} - e^{i\theta_j}\right|^2 + y^2\right),$$

where the numbers $|e^{i\varphi} - e^{\pm i\theta_j}|^2$ are simply a rearrangement of the numbers $|e^{i\theta} - e^{\pm i\theta_j}|^2$. Use part a] to show that

$$|T_n(x+iy)|^2 \le |T_n(x^*+iy)|^2.$$

Note that $\cos\theta_1 \le x^* \le 1$, where $\cos\theta_1$ is the right most zero of T_n. Hence

$$|T_n(x^*+iy)|^2 \le |T_n(1+iy)|^2,$$

and the proof is finished. □

E.2 Markov's Inequality for Higher Derivatives for $\mathcal{P}_n^c$.

a] Show that if p and q satisfy the conditions of Lemma 5.2.3 with $p \in \mathcal{P}_n$ replaced by $p \in \mathcal{P}_n^c$, then

$$|p^{(m)}(x)| \le |q^{(m)}(b)|, \qquad x \in [a,b]$$

for every $m \in \mathbb{N}$.

Hint: After differentiating (5.2.2) $m-1$ times, we obtain

$$p^{(m)}(x) = \sum_{j=1}^{n} \delta_j \frac{d^{m-1}}{dx^{m-1}} \left(\frac{q(x)}{x-\lambda_j}\right),$$

where $|\delta_j| = |p'(\lambda_j)/q'(\lambda_j)| \le 1$. It is evident that if $x \in (a,b)$ is fixed, then $|p^{(m)}(x)|$ attains its maximum when $\delta_j = \pm 1$ for each j, in which case p' has real coefficients. Now use Lemma 5.2.3. □

b] Show that if p satisfies the assumption of Theorem 5.2.1 with $p \in \mathcal{P}_n$ replaced by $p \in \mathcal{P}_n^c$, then

$$\|p^{(m)}\|_{[-1,1]} \leq |T_n^{(m)}(1)|$$

for every $m \in \mathbb{N}$. The equality can hold only if $p = cT_n$ for some $c \in \mathbb{C}$, $|c| = 1$.

Hint: Modify the proof of Theorem 5.2.1 by using part a]. □

c] Show that

$$T_n^{(m)}(1) = \frac{n^2(n^2-1)(n^2-2^2)\cdots(n^2-(m-1)^2)}{1\cdot 3\cdot 5\cdots(2m-1)}\,.$$

Hint: Differentiating the second-order differential equation for T_n (see E.3 e] of Section 2.1) $m-1$ times gives

$$(1-x^2)T_n^{(m+2)}(x) - (2m+1)xT_n^{(m+1)}(x) + (n^2-m^2)T_n^{(m)}(x) = 0$$

from which

$$(2m+1)\,T_n^{(m+1)}(1) = (n^2-m^2)\,T_n^{(m)}(1)$$

follows. Use induction and $T_n(1) = 1$ to finish the proof. □

d] The Main Inequality. Suppose $p \in \mathcal{P}_n^c$ satisfies

$$\left|p\left(\cos \tfrac{j\pi}{n}\right)\right| \leq 1\,, \quad j = 1, 2, \ldots, n\,.$$

Show that for $m = 1, 2, \ldots, n$,

$$\|p^{(m)}\|_{[-1,1]} \leq \frac{n^2(n^2-1)(n^2-2^2)\cdots(n^2-(m-1)^2)}{1\cdot 3\cdot 5\cdots(2m-1)}\,,$$

and the equality can occur only if $p = cT_n$ for some $c \in \mathbb{C}$, $|c| = 1$.

Hint: Combine parts c] and b]. □

A slightly weaker version of Markov's inequality for higher derivatives is much easier to prove.

E.3 A Weaker Version of Markov's Inequality. Show that

$$\|p^{(m)}\|_{[-1,1]} \leq 2^m n^{2m} \|p\|_{[-1,1]}$$

for every $p \in \mathcal{P}_n$.

Hint: First show by a variational method that the extremal problem

$$\max_{0 \neq p \in \mathcal{P}_n} \frac{|p'(\pm 1)|}{\|p\|_{[-1,1]}}$$

is solved by the Chebyshev polynomial T_n, hence

$$|p'(\pm 1)| \le n^2 \|p\|_{[-1,1]} \,.$$

Then, by using a linear transformation, show that

$$|p'(y)| \le \frac{2}{1-y} n^2 \|p\|_{[y,1]} \le 2n^2 \|p\|_{[-1,1]} \,, \qquad -1 \le y \le 0$$

and

$$|p'(y)| \le \frac{2}{1+y} n^2 \|p\|_{[-1,y]} \le 2n^2 \|p\|_{[-1,1]} \,, \qquad 0 \le y \le 1 \,.$$

Hence the inequality of the exercise is proved when $m = 1$. For larger values of m use induction. □

E.4 Weighted Bernstein and Markov Inequalities. Let $w \in C[-1,1]$ be strictly positive on $[-1,1]$.

a] Show that for every $\epsilon > 0$ there exists an n_0 depending on ϵ and w such that

$$\left\| p'(x) w(x) \sqrt{1-x^2} \right\|_{[-1,1]} \le n(1+\epsilon) \|pw\|_{[-1,1]}$$

for every $p \in \mathcal{P}_n$, $n \ge n_0$.

Proof. By the Weierstrass approximation theorem, for every $\eta > 0$ there is a $q \in \mathcal{P}_k$ such that

$$w(x) \le q(x) \le (1+\eta) w(x) \,, \qquad x \in [-1,1] \,.$$

Let $m := \min\{w(x) : x \in [-1,1]\}$. Applying Theorem 5.1.7 (Bernstein's inequality) to $pq \in \mathcal{P}_{n+k}$ and then to $q \in \mathcal{P}_k$, we obtain

$$\begin{aligned}
\left|p'(x)w(x)\sqrt{1-x^2}\right| &\le \left|p'(x)q(x)\sqrt{1-x^2}\right| \\
&\le \left|(pq)'(x)\sqrt{1-x^2}\right| + \left|p(x)q'(x)\sqrt{1-x^2}\right| \\
&\le (n+k)\|pq\|_{[-1,1]} + \|p\|_{[-1,1]} k \|q\|_{[-1,1]} \\
&\le (n+k)(1+\eta)\|pw\|_{[-1,1]} + \frac{1}{m}\|pw\|_{[-1,1]} k(1+\eta)\|w\|_{[-1,1]} \\
&\le n(1+\epsilon)\|pw\|_{[-1,1]}
\end{aligned}$$

for every $p \in \mathcal{P}_n$, provided $\eta > 0$ is sufficiently small and $n \ge n_0$. □

b] Show that for every $\epsilon > 0$ there exists an n_0 depending on ϵ and w such that

$$\|p'w\|_{[-1,1]} \leq n^2(1+\epsilon)\|pw\|_{[-1,1]}$$

for every $p \in \mathcal{P}_n, \ \ n \geq n_0$.

Hint: Use the idea given in the previous proof. □

Schaeffer and Duffin [38] prove an extension of Theorem 5.1.7 (Bernstein's inequality) to higher derivatives. They show that

$$\left|\frac{d^m}{dx^m} p(x)\right| \leq \left|\frac{d^m}{dx^m} \exp(in \arccos x)\right|, \qquad x \in (-1,1)$$

for every $p \in \mathcal{P}_n$. The following exercise gives a slightly weaker version of this, which is much simpler to prove. Some of this follows Lachance [84].

E.5 Bernstein's Inequality for Higher Derivatives.

a] Show that there exists a constant $c(m)$ depending only on m such that

$$|p^{(m)}(x)| \leq c(m)\left(\frac{n}{\sqrt{1-x^2}}\right)^m \|p\|_{[-1,1]}, \qquad x \in (-1,1)$$

for every $p \in \mathcal{P}_n^c$. (That we can choose $c(m) \leq 2^m$ is shown in parts c], d], and e].)

Hint: For $j = 1, 2, \dots, m$, let

$$a_j := x - \frac{(m-j)(1+x)}{m} \qquad \text{and} \qquad b_j := x + \frac{(m-j)(1-x)}{m}.$$

Use Corollary 5.1.5 to show that there are constants $c_j(m)$ depending only on m such that

$$\|p^{(j)}\|_{[a_j,b_j]} \leq c_j(m)\frac{n}{\sqrt{1-x^2}}\|p^{(j-1)}\|_{[a_{j-1},b_{j-1}]}$$

for every $p \in \mathcal{P}_n^c$ and $j = 1, 2, \dots, m$. □

b] Show that there exists a constant $c(m) > 0$ depending only on m such that

$$\sup_{0 \neq p \in \mathcal{P}_n} \frac{|p^{(m)}(x)|}{\|p\|_{[-1,1]}} \geq c(m)\left(\min\left\{n^2, \frac{n}{\sqrt{1-x^2}}\right\}\right)^m$$

for every $x \in [-1,1]$ and $m = 1, 2, \dots, n$.

Hint: First show that

$$\|T_n^{(m)}\|_{I(x)} \geq c(m)\left(\min\left\{n^2, \frac{n}{\sqrt{1-x^2}}\right\}\right)^m,$$

where $I(x) := [x - \frac{1}{2}(1-|x|), x + \frac{1}{2}(1-|x|)]$, then use a shift and a scaling. □

In the rest of the exercise we show that $c(m) = 2^m$ is a suitable choice for the constant in part a].

c] Let k be a positive integer. Then

$$\|p(x)(1-x^2)^{(k-1)/2}\|_{[-1,1]} \leq \frac{n+k}{k} \|p(x)(1-x^2)^{k/2}\|_{[-1,1]}$$

for every $p \in \mathcal{P}_n$.

Hint: Let $p \in \mathcal{P}_n$ be normalized so that $\|p(x)(1-x^2)^{k/2}\|_{[-1,1]} = 1$. Apply Theorem 5.1.3 (Bernstein-Szegő inequality) with

$$t(\theta) := p(\cos\theta)\sin^k\theta \in \mathcal{T}_{n+k}\,.$$

If $x_0 = \cos\theta_0$ denotes a relative extreme point for p on $(-1,1)$, then $p'(\cos\theta_0) = 0$, and after simplification we obtain that

$$(p(\cos\theta_0)\sin^{k-1}\theta_0)^2 \leq \frac{(n+k)^2}{((n+k)^2 - k^2)\sin^2\theta_0 + k^2} \leq \left(\frac{n+k}{k}\right)^2\,.$$

□

d] Let k be a positive integer. Then

$$\|p'(x)(1-x^2)^{(k+1)/2}\|_{[-1,1]} \leq 2(n+k)\|p(x)(1-x^2)^{k/2}\|_{[-1,1]}$$

for every $p \in \mathcal{P}_n$.

Proof. Let $p \in \mathcal{P}_n$ be normalized so that $\|p(x)(1-x^2)^{k/2}\|_{[-1,1]} = 1$. Applying Theorem 5.1.4 (Bernstein's inequality) with $m = 1$ and

$$t(\theta) := p(\cos\theta)\sin^k\theta \in \mathcal{T}_{n+k}\,,$$

we obtain

$$|p'(\cos\theta)\sin^{k+1}\theta + p(\cos\theta)k\sin^{k-1}\theta\cos\theta| \leq n+k\,.$$

Now the triangle inequality and part c] yield

$$\begin{aligned}|p'(\cos\theta)\sin^{k+1}\theta| &\leq (n+k) + k\left|p(\cos\theta)\sin^{k-1}\theta\right| \\ &\leq (n+k) + k\,\frac{n+k}{k} \leq 2(n+k)\,.\end{aligned}$$

□

e] Show that

$$|p^{(m)}(x)| \le \left(\frac{2n}{\sqrt{1-x^2}}\right)^m \|p\|_{[-1,1]}$$

for every $p \in \mathcal{P}_n$ and $x \in (-1,1)$.

Hint: Use induction on m, Theorem 5.1.7, and part d]. □

5.3 Inequalities for Norms of Factors

A typical result of this section is the following inequality due to Kneser [34].

Theorem 5.3.1. *Suppose $p = qr$, where $q \in \mathcal{P}_m^c$ and $r \in \mathcal{P}_{n-m}^c$. Then*

$$\|q\|_{[-1,1]}\|r\|_{[-1,1]} \le \tfrac{1}{2} C_{n,m} C_{n,n-m} \|p\|_{[-1,1]}\,,$$

where

$$C_{n,m} := 2^m \prod_{k=1}^{m} \left(1 + \cos \tfrac{(2k-1)\pi}{2n}\right).$$

Furthermore, for any n and $m \le n$ the inequality is sharp in the case that p is the Chebyshev polynomial T_n of degree n defined by (2.1.1), and the factor $q \in \mathcal{P}_m^c$ is chosen so that q vanishes at the m zeros of p closest to -1.

Before proving the above theorem, we establish an asymptotic formula for $C_{n,m}$ and formulate a corollary.

If $f \in C^2[a,b]$, then by the midpoint rule of numerical integration

$$\int_a^b f(x)\,dx = (b-a) f\left(\tfrac{1}{2}(a+b)\right) + \frac{(b-a)^3}{24} f''(\xi)$$

for some $\xi \in [a,b]$. Let $f(x) := \log(2 + 2\cos \pi x)$. Then

$$f''(x) = \frac{-\pi^2}{(1+\cos \pi x)^2}\,.$$

On applying the midpoint rule to the above f, we obtain

$$\int_0^{m/n} \log(2 + 2\cos \pi x)\,dx = \frac{1}{n} \sum_{k=1}^{m} \log\left(2 + 2\cos \tfrac{(2k-1)\pi}{2n}\right) + O\left(m \frac{1}{24} \frac{1}{n^3} \frac{n^4}{(n-m)^4}\right)$$

for all integers $0 \le m \le n$. Thus

$$C_{n,m} = \exp\left(\log \prod_{k=1}^{m}\left(2 + 2\cos\tfrac{(2k-1)\pi}{2n}\right)\right) \tag{5.3.1}$$
$$= \left(\exp\left(\frac{1}{n}\sum_{k=1}^{m}\log\left(2 + 2\cos\tfrac{(2k-1)\pi}{2n}\right)\right)\right)^n$$
$$= \exp\left(O\left(\frac{mn^2}{(n-m)^4}\right)\right)(\exp(I(n,m)))^n,$$

where

$$I(n,m) := \int_0^{m/n} \log(2 + 2\cos \pi x)\, dx.$$

So

$$\left(C_{n,\lfloor n/2 \rfloor}\right)^{1/n} \sim \exp\left(\int_0^{1/2}\log(2 + 2\cos\pi x)\, dx\right) = 1.7916\ldots \tag{5.3.2}$$

and

$$\left(C_{n,\lfloor 2n/3 \rfloor}\right)^{1/n} \sim \exp\left(\int_0^{2/3}\log(2 + 2\cos\pi x)\, dx\right) = 1.9081\ldots. \tag{5.3.3}$$

We use the notation $a_n \sim b_n$ and $a_n \lesssim b_n$ to mean $\lim\limits_{n\to\infty} a_n/b_n = 1$ and $\limsup\limits_{n\to\infty} a_n/b_n \le 1$, respectively.

On estimating $\frac{1}{2}C_{n,m}C_{n,n-m}$ in Theorem 5.3.1 and using (5.3.2), we obtain the following:

Corollary 5.3.2. *Let $p \in \mathcal{P}_n^c$ and suppose $p = qr$ for some polynomials q and r. Then*

$$\|q\|_{[-1,1]}\|r\|_{[-1,1]} \le 2^{n-1}\prod_{k=1}^{\lfloor n/2 \rfloor}\left(1 + \cos\tfrac{(2k-1)\pi}{2n}\right)^2 \|p\|_{[-1,1]}$$
$$\le \tfrac{1}{2}C_{n,\lfloor n/2 \rfloor}^2 \|p\|_{[-1,1]}$$

and equality holds when p is the Chebyshev polynomial T_n of degree n, and the factor $q \in \mathcal{P}_m^c$ is chosen so that $m := \lfloor n/2 \rfloor$ and q vanishes at the m zeros of T_n closest to -1. Here $C_{n,\lfloor n/2 \rfloor}^{2/n} \sim 3.20991\ldots$, hence

$$\left(\frac{\|q\|_{[-1,1]}\|r\|_{[-1,1]}}{\|p\|_{[-1,1]}}\right)^{1/n} \lesssim 3.20991\ldots.$$

The proof of Theorem 5.3.1 proceeds through a number of lemmas. For the remainder of the proof we assume that $0 < m < n$ are fixed. Now

$$\sup\{\|q\|_{[-1,1]}\|r\|_{[-1,1]} : \|qr\|_{[-1,1]} = 1\,,\ q \in \mathcal{P}_m^c\,,\ r \in \mathcal{P}_{n-m}^c\} \tag{5.3.4}$$

is attained for some $q \in \mathcal{P}_m^c$ and $r \in \mathcal{P}_{n-m}^c$. We proceed to show that there are extremal polynomials $q \in \mathcal{P}_m^c$ and $r \in \mathcal{P}_{n-m}^c$ such that $p := qr$ is the Chebyshev polynomial T_n of degree of n, and that the factors q and r are as advertised, that is,

$$p(x) = (qr)(x) = T_n(x) = \frac{1}{2}\prod_{k=1}^{n} 2\left(x - \cos\tfrac{(2k-1)\pi}{2n}\right),$$

and the extremal factors, q and r, are given by

$$q(x) := \frac{1}{\sqrt{2}}\prod_{k=1}^{m} 2\left(x - \cos\tfrac{(2k-1)\pi}{2n}\right)$$

and

$$r(x) := \frac{1}{\sqrt{2}}\prod_{k=m+1}^{n} 2\left(x - \cos\tfrac{(2k-1)\pi}{2n}\right),$$

respectively. Note that for the above q and r we have

$$\|q\|_{[-1,1]} = |q(-1)| = \frac{1}{\sqrt{2}}C_{n,m}$$

and

$$\|r\|_{[-1,1]} = |r(1)| = \frac{1}{\sqrt{2}}C_{n,n-m}\,.$$

First we show that there exist extremal polynomials $q \in \mathcal{P}_m^c$ and $r \in \mathcal{P}_{n-m}^c$ such that

$$|q(-1)| = \|q\|_{[-1,1]} \qquad \text{and} \qquad |r(1)| = \|r\|_{[-1,1]}\,. \tag{5.3.5}$$

To see this, choose $\alpha,\ \beta \in [-1,1]$ such that

$$|q(\alpha)| = \|q\|_{[-1,1]} \qquad \text{and} \qquad |r(\beta)| = \|r\|_{[-1,1]}\,,$$

where, considering $q(-z)$ and $r(-z)$ if necessary, we may assume that $\alpha \leq \beta$. Note that $\alpha = \beta$ cannot happen, so $\alpha < \beta$. We have

$$\frac{\|q\|_{[\alpha,\beta]}\|r\|_{[\alpha,\beta]}}{\|qr\|_{[\alpha,\beta]}} \geq \frac{\|q\|_{[-1,1]}\|r\|_{[-1,1]}}{\|qr\|_{[-1,1]}}$$

since the numerators are equal and

$$\|qr\|_{[\alpha,\beta]} \leq \|qr\|_{[-1,1]}\,.$$

Let $\widetilde{q} \in \mathcal{P}_m$ be defined by shifting q from $[\alpha,\beta]$ to $[-1,1]$ linearly so that $\alpha \to -1$. Let $\widetilde{r} \in \mathcal{P}_{n-m}^c$ be defined by shifting r from $[\alpha,\beta]$ to $[-1,1]$ linearly so that $\beta \to 1$. Then $\widetilde{q} \in \mathcal{P}_m^c$ and $\widetilde{r} \in \mathcal{P}_{n-m}^c$ are extremal polynomials for which (5.3.5) holds.

Lemma 5.3.3. *Suppose $q \in \mathcal{P}_m^c$ and $r \in \mathcal{P}_{n-m}^c$ are extremal polynomials for which (5.3.5) holds. Then there are extremal polynomials $\widetilde{q} \in \mathcal{P}_m^c$ and $\widetilde{r} \in \mathcal{P}_{n-m}^c$ having only real zeros for which (5.3.5) holds.*

Proof. Let $\widetilde{q}(z)$ be defined by replacing every factor $z - \alpha$ with nonreal α by $z - (|\alpha+1|-1)$ in the factorization of q. Let $\widetilde{r}(z)$ be defined by replacing every factor $z - \alpha$ with nonreal α by $z - (1 - |\alpha - 1|)$ in the factorization of r. Now it is elementary geometry to show that $\widetilde{q} \in \mathcal{P}_m^c$ and $\widetilde{r} \in \mathcal{P}_{n-m}^c$ are extremal polynomials for which (5.3.5) holds, and all the zeros of both $\widetilde{q}$ and $\widetilde{r}$ are real. □

Lemma 5.3.4. *Suppose $q \in \mathcal{P}_m^c$ and $r \in \mathcal{P}_{n-m}^c$ are extremal polynomials having only real zeros for which (5.3.5) holds. Then there are extremal polynomials $\widetilde{q} \in \mathcal{P}_m^c$ and $\widetilde{r} \in \mathcal{P}_{n-m}^c$ having all their zeros in $[-1, 1]$ for which (5.3.5) holds.*

Proof. Let $\widetilde{q}(z)$ be defined by replacing every factor $z - \alpha$ by $z - 1$ if $\alpha > 1$, and by 1 if $\alpha < -1$, in the factorization of q. Let $\widetilde{r}(z)$ be defined by replacing every factor $z - \alpha$ by $z + 1$ if $\alpha < -1$, and by 1 if $\alpha > 1$. Now it is again elementary geometry to show that $\widetilde{q} \in \mathcal{P}_m^c$ and $\widetilde{q} \in \mathcal{P}_{n-m}^c$ are extremal polynomials having all their zeros in $[-1, 1]$ for which (5.3.5) holds. □

So we now assume that $q \in \mathcal{P}_m^c$ and $r \in \mathcal{P}_{n-m}^c$ are extremal polynomials having all their zeros in $[-1, 1]$ for which (5.3.5) holds. We may also assume that $\deg(q) = m$ and $\deg(r) = n - m$, otherwise we would study

$$\widetilde{q}(z) := z^{m-\deg(q)} q(z) \in \mathcal{P}_m^c$$

and

$$\widetilde{r}(z) := z^{n-m-\deg(r)} r(z) \in \mathcal{P}_{n-m}^c ,$$

which are also extremal polynomials having all their zeros in $[-1, 1]$ for which (5.3.5) holds.

It is now clear that if q and r are extremal polynomials with the above properties, then the smallest zero of q is not less than the largest zero of r. Indeed, if there were numbers $-1 \leq \alpha < \beta \leq 1$ so that $q(\alpha) = r(\beta) = 0$, then the polynomials

$$\widetilde{q}(z) := q(z) \frac{z - \beta}{z - \alpha} \in \mathcal{P}_m^c$$

and

$$\widetilde{r}(z) := r(z) \frac{z - \alpha}{z - \beta} \in \mathcal{P}_{n-m}^c$$

would contradict the extremality of q and r since

$$\|\widetilde{q}\|_{[-1,1]} \geq |\widetilde{q}(-1)| > |q(-1)| = \|q\|_{[-1,1]},$$

$$\|\widetilde{r}\|_{[-1,1]} \geq |\widetilde{r}(1)| > |r(1)| = \|r\|_{[-1,1]},$$

and

$$\|\widetilde{q}\widetilde{r}\|_{[-1,1]} = \|qr\|_{[-1,1]}.$$

So now we have extremal polynomials $q \in \mathcal{P}_m^c$ and $r \in \mathcal{P}_{n-m}^c$ of the form

$$q(z) = \sqrt{a}\prod_{k=1}^{m}(z - \beta_k) \qquad \text{and} \qquad r(z) = \sqrt{a}\prod_{k=1}^{n-m}(z - \alpha_k)$$

satisfying

$$|q(-1)| = \|q\|_{[-1,1]} \qquad \text{and} \qquad |r(1)| = \|r\|_{[-1,1]},$$

where

$$-1 \leq \alpha_1 \leq \alpha_2 \leq \cdots \leq \alpha_{n-m} \leq \beta_1 \leq \beta_2 \leq \cdots \leq \beta_m \leq 1$$

and the constant $a > 0$ is chosen so that for $p := qr$ we have $\|p\|_{[-1,1]} = 1$. Now we are ready to prove Theorem 5.3.1.

Proof of Theorem 5.3.1. We show three properties of $p = qr$:

(1) $|p(-1)| = 1$ and $\|p(1)| = 1$.

(2) $\|p(x)\|_{[\alpha_i,\alpha_{i+1}]} = 1, \quad i = 1, 2, \ldots, n-m-1,$

$\|p(x)\|_{[\beta_i,\beta_{i+1}]} = 1, \quad i = 1, 2, \ldots, m-1.$

(3) $\|p\|_{[\alpha_{n-m},\beta_1]} = 1.$

These three facts show that p is indeed the Chebyshev polynomial $\pm T_n$ defined by (2.1.1) since $\pm T_n$ are the only polynomials of degree at most n that equioscillate $n+1$ times on $[-1,1]$ with uniform norm 1.

To prove (1), assume to the contrary that $|p(-1)| < 1$. Then there is a $\delta < -1$ such that

$$\|p\|_{[\delta,1]} = \|p\|_{[-1,1]}.$$

Since $|q|$ is strictly decreasing on $[\delta, -1]$,

$$\|q\|_{[\delta,1]} \geq |q(\delta)| > |q(-1)| = \|q\|_{[-1,1]}$$

and, of course,

$$\|r\|_{[\delta,1]} \geq \|r\|_{[-1,1]}.$$

Let $\widetilde{q} \in \mathcal{P}_m^c$ and $\widetilde{r} \in \mathcal{P}_{n-m}^c$ be the polynomials q and r shifted linearly from $[\delta, 1]$ to $[-1, 1]$ so that $1 \mapsto 1$. By the previous observations, these $\widetilde{q}$

and $\widetilde{r}$ contradict the extremality of q and r, hence $|p(-1)| = 1$, and we are finished. A similar argument shows that $|p(1)| = 1$.

To prove (2) let

$$s(z) := (z - \alpha_i)(z - \alpha_{i+1}) \, .$$

Given $\epsilon > 0$, we can find $0 < \delta_1, \delta_2 < \epsilon$ such that

$$t(z) := (z - (\alpha_i - \delta_1))(z - (\alpha_{i+1} + \delta_2))$$

satisfies

$$t(1) = s(1) \, , \qquad \|t - s\|_{[-1,1]} < \epsilon \, ,$$

and

$$|t(x)| < |s(x)| \, , \qquad x \in [-1, \alpha_i - \delta_1] \cup [\alpha_{i+1} + \delta_2, 1) \, .$$

Suppose $\|p\|_{[\alpha_i, \alpha_{i+1}]} < 1$. Let

$$\widehat{q}(z) := q(z) \in \mathcal{P}_m^c$$

and

$$\widehat{r}(z) := r(z) \frac{t(z)}{s(z)} \in \mathcal{P}_{n-m}^c \, .$$

If $\epsilon > 0$ is sufficiently small, then

$$\|\widehat{q}\widehat{r}\|_{[-1,1]} \leq \|qr\|_{[-1,1]} \qquad \text{and} \qquad |(\widehat{q}\widehat{r})(-1)| < 1 \, .$$

The second inequality guarantees that there exists a $\delta < -1$ such that

$$\|\widehat{q}\widehat{r}\|_{[\delta,1]} = \|\widehat{q}\widehat{r}\|_{[-1,1]} \, .$$

Since $|\widehat{q}|$ is (strictly) decreasing on $(-\infty, 1]$,

$$\|\widehat{q}\|_{\delta,1]} \geq |\widehat{q}(\delta)| > |\widehat{q}(-1)| = |q(-1)| = \|q\|_{[-1,1]} \, .$$

Also

$$\|\widehat{r}\|_{[\delta,1]} \geq \|\widehat{r}\|_{[-1,1]} \geq |\widehat{r}(1)| = |r(1)| = \|r\|_{[-1,1]} \, .$$

Now let $\widetilde{q} \in \mathcal{P}_m^c$ and $\widetilde{r} \in \mathcal{P}_{n-m}^c$ be the polynomials $\widehat{p}$ and $\widehat{q}$ shifted linearly from $[\delta, 1]$ to $[-1, 1]$ so that $-1 \to -1$. By the previous observations, these $\widetilde{q}$ and $\widetilde{r}$ contradict the extremality of q and r. Hence $\|p\|_{[\alpha_i, \alpha_{i+1}]} = 1$, and the proof is finished. The proof of $\|p\|_{[\beta_i, \beta_{i+1}]} = 1$ is identical.

To prove (3) assume that $\|p\|_{[\alpha_{n-m}, \beta_1]} < 1$. Let

$$\widetilde{q}(z) := \sqrt{a}(z - (\beta_1 + \epsilon)) \prod_{k=2}^{m} (z - \beta_k)$$

and

$$\widetilde{r}(z) := \sqrt{a}(z - (\alpha_{n-m} - \epsilon)) \prod_{k=1}^{n-m-1} (z - \alpha_k)\,.$$

If $\epsilon > 0$ is sufficiently small, then

$$\|\widetilde{q}\|_{[-1,1]} \geq |\widetilde{q}(-1)| > |q(-1)| = \|q\|_{[-1,1]}\,,$$

$$\|\widetilde{r}\|_{[-1,1]} \geq |\widetilde{r}(1)| > |r(1)| = \|r\|_{[-1,1]}\,,$$

and

$$\|\widetilde{q}\widetilde{r}\|_{[-1,1]} \leq \|qr\|_{[-1,1]}\,,$$

which contradicts the extremality of q and r. Hence $\|p\|_{[\alpha_{n-m},\beta_1]} = 1$, indeed. □

Theorem 5.3.5. *Suppose $p \in \mathcal{P}_n^c$ is monic and $q \in \mathcal{P}_m^c$ is a monic factor of p. Then*

$$|q(-\beta)| \leq \beta^{m-n} 2^{n-1} \prod_{k=1}^{m} \left(1 + \cos\tfrac{(2k-1)\pi}{2n}\right) \|p\|_{[-\beta,\beta]}$$

for every $\beta > 0$. Equality holds if p is the Chebyshev polynomial $T_{n,\beta}$ of degree n on $[-\beta,\beta]$ (normalized to be monic), and the monic factor $q \in \mathcal{P}_m^c$ is chosen so that q vanishes at the m zeros of $T_{n,\beta}$ closest to β. Note that $T_{n,\beta}(x) = \beta^n T_n(x/\beta)$, where T_n is defined by (2.1.1).

The proof of Theorem 5.3.5 is outlined in E.1.

Corollary 5.3.6. *Suppose $p \in \mathcal{P}_n^c$ is monic and $q \in \mathcal{P}_m^c$ is a monic factor of p. Then*

$$|q(-2)| \leq 2^{m-1} \prod_{k=1}^{m} \left(1 + \cos\tfrac{(2k-1)\pi}{2n}\right) \|p\|_{[-2,2]} = \frac{1}{2} C_{n,m} \|p\|_{[-2,2]}$$

and the inequality is sharp for all $m \leq n$. Here, for all $m \leq n$,

$$C_{n,m}^{1/n} \leq C_{n,\lfloor 2n/3 \rfloor}^{1/n} \sim 1.9081\ldots\,,$$

and hence

$$\left(\frac{|q(-2)|}{\|p\|_{[-2,2]}}\right)^{1/n} \lesssim 1.9081\ldots\,.$$

Proof. Take $\beta = 2$ in Theorem 5.3.5. Note that

$$2^m \prod_{k=1}^{m} \left(1 + \cos \tfrac{(2k-1)\pi}{2n}\right) \le 2^{\gamma(n)} \prod_{k=1}^{\gamma(n)} \left(1 + \cos \tfrac{(2k-1)\pi}{2n}\right),$$

where $\gamma(n)+1 := \lfloor \frac{2}{3}n + \frac{3}{2} \rfloor$ is the smallest $k \in \mathbb{N}$ for which $\cos \frac{(2k-1)\pi}{2n} \le -\frac{1}{2}$. So, for every $m = 0, 1, \ldots, n$,

$$C_{n,m} \le C_{n,\gamma(n)}$$

and by (5.3.3)

$$C_{n,m}^{1/n} \lesssim C_{n,\lfloor 2n/3 \rfloor} \lesssim 1.9081\ldots .$$

□

Theorem 5.3.7. *Suppose $p \in \mathcal{P}_n^c$ is monic and has a monic factor of the form qr, where $q \in \mathcal{P}_{m_1}^c$ and $r \in \mathcal{P}_{m_2}^c$. Then*

$$|q(-\beta)||r(\beta)| \le \beta^{m_1+m_2-n} 2^{n-1} \times \prod_{k=1}^{m_1} \left(1 + \cos \tfrac{(2k-1)\pi}{2n}\right) \prod_{k=1}^{m_2} \left(1 + \cos \tfrac{(2k-1)\pi}{2n}\right) \|p\|_{[-\beta,\beta]} \,.$$

Equality holds if p is the Chebyshev polynomial $T_{n,\beta}$ of degree n on $[-\beta, \beta]$ normalized to be monic, and the factors $q \in \mathcal{P}_{m_1}^c$ and $r \in \mathcal{P}_{m_2}^c$ are chosen so that q vanishes at the m_1 zeros of $T_{n,\beta}$ closest to β, while r vanishes at the m_2 zeros of $T_{n,\beta}$ closest to $-\beta$.

The proof of Theorem 5.3.7 is analogous to the proof of Theorem 5.3.1 and is left as an exercise (see E.2).

Theorem 5.3.8. *Suppose $p \in \mathcal{P}_n^c$ is monic and has a monic factor of the form $q_1 q_2 \cdots q_j$, where $q_i \in \mathcal{P}_{m_i}^c$ and $m := m_1 + m_2 + \cdots + m_j \le n$. Then*

$$\prod_{i=1}^{j} \|q_i\|_{[-\beta,\beta]} \le \beta^{m-n} 2^{n-1} \times \prod_{k=1}^{\lfloor m/2 \rfloor} \left(1 + \cos \tfrac{(2k-1)\pi}{2n}\right) \prod_{k=1}^{\lceil m/2 \rceil} \left(1 + \cos \tfrac{(2k-1)\pi}{2n}\right) \|p\|_{[-\beta,\beta]}$$

for every $\beta > 0$. Equality holds if p is the Chebyshev polynomial $T_{n,\beta}$ of degree n on $[-\beta, \beta]$ normalized to be monic, $j = 2$, and the factors $q_1 \in \mathcal{P}_{\lfloor m/2 \rfloor}^c$ and $q_2 \in \mathcal{P}_{\lceil m/2 \rceil}^c$ are chosen so that q_1 vanishes at the $\lfloor m/2 \rfloor$ zeros of $T_{n,\beta}$ closest to β, while q_2 vanishes at the $\lceil m/2 \rceil$ zeros of $T_{n,\beta}$ closest to $-\beta$.

Proof. For each q_i, write

$$q_i = r_i s_i \,,$$

where r_i is the monic factor of q_i composed of the roots of q_i with negative real part, while s_i is the monic factor of q_i composed of the roots of q_i with nonnegative real part. So

$$\|r_i\|_{[-\beta,\beta]} = |r_i(\beta)| \qquad \text{and} \qquad \|s_i\|_{[-\beta,\beta]} = |s_i(-\beta)| \,.$$

Thus

$$\prod_{i=1}^{j} \|q_i\|_{[-\beta,\beta]} \leq \left|\prod_{i=1}^{j} r_i(\beta)\right| \cdot \left|\prod_{i=1}^{j} s_i(-\beta)\right| .$$

We now apply Theorem 5.3.7 to the two factors $\prod_{i=1}^{j} r_i$ and $\prod_{i=1}^{j} s_i$ to finish the proof. □

As before, let $D := \{z \in \mathbb{C} : |z| < 1\}$. We now derive inequalities on the disk from those on the interval. A continuous function on $\overline{D}$ has the same uniform norm on both D and $\overline{D}$ and it is notationally convenient to state the remaining theorems over D. Suppose $t \in \mathcal{P}_n^c$, $s \in \mathcal{P}_m^c$, and $v \in \mathcal{P}_{n-m}^c$ are monic, and $t = sv$. By the maximum principle, t, s, and v achieve their maximum on $\overline{D}$ somewhere on ∂D. Now consider

$$p(x) := t(z)t(z^{-1}) \,,$$

$$q(x) := s(z)s(z^{-1}) \,, \qquad \text{and} \qquad r(x) := v(z)v(z^{-1})$$

with

$$x := z + z^{-1} \,.$$

The effect of this transformation on linear factors is

$$(z-\alpha)(z^{-1}-\alpha) = -\alpha x + 1 + \alpha^2 \,,$$

so $p \in \mathcal{P}_n^c$, $q \in \mathcal{P}_m^c$, $r \in \mathcal{P}_{n-m}^c$, and $p = qr$. Also

$$\|p\|_{[-2,2]} \leq \|t\|_D^2 \,.$$

If $t(0) \neq 0$, then the modulus of the leading coefficient of p is $|t(0)|$, while the modulus of the leading coefficient of q is $|s(0)|$, and the modulus of the leading coefficient of r is $|v(0)|$.

From these transformations and the interval inequalities we can deduce the next three theorems.

Theorem 5.3.9. *Let $t \in \mathcal{P}_n^c$ be monic and suppose $t = sv$, where $s \in \mathcal{P}_m^c$ and $v \in \mathcal{P}_{n-m}^c$. Then*

$$|v(0)|^{1/2}\|s\|_D \le \left(\tfrac{1}{2}C_{n,m}\right)^{1/2}\|t\|_D\,,$$

where $C_{n,m}$ is the same as in Theorem 5.3.1. This bound is attained when $m \le n$ are even, $t(z) = z^n + 1$, and $s \in \mathcal{P}_m^c$ vanishes at m adjacent zeros of t on the unit circle.

Proof. We may assume, by performing an initial rotation if necessary, that

$$\|s\|_D = |s(-1)|\,.$$

So from Corollary 5.3.6 we deduce that

$$\begin{aligned}
(5.3.6)\qquad \|s\|_D^2 &= |s(-1)|^2 = |q(-2)| \\
&\le |s(0)/t(0)|2^{m-1}\prod_{k=1}^{m}\left(1+\cos\tfrac{(2k-1)\pi}{2n}\right)\|p\|_{[-2,2]} \\
&\le |s(0)/t(0)|2^{m-1}\prod_{k=1}^{m}\left(1+\cos\tfrac{(2k-1)\pi}{2n}\right)\|t\|_D^2\,,
\end{aligned}$$

where $s(0)/t(0) = 1/v(0)$. □

Theorem 5.3.10. *Suppose $t = sv$, where $s \in \mathcal{P}_m$ and $r \in \mathcal{P}_{n-m}$. Then*

$$\|s\|_D\|v\|_D \le \left(\tfrac{1}{2}C_{n,m}C_{n,n-m}\right)^{1/2}\|t\|_D\,,$$

where $C_{n,m}$ is the same as in Theorem 5.3.1 and

$$(C_{n,m}C_{n,n-m})^{1/(2n)} \le C_{n,\lfloor n/2\rfloor}^{1/n} \sim 1.7916\ldots\,.$$

This bound is attained when $m \le n$ are even, $t(z) = z^n + 1$, and $s \in \mathcal{P}_m^c$ vanishes at the m zeros of t closest to 1 and $v \in \mathcal{P}_{n-m}$ vanishes at the $n-m$ zeros of t closest to -1.

Proof. From Theorem 5.3.1 we can deduce that if $a,\, b \in \partial D$, then

$$\begin{aligned}
|s(a)|^2|v(b)|^2 &= |s(a)s(a^{-1})||v(b)v(b^{-1})| \\
&= |q(a+a^{-1})||r(b+b^{-1})| \\
&\le \tfrac{1}{2}C_{n,m}C_{n,n-m}\|p\|_{[a+a^{-1},b+b^{-1}]} \\
&\le \tfrac{1}{2}C_{n,m}C_{n,n-m}\|p\|_{[-2,2]} \\
&\le \tfrac{1}{2}C_{n,m}C_{n,n-m}\|t\|_D^2\,,
\end{aligned}$$

where, without loss of generality, we may assume that $a + a^{-1} \le b + b^{-1}$. The result now follows on choosing a and b to be points on ∂D where s and v, respectively, achieve their uniform norm on D. □

In the multifactor case we have the following theorem:

Theorem 5.3.11. *Suppose $t \in \mathcal{P}_n$ is of the form $t = vs_1s_2\cdots s_j$, where $s_i \in \mathcal{P}_{m_i}$ and $v \in \mathcal{P}_{n-m}$ with $m := m_1 + m_2 + \cdots + m_j \le n$. Then*

$$|v(0)|^{1/2}\prod_{i=1}^{j}\|s_i\|_D \le 2^{(m-1)/2}$$

$$\times\left(\prod_{k=1}^{\lfloor m/2\rfloor}\left(1+\cos\frac{(2k-1)\pi}{2n}\right)\prod_{k=1}^{\lceil m/2\rceil}\left(1+\cos\frac{(2k-1)\pi}{2n}\right)\right)^{1/2}\|t\|_D\,.$$

Equality holds if $t(z) = z^n + 1$, $j = 2$, $m_1 = m_2 := m/2$ and n are even, and the factors $q_1 \in \mathcal{P}_{m/2}$ and $q_2 \in \mathcal{P}_{m/2}$ are chosen so that q_1 vanishes at the $m/2$ zeros of t closest to 1 and q_2 vanishes at the $m/2$ zeros of t closest to -1.

Proof. This follows from Theorem 5.3.8 in exactly the same way as Theorem 5.3.10 follows from Theorem 5.3.1. □

Comments, Exercises, and Examples.

The first result of this section is due to Kneser [34] and in part to Aumann [33]. The proof follows Borwein [94], as does most of the section. There are many variations and generalizations. See Boyd [92], [93a], [93b], [94a], and [94b]; Beauzamy and Enflo [85]; Beauzamy, Bombieri, Enflo, and Montgomery [90]; Gel'fond [60]; Glesser [90]; Granville [90]; Mahler [60], [62], and [64]; and Mignotte [82]. Some of these are presented in the exercises. In particular, E.6 reproduces a very pretty proof of Boyd [92] that

$$\|q\|_D\|r\|_D \le (1.7916\ldots)^n\|p\|_D\,,$$

where $p \in \mathcal{P}_n^c$ and $p = qr$ with some $q \in \mathcal{P}_m^c$ and $r \in \mathcal{P}_{n-m}^c$. (Note that we have not assumed real coefficients unlike in Theorem 5.3.10, and we have $\le$ instead of $\lesssim$.)

E.1 Proof of Theorem 5.3.5.

Outline. Let $m < n$ and $\beta > 0$ be fixed. The value

$$\sup\left\{\frac{|q(-\beta)|}{\|p\|_{[-\beta,\beta]}} : q \in \mathcal{P}_m^c \text{ and } p \in \mathcal{P}_n^c \text{ are monic and } q \text{ divides } p\right\}$$

is attained for some monic $q \in \mathcal{P}_m^c$ and $p \in \mathcal{P}_n^c$. We can now argue, exactly as in the proof of Lemma 5.3.3, that there are extremal polynomials $p \in \mathcal{P}_n^c$ and $q \in \mathcal{P}_m^c$ such that all the zeros of p are real and lie in $[-\beta, \infty)$. Arguing as in Lemma 5.3.4 gives that p has all its roots in $[-\beta, \beta]$. Thus q must be composed of the m roots of p closest to β.

The argument of the proof of Theorem 5.3.1 now applies essentially verbatim and proves that an extremal $p \in \mathcal{P}_n^c$ can be chosen to be the Chebyshev polynomial on $[-\beta, \beta]$ normalized to be monic. Thus on $[-1, 1]$

$$p(x) = \prod_{k=1}^{n} \left(x - \cos \tfrac{(2k-1)\pi}{2n} \right) \qquad \text{and} \qquad q(x) = \prod_{k=1}^{m} \left(x - \cos \tfrac{(2k-1)\pi}{2n} \right)$$

from which the result follows (on considering $\beta^n p(x/\beta)$ and $\beta^m q(x/\beta)$ on $[-\beta, \beta]$). □

E.2 Proof of Theorem 5.3.7.

Hint: Proceed as in the proof of Theorem 5.3.1 (or E.1). □

E.3 A Version of Theorem 5.3.10 for Complex Polynomials. Suppose $t = sv$, where $s \in \mathcal{P}_m^c$ and $v \in \mathcal{P}_{n-m}^c$. Then

$$|s(-1)||v(1)| \le \left(\tfrac{1}{2} C_{n,m} C_{n,n-m} \right)^{1/2} \|t\|_D$$

and if t is monic

$$|t(0)|^{1/2} \|s\|_D \|v\|_D \le \tfrac{1}{2} (C_{n,m} C_{n,n-m})^{1/2} \|t\|_D^2 .$$

Hint: The first inequality follows as in the proof of Theorem 5.3.10 with $a := -1$ and $b := 1$. The second part is immediate from Theorem 5.3.9. □

E.4 Mahler's Measure. Let $F : \mathbb{C}^k \to \mathbb{C}$, and let the *Mahler measure* of F be defined by

$$M_k(F) := \exp \left\{ \int_0^1 \cdots \int_0^1 \log |F(e^{2\pi i t_1}, \ldots, e^{2\pi i t_k})| \, dt_1 \cdots dt_k \right\}$$

if the integral exist.

a] Show that if

$$p(z) = c \prod_{i=1}^{n} (z - \alpha_i) , \qquad c, \, \alpha_i \in \mathbb{C} ,$$

then

$$M_1(p) = |c| \prod_{i=1}^{n} \max\{1, |\alpha_i|\} .$$

Hint: Use Jensen's formula (see E.10 c] of Section 4.2). □

b] Show that if $F := F(z_1, \ldots, z_k)$ and $G := G(z_1, \ldots, z_k)$, then

$$M_k(FG) = M_k(F) M_k(G) .$$

c] Show that

$$\begin{aligned} M_1(ax+b) &= \max(|a|,|b|)\,, \\ M_2(1+x+y) &= M_1(\max\{1,|1+x|\})\,, \\ M_2(1+x+y-xy) &= M_1(\max\{|1-x|,|1+x|\})\,. \end{aligned}$$

d] One can numerically check that

$$M_2(1+x+y) = 1.381356\ldots$$

and

$$M_2(1+x+y-xy) = 1.791622\ldots.$$

E.5 The Norm of a Factor of a $p \in \mathcal{P}_n^c$ on the Unit Disk. Suppose $p \in \mathcal{P}_n^c$ is monic and has a monic factor $q \in \mathcal{P}_m^c$. Then

$$\|q\|_D \le \beta^n \|p\|_D\,,$$

where $\beta := M_2(1+x+y) = 1.3813\ldots$.

Outline. Let

$$p(x) := \prod_{i=1}^{n}(x-\alpha_i) \qquad \text{and} \qquad q(x) := \prod_{i=1}^{m}(x-\alpha_i)\,, \qquad \alpha_i \in \mathbb{C}\,.$$

Suppose $\|q\|_D = |q(u)|$, where $u \in \partial D$. Then

$$\begin{aligned} \|q\|_D = |q(u)| &= \prod_{i=1}^{m}|u-\alpha_i| \le \prod_{i=1}^{n}\max\{|u-\alpha_i|,1\} \\ &= M_1(p(x+u)) \le M_1(\max\{1,|x+u|^n\}\|p\|_D)\,, \end{aligned}$$

where the last equality holds by E.4 a], and the last inequality follows because

$$|p(z)| \le \max\{1,|z|^n\}\cdot\|p\|_D \tag{5.3.7}$$

holds for every $p \in \mathcal{P}_n^c$ and $z \in \mathbb{C}$ by E.18 a] of Section 5.1. Now using E.4 b] and

$$M_1(\max\{1,|x+u|\}) = M_1(\max\{1,|x+1|\})\,,$$

we obtain

$$\begin{aligned} \|q\|_D &\le M_1(\max\{1,|x+u|^n\})\,\|p\|_D \\ &= M_1((\max\{1,|x+u|\})^n)\,\|p\|_D \\ &= M_1((\max\{1,|x+1|\})^n)\,\|p\|_D \\ &= (M_1(\max\{1,|x+1|\}))^n\,\|p\|_D \\ &= (M_2(1+x+y))^n\|p\|_D = \beta^n\|p\|_D\,. \end{aligned}$$

□

E.6 Another Inequality for the Factors of a $p \in \mathcal{P}_n^c$ on the Unit Disk. Let $p = qr$, where $q \in \mathcal{P}_m^c$ and $r \in \mathcal{P}_{n-m}^c$. Then

$$\|q\|_D \|r\|_D \le \delta^n \|p\|_D$$

where $\delta := M_2(1 + x + y - xy) = 1.7916\ldots$.

Outline. Without loss of generality we may assume that q and r are monic. Let

$$q(x) := \prod_{i=1}^{m}(x - \alpha_i) \qquad \text{and} \qquad r(x) := \prod_{i=m+1}^{n}(x - \alpha_i)\,, \qquad \alpha_i \in \mathbb{C}\,.$$

Choose $u \in \partial D$ and $v \in \partial D$ such that $|q(u)| = \|q\|_D$ and $|r(v)| = \|r\|_D$. Then, using E.4 b] and c], we obtain

$$\begin{aligned}
\|q\|_D \|r\|_D = |q(u)||r(v)| &= \prod_{i=1}^{m} |u - \alpha_i| \prod_{i=m+1}^{n} |v - \alpha_i| \\
&\le \prod_{i=1}^{n} \max\{|u - \alpha_i|, |v - \alpha_i|\} \\
&= \prod_{i=1}^{n} |u - \alpha_i| \max\left\{1, \left|\frac{v - \alpha_i}{u - \alpha_i}\right|\right\} \\
&= M_1\left((x-1)^n p\left(\frac{ux - v}{x - 1}\right)\right).
\end{aligned}$$

Now, by (5.3.7),

$$\left|(x-1)^n p\left(\frac{xu - v}{x - 1}\right)\right| \le \left(\max\left\{|x - 1|, \left|x - \frac{v}{u}\right|\right\}\right)^n \|p\|_D\,,$$

hence

$$\begin{aligned}
\|q\|_D \|r\|_D &\le M_1((\max\{|x-1|, |x - v/u|\})^n)\|p\|_D \\
&= (M_1(\max\{|x-1|, |x - v/u|\}))^n \|p\|_D \\
&\le (M_1(\max\{|1-x|, |1+x|\}))^n \|p\|_D \\
&= (M_2(1 + x + y - xy))^n \|p\|_D \\
&= (1.7916\ldots)^n \|p\|_D\,.
\end{aligned}$$

□

E.7 Bombieri's Norm. For $Q(z) := \sum_{k=0}^n a_k z^k$ the *Bombieri p norm* is defined by

$$[Q]_p := \left(\sum_{k=0}^{n} \binom{n}{k}^{1-p} |a_k|^p \right)^{1/p} .$$

Note that this is a norm on $\mathcal{P}_n^c$ for every $p \in [1, \infty)$, but it varies with varying n. The following remarkable inequality holds (see Beauzamy et al. [90]). If $Q = RS$ with $Q \in \mathcal{P}_n^c$, $R \in \mathcal{P}_m^c$, and $S \in \mathcal{P}_{n-m}^c$, then

$$[R]_2 [S]_2 \leq \binom{n}{m}^{1/2} [Q]_2$$

and this is sharp.

One feature of this inequality is that it extends naturally to the multivariate case. See Beauzamy, Enflo, and Wang [94] and Reznick [93] for further discussion.

6

Inequalities in Müntz Spaces

Overview

Versions of Markov's inequality for Müntz spaces, both in $C[a,b]$ and $L_p[0,1]$, are given in the first section of this chapter. Bernstein- and Nikolskii-type inequalities are treated in the exercises, as are various other inequalities for Müntz polynomials and exponential sums. The second section provides inequalities, including most significantly a Remez-type inequality, for nondense Müntz spaces.

6.1 Inequalities in Müntz Spaces

We first present a simplified version of Newman's beautiful proof of an essentially sharp Markov-type inequality for Müntz polynomials. This simplification allows us to prove the L_p analogs of Newman's inequality. Then, using the results of Section 3.4 on orthonormal Müntz-Legendre polynomials, we prove an L_2 version of Newman's inequality for Müntz polynomials with complex exponents. Some Nikolskii-type inequalities for Müntz polynomials are studied. The exercises treat a number of other inequalities for Müntz polynomials and exponential sums. Throughout this section we use the notation introduced in Section 3.4. Unless stated otherwise, the span always denotes the linear span over $\mathbb{R}$.

Theorem 6.1.1 (Newman's Inequality). *Let $\Lambda := (\lambda_i)_{i=0}^\infty$ be a sequence of distinct nonnegative real numbers. Then*

$$\frac{2}{3}\sum_{j=0}^n \lambda_j \leq \sup_{0 \neq p \in M_n(\Lambda)} \frac{\|xp'(x)\|_{[0,1]}}{\|p\|_{[0,1]}} \leq 9\sum_{j=0}^n \lambda_j$$

for every $n \in \mathbb{N}$, where $M_n(\Lambda) := \text{span}\{x^{\lambda_0}, x^{\lambda_1}, \dots, x^{\lambda_n}\}$.

Proof. It is equivalent to prove that

$$\frac{2}{3}\sum_{j=0}^n \lambda_j \leq \sup_{0 \neq P \in E_n(\Lambda)} \frac{\|P'\|_{[0,\infty)}}{\|P\|_{[0,\infty)}} \leq 9\sum_{j=0}^n \lambda_j\,, \tag{6.1.1}$$

where $E_n(\Lambda) := \text{span}\{e^{-\lambda_0 t}, e^{-\lambda_1 t}, \dots, e^{-\lambda_n t}\}$. Without loss of generality we may assume that $\lambda_0 := 0$. By a change of scale we may also assume that $\sum_{j=0}^n \lambda_j = 1$. We begin with the first inequality. We define the Blaschke product

$$B(z) := \prod_{j=1}^n \frac{z-\lambda_j}{z+\lambda_j}$$

and the function

$$T(t) := \frac{1}{2\pi i}\int_\Gamma \frac{e^{-zt}}{B(z)}\,dz\,, \qquad \Gamma := \{z \in \mathbb{C} : |z-1| = 1\}\,. \tag{6.1.2}$$

By the residue theorem

$$T(t) := \sum_{j=1}^n (B'(\lambda_j))^{-1} e^{-\lambda_j t}\,,$$

and hence $T \in E_n(\Lambda)$. We claim that

$$|B(z)| \geq \frac{1}{3}\,, \qquad z \in \Gamma\,. \tag{6.1.3}$$

Indeed, it is easy to see that $0 \leq \lambda_j \leq 1$ implies

$$\left|\frac{z-\lambda_j}{z+\lambda_j}\right| \geq \frac{2-\lambda_j}{2+\lambda_j} = \frac{1-\frac{1}{2}\lambda_j}{1+\frac{1}{2}\lambda_j}\,, \qquad z \in \Gamma\,.$$

So, for $z \in \Gamma$,

$$|B(z)| \geq \prod_{j=1}^n \frac{1-\frac{1}{2}\lambda_j}{1+\frac{1}{2}\lambda_j} \geq \frac{1-\frac{1}{2}\sum_{j=1}^n \lambda_j}{1+\frac{1}{2}\sum_{j=1}^n \lambda_j} = \frac{1-\frac{1}{2}}{1+\frac{1}{2}} = \frac{1}{3}\,.$$

Here the inequality

$$\frac{1-x}{1+x}\,\frac{1-y}{1+y} = \frac{1-(x+y)}{1+(x+y)} + \frac{2xy(x+y)}{(1+x)(1+y)(1+(x+y))}$$
$$\geq \frac{1-(x+y)}{1+(x+y)}\,, \qquad x, y \geq 0$$

is used. From (6.1.2) and (6.1.3) we can deduce that

$$(6.1.4) \qquad |T(t)| \leq \frac{1}{2\pi} \int_\Gamma \left| \frac{e^{-zt}}{B(z)} \right| |dz| \leq \frac{1}{2\pi} 3(2\pi) = 3\,, \quad t \geq 0\,.$$

Also

$$T'(t) = \frac{1}{2\pi i} \int_\Gamma \frac{-ze^{-zt}}{B(z)}\, dz$$

and

$$(6.1.5) \qquad T'(0) = -\frac{1}{2\pi i} \int_\Gamma \frac{z}{B(z)}\, dz = -\frac{1}{2\pi i} \int_{|z|=1} \frac{z}{B(z)}\, dz\,.$$

Now, for $|z| > \max_{1 \leq j \leq n} \lambda_j$, we have the Laurent series expansion

$$(6.1.6) \quad \frac{z}{B(z)} = z \prod_{j=1}^n \frac{1+\lambda_j/z}{1-\lambda_j/z} = z \prod_{j=1}^n \left(1 + 2 \sum_{k=1}^\infty \left(\frac{\lambda_j}{z} \right)^k \right)$$
$$= z \left(1 + 2 \left(\sum_{j=1}^n \lambda_j \right) z^{-1} + 2 \left(\sum_{j=1}^n \lambda_j \right)^2 z^{-2} + \cdots \right)$$
$$= z + 2 + 2z^{-1} + \cdots\,,$$

which, together with (6.1.5), yields that $T'(0) = -2$. Hence, by (6.1.4),

$$\frac{|T'(0)|}{\|T\|_{[0,\infty)}} \geq \frac{2}{3} = \frac{2}{3} \sum_{j=1}^n \lambda_j\,,$$

so the lower bound of the theorem is proved.

To prove the upper bound in (6.1.1), first we show that if

$$U(t) := \frac{1}{2\pi i} \int_\Gamma \frac{e^{-zt}}{(1-z)B(z)}\, dz\,, \qquad \Gamma := \{z \in \mathbb{C} : |z-1| = 1\}\,,$$

then

$$(6.1.7) \qquad \int_0^\infty |U''(t)|\, dt \leq 6\,.$$

Indeed, observe that if $z = 1 + e^{i\theta}$, then $|z|^2 = 2 + 2\cos\theta$, so (6.1.3) and Fubini's theorem yield that

$$\begin{aligned}
\int_0^\infty |U''(t)|\,dt &= \int_0^\infty \frac{1}{2\pi}\left|\int_\Gamma \frac{z^2 e^{-zt}}{(1-z)B(z)}\,dz\right|\,dt \\
&\le \frac{1}{2\pi}\int_0^\infty \int_0^{2\pi} \frac{|z|^2|e^{-zt}|}{|B(z)|}\,d\theta\,dt \\
&\le \frac{3}{2\pi}\int_0^\infty \int_0^{2\pi} (2+2\cos\theta)e^{-(1+\cos\theta)t}\,d\theta\,dt \\
&= \frac{3}{2\pi}\int_0^{2\pi} (2+2\cos\theta)\,\frac{1}{1+\cos\theta}\,d\theta = 6\,.
\end{aligned}$$

Now we show that

$$\int_0^\infty e^{-\lambda_j t}U''(t)\,dt = \lambda_j - 3\,. \tag{6.1.8}$$

To see this we write the left-hand side as

$$\begin{aligned}
\int_0^\infty e^{-\lambda_j t}U''(t)\,dt &= \int_0^\infty e^{-\lambda_j t}\frac{1}{2\pi i}\int_\Gamma \frac{z^2 e^{-zt}}{(1-z)B(z)}\,dz\,dt \\
&= \frac{1}{2\pi i}\int_0^\infty \int_\Gamma \frac{z^2 e^{-(z+\lambda_j)t}}{(1-z)B(z)}\,dz\,dt = \frac{1}{2\pi i}\int_\Gamma \frac{z^2}{(z+\lambda_j)(1-z)B(z)}\,dz \\
&= \frac{1}{2\pi i}\int_{|z|=2} \frac{z}{z+\lambda_j}\,\frac{z}{1-z}\,\frac{1}{B(z)}\,dz\,,
\end{aligned}$$

where in the third equality Fubini's theorem is used again. Here, for $|z| > 1$, we have the Laurent series expansions

$$\begin{aligned}
\frac{z}{z+\lambda_j} &= 1 - \lambda_j z^{-1} + \lambda_j^2 z^{-2} + \cdots\,, \\
\frac{z}{1-z} &= -1 - z^{-1} - z^{-2} - \cdots\,,
\end{aligned}$$

and, as in (6.1.6),

$$\frac{1}{B(z)} = 1 + 2z^{-1} + 2z^{-2} + \cdots\,.$$

Now (6.1.8) follows from the residue theorem (see, for example, Ash [71]). Let $P \in E_n(\Lambda)$ be of the form

$$P(t) = \sum_{j=0}^n c_j e^{-\lambda_j t}\,, \qquad c_j \in \mathbb{R}\,.$$

Then

$$\int_0^\infty P(t+a)U''(t)\,dt = \int_0^\infty \sum_{j=0}^n c_j e^{-\lambda_j a} e^{-\lambda_j t} U''(t)\,dt$$
$$= \sum_{j=0}^n c_j e^{-\lambda_j a} \int_0^\infty e^{-\lambda_j t} U''(t)\,dt = \sum_{j=0}^n c_j(\lambda_j - 3)e^{-\lambda_j a}$$
$$= -P'(a) - 3P(a)$$

and so

$$|P'(a)| \le 3|P(a)| + \int_0^\infty |P(t+a)\,U''(t)|\,dt\,. \tag{6.1.9}$$

Combining this with (6.1.7) gives

$$\|P'\|_{[0,\infty)} \le 3\,\|P\|_{[0,\infty)} + 6\,\|P\|_{[0,\infty)} = 9\,\|P\|_{[0,\infty)}\,,$$

and the theorem is proved. □

The next theorem establishes an L_p extension of Newman's inequality.

Theorem 6.1.2 (Newman's Inequality in L_p). *Let $p \in [1,\infty)$. If $\Lambda := (\lambda_i)_{i=0}^\infty$ is a sequence of distinct real numbers greater than $-1/p$, then*

$$\|xP'(x)\|_{L_p[0,1]} \le \left(\frac{1}{p} + 12\sum_{j=0}^n \left(\lambda_j + \frac{1}{p}\right)\right) \|P\|_{L_p[0,1]}$$

for every $P \in M_n(\Lambda) := \operatorname{span}\{x^{\lambda_0}, x^{\lambda_1}, \dots, x^{\lambda_n}\}$.

If $\Gamma := (\gamma_i)_{i=0}^\infty$ is a sequence of distinct positive real numbers, then

$$\|P'\|_{L_p[0,\infty)} \le 12\left(\sum_{j=0}^n \gamma_j\right) \|P\|_{L_p[0,\infty)}$$

for every $P \in E_n(\Gamma) := \operatorname{span}\{e^{-\gamma_0 t}, e^{-\gamma_1 t}, \dots, e^{-\gamma_n t}\}$.

Proof. First we show that the first statement of the theorem follows from the second. Indeed, if $(\lambda_i)_{i=0}^\infty$ is a sequence of distinct real numbers greater than $-1/p$ and $\gamma_i := \lambda_i + \frac{1}{p}$ for each i, then $(\gamma_i)_{i=1}^\infty$ is a sequence of distinct positive real numbers. Let $Q \in M_n(\Lambda)$. Applying the second inequality with

$$P(t) := Q(e^{-t})e^{-t/p} \in E_n(\Gamma)$$

and using the substitution $x = e^{-t}$, we obtain

$$\left(\int_0^1 \left| x \left(x^{1/p} Q(x) \right)' \right|^p x^{-1} \, dx \right)^{1/p} \leq 12 \left(\sum_{j=0}^n \left(\lambda_j + \frac{1}{p} \right) \right) \|Q\|_{L_p[0,1]} \, .$$

Now the product rule of differentiation and Minkowski's inequality yield

$$\|xQ'(x)\|_{L_p[0,1]} \leq \left(\frac{1}{p} + 12 \sum_{j=0}^n \left(\lambda_j + \frac{1}{p} \right) \right) \|Q\|_{L_p[0,1]} \, ,$$

which is the first statement of the theorem.

We prove the second statement. Let $P \in E_n(\Gamma)$ and $p \in [1, \infty)$ be fixed. As in the proof of Theorem 6.1.1, by a change of scale, without loss of generality we may assume that $\sum_{j=0}^n \gamma_j = 1$. It follows from (6.1.9) and Hölder's inequality (see E.7 a] of Section 2.2) that

$$\begin{aligned} |P'(a)|^p &\leq 2^{p-1} \left(3^p |P(a)|^p + \left(\int_0^\infty |P(t+a)| |U''(t)| \, dt \right)^p \right) \\ &\leq 6^p |P(a)|^p \\ &+ 2^{p-1} \left(\left(\int_0^\infty |P(t+a)|^p |U''(t)| \, dt \right)^{1/p} \left(\int_0^\infty |U''(t)| \, dt \right)^{1/q} \right)^p \end{aligned}$$

for every $a \in [0, \infty)$, where $q \in (1, \infty]$ is the conjugate exponent to p defined by $p^{-1} + q^{-1} = 1$. Combining the above inequality with (6.1.7), we obtain

$$|P'(a)|^p \leq 6^p |P(a)|^p + 2^{p-1} 6^{p/q} \int_0^\infty |P(t+a)|^p |U''(t)| \, dt$$

for every $a \in [0, \infty)$. Integrating with respect to a, then using Fubini's theorem and (6.1.7), we conclude that

$$\begin{aligned} \|P'\|_{L_p[0,\infty)}^p &\leq 6^p \|P\|_{L_p[0,\infty)}^p + 2^{p-1} 6^{p/q} \int_0^\infty \int_0^\infty |P(t+a)|^p |U''(t)| \, dt \, da \\ &\leq 6^p \|P\|_{L_p[0,\infty)}^p + 2^{p-1} 6^{p/q} \int_0^\infty \int_0^\infty |P(t+a)|^p |U''(t)| \, da \, dt \\ &\leq 6^p \|P\|_{L_p[0,\infty)}^p + 2^{p-1} 6^{p/q} \|P\|_{L_p[0,\infty)}^p \int_0^\infty |U''(t)| \, dt \\ &\leq 6^p \|P\|_{L_p[0,\infty)}^p + 2^{p-1} 6^{p/q+1} \|P\|_{L_p[0,\infty)}^p \\ &= (6^p + 2^{p-1} 6^p) \|P\|_{L_p[0,\infty)}^p \leq 12^p \|P\|_{L_p[0,\infty)}^p \end{aligned}$$

and the proof is finished. □

The following Nikolskii-type inequality follows from Theorem 6.1.1 quite simply:

Theorem 6.1.3 (Nikolskii-Type Inequality). *Suppose* $0 < q < p \le \infty$. *If* $\Lambda := (\lambda_i)_{i=0}^\infty$ *is a sequence of distinct real numbers greater than* $-1/q$, *then*

$$\|y^{1/q-1/p}P(y)\|_{L_p[0,1]} \le \left(18 \cdot 2^q \sum_{j=0}^{n} \left(\lambda_j + \frac{1}{q}\right)\right)^{1/q-1/p} \|P\|_{L_q[0,1]}$$

for every $P \in M_n(\Lambda) := \operatorname{span}\{x^{\lambda_0}, x^{\lambda_1}, \dots, x^{\lambda_n}\}$.

If $\Gamma := (\gamma_i)_{i=0}^\infty$ *is a sequence of distinct positive real numbers, then*

$$\|P\|_{L_p[0,\infty)} \le \left(18 \cdot 2^q \sum_{j=0}^{n} \gamma_j\right)^{1/q-1/p} \|P\|_{L_q[0,\infty)} \tag{6.1.10}$$

for every $P \in E_n(\Gamma) := \operatorname{span}\{e^{-\gamma_0 t}, e^{-\gamma_1 t}, \dots, e^{-\gamma_n t}\}$.

Proof. First we show that the first statement of the theorem follows from the second. If $(\lambda_i)_{i=0}^\infty$ is a sequence of distinct real numbers greater than $-1/q$ and $\gamma_i := \lambda_i + 1/q$ for each i, then $(\gamma_i)_{i=1}^\infty$ is a sequence of distinct positive real numbers. Let $Q \in M_n(\Lambda)$. Applying (6.1.10) with

$$P(t) := Q(e^{-t})e^{-t/q} \in E_n(\Gamma)$$

and using the substitution $x = e^{-t}$, we obtain

$$\begin{aligned}&\|y^{1/q-1/p}Q(y)\|_{L_p[0,1]}\\ &\qquad \le (18 \cdot 2^q)^{1/q-1/p} \left(\sum_{j=0}^{n} \left(\lambda_j + \frac{1}{q}\right)\right)^{1/q-1/p} \|Q\|_{L_q[0,1]},\end{aligned}$$

which is the first statement of the theorem.

It is sufficient to prove (6.1.10) when $p = \infty$, and then a simple argument gives the desired result for arbitrary $0 < q < p < \infty$. To see this, assume that there is a constant C so that

$$\|P\|_{[0,\infty)} \le C^{1/q}\|P\|_{L_q[0,\infty)}$$

for every $P \in E_n(\Gamma)$ and $0 < q < \infty$. Then

$$\begin{aligned}\|P\|_{L_p[0,\infty)}^p &= \int_0^\infty |P(t)|^{p-q+q}\, dt \le \|P\|_{[0,\infty)}^{p-q}\|P\|_{L_q[0,\infty)}^q\\ &\le C^{p/q-1}\|P\|_{L_q[0,\infty]}^{p-q}\|P\|_{L_q[0,\infty)}^q\end{aligned}$$

and therefore

$$\|P\|_{L_p[0,\infty)} \le C^{1/q-1/p}\|P\|_{L_q[0,\infty)}$$

for every $f \in E_n(\Gamma)$ and $0 < q < p \le \infty$.

When $p = \infty$, (6.1.10) can be proven as follows. Let $P \in E_n(\Gamma)$, and let $y \in [0, \infty)$ be chosen so that $|P(y)| = \|P\|_{[0,\infty)}$. From Theorem 6.1.1 and the Mean Value Theorem, we can deduce that $|P(t)| > \frac{1}{2}\|P\|_{[0,\infty)}$ for every

$$t \in I := \left[y, y + (18\gamma)^{-1}\right], \quad \text{where} \quad \gamma := \sum_{j=0}^{n} \gamma_j \,.$$

Thus

$$\|P\|^q_{L_q[0,\infty]} \geq \int_I |P(t)|^q \, dt \geq \left(18 \sum_{j=0}^{n} \gamma_j\right)^{-1} 2^{-q} \|P\|^q_{[0,\infty)} \,,$$

and the result follows. □

Theorem 6.1.3 immediately implies the following result, which is a special case of Theorem 4.2.4:

Theorem 6.1.4 (Müntz-Type Theorem in L_p). *Let $p \in [1, \infty)$. Let $(\lambda_i)_{i=0}^\infty$ be a sequence of distinct real numbers greater than $-1/p$ satisfying*

$$\sum_{j=0}^{\infty} \left(\lambda_j + \frac{1}{p}\right) < \infty \,.$$

Then $\operatorname{span}\{x^{\lambda_0}, x^{\lambda_1}, \dots\}$ *is not dense in* $L_p[0, 1]$.

The next theorem offers an L_2 analog of Theorem 6.1.1 even for complex exponents. It also improves the multiplicative constant 12 in the L_2 inequality of Theorem 6.1.2 and shows that the L_2 inequality of Theorem 6.1.2 is essentially sharp.

Theorem 6.1.5. *If $\Lambda := (\lambda_i)_{i=0}^\infty$ is a sequence of distinct complex numbers with* $\operatorname{Re}(\lambda_i) > -1/2$ *for each i, then*

$$\sup_{0 \neq p \in M_n(\Lambda)} \frac{\|xp'(x)\|_{L_2[0,1]}}{\|p\|_{L_2[0,1]}}$$
$$\leq \left(\sum_{j=0}^{n} |\lambda_j|^2 + \sum_{j=0}^{n} (1 + 2\operatorname{Re}(\lambda_j)) \sum_{k=j+1}^{n} (1 + 2\operatorname{Re}(\lambda_k))\right)^{1/2}$$

for every $n \in \mathbb{N}$, where $M_n(\Lambda)$ denotes the linear span of $\{x^{\lambda_0}, x^{\lambda_1}, \dots, x^{\lambda_n}\}$ over $\mathbb{C}$.

If $\Lambda := (\lambda_i)_{i=0}^\infty$ is a sequence of distinct nonnegative real numbers, then

$$\frac{1}{2\sqrt{30}} \sum_{j=0}^{n} \lambda_j \leq \sup_{0 \neq p \in M_n(\Lambda)} \frac{\|xp'(x)\|_{L_2[0,1]}}{\|p\|_{L_2[0,1]}} \leq \frac{1}{\sqrt{2}} \sum_{j=0}^{n} (1 + 2\lambda_j)$$

for every $n \in \mathbb{N}$, where $M_n(\Lambda)$ denotes the linear span of $\{x^{\lambda_0}, x^{\lambda_1}, \dots, x^{\lambda_n}\}$ over $\mathbb{R}$.

Proof. Let $p \in M_n(\Lambda)$ with $\|p\|_{L_2[0,1]} = 1$. Then

$$p(x) = \sum_{k=0}^{n} a_k L_k^*(x) \qquad \text{with} \qquad \sum_{k=0}^{n} |a_k|^2 = 1$$

and

$$xp'(x) = \sum_{k=0}^{n} a_k x L_k^{*\prime}(x)\,,$$

where $L_k^* \in M_k(\Lambda)$ denotes the kth orthonormal Müntz-Legendre polynomials on $[0,1]$. Using the recurrence formula of Corollary 3.4.5 b] for the terms $xL_k^{*\prime}(x)$ in the above sum, we obtain

$$xp'(x) = \sum_{j=0}^{n} \left(a_j \lambda_j + \sqrt{1 + \lambda_j + \overline{\lambda}_j} \sum_{k=j+1}^{n} a_k \sqrt{1 + \lambda_k + \overline{\lambda}_k} \right) L_j^*(x)\,.$$

Hence

$$\|xp'(x)\|_{L_2[0,1]}^2 = \sum_{j=0}^{n} \left| a_j \lambda_j + \sqrt{1 + \lambda_j + \overline{\lambda}_j} \sum_{k=j+1}^{n} a_k \sqrt{1 + \lambda_k + \overline{\lambda}_k} \right|^2.$$

If we apply the Cauchy-Schwarz inequality to each term in the first sum and recall that $\sum_{k=0}^{n} |a_k|^2 = 1$, we see that

$$\begin{aligned}
\|xp'(x)\|_{L_2[0,1]}^2 &\leq \sum_{j=0}^{n} \left(|\lambda_j|^2 + (1 + \lambda_j + \overline{\lambda}_j) \right) \sum_{k=j+1}^{n} (1 + \lambda_k + \overline{\lambda}_k) \\
&\leq \frac{1}{2} \left(\sum_{j=0}^{n} (1 + 2|\lambda_j|) \right)^2,
\end{aligned}$$

which proves the first part and the upper bound in the second part of the theorem.

Now we prove the lower estimate in the second part of the theorem. With the sequence $\Lambda := (\lambda_i)_{i=0}^{\infty}$ of distinct nonnegative real numbers, we associate

$$q(x) := \sum_{k=0}^{n} \sqrt{\lambda_k} \left(\sum_{j=0}^{k} \lambda_j \right) L_k^*(x) \in M_n(\Lambda)\,.$$

Since the system $(L_k^*)_{k=0}^{\infty}$ is orthonormal on [0,1], we have

$$\|q\|_{L_2[0,1]}^2 = \sum_{k=0}^{n} \lambda_k \left(\sum_{j=0}^{k} \lambda_j \right)^2 \leq \left(\sum_{j=0}^{n} \lambda_j \right)^3. \tag{6.1.11}$$

Furthermore

$$xq'(x) = \sum_{k=0}^{n} \sqrt{\lambda_k} \left(\sum_{j=0}^{k} \lambda_j \right) x L_k^{*\prime}(x) = \sum_{m=0}^{n} b_m L_m^*(x) \,,$$

where, by the recurrence formula of Corollary 3.4.5 b],

$$\begin{aligned} b_m &:= \lambda_m \sqrt{\lambda_m} \sum_{j=0}^{m} \lambda_j + \sqrt{1+2\lambda_m} \sum_{k=m+1}^{n} \sqrt{\lambda_k(1+2\lambda_k)} \sum_{j=0}^{k} \lambda_j \\ &\geq \sqrt{\lambda_m} \sum_{k=m}^{n} \lambda_k \sum_{j=0}^{k} \lambda_j \,. \end{aligned}$$

Hence

$$\begin{aligned} \|xq'(x)\|_{L_2[0,1]}^2 &= \sum_{m=0}^{n} b_m^2 \geq \sum_{m=0}^{n} \lambda_m \left(\sum_{k=m}^{n} \lambda_k \sum_{j=0}^{k} \lambda_j \right)^2 \\ &= \sum_{\substack{0 \leq m \leq n \\ m \leq k, k' \leq n}} \sum_{\substack{0 \leq j \leq k \\ 0 \leq j' \leq k'}} \lambda_m \lambda_k \lambda_j \lambda_{k'} \lambda_{j'} \\ &\geq \sum_{0 \leq m \leq j \leq j' \leq k \leq k' \leq n} \lambda_m \lambda_k \lambda_j \lambda_{k'} \lambda_{j'} \geq \frac{1}{5!} \left(\sum_{j=0}^{n} \lambda_j \right)^5 . \end{aligned}$$

This, together with (6.1.11), yields the lower bound in the second part of the theorem. □

Comments, Exercises, and Examples.

Theorem 6.1.1 is due to Newman [76]. We presented a modified version of Newman's original proof of Theorem 6.1.1. He worked with T instead of U, and instead of (6.1.9) he established a more complicated identity involving the second derivative of P. Therefore, he needed an application of Kolmogorov's inequality (see E.1) to finish his proof. It can be proven that if the exponents λ_j are distinct nonnegative integers, then $\|xp'(x)\|_{[0,1]}$ in Theorem 6.1.1 can be replaced by $\|p'\|_{[0,1]}$ (see E.3). Theorems 6.1.2 to 6.1.4 were proved by Borwein and Erdélyi [to appear 6], while Theorem 6.1.5 is due to Borwein, Erdélyi, and J. Zhang [94b]. It is shown in E.8 that Theorem 6.1.2 is essentially sharp for every Λ with a gap condition, and for every $p \in [2, \infty)$.

The interval $[0,1]$ plays a special role in this section, analogs of the results on $[a,b]$, $a > 0$, cannot be obtained by a linear transformation. E.10 deals with the nontrivial extension of Newman's inequality to intervals $[a,b]$, $a > 0$.

A conjecture of Lorentz about the "right" Bernstein-type inequality for exponential sums with n terms is settled in E.4.

E.1 Kolmogorov's Inequality. Show that

$$\|f'\|^2_{[0,\infty)} \leq 4\|f\|_{[0,\infty)}\|f''\|_{[0,\infty)}$$

for every $f \in C^2[0,\infty)$.

Hint: By Taylor's theorem

$$f(x+h) = f(x) + f'(x)h + \tfrac{1}{2}f''(\xi)h^2\,, \qquad h > 0$$

with some $\xi \in (x, x+h)$. Hence

$$\|f'\|_{[0,\infty)} \leq 2h^{-1}\|f\|_{[0,\infty)} + (h/2)\|f''\|_{[0,\infty)}\,.$$

Now minimize the right-hand side by taking

$$h := 2\left(\|f\|_{[0,\infty)}\|f''\|^{-1}_{[0,\infty)}\right)^{1/2}.$$

□

The constant 4 in E.1 is not the best possible. Kolmogorov [62] proved that

$$\|f^{(k)}\|_{\mathbb{R}} \leq K(n,k)\|f\|_{\mathbb{R}}^{1-k/n}\|f^{(n)}\|_{\mathbb{R}}^{k/n}$$

for every $f \in C^n(\mathbb{R})$ and $0 < k < n$ and found the best possible constants $K(n,k)$; see also DeVore and Lorentz [93]. This generalizes a result of Landau, who proved the above inequality for $n = 2$, $k = 1$, and showed that $K(2,1) = \sqrt{2}$. Various multivariate extensions of Kolmogorov's inequality have also been established; see, for example, Ditzian [89].

E.2 Nikolskii-Type Inequalities.

a] Suppose $(V, \|\cdot\|)$ is an $(n+1)$-dimensional real or complex Hilbert space, $(p_k)_{k=0}^n \subset V$ is an orthonormal system, and $\varphi \neq 0$ is a linear functional on V. Then

$$|\varphi(p)| \leq \left(\sum_{k=0}^n |\varphi(p_k)|^2\right)^{1/2} \|p\|$$

for every $p \in V$. Equality holds if and only if

$$p = c\sum_{j=0}^n \overline{\varphi(p_k)}p_k\,, \qquad c \in \mathbb{R} \quad \text{or} \quad c \in \mathbb{C}\,.$$

Hint: Write p as a linear combination of the orthonormal elements p_k, use the linearity of φ, then apply the Cauchy-Schwarz inequality. □

b] Suppose $\Lambda := (\lambda_i)_{i=0}^\infty$ is a set of distinct complex numbers satisfying $\mathrm{Re}(\lambda_i) > -1/2$ for each i, and $y \in [0, \infty)$ is fixed. Show that

$$|p^{(m)}(y)| \le \left(\sum_{k=0}^{n} |L_k^{*(m)}(y)|^2 \right)^{1/2} \|p\|_{L_2[0,1]}$$

for every $p \in M_n(\Lambda)$, where $M_n(\Lambda)$ denotes the linear span of

$$\{x^{\lambda_0}, x^{\lambda_1}, \dots, x^{\lambda_n}\}$$

over $\mathbb{C}$, and $L_k^* \in M_k(\Lambda)$ is the kth orthonormal Müntz-Legendre polynomial on $[0,1]$. Show that if there exists a $q \in M_n(\Lambda)$ with $q^{(m)}(y) \neq 0$, then equality holds if and only if

$$p = c \sum_{k=0}^{n} \overline{L_k^{*(m)}(y)} L_k^* \,, \qquad c \in \mathbb{C}\,.$$

c] Under the assumptions of part b] show that

$$\frac{|y^{1/2} p(y)|}{\|p\|_{L_2[0,1]}} \le \left(\sum_{j=0}^{n} (1 + 2\,\mathrm{Re}(\lambda_j)) \right)^{1/2}$$

and

$$\frac{|y^{3/2} p'(y)|}{\|p\|_{L_2[0,1]}} \le \left(\sum_{k=0}^{n} (1 + 2\,\mathrm{Re}(\lambda_k)) \left| \lambda_k + \sum_{j=0}^{k-1} (1 + 2\,\mathrm{Re}(\lambda_j)) \right|^2 \right)^{1/2}$$

for every $0 \neq p \in M_n(\Lambda)$ and $y \in [0,1]$.

Hint: When $y = 1$, use part b] and substitute the explicit values of $L_k^*(1)$ and $L_k^{*\prime}(1)$ (see Corollary 3.4.6 and formula (3.4.8)). If $0 < y < 1$, then the scaling $x \to yx$ reduces the inequality to the case $y = 1$. □

d] Show that if $n \ge 1$ and $p \neq 0$, then equality holds in the inequalities of part b] if and only if $y = 1$ and

$$p = c \sum_{k=0}^{n} L_k^*(1) L_k^* \qquad \text{or} \qquad p = c \sum_{k=0}^{n} L_k^{*\prime}(1) L_k^* \,,$$

respectively, with some $0 \neq c \in \mathbb{C}$.

e] Show that

$$\sup_{0\neq p\in\mathcal{P}_n} \frac{\|p\|_{[-1,1]}}{\|p\|_{L_2[-1,1]}} = \frac{n+1}{\sqrt{2}}\,.$$

Hint: Show that there is an extremal polynomial $\widetilde{p}$ for the above extremal problem for which $\|\widetilde{p}\|_{[-1,1]}$ is achieved at 1. Now use parts c] and d] to show that

$$|\widetilde{p}(1)|^2 = \frac{1}{2}\left(\sum_{k=0}^{n}(1+2k)\right)\int_{-1}^{1}\widetilde{p}^2(t)\,dt = \frac{1}{2}\,(n+1)^2\int_{-1}^{1}\widetilde{p}^2(t)\,dt$$

and the result follows. □

E.3 An Improvement of Newman's Inequality.

a] Suppose $\Lambda := (\lambda_k)_{k=0}^{\infty}$ is a sequence with $\lambda_0 = 0$ and $\lambda_{k+1} - \lambda_k \geq 1$ for each k. Show that

$$\|p'\|_{[0,1]} \leq 18\left(\sum_{j=1}^{n}\lambda_j\right)\|p\|_{[0,1]}$$

for every $p \in M_n(\Lambda) := \operatorname{span}\{x^{\lambda_0}, x^{\lambda_1}, \dots, x^{\lambda_n}\}$.

Hint: Let $y \in [0,1]$. To estimate $|p'(y)|$ distinguish two cases. If $\frac{1}{2} \leq y \leq 1$, use Theorem 6.1.1 (Newman's inequality), and if $0 \leq y \leq \frac{1}{2}$, use E.3 f] of Section 3.3 and Theorem 5.1.8 (Markov's inequality) transformed to $[y,1]$ to show that

$$|p'(y)| \leq \frac{2n^2}{1-y}\|p\|_{[y,1]} \leq 8\left(\sum_{j=1}^{n}\lambda_j\right)\|p\|_{[0,1]}$$

for every $p \in M_n(\Lambda)$. □

The next exercise is based on an example given by Bos.

b] Show that for every $\delta \in (0,1)$ there exists a sequence $\Lambda := (\lambda_k)_{k=0}^{\infty}$ with $\lambda_0 = 0$, $\lambda_1 \geq 1$, and

$$\lambda_{k+1} - \lambda_k \geq \delta\,, \qquad i = 0,1,2,\dots$$

such that

$$\lim_{n\to\infty}\ \sup_{0\neq p\in M_n(\Lambda)} \frac{|p'(0)|}{\left(\sum_{j=0}^{n}\lambda_j\right)\|p\|_{[0,1]}} = \infty\,.$$

Outline. Let Q_n be the Chebyshev polynomial T_n transformed linearly from $[-1,1]$ to $[0,1]$, that is,

$$Q_n(x) = \cos(n\arccos(2x-1))\,, \qquad x \in [0,1]\,.$$

Choose natural numbers u and v so that $\delta < u/v < 1$. Let $\Lambda := (\lambda_k)_{k=0}^{\infty}$ be defined by $\lambda_0 := 0$, $\lambda_1 := 1$, and

$$\lambda_k := 1 + \frac{ku}{v}, \qquad k = 1, 2, \dots .$$

Let

$$p_n(x) := x^{1-u}\left(Q_n(x^{u/v}) - (-1)^n\right)^v \in M_{nv-v}(\Lambda).$$

Then

$$|p_n'(0)| = \left(2n^2\right)^v.$$

For the sake of brevity let

$$q_n(x) := Q_n(x^{u/v}) - (-1)^n .$$

Use Theorem 5.1.8 (Markov's inequality) and the Mean Value Theorem to show that

$$\|q_n\|_{[y,1]} \geq \tfrac{1}{2}\|q_n\|_{[0,1]}$$

with

$$y := \left(2n^2\right)^{-v/u}.$$

Thus, if n is odd, then

$$\begin{aligned}\|p_n\|_{[0,1]} &\geq \|p_n\|_{[y,1]} \geq y^{1-u}(\|q_n\|_{[y,1]})^v \\ &\geq y^{1-u}\left(\tfrac{1}{2}\|q_n\|_{[0,1]}\right)^v = \left(2n^2\right)^{(u-1)v/u}\end{aligned}$$

and

$$\begin{aligned}\frac{|p_n'(0)|}{\left(\sum_{j=0}^{nv-v}\lambda_j\right)\|p_n\|_{[0,1]}} &= \frac{(2n^2)^v}{\left(\sum_{j=0}^{nv-v}\left(1+j\frac{u}{v}\right)\right)\left(2n^2\right)^{(u-1)v/u}} \\ &\geq \frac{\left(2n^2\right)^{v/u}}{(1+nu)nv} \geq \frac{\left(2n^2\right)^{v/u-1}}{uv} \underset{n\to\infty}{\longrightarrow} \infty .\end{aligned}$$

□

In his book *Nonlinear Approximation Theory*, Braess [86] writes the following: "The rational functions and exponential sums belong to those concrete families of functions which are the most frequently used in nonlinear approximation theory. The starting point of consideration of exponential sums is an approximation problem often encountered for the analysis of decay processes in natural sciences. A given empirical function on a real interval is to be approximated by sums of the form

$$\sum_{j=1}^{n} a_j e^{\lambda_j t},$$

where the parameters a_j and λ_j are to be determined, while n is fixed."

The next exercise treats inequalities for exponential sums of $n+1$ terms.

E.4 Nikolskii- and Bernstein-Type Inequalities for Exponential Sums. Let

$$\Lambda := (\lambda_i)_{i=1}^{\infty}$$

be a sequence of distinct nonzero real numbers. Let

$$E_n(\Lambda) := \left\{ f : f(t) = a_0 + \sum_{j=1}^{n} a_j e^{\lambda_j t}, \quad a_j \in \mathbb{R} \right\}$$

and

$$E_n := \bigcup_{\Lambda} E_n(\Lambda) = \left\{ f : f(t) = a_0 + \sum_{j=1}^{n} a_j e^{\lambda_j t}, \quad a_j, \lambda_j \in \mathbb{R} \right\},$$

that is, E_n is the collection of all $(n+1)$-term exponential sums with constant first term. Schmidt [70] proved that there is a constant $c(n)$ depending only on n such that

$$\|f'\|_{[a+\delta, b-\delta]} \leq c(n)\delta^{-1}\|f\|_{[a,b]}$$

for every $p \in E_n$ and $\delta \in \left(0, \frac{1}{2}(b-a)\right)$. Lorentz [89] improved Schmidt's result by showing that for every $\alpha > \frac{1}{2}$, there is a constant $c(\alpha)$ depending only on α such that $c(n)$ in the above inequality can be replaced by $c(\alpha)n^{\alpha \log n}$, and he speculated that there may be an absolute constant c such that Schmidt's inequality holds with $c(n)$ replaced by cn. Part d] of this exercise shows that Schmidt's inequality holds with $c(n) = 2n-1$. A weaker version of this showing that Schmidt's inequality holds with $c(n) = 8(n+1)^2$ is obtained in part b] and uses a Nikolskii-type inequality for exponential sums established in part a]. Part e] shows that the result of part d] is sharp up to a multiplicative absolute constant.

a] Let $p \in (0, 2]$. Show that

$$\|f\|_{[a+\delta, b-\delta]} \leq 2^{2/p^2} \left(\frac{n+1}{\delta} \right)^{1/p} \|f\|_{L_p[a,b]}$$

for every $f \in E_n$ and $\delta \in \left(0, \frac{1}{2}(b-a)\right)$.

Proof. Take the orthonormal sequence $(L_k)_{k=0}^{n}$ on $\left[-\frac{1}{2}, \frac{1}{2}\right]$, that is,

(1) $L_k \in \operatorname{span}\{1, e^{\lambda_1 t}, e^{\lambda_2 t}, \dots, e^{\lambda_k t}\}, \qquad k = 0, 1, \dots, n,$

and

(2) $\displaystyle\int_{-1/2}^{1/2} L_i L_j = \delta_{i,j}, \qquad 0 \leq i \leq j \leq n,$

where $\delta_{i,j}$ is the Kronecker symbol. On writing $f \in E_n(\Lambda)$ as a linear combination of $L_0, L_1, \dots, L_n$, and using the Cauchy-Schwarz inequality and the orthonormality of $(L_k)_{k=0}^n$ on $\left[-\frac{1}{2}, \frac{1}{2}\right]$, we obtain in a standard fashion that

$$\max_{0 \neq f \in E_n(\Lambda)} \frac{|f(t_0)|}{\|f\|_{L_2[-1/2,1/2]}} = \left(\sum_{k=0}^n L_k(t_0)^2\right)^{1/2}, \qquad t_0 \in \mathbb{R}.$$

Since

$$\int_{-1/2}^{1/2} \textstyle\sum_{k=0}^n L_k^2(x)\, dx = n+1,$$

there exists a $t_0 \in \left[-\frac{1}{2}, \frac{1}{2}\right]$ such that

$$\max_{0 \neq p \in E_n(\Lambda)} \frac{|f(t_0)|}{\|f\|_{L_2[-1/2,1/2]}} = \left(\sum_{k=0}^n L_k^2(t_0)\right)^{1/2} \leq \sqrt{n+1}.$$

Observe that if $f \in E_n(\Lambda)$, then $g(t) := f(t - t_0) \in E_n(\Lambda)$, so

$$\max_{0 \neq f \in E_n(\Lambda)} \frac{|f(0)|}{\|f\|_{L_2[-1,1]}} \leq \sqrt{n+1}.$$

Let

$$C := \max_{0 \neq f \in E_n(\Lambda)} \frac{|f(0)|}{\|f\|_{L_p[-2,2]}}.$$

Then

$$\max_{0 \neq f \in E_n(\Lambda)} \frac{|f(y)|}{\|f\|_{L_p[-2,2]}} \leq C\left(\frac{2}{2-|y|}\right)^{1/p} \leq 2^{1/p}C, \qquad y \in [-1,1].$$

Therefore, for every $f \in E_n(\Lambda)$,

$$\begin{aligned}
|f(0)| &\leq \sqrt{n+1}\, \|f\|_{L_2[-1,1]} \\
&\leq \sqrt{n+1}\left(\|f\|_{L_p[-1,1]}^p \|f\|_{[-1,1]}^{2-p}\right)^{1/2} \\
&\leq \sqrt{n+1}\left(\|f\|_{L_p[-1,1]}^p \left(2^{1/p}C\right)^{2-p} \|f\|_{L_p[-2,2]}^{2-p}\right)^{1/2} \\
&\leq \sqrt{n+1}\left(2^{1/p}C\right)^{1-p/2} \|f\|_{L_p[-2,2]} \\
&= 2^{1/p-1/2}\sqrt{n+1}\, C^{1-p/2} \|f\|_{L_p[-2,2]}.
\end{aligned}$$

Hence

$$C = \max_{0 \neq f \in E_n(\Lambda)} \frac{|f(0)|}{\|f\|_{L_p[-2,2]}} \leq 2^{1/p-1/2}\sqrt{n+1}\, C^{1-p/2}$$

and we conclude that $C \leq 2^{2/p^2-1/p}(n+1)^{1/p}$. So

$$|f(0)| \le 2^{2/p^2 - 1/p}(n+1)^{1/p} \|f\|_{L_p[-2,2]}$$

for every $f \in E_n(\Lambda)$. Now let $f \in E_n(\Lambda)$ and $t_0 \in [a+\delta, b-\delta]$. If we apply the above inequality to

$$g(t) := f\left(\tfrac{1}{2}\delta t + t_0\right) \in E_n(\Lambda)\,,$$

we obtain

$$\|f\|_{[a+\delta, b-\delta]} \le 2^{2/p^2 - 1/p}(n+1)^{1/p} \left(\frac{2}{\delta}\right)^{1/p} \|f\|_{L_p[a,b]}$$

and the result follows. □

The following Bernstein-type inequality can be obtained as a simple corollary of part a]:

b] Show that

$$\|f'\|_{[a+\delta, b-\delta]} \le 8(n+1)^2 \delta^{-1} \|f\|_{[a,b]}$$

for every $f \in E_n$ and $\delta \in (0, \frac{1}{2}(b-a))$.

Proof. Note that $f \in E_n(\Lambda)$ implies $f' \in E_n(\Lambda)$. Applying part a] to f' with $p = 1$, we obtain

$$|f'(0)| \le 2(n+1)\|f'\|_{L_1[-2,2]} = 2(n+1)\mathrm{Var}_{[-2,2]}(f) \le 4(n+1)^2 \|f\|_{[-2,2]}$$

for every $f \in E_n(\Lambda)$. If $f \in E_n(\Lambda)$ and $t_0 \in [a+\delta, b-\delta]$, then on applying the above inequality to

$$g(t) := f\left(\tfrac{1}{2}\delta t + t_0\right) \in E_n(\Lambda)\,,$$

we obtain the desired result. □

c] Lorentz's Conjecture. Show that

$$\sup_{0 \ne f \in \widetilde{E}_{2n}} \frac{|f'(0)|}{\|f\|_{[-1,1]}} = 2n - 1\,,$$

where

$$\widetilde{E}_{2n} := \left\{ f : f(t) = a_0 + \sum_{j=1}^{n} \left(a_j e^{\lambda_j t} + b_j e^{-\lambda_j t}\right), \quad a_j, b_j, \lambda_j \in \mathbb{R} \right\}.$$

Proof. First we prove that

$$|f'(0)| \le (2n-1)\,\|f\|_{[-1,1]}$$

for every $f \in \widetilde{E}_{2n}$. So let

$$f \in \text{span}\{1, e^{\pm\lambda_1 t}, e^{\pm\lambda_2 t}, \dots, e^{\pm\lambda_n t}\}$$

with some nonzero real numbers $\lambda_1, \lambda_2, \dots, \lambda_n$, where, without loss of generality, we may assume that

$$0 < \lambda_1 < \lambda_2 < \cdots < \lambda_n .$$

Let

$$g(t) := \tfrac{1}{2}(f(t) - f(-t)) .$$

Observe that

$$g \in \text{span}\{\sinh \lambda_1 t\,, \sinh \lambda_2 t\,, \dots, \sinh \lambda_n t\} .$$

It is also straightforward that

$$g'(0) = f'(0) \qquad \text{and} \qquad \|g\|_{[0,1]} \leq \|f\|_{[-1,1]} .$$

For a given $\epsilon > 0$, let

$$H_{n,\epsilon} := \text{span}\{\sinh \epsilon t\,, \sinh 2\epsilon t\,, \dots, \sinh n\epsilon t\}$$

and

$$K_{n,\epsilon} := \sup\left\{|h'(0)| : h \in H_{n,\epsilon}\,, \ \|h\|_{[0,1]} = 1\right\} .$$

The inequality of E.5 e] in Section 3.3 is the key to the proof. It shows that it is sufficient to prove that

$$\inf\{K_{n,\epsilon} : \epsilon > 0\} \leq 2n - 1 .$$

Observe that every $h \in H_{n,\epsilon}$ is of the form

$$h(t) = e^{-n\epsilon t} P(e^{\epsilon t})\,, \qquad P \in \mathcal{P}_{2n} .$$

Therefore, using E.23 c] of Section 5.1, we obtain for every $h \in H_{n,\epsilon}$ that

$$\begin{aligned}
|h'(0)| &= |\epsilon P'(1) - n\epsilon P(1)| \\
&\leq \frac{\epsilon(2n-1)}{1 - e^{-\epsilon}} \|P\|_{[e^{-\epsilon}, e^{\epsilon}]} + n\epsilon \|P\|_{[e^{-\epsilon}, e^{\epsilon}]} \\
&\leq \left(\frac{\epsilon(2n-1)}{1 - e^{-\epsilon}} + n\epsilon\right) e^{n\epsilon} \|h\|_{[-1,1]} .
\end{aligned}$$

It follows that

$$K_{n,\epsilon} \leq \left(\frac{\epsilon(2n-1)}{1 - e^{-\epsilon}} + n\epsilon\right) e^{n\epsilon} .$$

So $\inf\{K_{n,\epsilon} : \epsilon > 0\} \leq 2n - 1$, and the upper bound follows.

Now we prove that

$$\sup_{0\neq f\in \widetilde{E}_{2n}} \frac{|f'(0)|}{\|f\|_{[-1,1]}} \geq 2n-1\,.$$

Let $\epsilon > 0$ be fixed. We define

$$Q_{2n,\epsilon}(t) := e^{-n\epsilon t} T_{2n-1}\left(\frac{e^{\epsilon t}}{e^{\epsilon}-1} - \frac{1}{e^{\epsilon}-1}\right),$$

where T_{2n-1} denotes the Chebyshev polynomial of degree $2n-1$ defined by

$$T_{2n-1}(x) = \cos((2n-1)\arccos x), \quad x \in [-1,1]\,.$$

It is simple to check that $Q_{2n,\epsilon} \in \widetilde{E}_{2n}$,

$$\|Q_{2n,\epsilon}\|_{[-1,1]} \leq e^{n\epsilon t}$$

and

$$|Q'_{2n,\epsilon}(0)| \geq 2n-1-n\epsilon\,.$$

Now the result follows by letting ϵ decrease to 0. □

d] Show that

$$\|f'\|_{[a+\delta,b-\delta]} \leq (2n-1)\delta^{-1}\|f\|_{[a,b]}$$

for every $f \in E_n$ and $\delta \in \left(0, \frac{1}{2}(b-a)\right)$.

Proof. Observe that $E_n \subset \widetilde{E}_{2n}$. Hence the result follows from part c] by a linear substitution. □

e] Let $a < b$ and $y \in (a,b)$. Suppose that $n \in \mathbb{N}$ is odd. Let T_n be the Chebyshev polynomial of degree n defined by (2.1.1). Let

$$Q_n(t) := Q_{n,y}(t) := T_n\left(\frac{e}{e-1}\exp\left(\frac{t-b}{b-y}\right) - \frac{1}{e-1}\right)$$

and

$$R_n(t) := R_{n,y}(t) := T_n\left(\frac{e}{e-1}\exp\left(\frac{t-a}{a-y}\right) - \frac{1}{e-1}\right).$$

Show that $Q_n, R_n \in E_n$ and

$$\frac{|Q'_n(y)|}{\|Q_n\|_{[a,b]}} = \frac{1}{e-1}\frac{n}{b-y} \qquad \text{and} \qquad \frac{|R'_n(y)|}{\|R_n\|_{[a,b]}} = \frac{1}{e-1}\frac{n}{y-a}$$

for every $y \in (a,b)$.

f] Let $a < b$ and $y \in (a,b)$. Conclude that

$$\frac{1}{e-1}\frac{n-1}{\min\{y-a,b-y\}} \leq \sup_{0 \neq f \in E_n} \frac{|f'(y)|}{\|f\|_{[a,b]}} \leq \frac{2n-1}{\min\{y-a,b-y\}}\,.$$

In the rest of the exercise let

$$E_n^* := \left\{ f : f(t) = \sum_{j=1}^{l} P_{k_j}(t)e^{\lambda_j t}, \ \lambda_j \in \mathbb{R}\,, \ P_{k_j} \in \mathcal{P}_{k_j}\,, \ \sum_{j=1}^{l}(k_j+1) = n \right\}.$$

g] Extend the inequalities of parts a], c], and f] to E_n^*.

h] Let $[a,b]$ be a finite interval. Let $g \in C[a,b]$. Show that the value

$$\inf\{\|g-f\|_{[a,b]} : f \in E_n^*\}$$

is attained by an $\widetilde{f} \in E_n^*$.

Hint: Use Schmidt's inequality (or its improved form given by part c]). For the nontrivial details, see Braess [86]. □

i] Let $[a,b]$ be a finite interval. Let $p \in [1,\infty)$ and $g \in L_p[a,b]$. Show that the value

$$\inf\{\|g-f\|_{L_p[a,b]} : f \in E_n^*\}$$

is attained by an $\widetilde{f} \in E_n^*$.

Hint: Use part a] with $p = 1$ and Hölder's inequality. Once again, for the details, see Braess [86]. □

The following result is from Borwein and Erdélyi [95c]:

E.5 Upper Bound for the Derivative of Exponential Sums with Nonnegative Exponents. The equality

$$\sup_p \frac{|p'(a)|}{\|p\|_{[a,b]}} = \frac{2n^2}{b-a}$$

holds for every $a < b$, where the supremum is taken over all exponential sums $0 \neq p \in E_n$ with nonnegative exponents. The equality

$$\sup_p \frac{|p'(a)|}{\|p\|_{[a,b]}} = \frac{2n^2}{a(\log b - \log a)}$$

also holds for every $0 < a < b$, where the supremum is taken over all Müntz polynomials $0 \neq p$ of the form

$$p(x) = a_0 + \sum_{j=1}^{n} a_j x^{\lambda_j}\,, \qquad a_j \in \mathbb{R}\,, \quad \lambda_j > 0\,.$$

Outline. It is sufficient to prove only the second statement, the first can be obtained from it by the change of variable $x = e^t$. For $\epsilon > 0$, define $\Lambda_\epsilon := (j\epsilon)_{j=0}^\infty$. Let

$$T_{n,\epsilon} := T_n\{1, x^\epsilon, x^{2\epsilon}, \dots, x^{n\epsilon}; [a,b]\}$$

be the Chebyshev polynomial for $M_n(\Lambda_\epsilon)$ on $[a,b]$. Use E.3 b] and E.3 f] of Section 3.3 to show that

$$\frac{|p'(a)|}{\|p\|_{[a,b]}} \leq \lim_{\epsilon \to 0+} \frac{|T'_{n,\epsilon}(a)|}{\|T_{n,\epsilon}\|_{[a,b]}} = \lim_{\epsilon \to 0+} |T'_{n,\epsilon}(a)|$$

for every p of the form

$$p(x) = a_0 + \sum_{j=1}^n a_j x^{\lambda_j}, \qquad a_j \in \mathbb{R}, \quad \lambda_j > 0.$$

From the definition and uniqueness of $T_{n,\epsilon}$ it follows that

$$T_{n,\epsilon}(x) = T_n\left(\frac{2}{b^\epsilon - a^\epsilon} x^\epsilon - \frac{b^\epsilon + a^\epsilon}{b^\epsilon - a^\epsilon}\right)$$

where $T_n(y) = \cos(n \arccos y)$, $y \in [-1,1]$. Therefore

$$\begin{aligned} |T'_{n,\epsilon}(a)| &= |T'_n(-1)| \frac{2}{b^\epsilon - a^\epsilon} \epsilon a^{\epsilon - 1} \\ &= \frac{2n^2}{\epsilon^{-1}(b^\epsilon - 1) - \epsilon^{-1}(a^\epsilon - 1)} a^{\epsilon-1} \underset{\epsilon \to 0+}{\longrightarrow} \frac{2n^2}{a(\log b - \log a)} \end{aligned}$$

and the proof is finished. □

The next exercise follows Turán [84].

E.6 Turán's Inequalities for Exponential Sums.

a] Let

$$g(\nu) := \sum_{j=1}^n b_j z_j^\nu, \qquad b_j, z_j \in \mathbb{C}.$$

Suppose

$$|z_j| \geq 1, \qquad j = 1, 2, \dots, n.$$

Then

$$\max_{\nu = m+1, \dots, m+n} |g(\nu)| \geq \left(\frac{n}{2e(m+n)}\right)^2 |b_1 + b_2 + \dots + b_n|$$

for every nonnegative integer m.

Proof. Let

$$f(z) := \prod_{j=1}^{n} \left(1 - \frac{z}{z_j}\right) =: \sum_{\nu=0}^{n} \alpha_\nu z^\nu \,. \tag{6.1.12}$$

Since $f(z)$ has all its zeros outside the open unit disk, $g := 1/f$ is of the form

$$g(z) = \sum_{\nu=0}^{\infty} \beta_\nu z^\nu \,, \qquad |z| < 1 \,. \tag{6.1.13}$$

Let

$$g_m(z) := \sum_{\nu=0}^{m} \beta_\nu z^\nu \tag{6.1.14}$$

and

$$h := 1 - f g_m \in \mathcal{P}_{n+m}^c \,. \tag{6.1.15}$$

Note that

$$\begin{aligned} h(z) &= 1 - f(z) \left(g(z) - \sum_{\nu=m}^{\infty} \beta_\nu z^\nu \right) \\ &= f(z) \sum_{\nu=m+1}^{\infty} \beta_\nu z^\nu \end{aligned} \tag{6.1.16}$$

so h is of the form

$$h(z) = \sum_{\nu=m+1}^{m+n} \gamma_\nu z^\nu \,. \tag{6.1.17}$$

Observe that $f(z_j) = 0$ and (6.1.13) imply $h(z_j) = 1$, that is,

$$\sum_{\nu=m+1}^{m+n} \gamma_\nu z_j^\nu = 1 \,, \qquad j = 1, 2, \ldots, n \,.$$

Multiplication with b_j and summation over j yield the fundamental identity

$$\sum_{\nu=m+1}^{m+n} \gamma_\nu g(\nu) = \sum_{j=1}^{n} b_j \,. \tag{6.1.18}$$

This immediately gives

(6.1.19)
$$\max_{\nu=m+1,\dots,m+n} |g(\nu)| \left(\sum_{\nu=m+1}^{n} |\gamma_\nu| \right) \geq \left| \sum_{j=1}^{n} b_j \right| .$$

It follows from (6.1.15), (6.1.17), (6.1.12), and (6.1.14) that

(6.1.20)
$$\sum_{\nu=m+1}^{m+n} |\gamma_\nu| \leq \left(\sum_{\nu=0}^{n} |\alpha_\nu| \right) \left(\sum_{\nu=0}^{m} |\beta_\nu| \right) .$$

Since each z_j is of modulus at least 1, (6.1.12) yields that

(6.1.21)
$$\sum_{\nu=0}^{n} |\alpha_\nu| \leq 2^n .$$

Also, (6.1.13) implies that

$$\beta_\nu = \sum_{i_1+\cdots+i_n=\nu} \frac{1}{z_1^{i_1} z_2^{i_2} \cdots z_n^{i_n}} .$$

Again using that each z_j is of modulus at least 1, we obtain

$$|\beta_\nu| \leq \sum_{i_1+\cdots+i_n=\nu} 1 = \binom{\nu+n-1}{n-1} .$$

Hence

(6.1.22)
$$\sum_{\nu=0}^{m} |\beta_\nu| \leq \binom{m+n}{n} \leq \left(e \, \frac{m+n}{n} \right)^n .$$

By (6.1.20) to (6.1.22) we conclude that

$$\sum_{\nu=m+1}^{m+n} |\gamma_\nu| \leq \left(2e \, \frac{m+n}{n} \right)^n ,$$

which, together with (6.1.19), finishes the proof. □

b] Let

$$f(t) := \sum_{j=1}^{n} b_j e^{\lambda_j t} , \qquad b_j \, , \lambda_j \in \mathbb{C} .$$

Suppose

$$\mathrm{Re}(\lambda_j) \geq 0 , \qquad j = 1, 2, \dots , n .$$

Show that

$$|f(0)| \leq \left(\frac{2e(a+d)}{d} \right)^n \|f\|_{[a,a+d]}$$

for every $a > 0$ and $d > 0$.

Hint: First observe that the result of part a] can be formulated as

$$\max_{\substack{m \le \nu \le m+n \\ \nu \in \mathbb{N}}} |g(\nu)| \ge \left(\frac{n}{2e(m+n)} \right)^n |b_1 + b_2 + \cdots + b_n| \,,$$

where m is an arbitrary positive (not necessarily integer) number. Now apply the above inequality with

$$m := \frac{an}{d} \,, \qquad z_j := \exp\left(\frac{d\lambda_j}{n} \right) , \qquad j = 1, 2, \ldots, n \,.$$

□

c] Let

$$p(z) := \sum_{j=1}^{n} b_j z^{\lambda_j} \,, \qquad b_j \in \mathbb{C} \,, \quad \lambda_j \in \mathbb{R} \,, \quad z = e^{i\theta} \,.$$

Show that

$$\max_{|z|=1} |p(z)| \le \left(\frac{4e\pi}{\delta} \right)^n \max_{\substack{|z|=1 \\ \alpha \le \arg(z) \le \alpha+\delta}} |p(z)|$$

for every $0 \le \alpha < \alpha + \delta \le 2\pi$.

Hint: Use part b]. □

The inequalities of the above exercise and their variants play a central role in the book of Turán [83], where many applications are also presented. The main point in these inequalities is that the exponent on the right-hand side is only the number of terms n, and so it is independent of the numbers λ_j. An inequality, say in part c], of type

$$\max_{|z|=1} |p(z)| \le c(\delta)^{\lambda_n} \max_{\substack{|z|=1 \\ \alpha \le \arg(z) \le \alpha+\delta}} |p(z)| \,,$$

where $0 \le \lambda_1 < \lambda_2 < \cdots < \lambda_n$ are integers and $c(\delta)$ depends only on δ, could be obtained by a simple direct argument, but it is much less useful than the inequality of E.6 c].

E.7 Nikolskii-Type Inequality for Müntz Polynomials. Suppose that $\Lambda := (\lambda_i)_{i=0}^{\infty}$ is a sequence with $\lambda_0 := 0$ and $\lambda_{i+1} - \lambda_i \ge 1$ for each i. Show that

$$\|P\|_{L_p[0,1]} \le \left(18 \cdot 2^q \sum_{j=0}^{n} \lambda_j \right)^{1/q-1/p} \|P\|_{L_q[0,1]}$$

for every $P \in M_n(\Lambda) := \operatorname{span}\{x^{\lambda_0}, x^{\lambda_1}, \ldots, x^{\lambda_n}\}$ and $0 < q < p \le \infty$.

Proof. It is sufficient to study the case when $p = \infty$ (see the comment in the proof of Theorem 6.1.3). Let $P \in M_n(\Lambda)$ and let $x_0 \in [0,1]$ be chosen so that $|P(x_0)| = \|P\|_{[0,1]}$. Combining E.3 a] and the Mean Value Theorem, we obtain

$$|P(x)| \geq \frac{1}{2}\|P\|_{[0,1]}\,, \qquad x \in I\,,$$

where

$$I := \left[x_0 - (36\lambda)^{-1}\,, \, x_0 + (36\lambda)^{-1}\right] \quad \text{with} \quad \lambda := \sum_{j=0}^{n} \lambda_j\,.$$

So

$$\left(\int_0^1 |P(x)|^q\,dx\right)^{1/q} \geq \left(\int_I |P(x)|^q\,dx\right)^{1/q} \geq \left((18\lambda)^{-1}\left(\frac{1}{2}\|P\|_{[0,1]}\right)^q\right)^{1/q}$$

and the result follows. □

E.8 Sharpness of Theorem 6.1.2. Suppose $\Lambda := (\lambda_i)_{i=0}^{\infty}$ is a sequence with $\lambda_0 := 0$ and $\lambda_{i+1} - \lambda_i \geq 1$ for each i. Show that there exists an absolute constant $c > 0$ (independent of Λ and p) such that

$$\sup_{P \in M_n(\Lambda)} \frac{\|xP'(x)\|_{L_p[0,1]}}{\|P\|_{L_p[0,1]}} \geq c\sum_{k=0}^{n} \lambda_k$$

for every $p \in [2, \infty)$, where $M_n(\Lambda) := \text{span}\{x^{\lambda_0}, x^{\lambda_1}, \dots, x^{\lambda_n}\}$.

Proof. Let $L_k^* \in M_k(\Lambda)$ be the kth orthonormal Müntz-Legendre polynomial on $[0,1]$. Let $p \in [2, \infty)$ and

$$P := \sum_{k=0}^{n} L_k^{*\prime}(1) L_k^*\,.$$

For the sake of brevity let

$$\lambda := \sum_{j=0}^{n} \lambda_j\,.$$

Using Theorem 3.4.3 (orthogonality), (3.4.8), and Corollary 3.4.6, we obtain

$$\begin{aligned} (6.1.23) \qquad P'(1) &= \sum_{k=0}^{n} (L_k^{*\prime}(1))^2 \geq \sum_{k=0}^{n} (2\lambda_k + 1)\left(\sum_{j=0}^{k} \lambda_j\right)^2 \\ &\geq 2\lambda\left(\sum_{j=0}^{k} \lambda_j\right)^2 \geq \tfrac{1}{8}\lambda^3 \end{aligned}$$

and

$$\text{(6.1.24)} \qquad \|P\|_{L_2[0,1]} = \left(\sum_{k=0}^{n} (L_k^{*\prime}(1))^2\right)^{1/2}$$
$$\leq \left(\sum_{k=0}^{n} (2\lambda_k + 1)\left(\sum_{j=0}^{k} (2\lambda_j + 1)\right)^2\right)^{1/2}$$
$$\leq \sqrt{32}\,\lambda^{3/2}\,.$$

A combination of E.7 and (6.1.24) yields

$$\text{(6.1.25)} \qquad \|P\|_{L_p[0,1]} \leq (72\,\lambda)^{1/2-1/p} \|P\|_{L_2[0,1]}$$
$$\leq 72^{1/2-1/p}\sqrt{32}\,\lambda^{2-1/p} \leq 48\,\lambda^{2-1/p}\,.$$

From E.3 a], E.7, and (6.1.24) we can deduce that

$$\text{(6.1.26)} \qquad \|P'\|_{[0,1]} \leq 18\,\lambda\,\|P\|_{[0,1]}$$
$$\leq 18\,\lambda(72\,\lambda)^{1/2}\|P\|_{L_2[0,1]}$$
$$\leq 18\,\lambda(72\,\lambda)^{1/2}\sqrt{32}\,\lambda^{3/2}$$
$$\leq 18 \cdot 48\,\lambda^3\,.$$

Applying E.3 a] with P', we get

$$\text{(6.1.27)} \qquad \|P''\|_{[0,1]} \leq 18\,\lambda\,\|P'\|_{[0,1]} \leq 18^2 \cdot 48\,\lambda^4\,.$$

Now (6.1.23), (6.1.27), and the Mean Value Theorem give

$$|P'(x)| \geq \tfrac{1}{16}\lambda^3\,, \qquad x \in I\,,$$

where

$$I := \left[1 - \left(18^2 \cdot 48 \cdot 16\,\lambda\right)^{-1}, 1\right].$$

So

$$\text{(6.1.28)} \qquad \|xP'(x)\|_{L_p[0,1]} \geq \left(\int_I |xP'(x)|^p\,dx\right)^{1/p}$$
$$\geq \left(\left(18^2 \cdot 48 \cdot 16\,\lambda\right)^{-1}\left(\tfrac{1}{32}\lambda^3\right)^p\right)^{1/p}$$
$$\geq \left(18^2 \cdot 48 \cdot 16\right)^{-1/p} 32^{-1/2}\lambda^{3-1/p}$$
$$\geq \left(128 \cdot 9\sqrt{3}\right)^{-1}\lambda^{3-1/p}\,.$$

Combining (6.1.25) and (6.1.28), we obtain the required result with a constant $c = (128 \cdot 9\sqrt{3})^{-1}$. □

E.9 On the Interval Where the Sup Norm of a Müntz Polynomial Lives. Let $\Lambda := (\lambda_j)_{j=0}^\infty$ be an increasing sequence of nonnegative real numbers. Let $0 \neq p \in M_n(\Lambda) := \operatorname{span}\{x^{\lambda_0}, x^{\lambda_1}, \dots x^{\lambda_n}\}$, and let $q(x) := x^{k\alpha}p(x)$, where $\alpha > 0$ and k is a nonnegative integer. Let $\xi \in [0,1]$ be chosen so that $|q(\xi)| = \|q\|_{[0,1]}$.

a] Suppose $\alpha = 1$ and each λ_j is an integer. Then

$$\left(\frac{k}{k+n}\right)^2 < \xi\,.$$

Proof. See Saff and Varga [81]. □

The above result is sharp in a certain limiting sense, which is described in detail in Saff and Varga [78].

b] Suppose $\lambda_j = \alpha j$ for each j. Use part a] to show that

$$\left(\frac{k}{k+n}\right)^{2/\alpha} < \xi\,.$$

c] Suppose $\lambda_j = \alpha j$ for each j. Use E.11 of Section 5.1 to show, without using part a], that there exists an absolute constant $c > 0$ such that

$$\left(\frac{ck}{k+n}\right)^{2/\alpha} < \xi\,.$$

d] Suppose $\lambda_j \geq \alpha j$ for each j. Use part b] and E.3 g] of Section 3.3 to show that the conclusion of part b] remains valid.

e] Suppose $\lambda_j \geq \alpha j$ for each j. Use part c] and E.3 g] of Section 3.3 to show that the conclusion of part c] remains valid.

The following extension of Newman's inequality is in Borwein and Erdélyi [to appear 3].

E.10 Newman's Inequality on $[a,b] \subset (0,\infty)$.

a] Let $\Lambda := (\lambda_j)_{j=0}^\infty$ be an increasing sequence of nonnegative real numbers. Assume that there exists an $\alpha > 0$ such that $\lambda_j \geq \alpha j$ for each j. Suppose that $[a,b] \subset (0,\infty)$. Show that there exists a constant $c(a,b,\alpha)$ depending only on a, b, and α such that

$$\|p'\|_{[a,b]} \leq c(a,b,\alpha)\left(\sum_{j=0}^n \lambda_j\right)\|p\|_{[a,b]}$$

for every $p \in M_n(\Lambda) := \operatorname{span}\{x^{\lambda_0}, x^{\lambda_1}, \dots x^{\lambda_n}\}$.

Proof. We base the proof on E.9 d], however; it may also be based on E.9 e]. Let $p \in M_n(\Lambda)$. We want to estimate $p'(y)$ for every $y \in [a,b]$. First let $y \in \left[\frac{1}{2}(a+b), b\right]$. We define $q(x) := x^{mn\alpha}p(x)$, where m is the smallest positive integer satisfying

$$a \le \frac{a+b}{2}\left(\frac{m}{m+1}\right)^{2/\alpha}.$$

Scaling Newman's inequality from $[0,1]$ to $[0,y]$, then using E.9 d], we obtain

$$\begin{aligned}
|q'(y)| &\le \frac{9}{y}\sum_{j=0}^{n}(\lambda_j + mn\alpha)\|q\|_{[0,y]} \\
&= \frac{9}{y}\sum_{j=0}^{n}(\lambda_j + mn\alpha)\|q\|_{\left[y\left(\frac{m}{m+1}\right)^{2/\alpha},y\right]} \\
&\le c_1(a,b,\alpha)\left(\sum_{j=0}^{n}\lambda_j\right)\|q\|_{[a,y]}
\end{aligned}$$

with a constant $c_1(a,b,\alpha)$ depending only on a, b, and α. Hence

$$\begin{aligned}
|p'(y)| &\le \left|q'(y)y^{-mn\alpha}\right| + \frac{mn\alpha}{y}|p(y)| \\
&\le y^{-mn\alpha}c_1(a,b,\alpha)\left(\sum_{j=0}^{n}\lambda_j\right)\|q\|_{[a,y]} + \frac{mn}{y}\|p\|_{[a,y]} \\
&\le c_2(a,b,\alpha)\left(\sum_{j=0}^{n}\lambda_j\right)\|p\|_{[a,y]} \\
&\le c_2(a,b,\alpha)\left(\sum_{j=0}^{n}\lambda_j\right)\|p\|_{[a,b]}
\end{aligned}$$

with a constant $c_2(a,b,\alpha)$ depending only on a, b, and α.

Now let $y \in \left[a, \frac{1}{2}(a+b)\right]$. Then, by E.3 b] and f] of Section 3.3, we can deduce that

$$\begin{aligned}
|p'(y)| &\le \left|T_n'\{x^0, x^\alpha, x^{2\alpha}, \dots, x^{n\alpha}; [y,b]\}(y)\right| \, \|p\|_{[y,b]} \\
&= \frac{2\alpha y^{\alpha-1}}{b^\alpha - y^\alpha}n^2\|p\|_{[y,b]} \le c_3(a,b,\alpha)n^2\|p\|_{[y,b]} \\
&\le c_4(a,b,\alpha)\left(\sum_{j=0}^{n}\lambda_j\right)\|p\|_{[y,b]}
\end{aligned}$$

with constants $c_3(a,b,\alpha)$ and $c_4(a,b,\alpha)$ depending only on a, b, and α. This finishes the proof. □

b] Show that if the gap condition $\lambda_j \geq j\alpha$ in part a] is dropped, then the conclusion of part a] fails to hold with $c(a,b,\alpha)$ replaced by a constant $c(a,b)$ depending only on a and b.

Hint: Study
$$T'_n\left\{x^0, x^\alpha, x^{2\alpha}, \dots, x^{n\alpha}; \left[\tfrac{1}{2}, 1\right]\right\}$$
and let $\alpha \to 0+$. □

6.2 Nondense Müntz Spaces

Throughout this section we assume that $\Lambda := (\lambda_i)_{i=0}^\infty$ is a sequence of distinct nonnegative real numbers, $M_n(\Lambda)$ denotes the linear span of $\{x^{\lambda_0}, x^{\lambda_1}, \dots, x^{\lambda_n}\}$ over $\mathbb{R}$, and

$$M(\Lambda) := \bigcup_{n=0}^\infty M_n(\Lambda) = \operatorname{span}\{x^{\lambda_0}, x^{\lambda_1}, \dots\}.$$

If $A \subset [0,1]$ is compact, then a combination of Tietze's and Müntz's theorems yields that $M(\Lambda)$ is dense in $C(A)$ whenever $\sum_{i=1}^\infty 1/\lambda_i = \infty$. (Recall that Tietze's theorem guarantees that if $A \subset [0,1]$ is compact, then for every $f \in C(A)$ there exists an $\widetilde{f} \in C[0,1]$ such that $\widetilde{f}(x) = f(x)$ for all $x \in A$.) If the Lebesgue measure $m(A)$ of A is positive, then the converse is also true. More precisely, we have the following.

Theorem 6.2.1 (Müntz Theorem on Compact Sets of Positive Measure). *Suppose $\lambda_0 := 0$ and $\sum_{i=1}^\infty 1/\lambda_i < \infty$. Let $A \subset [0,1]$ be a compact set with positive Lebesgue measure. Then $M(\Lambda)$ is not dense in $C(A)$. Moreover, if the gap condition*
$$\inf\{\lambda_i - \lambda_{i-1} : i \in \mathbb{N}\} > 0$$
holds and
$$r_A := \sup\{x \in [0,\infty) : m(A \cap (x,\infty)) > 0\},$$
then every function $f \in C(A)$ from the uniform closure of $M(\Lambda)$ on A is of the form
$$f(x) = \sum_{j=0}^\infty a_j x^{\lambda_j}, \qquad x \in A \cap [0, r_A).$$
If the above gap condition does not hold, then every function $f \in C(A)$ from the uniform closure of $M(\Lambda)$ on A can still be extended analytically throughout the region
$$\{z \in \mathbb{C} \setminus (-\infty, 0] : |z| < r_A\}.$$

The proof of the above theorem rests on the following bounded Remez-type inequality:

Theorem 6.2.2 (Remez-Type Inequality for Nondense Müntz Spaces). *Let $\lambda_0 := 0$ and $\sum_{i=0}^{\infty} 1/\lambda_i < \infty$. Then there exists a constant c depending only on Λ and s (and not on ϱ, A, or the number of terms in p) such that*

$$\|p\|_{[0,\varrho]} \leq c\,\|p\|_A$$

for every $p \in \operatorname{span}\{x^{\lambda_0}, x^{\lambda_1}, \dots\}$ and for every compact set $A \subset [\varrho, 1]$ of Lebesgue measure at least $s > 0$.

To prove Theorem 6.2.2 we need a few lemmas. We use the notation introduced in Section 3.3. However, for notational convenience, we let

$$T_n\{\lambda_0, \lambda_1, \dots, \lambda_n; A\} := T_n\{x^{\lambda_0}, x^{\lambda_1}, \dots, x^{\lambda_n}; A\}$$

for compact sets $A \subset [0, \infty)$.

By the unique interpolation property of Chebyshev spaces (see Proposition 3.1.2), associated with

$$0 < x_0 < x_1 < \cdots < x_n\,,$$

we can define

$$\ell_k\{x_0, x_1, \dots, x_n\} \in M_n(\Lambda)\,, \qquad k = 0, 1, \dots, n$$

such that

$$\ell_k\{x_0, x_1, \dots, x_n\}(x_j) = \delta_{j,k} := \begin{cases} 1 & \text{if } j = k \\ 0 & \text{if } j \neq k\,. \end{cases}$$

Lemma 6.2.3. *Let*

$$0 < x_0 < x_1 < \cdots < x_n \qquad \textit{and} \qquad 0 < \widetilde{x}_0 < \widetilde{x}_1 < \cdots < \widetilde{x}_n\,.$$

Suppose $0 \leq k \leq n$ and

$$\begin{aligned} x_j &\leq \widetilde{x}_j \quad \text{if } j = 0, 1, \dots, k-1\,; \\ x_j &= \widetilde{x}_j \quad \text{if } j = k\,; \\ x_j &\geq \widetilde{x}_j \quad \text{if } j = k+1, k+2, \dots, n\,. \end{aligned}$$

For notational convenience, let

$$\ell_k := \ell_k\{x_0, x_1, \dots, x_n\} \qquad \textit{and} \qquad \widetilde{\ell}_k := \ell_k\{\widetilde{x}_0, \widetilde{x}_1, \dots, \widetilde{x}_n\}\,.$$

Then

$$|\ell_k(0)| \leq |\widetilde{\ell}_k(0)|\,.$$

Proof. It is sufficient to prove the lemma in the case where there is an index m such that $1 \leq m \leq n$, $m \neq k$, and

$$\begin{aligned} x_j &= \widetilde{x}_j \quad \text{if } j = 0, 1, \dots, n, \;\; j \neq m\,; \\ x_m &< \widetilde{x}_m \quad \text{if } m < k\,; \\ x_m &> \widetilde{x}_m \quad \text{if } m > k\,. \end{aligned}$$

The general case of the lemma then follows from repeated applications of the above special cases. Note that in the above special cases

$$\ell_k - \widetilde{\ell}_k \in M_n(\Lambda)$$

has a zero at each of the points

$$x_j\,, \quad j = 0, 1, \dots, n\,, \;\; j \neq m\,;$$

hence it changes sign at each of these points, and it has no other zero in $[0, \infty)$ (see E.10 of Section 3.1). It is also obvious that

$$\operatorname{sign}(\ell_k(x)) = \operatorname{sign}(\widetilde{\ell}_k(x))\,, \qquad x \in (0, x_0)\,,$$

which, together with the previous observation and the inequality $x_0 \leq \widetilde{x}_0$, yields that

$$|\ell_k(0)| \leq |\widetilde{\ell}_k(0)|\,,$$

which finishes the proof. □

By a simple scaling we can extend Lemma 6.2.3 as follows. We use the notation introduced in Lemma 6.2.3.

Lemma 6.2.4. *Let*

$$0 < x_0 < x_1 < \cdots < x_n \qquad \textit{and} \qquad 0 < \widetilde{x}_0 < \widetilde{x}_1 < \cdots < \widetilde{x}_n\,.$$

Suppose $0 \leq k \leq n$, $\gamma \geq 0$, *and*

$$\begin{aligned} x_j &\leq \widetilde{x}_j - \gamma \quad \textit{if} \;\; j = 0, 1, \dots, k-1\,; \\ x_j &= \widetilde{x}_j - \gamma \quad \textit{if} \;\; j = k\,; \\ x_j &\geq \widetilde{x}_j - \gamma \quad \textit{if} \;\; j = k+1, k+2, \dots, n\,. \end{aligned}$$

Then

$$|\ell_k(0)| \leq |\widetilde{\ell}_k(0)|\,.$$

Proof. If $\gamma = 0$, then Lemma 6.2.3 yields the lemma. So we may assume that $\gamma > 0$. Let

$$\beta := \frac{x_k}{\widetilde{x}_k} = \frac{\widetilde{x}_k - \gamma}{\widetilde{x}_k}\,,$$
$$x_j^* := \beta \widetilde{x}_j\,, \qquad j = 0, 1, \dots, n\,,$$

and

$$\ell_k^* := \ell_k\{x_0^*, x_1^*, \dots, x_n^*\}\,, \qquad k = 0, 1, \dots, n\,.$$

Obviously

$$\widetilde{\ell}_k(\beta x) = \ell_k^*(x)\,, \qquad x \in [0, \infty)$$

and

$$\begin{aligned} x_j &\le x_j^* \quad \text{if } j = 0, 1, \dots, k-1\,,\\ x_j &= x_j^* \quad \text{if } j = k\,,\\ x_j &\ge x_j^* \quad \text{if } j = k+1, k+2, \dots, n\,. \end{aligned}$$

Hence Lemma 6.2.3 implies that

$$|\ell_k(0)| \le |\ell_k^*(0)| = |\widetilde{\ell}_k(0)|$$

which finishes the proof. □

Lemma 6.2.5. *Let A be a closed subset of $[0, 1]$ with Lebesgue measure at least $s \in (0, 1)$. Then*

$$|p(0)| \le |T_n\{\lambda_0, \lambda_1, \dots, \lambda_n; [1-s, 1]\}(0)| \cdot \|p\|_A$$

for every $p \in M(\Lambda)$.

Proof. If $0 \in A$, then the statement is trivial. So assume that $0 \notin A$. Let

$$\widetilde{x}_0 < \widetilde{x}_1 < \cdots < \widetilde{x}_n$$

denote the extreme points of

$$T_n := T_n\{\lambda_0, \lambda_1, \dots, \lambda_n; [1-s, 1]\}$$

in $[1-s, 1]$, that is,

$$T_n(\widetilde{x}_j) = (-1)^{n-j}\,, \qquad j = 0, 1, \dots, n\,.$$

Let $x_j \in A,\ \ j = 0, 1, \dots, n$, be defined so that

$$m([x_j, 1] \cap A) = m([\widetilde{x}_j, \widetilde{x}_n]) = \widetilde{x}_n - \widetilde{x}_j\,.$$

Since A is a closed subset of $[0, 1]$ with $m(A) \ge s$, such points $x_j \in A$ exist. Let $p \in M_n(\Lambda)$. Then, by Lemma 6.2.4, we can deduce that

$$
\begin{aligned}
|p(0)| &= \left| \sum_{k=0}^{n} p(x_k)\ell_k(0) \right| \\
&\leq \left(\sum_{k=0}^{n} |\ell_k(0)| \right) \|p\|_A \\
&\leq \left(\sum_{k=0}^{n} |\widetilde{\ell}_k(0)| \right) \|p\|_A \\
&= \left(\sum_{k=0}^{n} (-1)^{n-k} \widetilde{\ell}_k(0) \right) \|p\|_A \\
&= \left(\sum_{k=0}^{n} T_n(\widetilde{x}_k)\widetilde{\ell}_k(0) \right) \|p\|_A \\
&= |T_n(0)| \cdot \|p\|_A
\end{aligned}
$$

which proves the lemma. □

Lemma 6.2.6. *Suppose* $\lambda_0 := 0$. *Let* A *be a closed subset of* $[0,1]$ *with Lebesgue measure at least* $s \in (0,1)$. *Then*

$$|p(y)| \leq |T_n\{\lambda_0, \lambda_1, \ldots, \lambda_n; [1-s, 1]\}(0)| \cdot \|p\|_A$$

for every $p \in M_n(\Lambda)$ *and* $y \in [0, \inf A]$.

Proof. For notational convenience, let

$$T_{n,A} := T_n\{\lambda_0, \lambda_1, \ldots, \lambda_n; A\}.$$

Note that $\lambda_0 = 0$ implies that $|T_{n,A}|$ is decreasing on $[0, \inf A]$, otherwise

$$T'_{n,A} \in \operatorname{span}\{x^{\lambda_1 - 1}, x^{\lambda_2 - 1}, \ldots, x^{\lambda_n - 1}\}$$

would have at least $n+1$ zeros in $(0,1]$, which is impossible. Hence, it follows from E.3 and Lemma 6.2.5 that

$$
\begin{aligned}
\frac{|p(y)|}{\|p\|_A} \leq \frac{|T_{n,A}(y)|}{\|T_{n,A}\|_A} &= |T_{n,A}(y)| \leq |T_{n,A}(0)| \\
&\leq |T_n\{\lambda_0, \lambda_1, \ldots, \lambda_n; [1-s, 1]\}(0)|
\end{aligned}
$$

for every $0 \neq p \in M_n(\Lambda)$. This finishes the proof. □

Proof of Theorem 6.2.2. Lemma 6.2.6 and E.5 a] of Section 4.2 yield the theorem. □

Proof of Theorem 6.2.1. The theorem follows from E.5 of this section and E.3 e] and E.8 b] of Section 4.2. □

Our next theorem is an interesting characterization of lacunary sequences.

Theorem 6.2.7 (Characterization of Lacunary Müntz Spaces). *Suppose $\Lambda := (\lambda_i)_{i=0}^\infty$ with $0 \le \lambda_0 < \lambda_1 < \cdots$. There exists a constant c depending only on Λ such that*

$$|a_{j,n}| \le c\,\|p\|_{[0,1]}\,, \qquad j = 0, 1, \dots, n\,, \quad n \in \mathbb{N}$$

for every $p \in M(\Lambda)$ of the form

$$p(x) = \sum_{j=0}^{n} a_{j,n} x^{\lambda_j}$$

if and only if Λ is lacunary, that is, if and only if the elements λ_i of Λ satisfy

$$\inf\{\lambda_{i+1}/\lambda_i : i \in \mathbb{N}\} > 1\,.$$

To prove Theorem 6.2.7 we need the following result of Hardy and Littlewood [26] whose proof we do not reproduce:

Theorem 6.2.8. *Suppose $0 = \gamma_0 < \gamma_1 < \cdots$ is a lacunary sequence, that is,*

$$\inf\{\gamma_{i+1}/\gamma_i : i \in \mathbb{N}\} > 1\,.$$

Suppose the function f is of the form

$$f(x) = \sum_{i=0}^{\infty} a_i x^{\gamma_i}\,, \qquad a_i \in \mathbb{R}\,, \quad x \in [0,1)$$

and $A := \lim_{x\to 1-} f(x)$ exists and is finite. Then $\sum_{i=0}^\infty a_i = A$.

Proof of Theorem 6.2.7. Suppose Λ is lacunary and suppose there exists a sequence $(P_k)_{k=1}^\infty \subset M(\Lambda)$ such that if P_k is of the form

$$P_k(x) = \sum_{j=0}^{n_k} a_{j,n_k} x^{\lambda_j}\,, \qquad a_{j,n_k} \in \mathbb{R}\,,$$

then

$$\text{(6.2.1)} \quad \|P_k\|_{[0,1]} = 1 \qquad \text{and} \qquad \max_{0\le j\le n_k} |a_{j,n_k}| \ge k^2\,, \qquad k = 1, 2, \dots\,.$$

We may assume, without loss of generality, that $\lambda_0 = 0$. Choose a sequence $(\alpha_k)_{k=1}^\infty$ of positive integers such that

$$\alpha_1 = 1\,, \qquad \lambda_1\alpha_{k+1} > 2\alpha_k\lambda_{n_k}\,, \qquad k = 1, 2, \dots\,.$$

Now let the function f be defined by

$$f(x) = \sum_{k=1}^{\infty} k^{-2} P_k(x^{\alpha_k}) \le \sum_{k=1}^{\infty} k^{-2} < \infty\,, \qquad 0 \le x \le 1\,. \tag{6.2.2}$$

Note that $f \in C[0,1]$ by the Weierstrass M-test. For notational convenience, let

$$m_0 := 0, \qquad m_k := \sum_{i=1}^{k} n_i\,, \qquad k = 1, 2, \dots\,.$$

Further, let

$$\gamma_0 := 0 \qquad \text{and} \qquad a_0 := \sum_{k=1}^{\infty} k^{-2} a_{0,n_k} = \sum_{k=1}^{\infty} k^{-2} P_k(0)$$

and

$$\gamma_{m_{k-1}+j} := \alpha_k \lambda_j\,, \quad j = 1, 2, \dots, n_k\,, \quad k = 1, 2, \dots\,.$$

Observe that $a_0 \in \mathbb{R}$ is well-defined since $|P_k(0)| \le 1$ for each $k \in \mathbb{N}$. Also

$$\inf\{\gamma_{i+1}/\gamma_i : i \in \mathbb{N}\} \ge \min\{2, \inf\{\lambda_{i+1}/\lambda_i : i \in \mathbb{N}\}\} > 1\,.$$

Let $\Gamma := (\gamma_i)_{i=0}^{\infty}$. Then $f \in C[0,1]$ defined by (6.2.2) is in the uniform closure of $M(\Gamma)$ on $[0,1]$; hence, by the Clarkson-Erdős theorem (see E.3 e] of Section 4.2), f is of the form

$$f(x) = \sum_{i=0}^{\infty} a_i x^{\gamma_i}\,, \qquad x \in [0,1)\,.$$

Since $f \in C[0,1]$, Theorem 6.2.8 implies that $A := \sum_{i=0}^{\infty} a_i$ exists and is finite. Recalling (6.2.2) and the choice of α_k, and using E.3 e] of Section 4.2, we can deduce that each

$$k^{-2} a_{j,n_k}\,, \qquad j = 1, 2, \dots, n_k\,, \quad k = 1, 2, \dots$$

is equal to one of the coefficients $a_1, a_2 \dots$. Since $|a_{0,n_k}| = |P_k(0)| \le 1$ for each $k \in \mathbb{N}$, from (6.2.1) and (6.2.2) we see that $|a_i| \ge 1$ holds for infinitely many $i \in \mathbb{N}$, which contradicts the fact that $\sum_{i=0}^{\infty} a_i$ converges. This finishes the *if* part of the theorem.

Now assume that Λ is not lacunary. Then for every $\epsilon > 0$ there is an $n \in \mathbb{N}$ such that $\lambda_{n-1}/\lambda_n > 1 - \epsilon$. Observe that $P_n(x) := x^{\lambda_n} - x^{\lambda_{n-1}}$ achieves its maximum modulus on $[0,1]$ at

$$x = \left(\frac{\lambda_{n-1}}{\lambda_n}\right)^{1/(\lambda_n - \lambda_{n-1})}$$

and hence

$$\|P_n\|_{[0,1]} \le \left(\frac{\lambda_{n-1}}{\lambda_n}\right)^{\lambda_{n-1}/(\lambda_n-\lambda_{n-1})}\left(1-\frac{\lambda_{n-1}}{\lambda_n}\right) \le 1-\frac{\lambda_{n-1}}{\lambda_n} < \epsilon\,,$$

which shows that the lead coefficient $a_{n,n}$ of

$$T_n\{\lambda_0, \lambda_1, \dots, \lambda_n; [0,1]\}$$

is at least $1/\epsilon$, otherwise $a_{n,n}^{-1}T_n - P_n \in M_{n-1}(\Lambda)$ would have at least n zeros on $(0,1)$, which is a contradiction. The proof of the *only if* part of the theorem is now finished. □

From the above proof it also follows that under the assumptions of Theorem 6.2.7, the Chebyshev polynomials

$$T_n\{\lambda_0, \lambda_1, \dots, \lambda_n; [0,1]\}$$

have uniformly bounded coefficients if and only if Λ is lacunary.

As an application of Theorem 6.2.7 we derive the following Bernstein-type inequality.

Theorem 6.2.9 (Bernstein-Type Inequality). *Suppose* $\lambda_0 := 0$, $\lambda_1 \ge 1$, *and suppose* $\Lambda := (\lambda_i)_{i=0}^\infty$ *is lacunary, that is,*

$$\inf\{\lambda_{i+1}/\lambda_i : i \in \mathbb{N}\} > 1\,.$$

Then there exists a constant c depending only on Λ (and not on y or the number of terms in p) such that

$$|p'(y)| \le \frac{c}{1-y}\,\|p\|_{[0,1]}$$

for every $p \in M(\Lambda) = \mathrm{span}\{x^{\lambda_0}, x^{\lambda_1}, \dots\}$ *and for every* $y \in [0,1)$.

Proof. Let $p \in M(\Lambda)$ be of the form

$$p(x) = a_{0,n} + \sum_{j=1}^n a_{j,n}x^{\lambda_j}\,, \qquad \|p\|_{[0,1]} = 1\,.$$

Theorem 6.2.7 and the assumptions on Λ yield

$$\begin{aligned}|p'(y)| &\le \left|\sum_{j=1}^n a_{j,n}\lambda_j y^{\lambda_j-1}\right| \le \sum_{j=1}^n |a_{j,n}|\,\lambda_j y^{\lambda_j-1}\\ &\le c_1\sum_{j=1}^n \lambda_j y^{\lambda_j-1} \le c_2\sum_{j=0}^\infty y^j = \frac{c_2}{1-y}\,,\end{aligned}$$

where c_1 and c_2 depend only on Λ, and the theorem is proved. □

Comments, Exercises, and Examples.

The results of this section have been obtained by Borwein and Erdélyi [91, 93, 95b, to appear 1]. In E.1 we present some of the several important consequences of our central result, Theorem 6.2.2. In E.6 we offer another proof of the first part of Theorem 6.2.1 when Λ is lacunary, while E.9 shows that the Bernstein-type inequality of Theorem 6.2.9 "almost" characterizes the lacunary Müntz spaces. Note that if $A \subset [0,1]$ contains an interval, then the first part of Theorem 6.2.1 follows immediately from E.3 of Section 4.2. A typical case that does not follow from that exercise is when $A \subset [0,1]$ is a "fat" Cantor-type set of positive measure.

E.1 Some Consequences of Theorem 6.2.2. Let $A \subset [0,\infty)$ be a set of positive Lebesgue measure, and let r_A be the essential supremum of A as defined in Theorem 6.2.1. Suppose $q \in (0,\infty)$ and suppose w is a nonnegative-valued, integrable weight function on A with $\int_A w > 0$. Let $L_q(w) := L_q(\mu)$, where $d\mu = w\,dt$, and where $L_q(\mu)$ is defined in E.7 of Section 2.2. Let $\Lambda := (\lambda_i)_{i=0}^\infty$ be a sequence of distinct nonnegative real numbers with $\lambda_i \neq 0$ for each $i = 1, 2, \ldots$.

a] Suppose $\sum_{i=1}^\infty 1/\lambda_i < \infty$. Then $M(\Lambda)$ is not dense in $L_q(w)$. Moreover, if the gap condition

$$\inf\{\lambda_i - \lambda_{i-1} : i \in \mathbb{N}\} > 0$$

holds, then every function $f \in L_q(w)$ belonging to the $L_q(w)$ closure of $M(\Lambda)$ can be represented as

$$f(x) = \sum_{i=0}^\infty a_i x^{\lambda_i}, \qquad a_i \in \mathbb{R}, \quad x \in A \cap [0, r_w),$$

where

$$r_w := \sup\left\{ y \in [0,\infty) : \int_{A\cap(y,\infty)} w(x)\,dx > 0 \right\}.$$

If the above gap condition does not hold, then every function $f \in L_q(w)$ belonging to the $L_q(w)$ closure of $M(\Lambda)$ can still be represented as an analytic function on

$$\{z \in \mathbb{C} \setminus (-\infty, 0] : |z| < r_w\}$$

restricted to A.

Proof. Suppose $f \in L_q(w)$ and suppose there is a sequence $(p_i)_{i=1}^\infty \subset M(\Lambda)$ such that

$$\lim_{i\to\infty} \|f - p_i\|_{L_q(w)} = 0.$$

Minkowski's inequality (see E.7 b] and E.7 i] of Section 2.2.) yields that $(p_i)_{i=1}^\infty$ is a Cauchy sequence in $L_q(w)$. The assumptions on w imply that for every $\delta \in (0, r_w)$ there exists an $\alpha > 0$ such that

$$B := \{x \in A \cap (\delta, \infty) : w(x) > \alpha\}$$

is of positive Lebesgue measure. Let $s := m(B) > 0$. Observe that if $\|p\|_{L_q(w)} < \varepsilon$, then

$$m\left(\left\{x \in B : |p(x)| \geq \left(\frac{2\varepsilon}{\alpha s}\right)^{1/q}\right\}\right) \leq \frac{s}{2},$$

so

$$m\left(\left\{x \in B : |p(x)| < \left(\frac{2\varepsilon}{\alpha s}\right)^{1/q}\right\}\right) > \frac{s}{2}.$$

Hence, by Theorem 6.2.2, $(p_i)_{i=1}^\infty$ is uniformly Cauchy on $[0, \delta]$. The proof can now be finished as that of Theorem 6.2.1. □

b] Müntz-Type Theorem in $L_q(w)$. *$M(\Lambda)$ is dense in $L_q(w)$ if and only if $\sum_{i=1}^\infty 1/\lambda_i = \infty$.*

Proof. Suppose $\sum_{i=1}^\infty 1/\lambda_i = \infty$. Let $f \in L_q(w)$. It is standard measure theory to show that for every $\varepsilon > 0$ there exists a $g \in C[0,1]$ such that $g(0) = 0$ and

$$\|f - g\|_{L_q(w)} < \frac{\varepsilon}{2}.$$

Now Theorem 4.2.1 (full Müntz theorem in $C[0,1]$) implies that there exists a $p \in M(\Lambda)$ such that

$$\|g - p\|_{L_q(w)} \leq \|g - p\|_A \left(\int_A w\right)^{1/q} \leq \|g - p\|_{[0,1]} \left(\int_A w\right)^{1/q} < \frac{\varepsilon}{2}.$$

Therefore $M(\Lambda)$ is dense in $L_q(w)$.

Suppose now that $\sum_{i=1}^\infty 1/\lambda_i < \infty$. Then part a] yields that $M(\Lambda)$ is not dense in $L_q(w)$. □

c] Convergence in $M(\Lambda)$. Suppose $\sum_{i=1}^\infty 1/\lambda_i < \infty$, $(p_i)_{i=1}^\infty \subset M(\Lambda)$, and

$$p_i(x) \to f(x), \qquad x \in A.$$

Then $(p_i)_{i=1}^\infty$ converges uniformly on every closed subinterval of $[0, r_A)$.

Proof. Let $\delta \in (0, r_A)$ be fixed. Egoroff's theorem (see, for example, Royden [88]) and the definition of r_A imply the existence of a set $B \subset A \cap (\delta, \infty)$ of positive Lebesgue measure so that $(p_i)_{i=1}^\infty$ converges uniformly on B and hence is uniformly Cauchy on B. Now Theorem 6.2.2 yields that $(p_i)_{i=1}^\infty$ is uniformly Cauchy on $[0, \delta]$, and the result follows. □

d] Suppose $\sum_{i=1}^{\infty} 1/\lambda_i = \infty$. Show that there is a sequence $(p_i)_{i=1}^{\infty} \subset M(\Lambda)$ that converges pointwise on $[0, \infty)$ but does not converge uniformly on $A \cap [0, a]$ for some $a \in (0, r_A)$.

Hint: Use Theorem 4.2.1 (Müntz's theorem). □

e] Suppose $\sum_{i=1}^{\infty} 1/\lambda_i < \infty$ and

$$\inf\{\lambda_i - \lambda_{i-1} : i \in \mathbb{N}\} > 0\,.$$

Let $P(\Lambda)$ denote the collection of all real-valued functions f defined on $[0,1)$ by a power series

$$f(x) = \sum_{i=0}^{\infty} a_i x^{\lambda_i}\,, \qquad a_i \in \mathbb{R}\,, \quad x \in [0,1)\,.$$

Suppose that $A \subset [0,1]$ with $r_A = 1$. Show that if $(f_i)_{i=1}^{\infty} \subset P(\Lambda)$ and

$$f_i(x) \to f(x)\,, \qquad x \in A\,,$$

then

$$f_i(x) \to \widetilde{f}(x)\,, \qquad x \in [0,1)\,,$$

where $\widetilde{f} \in P(\Lambda)$.

Hint: Use part c] and E.3 e] of Section 4.2. □

E.2 On the Smallest Zero of Chebyshev Polynomials in Nondense Müntz Spaces. Suppose $\lambda_0 := 0$ and $\sum_{i=1}^{\infty} 1/\lambda_i < \infty$. Show that there exists a constant $c > 0$ depending only on $\Lambda := (\lambda_i)_{i=0}^{\infty}$ (and not on n) such that the smallest positive zero of

$$T_n\{0, \lambda_1, \lambda_2, \dots, \lambda_n; [0,1]\}\,, \qquad n = 1, 2, \dots$$

is greater than c.

Hint: If $\lambda_1 \geq 1$, then use the Mean Value Theorem, E.1 a] of Section 3.3, and E.5 b] of Section 4.2. If $0 < \lambda_1 < 1$, then the scaling $x \to x^{1/\lambda_1}$ reduces the problem to the case $\lambda_1 = 1$. □

E.3 Extremal Functions for the Remez-Type Inequality of Theorem 6.2.2. Suppose $0 \leq \lambda_0 < \lambda_1 < \cdots < \lambda_n$, $0 < \varrho$, $A \subset [\varrho, \infty)$ is a compact set containing at least $n+1$ points, and $y \in (0, \varrho)$ is fixed. Let

$$M_n(\Lambda) := \operatorname{span}\{x^{\lambda_0}, x^{\lambda_1}, \dots, x^{\lambda_n}\}\,.$$

a] Show that there is a $0 \neq p^* \in M_n(\Lambda)$ such that

$$\frac{|p^*(y)|}{\|p^*\|_A} = \sup_{0 \neq p \in M_n(\Lambda)} \frac{|p(y)|}{\|p\|_A}\,.$$

Hint: Use a compactness argument. □

b] Show that $p^* = cT_n\{\lambda_0, \lambda_1, \dots, \lambda_n; A\}$ for some $c \in \mathbb{R}$.

Hint: Use a perturbation argument. □

E.4 A Lexicographic Property of Chebyshev Polynomials in Different Müntz Spaces. Let $0 := \lambda_0 < \lambda_1 < \cdots < \lambda_n$, $0 := \gamma_0 < \gamma_1 < \cdots < \gamma_n$, and

$$\lambda_j \leq \gamma_j\,, \qquad j = 0, 1, \ldots, n\,.$$

Let $\varrho > 0$ and let $A \subset [\varrho, \infty)$ be compact containing at least $n+1$ points, and let

$$T_{n,\Lambda} := T_n\{\lambda_0, \lambda_1, \ldots, \lambda_n; A\} \quad \text{and} \quad T_{n,\Gamma} := T_n\{\gamma_0, \gamma_1, \ldots, \gamma_n; A\}\,.$$

a] Show that $|T_{n,\Gamma}(y)| \leq |T_{n,\Lambda}(y)|$ for every $y \in [0, \varrho)$.

Hint: Suppose, without loss of generality, that there is an index m, $1 \leq m \leq n$, such that $\lambda_m < \gamma_m$ and $\lambda_j = \gamma_j$ if $j \neq m$. We choose an $R_{n,\Lambda} \in M_n(\Lambda)$ that interpolates $T_{n,\Gamma}$ at the n zeros of $T_{n,\Gamma}$ and is normalized so that $R_{n,\Lambda}(0) = T_{n,\Gamma}(0)$. Use Theorem 3.2.5 to show that $|R_{n,\Lambda}(x)| \leq |T_{n,\Gamma}(x)|$ for every $x \in [0, \infty)$, in particular for every $x \in A$. Now use E.3 to show that $|T_{n,\Gamma}(0)| = |R_{n,\Lambda}(0)| \leq |T_{n,\Lambda}(0)|$, which gives the desired result for $y = 0$. Using this, we can deduce that $|T_{n,\Gamma}(y)| \leq |T_{n,\Lambda}(y)|$ for every $y \in [0, \varrho)$, otherwise

$$T_{n,\Lambda} - T_{n,\Gamma} \in \operatorname{span}\{x^{\lambda_0}, x^{\lambda_1}, \ldots, x^{\lambda_n}, x^{\gamma_m}\}$$

would have at least $(n+2)$ zeros in $(0, \infty)$, which is a contradiction. □

b] Show that

$$\max_{p \in M_n(\Gamma)} \frac{|p(y)|}{\|p\|_A} \leq \max_{p \in M_n(\Lambda)} \frac{|p(y)|}{\|p\|_A}$$

for every $y \in [0, \varrho)$, where

$$M_n(\Gamma) := \operatorname{span}\{x^{\gamma_0}, x^{\gamma_1}, \ldots, x^{\gamma_n}\}$$

and

$$M_n(\Lambda) := \operatorname{span}\{x^{\lambda_0}, x^{\lambda_1}, \ldots, x^{\lambda_n}\}\,.$$

Hint: Combine part a] and E.4. □

E.5 Theorem 6.2.1 Follows from Theorem 6.2.2. Under the assumptions of Theorem 6.2.1 show that if $(p_j)_{j=1}^{\infty} \subset M(\Lambda)$ is uniformly Cauchy in $C(A)$, then it is uniformly Cauchy in $C[0, y]$ for every $y \in (0, r_A)$, where r_A is defined as in Theorem 6.2.1.

Hint: Use Theorem 6.2.2. □

E.6 Some Corollaries of Theorem 6.2.9 in the Lacunary Case. Suppose $\lambda_0 := 0$, $\lambda_1 \geq 1$, and $\Lambda := (\lambda_i)_{i=0}^\infty$ is lacunary.

a] Show that the Chebyshev polynomials

$$T_n := T_n\{\lambda_0, \lambda_1, \dots, \lambda_n : [0,1]\}, \qquad n = 1, 2, \dots$$

have the following property: There is a constant $c_1 \in (0,1)$ depending only on Λ (and not on n) such that if $y \in [0,1)$ and $|T_n(y)| = 1$, then $|T_n(x)| \geq \frac{1}{2}$ for every $x \in [y, y + c_1(1-y)]$.

Hint: Use the Mean Value Theorem and the Bernstein-type inequality of Theorem 6.2.9. □

b] Show that there is a constant $c_2 \in (0,1)$ depending only on Λ (and not on n) so that if $a < b$ are two consecutive zeros of T_n, then $1-b < c_2(1-a)$.

c] Let $\epsilon \in (0,1)$. Show that there is an $n_0 \in \mathbb{N}$ depending only on ϵ (and not on n) so that every T_n has at most n_0 zeros in $[0, 1-\epsilon]$.

d] Give a new proof of the first part of Theorem 6.2.1 based on parts a] and c].

Outline. By Lebesgue's density theorem (see Royden [88]), it may be supposed, without loss of generality, that the left-hand side Lebesgue density of A at 1 is 1. Choose $\epsilon \in (0,1)$ so that $A \cap [0, 1-\epsilon)$ contains infinitely many points and

$$\frac{m(A \cap [y,1])}{1-y} > c_1 \tag{6.2.5}$$

for every $y \in [1-\epsilon, 1]$, where $c_1 \in (0,1)$ is the same as in part a]. For this ϵ, choose n_0 according to part c]. Now define $g \in C(A)$ so that g alternates $n_0 + 3$ times in $A \cap [0, 1-\epsilon)$ between 2 and -2 and is identically zero on $[1-\epsilon, 1]$. Assume that there exists a $p \in M_n(\Lambda)$ such that $\|p - g\|_A \leq \frac{1}{4}$. Use part a] and (6.2.5) to show that $p - T_n \in M_n(\Lambda)$ has more than n distinct zeros in $[0,1]$, which is a contradiction. □

The following simple application of Theorem 6.2.9 was pointed out by Wojcieszyk:

e] Suppose $A \subset [0,1]$ is a measurable set and the left-hand side Lebesgue density of A at 1 is 1. Show that there is a constant $c > 0$ depending only on Λ and A so that

$$\|p\|_{[0,1]} \leq c\,\|p\|_A$$

for every $p \in M(\Lambda) = \mathrm{span}\{x^{\lambda_0}, x^{\lambda_1}, \dots\}$.

Hint: Use the Mean Value Theorem, the Bernstein-type inequality of Theorem 6.2.9, and the Chebyshev-type inequality of E.3 f] of Section 4.2. □

f] Use part e] to give another proof of Theorem 6.2.1.

The following exercise constructs quasi-Chebyshev polynomials P_n for $M_n(\Lambda)$ if the lacunarity constant of Λ is large:

E.7 Quasi-Chebyshev Polynomials in Very Lacunary Müntz Spaces. Let $\lambda_0 = 0$, $\lambda_1 = 2$, and $\lambda_{i+1}/\lambda_i \geq 16$ for $i = 1, 2, \dots$. Let

$$P_n(x) := 1 + 2\sum_{j=1}^{n} (-1)^j x^{\lambda_j}, \qquad n = 1, 2, \dots .$$

Let $y_i := (4\lambda_i)^{-1}$. Prove the following statements:

a] $\|P_n\|_{[0,1]} = 1$ and $P_n(1) = (-1)^n$.

b] P_n has exactly n zeros, $x_{1,n} < x_{2,n} < \cdots < x_{n,n}$, in $(0,1)$.

c] $|P_n'(\xi)| \leq 2\lambda_n$ for every $\xi \in [x_{n,n}, 1]$.

d] We have

$$P_n(x) \leq -\frac{1}{5} \quad \text{if} \quad 1 - y_{2k} \leq x \leq 1 - \frac{y_{2k}}{2} \quad \text{and} \quad 1 \leq 2k \leq n$$

and

$$P_n(x) \geq \frac{1}{5} \quad \text{if} \quad 1 - y_{2k+1} \leq x \leq 1 - \frac{y_{2k+1}}{2} \quad \text{and} \quad 1 \leq 2k+1 \leq n\,.$$

Hint: Part a] is obvious. Prove the rest together, by induction on n. □

The next exercise follows Borwein and Erdélyi [95b].

E.8 Products of Müntz Spaces. Associated with $\Lambda := (\lambda_j)_{j=0}^{\infty}$, let

$$M^k(\Lambda) := \left\{ p = \prod_{j=1}^{k} p_j : p_j \in M(\Lambda) \right\}, \qquad k = 1, 2, \dots .$$

Is $M^2(\Lambda)$ dense in $C[0,1]$ for $\Lambda := (j^2)_{j=0}^{\infty}$?

Note that $M^k(\Lambda)$, $k \geq 2$ is not the linear span of monomials, and Müntz's theorem does not give the answer. This exercise establishes Remez-, Bernstein-, and Nikolskii-type inequalities for $M^k(\Lambda)$. From any of these it follows immediately that if $\sum_{j=1}^{\infty} 1/\lambda_j < \infty$ and $A \subset [0,1]$ is a set of positive Lebesgue measure, then $M^k(\Lambda)$ is not dense in $C(A)$.

Throughout parts a] to d] of the exercise we assume $0 = \lambda_0 < \lambda_1 < \cdots$, $\sum_{j=1}^{\infty} 1/\lambda_j < \infty$, and $s \in (0,1)$.

a] Remez-Type Inequality for $M^k(\Lambda)$. There exists a constant c depending only on Λ, s, and k (and not on ϱ or A) such that

$$\|p\|_{[0,\varrho]} \leq c\,\|p\|_A$$

for every $p \in M^k(\Lambda)$ and for every compact set $A \subset [\varrho, 1]$ of Lebesgue measure at least $s > 0$.

Proof. Theorem 6.2.2 implies that there exists a constant $\alpha > 0$ depending only on Λ, s, and k such that

$$m(\{x \in [y,1] : |p(x)| > \alpha^{-1}|p(y)|\}) \geq 1 - y - \frac{s}{2k}$$

for every $p \in M(\Lambda)$ and $y \in [0, 1-s]$. Now let $p \in M^k(\Lambda)$, that is,

$$p = \prod_{j=1}^{k} p_j\,, \qquad p_j \in M(\Lambda)\,.$$

Then, for every $y \in [0, 1-s]$,

$$\begin{aligned} &m(\{x \in [y,1] : |p(x)| > \alpha^{-k}|p(y)|\}) \\ &\qquad \geq m\left(\bigcap_{j=1}^{k}\{x \in [y,1] : |p_j(x)| > \alpha^{-1}|p_j(y)|\}\right) \\ &\qquad \geq 1 - y - k\frac{s}{2k} = 1 - y - \frac{s}{2}\,. \end{aligned}$$

Hence $y \in [0, \inf A]$ and $m(A) \geq s$ imply that

$$m(\{x \in A : |p(x)| > \alpha^{-k}|p(y)|\}) \geq \frac{s}{2} > 0$$

and the inequality follows with $c = \alpha^k$. □

b] Solution to Newman's Problem. Let $A \subset [0,1]$ be a set of positive Lebesgue measure. Then $M^k(\Lambda)$ is not dense in $C(A)$.

Proof. This follows from part a]. □

c] Bernstein-Type Inequality for $M^k(\Lambda)$. Suppose $\lambda_1 \geq 1$. There exists a constant c depending only on Λ, s, and k (and not on ϱ and A) such that

$$\|p'\|_{[0,\varrho]} \leq c\,\|p\|_A$$

for every $p \in M^k(\Lambda)$ and for every compact set $A \subset [\varrho, 1]$ of Lebesgue measure at least $s > 0$.

Hint: Use the product rule of differentiation, and estimate each term separately. Proceed as in the proof of part a]. Use Theorem 6.2.2 and E.5 a] and b] of Section 4.2. □

d] Nikolskii-Type Inequality for $M^k(\Lambda)$. There exists a constant c depending only on Λ, s, k, q, and w (and not on ϱ and A) such that

$$\|p\|_{[0,\varrho]}^q \leq c \int_A |p(x)|^q w(x)\,dx$$

for every $p \in M^k(\Lambda)$, for every compact set $A \subset [\varrho, 1]$ of Lebesgue measure at least $s > 0$, for every function w measurable and positive a.e. on $[0,1]$, and for every $q \in (0, \infty)$.

Hint: Use part a]. □

e] Associated with

$$\Lambda_j = (\lambda_{i,j})_{i=0}^{\infty}\,, \qquad j = 1, 2, \dots, k\,,$$

let

$$M(\Lambda_1, \Lambda_2, \dots, \Lambda_k) := \left\{ p = \prod_{j=1}^{k} p_j : p_j \in M(\Lambda_j) \right\}.$$

Formulate and prove the analogs of parts a] to d] for $M(\Lambda_1, \Lambda_2, \dots, \Lambda_k)$.

E.9 A Weak Converse of Theorem 6.2.9. Suppose $\Lambda := (\lambda_i)_{i=0}^{\infty}$ is a (strictly) increasing sequence of nonnegative real numbers with $\lambda_0 := 0$ and $\lambda_1 \geq 1$. Suppose also that there exists a constant c depending only on Λ (and not on y or the number of terms in p) such that

$$|p'(y)| \leq \frac{c}{1-y} \, \|p\|_{[0,1]}$$

for every $p \in M(\Lambda) = \text{span}\{x^{\lambda_0}, x^{\lambda_1}, \dots\}$ and for every $y \in [0,1)$. Show that there is a constant $\lambda > 1$ depending only on Λ such that $\lambda_n \geq \lambda^n$.

Outline. Let

$$T_n := T_n\{\lambda_0, \lambda_1, \dots, \lambda_n; [0,1]\}$$

and denote its zeros in $(0,1)$ by $x_{1,n} > x_{2,n} > \cdots > x_{n,n}$. Use the Mean Value Theorem and the assumed Bernstein-type inequality to show that there is a constant $\gamma \in (0,1)$ depending only on Λ such that

$$1 - x_{j,n} \leq \gamma(1 - x_{j+1,n})\,, \qquad j = 1, 2, \dots, n-1\,, \quad n \in \mathbb{N}\,;$$

hence $1 - x_{1,n} \leq \gamma^n$. On the other hand, use the Mean Value Theorem and Theorem 6.1.1 (Newman's inequality) to show that

$$1 - x_{1,n} \geq \left(1 + 9\sum_{j=1}^{n} \lambda_j\right)^{-1} \geq (9(n+1)\lambda_n)^{-1}\,.$$

Finally, combine the lower and upper bounds for $1 - x_{1,n}$ to conclude that

$$\lambda_n \geq \frac{\gamma^{-n}}{9(n+1)}\,.$$

□

E.10 Polynomials in x^{λ_n}. Given $n \in \mathbb{N}$ and $\lambda_n \in \mathbb{R}$, let

$$\mathcal{P}_n(\lambda_n) := \{p_n(x^{\lambda_n}) : p_n \in \mathcal{P}_n\}$$

(as in E.6 of Section 4.1). Suppose $\lambda_n \geq 1$ for all $n \in \mathbb{N}$. Let $\delta \in \mathbb{R}$ be defined by

$$\limsup_n \frac{\log n}{\lambda_n} = \frac{1}{2} \log \frac{1}{\delta} \,.$$

Suppose $\delta > 0$.

a] **Bounded Remez-Type Inequality.** Suppose $0 < \widetilde{\delta} < \delta$. Show that there exists a constant c depending only on $\widetilde{\delta}$ (and not on n, y, or A) such that

$$|p(y)| \leq c \, \|p\|_A$$

for every $p \in \cup_{n=1}^{\infty} \mathcal{P}_n(\lambda_n)$, for every $A \subset [0,1]$ of Lebesgue measure at least $1 - \widetilde{\delta}$, and for every $y \in [0, \inf A]$.

Hint: Use Lemma 6.2.6 and E.6 a] of Section 4.1. □

b] **Müntz-Type Theorem.** *If* $0 \leq \widetilde{\delta} < \delta$ *and* $A \subset [0,1]$ *is a set of Lebesgue measure at least* $1 - \widetilde{\delta}$, *then* $\cup_{n=1}^{\infty} \mathcal{P}_n(\lambda_n)$ *is not dense in* $C(A)$.

Hint: Use part a]. □

7

Inequalities for Rational Function Spaces

Overview

Precise Markov- and Bernstein-type inequalities are given for various classes of rational functions in the first section of this chapter. Extensions of the inequalities of Lax, Schur, and Russak are also presented, as are inequalities for self-reciprocal polynomials. The second section of the chapter is concerned with metric inequalities for polynomials and rational functions.

7.1 Inequalities for Rational Function Spaces

Sharp extensions of most of the polynomial inequalities of Section 5.1 are established for rational function spaces on $K := \mathbb{R} \pmod{2\pi}$, on the interval $[-1, 1]$, on the unit circle of $\mathbb{C}$, and on the real line. The classical inequalities of Section 5.1 are then recovered as limiting cases. A sharp extension of Lax's inequality is also given. Essentially sharp Markov- and Bernstein-type inequalities for self-reciprocal and antiself-reciprocal polynomials are presented in the exercises.

Let $D := \{z \in \mathbb{C} : |z| < 1\}$ and $\partial D := \{z \in \mathbb{C} : |z| = 1\}$, as before. We study the rational function spaces:

$$\mathcal{T}_n(a_1, a_2, \ldots, a_{2n}; K) := \left\{ \frac{t(\theta)}{\prod_{k=1}^{2n} |\sin((\theta - a_k)/2)|} : t \in \mathcal{T}_n \right\}$$

and

$$\mathcal{T}_n^c(a_1, a_2, \ldots, a_{2n}; K) := \left\{ \frac{t(\theta)}{\prod_{k=1}^{2n} \sin((\theta - a_k)/2)} : t \in \mathcal{T}_n^c \right\}$$

on K with $a_1, a_2, \ldots, a_{2n} \in \mathbb{C} \setminus \mathbb{R}$;

$$\mathcal{P}_n(a_1, a_2, \ldots, a_n; [-1,1]) := \left\{ \frac{p(x)}{\prod_{k=1}^{n} |x - a_k|} : p \in \mathcal{P}_n \right\}$$

and

$$\mathcal{P}_n^c(a_1, a_2, \ldots, a_n; [-1,1]) := \left\{ \frac{p(x)}{\prod_{k=1}^{n} (x - a_k)} : p \in \mathcal{P}_n^c \right\}$$

on $[-1,1]$ with $a_1, a_2, \ldots, a_n \in \mathbb{C} \setminus [-1,1]$;

$$\mathcal{P}_n^c(a_1, a_2, \ldots, a_n; \partial D) := \left\{ \frac{p(z)}{\prod_{k=1}^{n} (z - a_k)} : p \in \mathcal{P}_n^c \right\}$$

on ∂D with $a_1, a_2, \ldots, a_n \in \mathbb{C} \setminus \partial D$; and

$$\mathcal{P}_n(a_1, a_2, \ldots, a_n; \mathbb{R}) := \left\{ \frac{p(x)}{\prod_{k=1}^{n} |x - a_k|} : p \in \mathcal{P}_n \right\}$$

and

$$\mathcal{P}_n^c(a_1, a_2, \ldots, a_n; \mathbb{R}) := \left\{ \frac{p(z)}{\prod_{k=1}^{n} (z - a_k)} : p \in \mathcal{P}_n^c \right\}$$

on $\mathbb{R}$ with $a_1, a_2, \ldots, a_n \in \mathbb{C} \setminus \mathbb{R}$.

The Chebyshev polynomials $\widetilde{T}_n$, $\widetilde{U}_n$, and

$$V := (\cos \alpha)\widetilde{T}_n + (\sin \alpha)\widetilde{U}_n, \qquad \alpha \in K$$

for the rational function space $\mathcal{T}_n(a_1, a_2, \ldots, a_{2n}; K)$ are defined in E.3 of Section 3.5, and they play a central role in this section.

Theorem 7.1.1 (Bernstein-Szegő-Type Inequality on K). *Given*

$$(a_k)_{k=1}^{2n} \subset \mathbb{C} \setminus \mathbb{R}, \qquad \operatorname{Im}(a_k) > 0,$$

let

$$\widetilde{B}_n(\theta) := \frac{1}{2} \sum_{k=1}^{2n} \frac{1 - |e^{ia_k}|^2}{|e^{ia_k} - e^{i\theta}|^2}.$$

Then

$$f'(\theta)^2 + \widetilde{B}_n^2(\theta) f^2(\theta) \le \widetilde{B}_n^2(\theta) \|f\|_K^2, \qquad \theta \in K$$

for every $f \in \mathcal{T}_n(a_1, a_2, \dots, a_{2n}; K)$.

Equality holds if and only if either θ is a maximum point of $|f|$ (that is, $f(\theta) = \pm\|f\|_K$) or f is a linear combination of $\widetilde{T}_n$ and $\widetilde{U}_n$ (with real coefficients) as defined in E.3 of Section 3.5.

Corollary 7.1.2 (Bernstein-Type Inequality on K, Real Case). *Given*

$$(a_k)_{k=1}^{2n} \subset \mathbb{C} \setminus \mathbb{R}, \qquad \operatorname{Im}(a_k) > 0,$$

let the Bernstein factor $\widetilde{B}_n$ be defined as in Theorem 7.1.1. Then

$$|f'(\theta)| \le \widetilde{B}_n(\theta) \|f\|_K, \qquad \theta \in K$$

for every $f \in \mathcal{T}_n(a_1, a_2, \dots, a_{2n;} K)$.

Equality holds if and cnly if f is a linear combination of $\widetilde{T}_n$ and $\widetilde{U}_n$ (with real coefficients) as defined in E.3 of Section 3.5, and $f(\theta) = 0$.

Theorem 7.1.1 and Corollary 7.1.2 can be easily obtained from the extension of Theorem 3.5.3 given by E.3 of Section 3.5, which gives explicit formulas for the Chebyshev polynomials for these classes $\mathcal{T}_n(a_1, a_2, \dots, a_{2n}; K)$. The arguments are outlined in E.1.

The following two results can be obtained from Theorem 7.1.1 and Corollary 7.1.2 by the substitution $x = \cos\theta$; see E.2.

Corollary 7.1.3 (Bernstein-Szegő-Type Inequality on $[-1, 1]$). *Associated with $(a_k)_{k=1}^n \subset \mathbb{C} \setminus [-1, 1]$, let the Bernstein factor B_n be defined by*

$$B_n(x) := \sum_{k=1}^{n} \operatorname{Re}\left(\frac{\sqrt{a_k^2 - 1}}{a_k - x} \right),$$

where the choice of $\sqrt{a_k^2 - 1}$ is determined by $\left|a_k - \sqrt{a_k^2 - 1}\right| < 1$. Then

$$(1 - x^2) f'(x)^2 + B_n^2(x) f^2(x) \le B_n^2(x) \|f\|_{[-1,1]}^2, \qquad x \in [-1, 1]$$

for every $f \in \mathcal{P}_n(a_1, a_2, \dots, a_n; [-1, 1])$.

Equality holds if and only if either x is a maximum point of $|f|$ (that is, $f(x) = \pm\|f\|_{[-1,1]}$) or $f = cT_n$ with $c \in \mathbb{R}$, where T_n is defined as in Section 3.5.

Corollary 7.1.4 (Bernstein-Type Inequality on $[-1,1]$, Real Case). *Given $(a_k)_{k=1}^n \subset \mathbb{C} \setminus [-1,1]$, let the Bernstein factor B_n be defined as in Corollary 7.1.3. Then*

$$|f'(x)| \leq \frac{B_n(x)}{\sqrt{1-x^2}} \|f\|_{[-1,1]}, \qquad x \in (-1,1)$$

for every $f \in \mathcal{P}_n(a_1, a_2, \ldots, a_n; [-1,1])$.

Equality holds if and only if $f = cT_n$ with $c \in \mathbb{R}$, where T_n is defined as in Section 3.5, and $f(x) = 0$. (Note that $B_n(x) > 0$ for every $x \in (-1,1)$.)

Our next result follows from Theorem 7.1.1; see the hints to E.3.

Corollary 7.1.5 (Bernstein-Szegő-Type Inequality on $\mathbb{R}$). *Given*

$$(a_k)_{k=1}^n \subset \mathbb{C} \setminus \mathbb{R}, \qquad \operatorname{Im}(a_k) > 0,$$

let the Bernstein factor B_n be defined by

$$B_n(x) := \sum_{k=1}^n \frac{\operatorname{Im}(a_k)}{|x-a_k|^2}.$$

Then

$$f'(x)^2 + B_n^2(x) f^2(x) \leq B_n^2(x) \|f\|_{\mathbb{R}}^2, \qquad x \in \mathbb{R}$$

for every $f \in \mathcal{P}_n(a_1, a_2, \ldots, a_n; \mathbb{R})$.

Equality holds if and only if either x is a maximum point of $|f|$ (that is, $|f(x)| = \pm\|f\|_{\mathbb{R}}$) or f is a linear combination of T_n and U_n (with real coefficients) defined in E.5 of Section 3.5.

Corollary 7.1.6 (Bernstein-Type Inequality on $\mathbb{R}$, Real Case). *Associated with $(a_k)_{k=1}^n \subset \mathbb{C} \setminus \mathbb{R}$, let the Bernstein factor B_n be defined as in Corollary 7.1.5. Then*

$$|f'(x)| \leq B_n(x) \|f\|_{\mathbb{R}}, \qquad x \in \mathbb{R}$$

for every $f \in \mathcal{P}_n(a_1, a_2, \ldots, a_n; \mathbb{R})$.

Equality holds if and only if f is a linear combination of T_n and U_n (with real coefficients) defined in E.5 of Section 3.5, and $f(x) = 0$.

To formulate our next theorem we introduce some notation. For a polynomial

$$q(z) := \prod_{k=1}^n (z - a_k), \qquad a_k \in \mathbb{C},$$

we define

$$q^*(z) := \prod_{k=1}^{n} (1 - \overline{a}_k z) =: z^n \overline{q}_n(z^{-1}) .$$

Then

(7.1.1)
$$|q(z)| = |q^*(z)| , \qquad z \in \partial D .$$

The function

$$S_n(z) := \frac{q^*(z)}{q(z)} = \prod_{k=1}^{n} \frac{1 - \overline{a}_k z}{z - a_k}$$

is the Blaschke product associated with $(a_k)_{k=1}^n$.

Theorem 7.1.7 (Bernstein-Type Inequality on ∂D, Complex Case). *Given $(a_k)_{k=1}^n \subset \mathbb{C} \setminus \partial D$, let the Bernstein factor B_n be defined by*

$$B_n(z) := \max\{B_n^+(z), B_n^-(z)\}$$

with

$$B_n^+(z) := \sum_{\substack{k=1 \\ |a_k|>1}}^{n} \frac{|a_k|^2 - 1}{|a_k - z|^2} \qquad \text{and} \qquad B_n^-(z) := \sum_{\substack{k=1 \\ |a_k|<1}}^{n} \frac{1 - |a_k|^2}{|a_k - z|^2} .$$

Then

$$|f'(z)| \le B_n(z) \|f\|_{\partial D} , \qquad z \in \partial D$$

for every $f \in \mathcal{P}_n^c(a_1, a_2, \dots, a_n; \partial D)$.

If the first sum is not less than the second sum for a fixed $z \in \partial D$, then equality holds for $f = cS_n^+$ with $c \in \mathbb{C}$, where S_n^+ is the Blaschke product associated with those a_k for which $|a_k| > 1$. If the second sum is not less than the first sum for a fixed $z \in \partial D$, then equality holds for $f = cS_n^-$ with $c \in \mathbb{C}$, where S_n^- is the Blaschke product associated with those a_k for which $|a_k| < 1$.

Proof. For reasons of symmetry it is sufficient to prove the theorem only for $z = 1$. Without loss of generality we may assume that

(7.1.2)
$$\operatorname{Re}\left(\sum_{k=1}^{n} \frac{1}{1 - a_k}\right) \ne \frac{n}{2} ;$$

the remaining cases follow from this by a limiting argument. Let $Q := \partial D$ (equipped with the usual metric topology), $V := \mathcal{P}_n^c(a_1, a_2, \dots, a_n; \partial D)$, and $L(f) := f'(1)$ for $f \in V$. We show in this situation that $n + 1 \le r$ in Theorem A.3.3 (interpolation of linear functionals). Suppose to the contrary that $r \le n$. By Theorem A.3.3, there are distinct points $x_1, x_2, \dots, x_r$ on ∂D, and there are constants $c_1, c_2, \dots, c_r \in \mathbb{C}$ such that

$$\frac{p'(1)q(1) - q'(1)p(1)}{q(1)^2} = \sum_{k=1}^{r} c_k \frac{p(x_k)}{q(x_k)}, \qquad p \in \mathcal{P}_n^c, \tag{7.1.3}$$

where

$$q(z) := \prod_{k=1}^{n} (z - a_k). \tag{7.1.4}$$

We claim that $x_k \neq 1$ for each $k = 1, 2, \dots, r$. Indeed, if there is index k such that $x_k = 1$, then Theorem A.3.3 implies that

$$p(z) := (z+1)^{n-r} \prod_{k=1}^{r} (z - x_k) \in \mathcal{P}_n^c$$

has a zero at 1 with multiplicity at least two, which is a contradiction. Applying (7.1.3) with the above p, we obtain

$$p'(1)q(1) - q'(1)p(1) = 0$$

and since $p(1) \neq 0$ and $q(1) \neq 0$, this is equivalent to

$$\frac{q'(1)}{q(1)} = \frac{p'(1)}{p(1)},$$

that is, in terms of the zeros of p_n and q_n,

$$\sum_{k=1}^{n} \frac{1}{1 - a_k} = \frac{n-r}{2} + \sum_{k=1}^{r} \frac{1}{1 - x_k}. \tag{7.1.5}$$

Since $x_k \in \partial D$ and $x_k \neq 1$, we have

$$\operatorname{Re}\left(\frac{1}{1 - x_k}\right) = \frac{1}{2}, \qquad k = 1, 2, \dots, r. \tag{7.1.6}$$

It follows from (7.1.5) and (7.1.6) that

$$\operatorname{Re}\left(\sum_{k=1}^{n} \frac{1}{1 - a_k}\right) = \frac{n}{2},$$

which contradicts assumption (7.1.2). So $n + 1 \leq r$, indeed.

A compactness argument shows that there is a function $\widetilde{f} \in V$ such that

$$\|\widetilde{f}\|_{\partial D} = 1 \qquad \text{and} \qquad L(\widetilde{f}) = \|L\| := \max_{0 \neq f \in V} \frac{|L(f)|}{\|f\|_{\partial D}}.$$

Theorem A.3.3 implies $|\widetilde{f}(x_k)| = 1$ for every $k = 1, 2, \ldots, r$. Hence, if

$$\widetilde{f} = \frac{\widetilde{p}}{q}, \qquad \widetilde{p} \in \mathcal{P}_n^c, \qquad q(z) = \prod_{k=1}^{n} (z - a_k),$$

then

$$h(z) := |\widetilde{p}(z)|^2 - |q(z)|^2 \le 0, \qquad z \in \partial D \tag{7.1.7}$$

and

$$h(x_k) = 0, \qquad k = 1, 2, \ldots, r. \tag{7.1.8}$$

Note that $t(\theta) := h(e^{i\theta}) \in \mathcal{T}_n$ vanishes at each θ_k, where the numbers $\theta_k \in [-\pi, \pi)$ are defined by $x_k = e^{i\theta_k}$, $k = 1, 2, \ldots, r$. Because of (7.1.7), each of these zeros is of even multiplicity. Hence, $n + 1 \le r$ implies that $t \in \mathcal{T}_n$ has at least $2n + 2$ zeros and therefore $t = 0$. From this we can deduce that $h(z) = 0$ for every $z \in \partial D$, so

$$|\widetilde{p}(z)| = |q(z)|, \qquad z \in \partial D. \tag{7.1.9}$$

We now have

$$z^{-n}\widetilde{p}(z)\widetilde{p}^*(z) = |\widetilde{p}(z)|^2 = |q(z)|^2 = z^{-n}q(z)q^*(z), \qquad z \in \partial D,$$

so by the unicity theorem for analytic functions (see E.1 e] of Section 1.2)

$$\widetilde{p}\widetilde{p}^* = qq^*.$$

From this, it follows that there exists a constant $0 \ne c \in \mathbb{C}$ such that

$$\widetilde{f}(z) = \frac{\widetilde{p}(z)}{q(z)} = c \prod_{k=1}^{m} \frac{z - \overline{\alpha}_k^{-1}}{z - \alpha_k}, \qquad z \in \mathbb{C}, \quad q(z) \ne 0$$

with some $m \le n$ and

$$\alpha_k := a_{j_k}, \qquad k = 1, 2, \ldots, m, \qquad 1 \le j_1 < j_2 < \cdots < j_m \le n.$$

A straightforward calculation gives that

$$\begin{aligned} |\widetilde{f}'(1)| &= \left| \frac{\widetilde{f}'(1)}{\widetilde{f}(1)} \right| = \left| \sum_{k=1}^{m} \left(\frac{1}{1 - \overline{\alpha}_k^{-1}} - \frac{1}{1 - \alpha_k} \right) \right| \\ &= \left| \sum_{k=1}^{m} \frac{|\alpha_k|^2 - 1}{|\alpha_k - 1|^2} \right| \le \max\{B_n^+(z), B_n^-(z)\}, \end{aligned}$$

which finishes the proof. $\square$

Corollary 7.1.8 (Bernstein-Type Inequality on K, Complex Case). *Given $(a_k)_{k=1}^{2n} \subset \mathbb{C} \setminus \mathbb{R}$, let the Bernstein factor B_n be defined by*

$$\widetilde{B}_n(\theta) := \max\{\widetilde{B}_n^+(\theta), \widetilde{B}_n^-(\theta)\}$$

with

$$\widetilde{B}_n^+(\theta) := \sum_{\substack{k=1 \\ \operatorname{Im}(a_k)<0}}^{2n} \frac{|e^{ia_k}|^2 - 1}{|e^{ia_k} - e^{i\theta}|^2} \quad \text{and} \quad \widetilde{B}_n^-(\theta) := \sum_{\substack{k=1 \\ \operatorname{Im}(a_k)>0}}^{2n} \frac{1 - |e^{ia_k}|^2}{|e^{ia_k} - e^{i\theta}|^2}\,.$$

Then

$$|f'(\theta)| \leq B_n(\theta)\|f\|_K\,, \qquad \theta \in K$$

for every $f \in \mathcal{T}_n^c(a_1, a_2, \dots, a_{2n}; K)$.

If the first sum is not less than the second sum for a fixed $\theta \in K$, then equality holds for $f(\theta) = cS_{2n}^+(e^{i\theta})$ with $c \in \mathbb{C}$, where S_{2n}^+ is the Blaschke product associated with those e^{ia_k} for which $\operatorname{Im}(a_k) < 0$. If the second sum is not less than the first sum for a fixed $\theta \in K$, then equality holds for $f(\theta) = cS_{2n}^-(e^{i\theta})$ with $c \in \mathbb{C}$, where S_{2n}^- is the Blaschke product associated with those e^{ia_k} for which $\operatorname{Im}(a_k) > 0$. Note that

$$S_{2n}^{\pm}(e^{i\theta}) \in \mathcal{T}_n^c(a_1, a_2, \dots, a_{2n}; K)\,.$$

Proof. Observe that if

$$h(\theta) := \prod_{j=1}^{2n} \sin((\theta - a_j)/2) \in \mathcal{T}_n^c$$

and $t_n \in \mathcal{T}_n^c$, then there are $p \in \mathcal{P}_{2n}^c$ and $q \in \mathcal{P}_{2n}^c$ such that

$$\frac{t(\theta)}{h(\theta)} = \frac{p(e^{i\theta})e^{-in\theta}}{q(e^{i\theta})e^{-in\theta}} = \frac{p(e^{i\theta})}{q(e^{i\theta})}\,,$$

where q is of the form

$$q(z) = c\prod_{j=1}^{2n}(z - e^{ia_j})$$

with some $c \in \mathbb{C}$. So the corollary follows from Theorem 7.1.7. □

Corollary 7.1.9 (Bernstein-Type Inequality on $[-1,1]$, Complex Case). *Given $\{a_k\}_{k=1}^n \subset \mathbb{C} \setminus [-1,1]$, let the Bernstein factor $B_n(x)$ be defined by*

$$B_n(x) = \max\left\{\sum_{k=1}^n \frac{1 - |c_k|^2}{|c_k - z|^2}\,,\ \sum_{k=1}^n \frac{|c_k|^{-2} - 1}{|c_k^{-1} - z|^2}\right\},$$

where c_k and z are determined by

$$a_k := \tfrac{1}{2}(c_k + c_k^{-1}), \qquad |c_k| < 1;$$
$$x := \tfrac{1}{2}(z + z^{-1}), \qquad \mathrm{Im}(z) > 0.$$

Then

$$|f'(x)| \le \frac{B_n(x)}{\sqrt{1-x^2}} \|f\|_{[-1,1]}, \qquad x \in (-1,1)$$

for every $f \in \mathcal{P}_n^c(a_1, a_2, \dots, a_n; [-1,1])$. Note that

$$\sum_{k=1}^{n} \frac{1-|c_k|^2}{|c_k - z|^2} = \mathrm{Re}\left(\sum_{k=1}^{n} \frac{\sqrt{a_k^2 - 1}}{a_k - x} \right), \qquad x \in [-1,1],$$

where the choice of $\sqrt{a_k^2-1}$ is determined by $\left|a_k - \sqrt{a_k^2-1}\right| < 1$.

Proof. The corollary follows from Theorem 7.1.7 by the substitution $x = \frac{1}{2}(z + z^{-1})$. □

Bernstein's classical polynomial inequalities discussed in Section 5.1 are contained in Theorem 7.1.7 and Corollaries 7.1.8 and 7.1.9 as limiting cases. In Theorem 7.1.7 and Corollary 7.1.9 we take

$$(a_1^{(m)}, a_2^{(m)}, \dots, a_n^{(m)}) \subset \mathbb{C} \setminus D$$

so that

$$\lim_{m \to \infty} |a_k^{(m)}| = \infty, \qquad k = 1, 2, \dots, n.$$

In Corollary 7.1.8 we take

$$(a_1^{(m)}, a_2^{(m)}, \dots, a_{2n}^{(m)}) \subset \mathbb{C} \setminus \mathbb{R}$$

so that

$$a_{n+k}^{(m)} = \overline{a}_k^{(m)} \quad \text{and} \quad \lim_{m \to \infty} |\mathrm{Im}(a_k^{(m)})| = \infty, \qquad k = 1, 2, \dots, n.$$

To formulate our next result we introduce the Blaschke product

$$Q_n(z) := \prod_{k=1}^{n} \frac{z - \overline{a}_k}{z - a_k}$$

associated with $(a_1, a_2, \dots, a_n) \subset \mathbb{C} \setminus \mathbb{R}$. Obviously $|Q_n(z)| = 1$ for every $z \in \mathbb{R}$.

Corollary 7.1.10 (Bernstein-Type Inequality on $\mathbb{R}$, Complex Case). *Given $(a_k)_{k=1}^n \subset \mathbb{C} \setminus \mathbb{R}$, let the Bernstein factor $B_n(x)$ be defined by*

$$B_n(x) := \max\{B_n^+(x), B_n^-(x)\}$$

with

$$B_n^+(x) := \sum_{\substack{k=1 \\ \mathrm{Im}(a_k)>0}}^{n} \frac{2\,|\mathrm{Im}(a_k)|}{|x-a_k|^2} \quad \text{and} \quad B_n^-(x) := \sum_{\substack{k=1 \\ \mathrm{Im}(a_k)<0}}^{n} \frac{2\,|\mathrm{Im}(a_k)|}{|x-a_k|^2}$$

for every $x \in \mathbb{R}$. Then

$$|f'(x)| \le B_n(x)\|f\|_{\mathbb{R}}\,, \qquad x \in \mathbb{R}$$

for every $f \in \mathcal{P}_n^c(a_1, a_2, \dots, a_n; \mathbb{R})$.

If the first sum is not less than the second sum for a fixed $x \in \mathbb{R}$, then equality holds for $f = cQ_n^+$ with $c \in \mathbb{C}$, where Q_n^+ is the Blaschke product associated with the poles a_k lying in the open upper half-plane

$$H^+ := \{z \in \mathbb{C} : \mathrm{Im}(z) > 0\}\,.$$

If the second sum is not less than the first sum for a fixed $x \in \mathbb{R}$, then equality holds for $f = cQ_n^-$ with $c \in \mathbb{C}$, where Q_n^- is the Blaschke product associated with the poles a_k lying in the open lower half-plane

$$H^- := \{z \in \mathbb{C} : \mathrm{Im}(z) < 0\}\,.$$

Corollary 7.1.10 follows from Theorem 7.1.7; see E.4.

The next theorem improves the Bernstein-type inequality of Theorem 7.1.7 in the case when $\{a_k\}_{k=1}^n \subset \mathbb{C} \setminus \overline{D}$ and f has all its zeros in $\mathbb{C} \setminus D$. It extends Lax [44].

Theorem 7.1.11 (Lax-Type Inequality). *Given $(a_k)_{k=1}^n \subset \mathbb{C} \setminus \overline{D}$, let the Bernstein factor B_n be, as in Theorem 7.1.7, defined by*

$$B_n(z) := \sum_{k=1}^{n} \frac{|a_k|^2 - 1}{|a_k - z|^2}\,.$$

Then

$$|h'(z)| \le \tfrac{1}{2} B_n(z)\|h\|_{\partial D}\,, \qquad z \in \partial D$$

for every $h \in \mathcal{P}_n^c(a_1, a_2, \dots, a_n; \partial D)$ having all its zeros in $\mathbb{C} \setminus D$.

Equality holds for $h = c(S_n + 1)$ with $c \in \mathbb{C}$, where S_n is the Blaschke product associated with $(a_k)_{k=1}^n$.

Note that $B_n(z) = |S_n'(z)|$. *Note also that*

$$h := c(S_n + 1) \in \mathcal{P}_n^c(a_1, a_2, \dots, a_n; \partial D)\,, \qquad c \neq 0$$

has all its zeros on ∂D.

Proof. First assume that each zero of h is on ∂D, the general case can be reduced to this (see E.5). Thus let $h := p/q$, where $p \in \mathcal{P}_n^c$ has all its zeros on ∂D and where

$$q(z) := \prod_{k=1}^{n} (z - a_k)\,, \qquad |a_k| > 1\,.$$

Let

$$q^*(z) := \prod_{k=1}^{n} (1 - \overline{a}_k z)\,.$$

We study

$$u(\theta) := \frac{p(e^{2i\theta})e^{-in\theta}}{|q(e^{2i\theta})|} = \frac{p(e^{2i\theta})}{\sqrt{q(e^{2i\theta})}\sqrt{q^*(e^{2i\theta})}}$$

for $\theta \in \mathbb{R}$, where the square roots are taken so that $\sqrt{q}$ is analytic in a neighborhood of the closed unit disk, and $\sqrt{q^*}$ is analytic in a neighborhood of the complement of the open unit disk. Since $p \in \mathcal{P}_n^c$ has all its zeros on ∂D, there exists a $0 \neq \beta \in \mathbb{C}$ such that

$$t(\theta) := \beta p(e^{2i\theta})e^{-in\theta}$$

is a real trigonometric polynomial of degree at most n (see E.5 a]). Also

$$|q(e^{2i\theta})| = |q^*(e^{2i\theta})| = \prod_{k=1}^{n} |1 - \overline{a}_k e^{2i\theta}|$$

$$= \gamma \prod_{k=1}^{2n} |\sin((\theta - c_k)/2)|\,,$$

where $\gamma > 0$,

$$e^{ic_k} = \overline{a}_k^{\,-1/2}\,, \qquad \mathrm{Im}(c_k) > 0\,, \quad k = 1, 2, \dots, n$$

and

$$e^{ic_k} = -\overline{a}_k^{\,-1/2}\,, \qquad \mathrm{Im}(c_k) > 0\,, \quad k = n+1, n+2, \dots, 2n\,.$$

Applying Theorem 7.1.1 to

$$\beta^{-1}u \in \mathcal{T}_n(c_1, c_2, \dots, c_{2n}; K)\,,$$

we obtain

$$|u'(\theta) - i\widetilde{B}_n(\theta)u(\theta)| \leq \widetilde{B}_n(\theta)\|u\|_K\,, \qquad \theta \in K\,, \tag{7.1.10}$$

where

$$\begin{aligned}\widetilde{B}_n(\theta) &= \frac{1}{2}\sum_{k=1}^{n}\left(\frac{1-|a_k|^{-1}}{|\overline{a}_k^{-1/2} - e^{i\theta}|^2} + \frac{1-|a_k|^{-1}}{|-\overline{a}_k^{-1/2} - e^{i\theta}|^2}\right) \\ &= \sum_{k=1}^{n}\frac{1-|a_k|^{-2}}{|\overline{a}_k^{-1} - e^{2i\theta}|^2} = \sum_{k=1}^{n}\frac{|a_k|^2-1}{|a_k - e^{2i\theta}|^2}\,.\end{aligned} \tag{7.1.11}$$

Observe that

$$u(\theta) = \frac{p(e^{2i\theta})}{q(e^{2i\theta})}\,\frac{\sqrt{q(e^{2i\theta})}}{\sqrt{q^*(e^{2i\theta})}} = \frac{p(e^{2i\theta})}{q(e^{2i\theta})}f_n(e^{i\theta})\,, \tag{7.1.12}$$

where

$$f_n(z) := \frac{\sqrt{q(z^2)}}{\sqrt{q^*(z^2)}}\,. \tag{7.1.13}$$

A simple calculation (see E.4. of Section 3.5) shows that

$$\widetilde{B}_n(\theta) = e^{i\theta}\frac{f_n'(e^{i\theta})}{f_n(e^{i\theta})}\,, \qquad \theta \in K\,. \tag{7.1.14}$$

Also, since $|f_n(e^{i\theta})| = 1$ for every $\theta \in K$, we have

$$\|u\|_K = \left\|\frac{p}{q}\right\|_{\partial D} = \|h\|_{\partial D}\,. \tag{7.1.15}$$

Now (7.1.10) to (7.1.15) yield

$$\left|\frac{d}{d\theta}\left(\frac{p(e^{2i\theta})}{q(e^{2i\theta})}f_n(e^{i\theta})\right) - ie^{i\theta}\frac{f_n'(e^{i\theta})}{f_n(e^{i\theta})}\,\frac{p(e^{2i\theta})}{q(e^{2i\theta})}f_n(e^{i\theta})\right| \leq \widetilde{B}_n(\theta)\|h\|_{\partial D}\,.$$

So

$$\begin{aligned}&|2ie^{2i\theta}h'(e^{2i\theta})f_n(e^{2i\theta}) + ie^{i\theta}f_n'(e^{i\theta})h(e^{2i\theta}) \\ &\qquad - ie^{i\theta}f_n'(e^{i\theta})h(e^{2i\theta})| \leq \widetilde{B}_n(\theta)\|h\|_{\partial D}\,.\end{aligned}$$

Thus

$$2\,|h'(e^{2i\theta})| \leq \widetilde{B}_n(\theta)\|h\|_{\partial D}\,,$$

which, together with (7.1.11), finishes the proof. □

Comments, Exercises, and Examples.

Most of the results in this section have been proved in Borwein and Erdélyi [to appear 4] and in Borwein, Erdélyi, and Zhang [94a]. A weaker version of Corollary 7.1.9 has been obtained by Russak (see Petrushev and Popov [87]). Theorem 7.1.11 contains, as a limiting case, an inequality of Lax [44] conjectured by Erdős. Lax's inequality establishes the sharp Bernstein-type inequality on the unit disk for polynomials $p \in \mathcal{P}_n^c$ having no zeros in the open unit disk. That is,

$$\|p'\|_D \leq \frac{n}{2} \|p\|_D$$

for such polynomials. Various extensions of this inequality are given by Ankeny and Rivlin [55], Govil [73], Malik [69], and others. We discuss some of these in E.16 of Appendix 5.

E.1 Proof of Theorem 7.1.1 and Corollary 7.1.2. Given $(a_k)_{k=1}^{2n} \subset \mathbb{C} \setminus \mathbb{R}$, let

$$\mathcal{T}_{n,a} := \mathcal{T}_n(a_1, a_2, \dots, a_{2n}; K).$$

a] Show that $\mathcal{T}_{n,a}$ is a Hermite interpolation space. That is, if the points $x_1, x_2, \dots, x_k \in K$ are distinct, and $m_1, m_2, \dots, m_k$ are positive integers with $\sum_{i=1}^k m_i \leq 2n+1$, then for any choice of real numbers $y_{i,j}$, there is a function $f \in \mathcal{T}_{n,a}$ such that

$$f^{(j)}(x_i) = y_{i,j}, \qquad i = 1, 2, \dots, k, \quad j = 0, 1, \dots, m_i - 1.$$

Hint: See the hint to E.7 of Section 1.1. □

b] Show that for every fixed $\theta \in K$, the value

$$\max_{0 \neq f \in \mathcal{T}_{n,a}} \frac{f'(\theta)^2 + B_n^2(\theta) f^2(\theta)}{\|f\|_K^2}$$

is attained by an $\widetilde{f} \in \mathcal{T}_{n,a}$.

Hint: Use a compactness argument. □

c] Show that $\widetilde{f} = cV$, where $c \in \mathbb{R}$ and V is one of the Chebyshev polynomials for $\mathcal{T}_{n,a}$ defined in Theorem 3.5.3 and E.3 of Section 3.5.

Hint: Use a variational method with the help of part a]. □

d] Prove Theorem 7.1.1.

Hint: Use part c] and E.4 of Section 3.5. □

e] Prove Corollary 7.1.2.

E.2 Proof of Corollaries 7.1.3 and 7.1.4.

a] Prove the inequality of Corollary 7.1.3.

Hint: Use Theorem 7.1.1 with the substitution $x = \cos\theta$. Note that $f \in \mathcal{P}_n(a_1, a_2, \dots, a_n; [-1,1])$ implies

$$g(\theta) := f(\cos\theta) = \mathcal{T}_n(c_1, c_1, c_2, c_2, \dots, c_n, c_n; K),$$

where the numbers $c_k \in \mathbb{C}$ are defined by

$$e^{ic_k} = a_k - \sqrt{a_k^2 - 1}, \qquad a_k = \tfrac{1}{2}(e^{ic_k} + e^{-ic_k}), \quad \operatorname{Im}(c_k) > 0.$$

Verify that if $\widetilde{B}_n$ is the Bernstein factor given in Theorem 7.1.1 associated with

$$(c_1, c_1, c_2, c_2, \dots, c_n, c_n),$$

then

$$\widetilde{B}_n(\theta) = \sum_{k=1}^{n} \operatorname{Re}\left(\frac{\sqrt{a_k^2 - 1}}{a_k - \cos\theta}\right), \qquad \theta \in K,$$

where the choice of $\sqrt{a_k^2 - 1}$ is determined by $\left|a_k - \sqrt{a_k^2 - 1}\right| < 1$. □

b] Given $x \in [-1,1]$, prove that equality holds in the inequality of Corollary 7.1.3 if and only if either x is a maximum point of $|f|$ (that is, $f(x) = \pm\|f\|_{[-1,1]}$) or $f = cT_n$ with $c \in \mathbb{R}$, where T_n is defined in Section 3.5.

Hint: Observe that

$$V = (\cos\alpha)\widetilde{T}_n + (\sin\alpha)\widetilde{U}_n$$

is even if and only if $V = \pm\widetilde{T}_n$ and use Theorem 7.1.1. □

c] Prove Corollary 7.1.4.

E.3 Proof of Corollaries 7.1.5 and 7.1.6.

a] Prove Corollary 7.1.5.

Hint: Use Theorem 7.1.1 and the substitution $x = i\dfrac{e^{i\theta}+1}{e^{i\theta}-1}$, which maps K onto $\mathbb{R} \cup \{\infty\}$. □

b] Prove Corollary 7.1.6.

E.4 Proof of Corollary 7.1.10. Prove Corollary 7.1.10.

Hint: Use Theorem 7.1.7 and the substitution $x = i\dfrac{z+1}{z-1}$, which maps ∂D onto $\mathbb{R} \cup \{\infty\}$. □

E.5 Completion of the Proof of Theorem 7.1.11.

a] Show that if $p \in \mathcal{P}_n^c$ has all its zeros on the unit circle, then there is a $0 \neq \beta \in \mathbb{C}$ such that $g(\theta) := \beta e^{-in\theta} p(e^{2i\theta})$ is a real trigonometric polynomial of degree at most n.

For $f = p/q$ with

$$q(z) := \prod_{k=1}^{n} (z - a_k), \qquad a_k \in \mathbb{C}$$

and

$$p(z) := \gamma \prod_{k=1}^{m} (z - b_k), \qquad b_k \in \mathbb{C}, \quad \gamma \in \mathbb{C}, \quad m \le n,$$

we define $f^* := p^*/q$, where

$$p^*(z) := \overline{\gamma} \prod_{k=1}^{m} (1 - z\overline{b}_k).$$

Let $D := \{z \in \mathbb{C} : |z| < 1\}$.

b] Show that $|f^*(z)| = |f(z)|$ for every $z \in \partial D$.

In each of the remaining parts of the exercise suppose that $|a_k| > 1$ for each k.

c] Show that if $\epsilon \in \partial D$ and $|b_k| \ge 1$ for each k, then $f + \epsilon f^*$ has all its zeros on the unit circle.

d] Show that if $|b_k| \ge 1$ for each k, then

$$|f'(z)| \le |f^{*\prime}(z)|, \qquad z \in \partial D.$$

Hint: First observe that it is sufficient to study the case $z = 1$. We have

$$\begin{aligned}
\left|\operatorname{Re}\left(\frac{f'(1)}{f(1)}\right)\right| &= \left|\operatorname{Re}\left(\sum_{k=1}^{n} \frac{1}{1-b_k}\right) - \operatorname{Re}\left(\sum_{k=1}^{n} \frac{1}{1-a_k}\right)\right| \\
&\le \left|\operatorname{Re}\left(\sum_{k=1}^{n} \frac{1}{1-b_k}\right) - \frac{n}{2}\right| + \left|\frac{n}{2} - \operatorname{Re}\left(\sum_{k=1}^{n} \frac{1}{1-a_k}\right)\right| \\
&= \frac{n}{2} - \operatorname{Re}\left(\sum_{k=1}^{n} \frac{1}{1-b_k}\right) + \frac{n}{2} - \operatorname{Re}\left(\sum_{k=1}^{n} \frac{1}{1-a_k}\right) \\
&= n - \operatorname{Re}\left(\sum_{k=1}^{n} \frac{1}{1-b_k}\right) - \operatorname{Re}\left(\sum_{k=1}^{n} \frac{1}{1-a_k}\right) \\
&= \operatorname{Re}\left(\sum_{k=1}^{n} \frac{1}{1-b_k^{-1}}\right) - \operatorname{Re}\left(\sum_{k=1}^{n} \frac{1}{1-a_k}\right) \\
&= \operatorname{Re}\left(\sum_{k=1}^{n} \frac{1}{1-\overline{b}_k^{-1}}\right) - \operatorname{Re}\left(\sum_{k=1}^{n} \frac{1}{1-a_k}\right) \\
&= \left|\operatorname{Re}\left(\frac{f^{*\prime}(1)}{f^*(1)}\right)\right|
\end{aligned}$$

and

$$\begin{aligned}\left|\operatorname{Im}\left(\frac{f'(1)}{f(1)}\right)\right| &= \left|\operatorname{Im}\left(\sum_{k=1}^{n}\frac{1}{1-b_k}\right) - \operatorname{Im}\left(\sum_{k=1}^{n}\frac{1}{1-a_k}\right)\right| \\ &= \left|\operatorname{Im}\left(n - \sum_{k=1}^{n}\frac{1}{1-b_k}\right) - \operatorname{Im}\left(\sum_{k=1}^{n}\frac{1}{1-a_k}\right)\right| \\ &= \left|-\operatorname{Im}\left(\sum_{k=1}^{n}\frac{1}{1-\overline{b}_k^{-1}}\right) + \operatorname{Im}\left(\sum_{k=1}^{n}\frac{1}{1-a_k}\right)\right| \\ &= \left|\operatorname{Im}\left(\frac{f^{*\prime}(1)}{f^*(1)}\right)\right| .\end{aligned}$$

The result now follows from the combination of part b] and the above two inequalities. □

e] Prove that if $|b_k| \geq 1$ for each k, then

$$2|f'(z)| \leq |f'(z)| + |f^{*\prime}(z)| \leq B_n(z)\|f\|_{\partial D}, \qquad z \in \partial D,$$

where $B_n(z)$ is the Bernstein factor defined in Theorem 7.1.11.

Hint: Use parts c] and d] and the already proved part of Theorem 7.1.11 (when f has all its zeros on the unit circle). □

f] Show that if $|b_k| \geq 1$ for each k, then

$$|f'(z)| \leq \frac{1}{2}B_n(z)\left(\max_{z\in\partial D}|f(z)| - \min_{z\in\partial D}|f(z)|\right), \qquad z \in \partial D,$$

where $B_n(z)$ is the Bernstein factor defined in Theorem 7.1.7. This extends a result of Aziz and Dawood [88].

Hint: Assume that $\|f\|_{\partial D} = 1$. Let $m := \min_{z\in\partial D}|f(z)|$. Let α be a constant of modulus less than 1. Let $g(z) := f(z) - \alpha m$. Observe that the argument of α can be chosen so that

$$|g^{*\prime}(z)| = |f^{*\prime}(z)| - |\alpha| m B_n(z) .$$

By Rouché's theorem, g has no zeros in D. So parts d] and e] imply that

$$\begin{aligned}2\,|f^{*\prime}(z)| - 2\,|\alpha|\,mB_n(z) &= 2\,|g^{*\prime}(z)| \leq |g'(z)| + |g^{*\prime}(z)| \\ &= |f'(z)| + |f^{*\prime}(z)| - |\alpha| m B_n(z) \\ &\leq B_n(z) - |\alpha| m B_n(z) .\end{aligned}$$

Since $|\alpha|$ can be chosen arbitrarily close to 1, the result follows. □

E.6 Extensions of Russak's Inequalities.

a] Given $(a_k)_{k=1}^{2n} \subset \mathbb{C} \setminus \mathbb{R}$, show that

$$\|f'\|_{L_1(K)} \leq 2\pi n \, \|f\|_K$$

for every $f \in \mathcal{T}_n(a_1, a_2, \ldots, a_{2n}; K)$ and

$$\|f'\|_{L_1(K)} \leq 4\pi n \, \|f\|_K$$

for every $f \in \mathcal{T}_n^c(a_1, a_2, \ldots, a_{2n}; K)$.

b] Given $(a_k)_{k=1}^{n} \subset \mathbb{C} \setminus \mathbb{R}$, show that

$$\|f'\|_{L_1(\mathbb{R})} \leq \pi n \, \|f\|_{\mathbb{R}}$$

for every $f \in \mathcal{P}_n(a_1, a_2, \ldots, a_n; \mathbb{R},)$ and

$$\|f'\|_{L_1(\mathbb{R})} \leq 2\pi n \, \|f\|_{\mathbb{R}}$$

for every $f \in \mathcal{P}_n^c(a_1, a_2, \ldots, a_n; \mathbb{R})$.

Hint: Use Corollaries 7.1.2, 7.1.8, 7.1.6, and 7.1.10. Write the Bernstein factors in a form so that the integral (of each term in the maximum if the Bernstein factor is defined by a maximum) can be evaluated by the residue theorem (in part a]) and by finding the antiderivative (in part b]). □

c] Are any of the inequalities of parts a] and b] sharp? If so, in which cases?

E.7 Markov-Type Inequality. Given $(a_k)_{k=1}^{n} \subset \mathbb{R} \setminus [-1, 1]$, show that

$$\|f'\|_{[-1,1]} \leq \frac{n}{n-1} \left(\sum_{k=1}^{n} \frac{1 + |c_k|}{1 - |c_k|} \right)^2 \|f\|_{[-1,1]}$$

for every $f \in \mathcal{P}_n^c(a_1, a_2, \ldots, a_n; [-1, 1])$, where the numbers c_k are defined by

$$c_k := a_k - \sqrt{a_k^2 - 1}\,, \qquad a_k = \tfrac{1}{2}(c_k + c_k^{-1})\,, \qquad |c_k| < 1\,.$$

Proceed as follows:

a] Given $(a_k)_{k=1}^{n} \subset \mathbb{R} \setminus [-1, 1]$, let

$$a_k(y) := \begin{cases} \dfrac{2a_k}{1+y} + \dfrac{1-y}{1+y} & \text{if} \quad 0 \leq y \leq 1 \\[2ex] \dfrac{2a_k}{1-y} + \dfrac{1+y}{1-y} & \text{if} \quad -1 \leq y \leq 0 \end{cases}$$

and let $c_k(y)$ be defined by

$$a_k(y) =: \tfrac{1}{2}(c_k(y) + c_k(y)^{-1}), \qquad |c_k(y)| < 1.$$

Show that

$$|f'(y)| \le \frac{2}{1+|y|}\left(\sum_{k=1}^{n} \frac{1+c_k(y)}{1-c_k(y)}\right)^2 \|f\|_{[-1,1]}$$

for every $f \in \mathcal{P}_n^c(a_1, a_2, \dots, a_n; [-1,1])$.

Hint: Show by a variational method that

$$\max_f \frac{|f'(\pm 1)|}{\|f\|_{[-1,1]}} = |T_n'(\pm 1)|,$$

where the maximum is taken for all $0 \ne f \in \mathcal{P}_n(a_1, a_2, \dots, a_n; [-1,1])$, and T_n is the Chebyshev polynomial for $\mathcal{P}_n(a_1, a_2, \dots, a_n; [-1,1])$ defined in Section 3.5. Now the result follows from E.1 c] of Section 3.5 by a linear shift from $[-1,1]$ to $[-1,y]$ if $0 \le y \le 1$, or to $[y,1]$ if $-1 \le y \le 0$. □

b] Given $(a_k)_{k=1}^n \subset \mathbb{R} \setminus [-1,1]$, show that

$$|f'(y)| \le \frac{n}{1-|y|}\|f\|_{[-1,1]}, \qquad y \in (-1,1)$$

for every $f \in \mathcal{P}_n(a_1, a_2, \dots, a_n; [-1,1])$.

Hint: When $y = 0$, this follows from Corollary 7.1.3. When $y \in (-1,1)$ is arbitrary, use a linear shift from $[-1,1]$ to $[2y-1,1]$ if $0 \le y \le 1$, or to $[-1, 2y+1]$ if $-1 < y \le 0$. □

c] Prove the Markov-type inequality of the exercise.

Hint: Combine parts a] and b]. Note that

$$|a_k(y)| \ge |a_k| \qquad \text{and} \qquad |c_k(y)| \le |c_k| < 1, \qquad k = 1, 2, \dots, n$$

holds for every $y \in [-1,1]$. □

E.8 Schur-Type Inequality. Given $\{a_k\}_{k=1}^n \subset \mathbb{R} \setminus [-1,1]$, show that

$$\|f\|_{[-1,1]} \le \max\{|U_n(1)|, |U_n(-1)|\} \cdot \left\|f(x)\sqrt{1-x^2}\right\|_{[-1,1]}$$

for every $f \in \mathcal{P}_n(a_1, a_2, \dots, a_n; [-1,1])$, where U_n is the Chebyshev polynomial (of the second kind) for $\mathcal{P}_n(a_1, a_2, \dots, a_n; [-1,1])$ defined in Section 3.5, and

$$|U_n(\pm 1)| = \left|\sum_{k=1}^{n} \frac{\sqrt{a_k^2 - 1}}{a_k \mp 1}\right|$$

with the choice of $\sqrt{a_k^2-1}$ determined by $\left|a_k - \sqrt{a_k^2 - 1}\right| < 1$. Show that equality holds if and only if $f = cU_n$, $c \in \mathbb{R}$.

Hint: First show that if $\widetilde{f} \in \mathcal{P}_n(a_1, a_2, \dots, a_n; [-1,1])$ is extremal for

$$\max_f \frac{|f(\pm 1)|}{\left\| f(x)\sqrt{1-x^2} \right\|_{[-1,1]}},$$

where the maximum is taken for all $0 \neq f \in \mathcal{P}_n(a_1, a_2, \dots, a_n; [-1,1])$, then $f = cU_n$ with some $c \in \mathbb{R}$. Observe that $U_n(\pm 1)$ can be evaluated by L'Hospital's rule since

$$U_n(x)^2 = \frac{1 - T_n(x)^2}{1 - x^2};$$

hence $|U_n(\pm 1)|^2 = |T_n'(\pm 1)| = |B_n(\pm 1)| \cdot |U_n(\pm 1)|$ with the notation of Section 3.5. Thus $|U_n(\pm 1)| = |B_n(\pm 1)|$.

If $y \in [-1,1]$ is arbitrary, then use a linear shift from $[-1,1]$ to $[-y,y]$ (some caution must be exercised about the change of poles). □

E.9 Extension of Lax's Inequality on the Half-Plane. Associated with $(a_k)_{k=1}^n \subset H^+ := \{z \in \mathbb{C} : \operatorname{Im}(z) > 0\}$, let the Bernstein factor B_n be, as in Corollary 7.1.10, defined by

$$B_n(x) := \sum_{k=1}^n \frac{2\operatorname{Im}(a_k)}{|x - a_k|^2}.$$

Show that

$$|h'(x)| \leq \tfrac{1}{2} B_n(x) \|h\|_{\mathbb{R}}, \qquad x \in \mathbb{R}$$

for every $h \in \mathcal{P}_n^c(a_1, a_2, \dots, a_n; \mathbb{R})$ having all its zeros in H^+.

Equality holds for $h = c(\widetilde{S}_n + 1)$ with $c \in \mathbb{C}$, where $\widetilde{S}_n$ is the Blaschke product associated with $(a_k)_{k=1}^n$. Note that $B_n(z) = |\widetilde{S}_n'(z)|$. Note also that

$$h = c(\widetilde{S} + 1) \in \mathcal{P}_n^c(a_1, a_2, \dots, a_n; \mathbb{R}), \qquad c \neq 0$$

has all its zeros on $\mathbb{R}$.

Hint: Use Theorem 7.1.11 with the substitution $x = i\dfrac{z+1}{z-1}$. □

E.10 Remarks on Theorem 7.1.7 and Corollary 7.1.10.

a] Given $(a_k)_{k=1}^n \subset D$ and $z \in \partial D$, show that equality holds in the inequality of Theorem 7.1.7 if and only if $f = cS_n$ with $c \in \mathbb{C}$, where S_n is the Blaschke product associated with $(a_k)_{k=1}^n$.

Hint: Analyze the proof of Theorem 7.1.7. □

b] Given $(a_k)_{k=1}^n \subset H^+ = \{z \in \mathbb{C} : \operatorname{Im}(z) > 0\}$ and $x \in \mathbb{R}$, show that equality holds in the inequality of Corollary 7.1.10 if and only if $f = cQ_n$ with $c \in \mathbb{C}$, where Q_n is the Blaschke product associated with $(a_k)_{k=1}^n$.

Note that the *only if* parts of E.10 a] and E.10 b] above are not claimed in the general case of Theorem 7.1.7 and Corollary 7.1.10 (why?).

E.11 Markov-Bernstein-Type Inequality for SR_n^c and ASR_n^c. Let SR_n^c denote the set of all self-reciprocal polynomials $p \in \mathcal{P}_n^c$ satisfying

$$p(z) \equiv z^n p(z^{-1}) .$$

Let SR_n denote the set of all real self-reciprocal polynomials of degree at most n, that is, $\mathrm{SR}_n := \mathrm{SR}_n^c \cap \mathcal{P}_n$. For a polynomial $p \in \mathcal{P}_n^c$ of the form

$$p(z) = \sum_{j=0}^{n} c_j z^j , \qquad c_j \in \mathbb{C} , \tag{7.1.16}$$

$p \in \mathrm{SR}_n^c$ if and only if

$$c_j = c_{n-j} , \qquad j = 0, 1, \dots , n .$$

Let ASR_n^c denote the set of all antiself-reciprocal polynomials $p \in \mathcal{P}_n^c$ satisfying

$$p(z) \equiv -z^n p(z^{-1}) .$$

Let ASR_n denote the set of all real antiself-reciprocal polynomials of degree at most n, that is, $\mathrm{ASR}_n := \mathrm{ASR}_n^c \cap \mathcal{P}_n$. Let $\mathrm{ASR}_n := \mathrm{ASR}_n^c \cap \mathcal{P}_n$. For a polynomial $p \in \mathcal{P}_n^c$ of the form (7.1.16) $p \in \mathrm{ASR}_n^c$ if and only if

$$c_j = -c_{n-j} , \qquad j = 0, 1, \dots , n .$$

a] There exists an absolute constant c such that

$$|p'(x)| \leq cn \min \left\{ (1 + \log n), \ \log \left(\frac{e}{1 - x^2} \right) \right\} \|p\|_{[-1,1]}$$

holds for every $x \in [-1, 1]$ and for every $p \in \mathcal{P}_n^c$ satisfying

$$|p(x)| \leq (1 + |x|^n) \|p\|_{[-1,1]} , \qquad x \in \mathbb{R} , \tag{7.1.17}$$

in particular, for every $p \in \mathrm{SR}_n^c$ and for every $p \in \mathrm{ASR}_n^c$.

The inequality

$$|p'(x)| \leq cn(1 + \log n) \, \|p\|_{[-1,1]}$$

for all $p \in \mathrm{SR}_n$ and for all $p \in \mathrm{ASR}_n$ was first obtained by Kroó and Szabados [94a]. They also showed that up to the constant $c > 0$ the above inequality is sharp for both SR_n and ASR_n. Here we present a distinct proof. The sharpness is studied in E.12 f].

Outline. Suppose $p \in \mathcal{P}_n^c$ satisfies (7.1.17). Then

$$f(x) := \frac{p(x)}{1 + x^{2n}} \tag{7.1.18}$$

satisfies

$$\|f\|_{\mathbb{R}} \leq 2 \|p\|_{[-1,1]} \,. \tag{7.1.19}$$

Let

$$a_k := e^{i(2k-1)\pi/(2n)}\,, \qquad k = 1, 2, \ldots, 2n \tag{7.1.20}$$

be the zeros of the equation $z^{2n} + 1 = 0$. Now Corollary 7.1.10, together with (7.1.18) to (7.1.20), yields that

$$\begin{aligned} |f'(x)| &\leq 2 \left(\sum_{k=1}^{n} \frac{\operatorname{Im}(a_k)}{|x - a_k|^2} \right) \|f\|_{\mathbb{R}} \\ &\leq 4 \left(\sum_{k=1}^{n} \frac{\operatorname{Im}(a_k)}{|x - a_k|^2} \right) \|p\|_{[-1,1]}\,, \qquad x \in \mathbb{R}\,. \end{aligned} \tag{7.1.21}$$

Show that if $n \in \mathbb{N}$ and $x \in [-1, 1]$, then

$$4 \sum_{k=1}^{n} \frac{\operatorname{Im}(a_k)}{|x - a_k|^2} \sim n \min \left\{ (1 + \log n),\ \log \left(\frac{e}{1 - x^2} \right) \right\}, \tag{7.1.22}$$

where here the $\sim$ symbol means that there are absolute constants $c_1 > 0$ and $c_2 > 0$ (independent of $n \in \mathbb{N}$ and $x \in [-1, 1]$) such that the left-hand side is between c_1 times the right-hand side and c_2 times the right-hand side for every $n \in \mathbb{N}$ and $x \in [-1, 1]$. Combining (7.1.18), (7.1.21), and (7.1.22), we conclude that there is an absolute constant c_2 such that

$$\begin{aligned} &\left| \frac{p'(x)}{1 + x^{2n}} - \frac{2nx^{2n-1}}{1 + x^{2n}} \frac{p(x)}{1 + x^{2n}} \right| \\ &\qquad \leq c_2 n \min \left\{ (1 + \log n),\ \log \left(\frac{e}{1 - x^2} \right) \right\} \|p\|_{[-1,1]}\,, \qquad x \in \mathbb{R}\,. \end{aligned}$$

So if $x \in [-1, 1]$, then

$$|p'(x)| \leq \left(2c_2 n \min \left\{ (1 + \log n),\ \log \left(\frac{e}{1 - x^2} \right) \right\} + 2n \right) \|p\|_{[-1,1]}$$

and the proof is finished. □

b] There exists an absolute constant $c > 0$ such that

$$\left\| \frac{p'(x)}{1 + x^{2n}} \right\|_{\mathbb{R}} \leq cn(1 + \log n) \left\| \frac{p(x)}{1 + x^{2n}} \right\|_{\mathbb{R}}$$

for every $p \in \mathcal{P}_{2n}^c$.

Proof. Associated with $p \in \mathcal{P}_{2n}^c$, let

$$f(x) := \frac{p(x)}{1 + x^{2n}}.$$

Let

$$a_k := e^{i(2k-1)\pi/(2n)}, \qquad k = 1, 2, \dots, 2n$$

be the zeros of the equation $z^{2n} + 1 = 0$. Using Corollary 7.1.10, we can deduce that there are absolute constants c_1 and c_2 such that

$$\begin{aligned}\|f'\|_{\mathbb{R}} &\leq \left(\sum_{k=1}^{n} 2\,|\mathrm{Im}(a_k)|^{-1}\right)\|f\|_{\mathbb{R}} \leq c_1 \left(\sum_{k=1}^{n}\left(\frac{k}{n}\right)^{-1}\right)\|f\|_{\mathbb{R}} \\ &\leq c_2 n(1 + \log n)\|f\|_{\mathbb{R}}.\end{aligned}$$

Therefore

$$\left|\frac{p'(x)}{1 + x^{2n}} - \frac{2nx^{2n-1}}{1 + x^{2n}}\frac{p(x)}{1 + x^{2n}}\right| \leq c_2 n(1 + \log n)\left\|\frac{p(x)}{1 + x^{2n}}\right\|_{\mathbb{R}}$$

for every $x \in \mathbb{R}$, which implies

$$\left\|\frac{p'(x)}{1 + x^{2n}}\right\|_{\mathbb{R}} \leq (c_2 n(1 + \log n) + 2n)\left\|\frac{p(x)}{1 + x^{2n}}\right\|_{\mathbb{R}},$$

and the result follows. □

c] For every $m \in \mathbb{N}$, there exists a constant $c(m)$ depending only on m so that

$$\|p^{(m)}\|_{[-1,1]} \leq c(m)(n(1 + \log n))^m \|p\|_{[-1,1]}$$

for every $p \in \mathcal{P}_n^c$ satisfying

$$|p(x)| \leq (1 + |x|^n)\|p\|_{[-1,1]}. \tag{7.1.23}$$

Proof. Using part b] and induction on m, we see that there exists a constant $c_1(m)$ depending only on m such that

$$\left\|\frac{p^{(m)}(x)}{1 + x^{2n}}\right\|_{\mathbb{R}} \leq c_1(m)(n(1 + \log n))^m \left\|\frac{p(x)}{1 + x^{2n}}\right\|_{\mathbb{R}}$$

for every $p \in \mathcal{P}_{2n}^c$. Note that if $p \in \mathcal{P}_n^c$ satisfies (7.1.23), then

$$\left\|\frac{p(x)}{1 + x^{2n}}\right\|_{\mathbb{R}} \leq 2\left\|\frac{p(x)}{1 + |x|^n}\right\|_{\mathbb{R}} \leq 2\|p\|_{[-1,1]},$$

and the result follows. □

E.12 Quasi-Chebyshev Polynomials for SR_{2n} and ASR_{2n}. Let

$$a_k := e^{i(2k-1)\pi/(2n)}\,, \qquad k = 1, 2, \dots n$$

be the zeros of the equation $z^{2n} + 1 = 0$ in the upper half-plane. Let

$$M_{2n}(z) := \prod_{k=1}^{n} (z - a_k)^2\,, \qquad z \in \mathbb{C}\,,$$

$$P_{2n}(x) := \mathrm{Re}(M_{2n}(x))\,, \qquad x \in \mathbb{R}\,,$$

and

$$Q_{2n}(x) := \mathrm{Im}(M_{2n}(x))\,, \qquad x \in \mathbb{R}\,.$$

a] Show that if n is even, then $P_{2n} \in \mathrm{SR}_{2n}$ and $Q_{2n} \in \mathrm{ASR}_{2n}$, while if n is odd, then $Q_{2n} \in \mathrm{SR}_{2n}$ and $P_{2n} \in \mathrm{ASR}_{2n}$.

b] Show that

$$|M_{2n}(x)| = 1 + x^{2n}\,, \qquad x \in \mathbb{R}$$

and

$$P_{2n}(x)^2 + Q_{2n}(x)^2 = (1 + x^{2n})^2\,, \qquad x \in \mathbb{R}\,,$$

in particular

$$\|P_{2n}\|_{[-1,1]} \le 2 \qquad \text{and} \qquad \|Q_{2n}\|_{[-1,1]} \le 2\,.$$

Proof. Note that

$$\frac{M_{2n}(x)}{1 + x^{2n}} = \prod_{k=1}^{n} \frac{x - a_k}{x - \overline{a}_k}\,, \tag{7.1.24}$$

which implies the first equality. The rest is straightforward from the definitions. □

c] Show that there are extended real numbers

$$\infty = z_0 > z_1 > \cdots > z_{2n} = -\infty$$

such that

$$\frac{M_{2n}(z_j)}{1 + z_j^{2n}} = (-1)^j\,, \qquad j = 0, 1, \dots, 2n$$

(the value of the left-hand side at $\pm\infty$ is defined by taking the limit when $x \to \pm\infty$).

Hint: Use (7.1.24) and the argument principle (see, for example, Ash [71]). □

d] Let $n \in \mathbb{N}$ be even. Show that there exist

$$1 = x_0 > y_1 > x_1 > y_2 > \cdots > x_{n-1} > y_n > x_n = -1$$

such that

$$\frac{P_{2n}(x_j)}{1 + x_j^{2n}} = (-1)^{j+n/2}, \qquad Q_{2n}(x_j) = 0, \quad j = 0, 1, \ldots, n$$

and

$$\frac{Q_{2n}(y_j)}{1 + y_j^{2n}} = (-1)^{j+1+n/2}, \qquad P_{2n}(y_j) = 0, \quad j = 1, 2, \ldots n.$$

Formulate the analog statement when $n \in \mathbb{N}$ is odd.

e] Show that there exists an absolute constant $c > 0$ such that

$$|P'_{2n}(x)| + |Q'_{2n}(x)| \geq cn \min\left\{(1 + \log n), \ \log\left(\frac{e}{1 - x^2}\right)\right\}$$

for every $n \in \mathbb{N}$ and $x \in [-1, 1]$.

Proof. If $x \in [-1, 1]$, then

$$\begin{aligned} |P'_{2n}(x)| + |Q'_{2n}(x)| &\geq |M'_{2n}(x)| = |M'_{2n}(x)| \left|\frac{1 + x^{2n}}{M_{2n}(x)}\right| \\ &\geq \left|\frac{M'_{2n}(x)}{M_{2n}(x)}\right| = \left|\sum_{k=1}^{n} \frac{2}{x - a_k}\right| \geq 2 \sum_{k=1}^{n} \frac{\operatorname{Im}(a_k)}{|x - a_k|^2} \\ &\geq cn \min\left\{(1 + \log n), \ \log\left(\frac{e}{1 - x^2}\right)\right\} \end{aligned}$$

with an absolute constant $c > 0$, where the last inequality follows from (7.1.22). $\square$

The next part shows the sharpness of the inequality of E.11 a].

f] Let $c > 0$ be the same absolute constant as in part e]. For the sake of brevity let

$$\delta_n(x) := \left(cn \min\left\{(1 + \log n), \ \log\left(\frac{e}{1 - x^2}\right)\right\}\right)^{-1}.$$

Show that for every interval

$$I_{n,x} := [x, x + 8\delta_n(x)] \subset [0, 1],$$

there exist $y_1 \in I_{n,x}$ and $y_2 \in I_{n,x}$ such that

$$|P'_{2n}(y_1)| \geq \tfrac{1}{2}\delta_n(x)^{-1} \qquad \text{and} \qquad |Q'_{2n}(y_2)| \geq \tfrac{1}{2}\delta_n(x)^{-1} .$$

Proof. Suppose that

$$|P'_{2n}(y)| < \tfrac{1}{2}\delta_n(x)^{-1}$$

for every $y \in I_{n,x}$. Then, from part e] we can deduce that

$$|Q'_{2n}(y)| > \tfrac{1}{2}\delta_n(x)^{-1}$$

for every $y \in I_{n,x}$. Therefore

$$\|Q_{2n}\|_{[-1,1]} \geq \tfrac{1}{2}\int_{I_{n,x}} |Q'_{2n}(y)|\, dy > \tfrac{1}{2}8\delta_n(x)\tfrac{1}{2}\delta_n(x)^{-1} = 2 ,$$

which contradicts the last inequality of part b]. This finishes the proof of the first inequality. The second inequality can be proven in the same way. □

g] Show that

$$p'(1) = \tfrac{1}{2}n\, p(1)$$

for every $p \in \mathrm{SR}_n^c$.

By using the quasi Chebyshev polynomials for SR_{2n} and ASR_{2n}, it can be shown that the inequality of E.11 c] is essentially sharp for the classes SR_n and ASR_{2n} for every m. This has been pointed out to us by Szabados. The argument requires some more technical details than the proof in the $m = 1$ case discussed in the above exercise.

7.2 Inequalities for Logarithmic Derivatives

We derive a series of metric inequalities of the form

$$m\left(\left\{x \in \mathbb{R} : \frac{r'(x)}{r(x)} \geq \alpha\right\}\right) \leq \frac{cn}{\alpha}, \qquad \alpha > 0 ,$$

where r is a rational function of type (n, n) and c is a constant independent of n. Here m is the Lebesgue measure, although, since the sets in question are usually just finite unions of intervals, this is mostly a notational convenience. One of the interesting features of these inequalities is their easy extension from the polynomial case to the rational case.

The basic inequality is due to Loomis [46]. Note the invariance of the measure of the set in this case.

Theorem 7.2.1. *If $p \in \mathcal{P}_n$ has n real roots, then*

$$m\left(\left\{x \in \mathbb{R} : \frac{p'(x)}{p(x)} \geq \alpha\right\}\right) = \frac{n}{\alpha}, \qquad \alpha > 0.$$

Proof. By considering $\alpha^{-1}p$ instead of p, it is sufficient to prove the theorem for $\alpha = 1$. We first consider the case where p has distinct roots, which are denoted by $\alpha_1 < \alpha_2 < \cdots < \alpha_n$. Then

$$\frac{p'(x)}{p(x)} = \sum_{k=1}^{n} \frac{1}{x - \alpha_i}.$$

Let $\beta_1 < \beta_2 < \cdots < \beta_n$ be the roots of $p - p'$, which must all be real. Note that these are the points where $p'(x)/p(x) = 1$. It is now easy to see from the graph that

$$\left\{x \in \mathbb{R} : \frac{p'(x)}{p(x)} \geq 1\right\} = [\alpha_1, \beta_1] \cup [\alpha_2, \beta_2] \cup \cdots \cup [\alpha_n, \beta_n]$$

and

$$m\left(\left\{x \in \mathbb{R} : \frac{p'(x)}{p(x)} \geq 1\right\}\right) = \sum_{i=1}^{n} \beta_i - \sum_{i=1}^{n} \alpha_i.$$

However, if $p(x) := x^n + a_{n-1}x^{n-1} + \cdots + a_0$, then

$$\sum_{i=1}^{n} \alpha_i = -a_{n-1},$$

while

$$\sum_{i=1}^{n} \beta_i = -(a_{n-1} - n)$$

is -1 times the second coefficient of $p - p'$.

This gives the result for distinct roots. The case when some of the roots of p are repeated can be handled by an easy limiting argument. □

Corollary 7.2.2. *If $\alpha_i \in \mathbb{R}$, $c_i > 0$, $i = 1, 2, \ldots, n$, and $\sum_{i=1}^{n} c_i = 1$, then*

$$m\left(\left\{x \in \mathbb{R} : \sum_{i=1}^{n} \frac{c_i}{x - \alpha_i} \geq \alpha\right\}\right) = \frac{1}{\alpha}, \qquad \alpha > 0.$$

Proof. For c_i rational this follows immediately from Theorem 7.2.1 on clearing the denominators of c_i by multiplying by an integer. The real case is an obvious limiting argument. □

In order to extend Theorem 7.2.1 to arbitrary polynomials we need the following generalization of E.3 of Section 2.4 due to Videnskii [51]. The proof is indicated in the exercises.

Lemma 7.2.3. *If $p \in \mathcal{P}_n$ is positive on $[a,b]$, then there exists $q, s \in \mathcal{P}_n$ nonnegative on $[a,b]$ with all real roots (in $[a,b]$) so that*

$$p(x) = q(x) + s(x).$$

We now prove the unrestricted case of Theorem 7.2.1.

Theorem 7.2.4. *Let $p \in \mathcal{P}_n$. Then*

$$m\left(\left\{x \in \mathbb{R} : \frac{p'(x)}{p(x)} \geq \alpha\right\}\right) \leq \frac{2n}{\alpha}, \qquad \alpha > 0.$$

Proof. Let $\alpha > 0$ and let $p_n \in \mathcal{P}_n$. Choose a and b such that

$$\left\{x \in \mathbb{R} : \frac{p'(x)}{p(x)} \geq \alpha\right\} \subset [a,b].$$

By Lemma 7.2.3 we can find polynomials $q \in \mathcal{P}_{2n}$ and $s \in \mathcal{P}_{2n}$ such that

$$p^2(x) = q(x) + s(x)$$

where, for $x \in [a,b]$,

$$0 \leq q(x) \leq p^2(x) \qquad \text{and} \qquad 0 \leq s(x) \leq p^2(x)$$

and both q and s have only real roots. Now

$$\left\{x \in \mathbb{R} : \frac{p'(x)}{p(x)} \geq \alpha\right\} = \left\{x \in \mathbb{R} : \frac{(p^2)'(x)}{p^2(x)} \geq 2\alpha\right\}.$$

Also,

$$(p^2)'(x) \geq 2\alpha\, p^2(x)$$

holds exactly when

$$q'(x) + s'(x) \geq 2\alpha(q(x) + s(x)).$$

By Theorem 7.2.1

$$m\left(\left\{x \in \mathbb{R} : \frac{q'(x)}{q(x)} \geq 2\alpha\right\}\right) = m\left(\left\{x \in \mathbb{R} : \frac{s'(x)}{s(x)} \geq 2\alpha\right\}\right) = \frac{n}{\alpha}.$$

Since q and s are nonnegative on $[a,b]$, it follows that

$$m(\{x \in [a,b] : q'(x) + s'(x) \geq 2\alpha(q(x) + s(x))\}) \leq \frac{2n}{\alpha},$$

and the proof is finished. □

It can be shown that this inequality is asymptotically sharp to the extent that the constant 2 cannot be replaced by any smaller constant for large n; see Kristiansen [82].

Theorem 7.2.4 extends easily to rational functions.

Theorem 7.2.5. *If $r = p/q$, where $p, q \in \mathcal{P}_n$, then*

$$m\left(\left\{x \in \mathbb{R} : \frac{r'(x)}{r(x)} \geq \alpha\right\}\right) \leq \frac{8n}{\alpha}, \qquad \alpha > 0.$$

Proof. We have

$$\frac{r'}{r} = \frac{p'}{p} - \frac{q'}{q}$$

and for $\alpha > 0$,

$$\left\{x \in \mathbb{R} : \frac{r'(x)}{r(x)} \geq \alpha\right\} \subset \left\{x \in \mathbb{R} : \frac{p'(x)}{p(x)} \geq \frac{\alpha}{2}\right\} \cup \left\{x \in \mathbb{R} : \frac{q'(x)}{q(x)} \leq -\frac{\alpha}{2}\right\}.$$

By Theorem 7.2.4

$$m\left(\left\{x \in \mathbb{R} : \frac{p'(x)}{p(x)} \geq \frac{\alpha}{2}\right\}\right) \leq \frac{4n}{\alpha},$$

and with $s(x) := q(-x)$,

$$m\left(\left\{x \in \mathbb{R} : \frac{q'(x)}{q(x)} \leq -\frac{\alpha}{2}\right\}\right) = m\left(\left\{x \in \mathbb{R} : \frac{s'(x)}{s(x)} \geq \frac{\alpha}{2}\right\}\right) \leq \frac{4n}{\alpha}.$$

It follows that

$$m\left(\left\{x \in \mathbb{R} : \frac{r'(x)}{r(x)} \geq \alpha\right\}\right) \leq \frac{8n}{\alpha}$$

for every $\alpha > 0$. □

This inequality probably does not have the exact constant. It can be shown (see Borwein, Rakhmanov, and Saff [to appear]) that the constant 8 cannot be replaced by any constant less than or equal to 2π.

Comments, Exercises, and Examples.

Many variants on the inequalities of this section are presented in E.2, E.3, E.4, and E.5. Some of these are in Borwein [82].

E.5 explores some metric properties of the lemniscate

$$E(p) := \{z \in \mathbb{C} : |p(z)| \leq 1\}$$

of a monic polynomial $p \in \mathcal{P}_n^c$ of the form

$$p(z) = \prod_{i=1}^{n} (z - z_i), \qquad z_i \in \mathbb{C}.$$

The reader is referred to Erdős, Herzog, and Piranian [58] for many results and open problems concerning the lemniscate of monic polynomials. One in particular, which is still open, conjectures that for monic polynomials $p \in \mathcal{P}_n^c$, the length of the boundary of $E(p)$ is maximal for $p(z) := z^n - 1$ and so is $O(n)$. Pommerenke [61] has shown that this length is $O(n^2)$. Borwein [95] improves this to $O(n)$; see E.7. E.9 solves another problem of Erdős, namely, the diameter of $E(p)$ for a monic polynomial $p \in \mathcal{P}_n^c$ is always at least 2.

Erdős [76] contains several other related open problems.

E.1 Polynomials as Sums of Polynomials with Real Roots.

a] Suppose $p \in \mathcal{P}_{2n} \setminus \mathcal{P}_{2n-1}$, and suppose that $p > 0$ on $[a, b]$. Then

$$p(x) = (x-a)(b-x)u^2(x) + v^2(x)$$

for some $u \in \mathcal{P}_{n-1}$ and $v \in \mathcal{P}_n$, which have all their zeros in $[a, b]$.

b] Suppose $p \in \mathcal{P}_{2n+1} \setminus \mathcal{P}_{2n}$ and suppose that $p > 0$ on $[a, b]$. Then

$$p(x) = (b-x)u^2(x) + (x-a)v^2(x)$$

for some $u, v \in \mathcal{P}_n$, which have all their zeros in $[a, b]$.

Hint: Let p be a polynomial of degree $2n$ that is strictly positive on $[a, b]$. Let T_n be the Chebyshev polynomial for the Chebyshev system

$$\left\{ \frac{1}{\sqrt{p(x)}}, \frac{x}{\sqrt{p(x)}}, \dots, \frac{x^n}{\sqrt{p(x)}} \right\}.$$

Then $T_n(x) = v(x)/\sqrt{p(x)}$ with some $v \in \mathcal{P}_n$. Show that, for part a], v and u defined by

$$(x-a)(b-x)u^2(x) = p(x) - v^2(x)$$

are the required polynomials. Use a similar construction for part b]. □

E.2 Various Specializations. For the next exercises we use the notation $\mathcal{P}_n^+$ to denote the polynomial of degree at most n with nonnegative coefficients, and $\mathcal{P}_n^\uparrow$ to denote those elements of $\mathcal{P}_n$ that are nondecreasing on $[0, \infty)$.

a] If $r = p/q$, where $p, q \in \mathcal{P}_n$ and both p and q have only real roots, then

$$m\left(\left\{x \in \mathbb{R} : \frac{r'(x)}{r(x)} \geq \alpha\right\}\right) \leq \frac{4n}{\alpha}, \qquad \alpha > 0.$$

b] Let $r(x) := x^n/(4n-x)^n$. Then

$$m\left(\left\{x \in \mathbb{R} : \frac{r'(x)}{r(x)} \geq 1\right\}\right) = 4n\,.$$

c] If $r = p/q$, where $p \in \mathcal{P}_n$ and $q \in \mathcal{P}_n^{\uparrow}$, then

$$m\left(\left\{x \geq 0 : \frac{r'(x)}{r(x)} \geq \alpha\right\}\right) \leq \frac{2n}{\alpha}\,, \qquad \alpha > 0\,.$$

d] If $r = p/q$, where $p \in \mathcal{P}_n^+$ and $q \in \mathcal{P}_n^{\uparrow}$, then

$$m\left(\left\{x \geq 0 : \frac{r'(x)}{r(x)} \geq \alpha\right\}\right) \leq \frac{n}{\alpha}\,, \qquad \alpha > 0\,.$$

e] Let $r(x) := x^n$. Then

$$m\left(\left\{x \geq 0 : \frac{r'(x)}{r(x)} \geq \alpha\right\}\right) = \frac{n}{\alpha}\,, \qquad \alpha > 0\,.$$

E.3 Another Metric Inequality. If $p \in \mathcal{P}_n$ has n real roots lying in the interval (a,b), then

$$m\left(\left\{x \in \mathbb{R} : \left|\frac{p'(x)}{p(x)}\right| \leq \frac{\alpha}{|(x-a)(b-x)|}\right\}\right) = \frac{2n}{\alpha}\,, \qquad \alpha > 0\,.$$

Outline. Prove that

$$m\left(\left\{x \in \mathbb{R} : 0 \leq \frac{(x-a)(b-x)p'(x)}{p(x)} \leq \alpha\right\}\right) = \frac{\alpha}{n}$$

and

$$m\left(\left\{x \in \mathbb{R} : 0 \geq \frac{(x-a)(b-x)p'(x)}{p(x)} \geq -\alpha\right\}\right) = \frac{\alpha}{n}\,.$$

First consider the case when p has distinct zeros. Let $y_0 < y_1 < \cdots < y_n$ denote the $n+1$ roots of $(x-a)(b-x)p'(x)$, and let $x_0 < x_1 < \cdots < x_{n-1}$ denote the n roots of $p(x)$. Then

$$y_0 < x_0 \leq y_1 < \cdots x_{n-1} < y_n < x_n := \infty\,.$$

Since

$$\lim_{x \to x_i -} (x-a)(b-x)\frac{p'(x)}{p(x)} = -\infty\,,$$

we can deduce that for each interval (y_i, x_i) there exists a point $\delta_i \in (y_i, x_i)$ such that

$$(\delta_i - a)(b - \delta_i)p'(\delta_i) = -\alpha p(\delta_i)\,.$$

Since the above equation can have at most $n+1$ solutions, we have

$$m\left(\left\{x \in \mathbb{R} : 0 \geq \frac{(x-a)(b-x)p'(x)}{p(x)} \geq -\alpha\right\}\right) = \sum_{i=0}^{n}(\delta_i - y_i)\,.$$

Now proceed as in the proof of Theorem 7.2.1. □

E.4 Extensions to $\mathcal{P}_n^c$.

a] If $p \in \mathcal{P}_n^c$, then

$$m\left(\left\{x \in \mathbb{R} : \left|\frac{p'(x)}{p(x)}\right| \geq \alpha\right\}\right) \leq \frac{8\sqrt{2}\,n}{\alpha}, \qquad \alpha > 0.$$

Hint: Write p as the sum of its real and imaginary parts. □

b] If $r = p/q$ with $p, q \in \mathcal{P}_n^c$, then

$$m\left(\left\{x \in \mathbb{R} : \left|\frac{r'(x)}{r(x)}\right| \geq \alpha\right\}\right) \leq \frac{32\sqrt{2}\,n}{\alpha}, \qquad \alpha > 0.$$

E.5 On the Lemniscate $E(p)$. Let

$$p(z) := \prod_{i=1}^{n}(z - z_i), \qquad z_i \in \mathbb{C}$$

and let

$$E(p) := \{z \in \mathbb{C} : |p(z)| \leq 1\}.$$

a] Show that

$$m(E(p) \cap \mathbb{R}) \leq 4 \cdot 2^{-1/n}$$

with equality only for the Chebyshev polynomial of degree n normalized to have lead coefficient 1; see Pólya [28].

Hint: Analogously to the proof of the Remez inequality of Section 5.1, show that the Chebyshev polynomial transformed to an interval of length 4 is extremal for this problem. □

b] Let $m_2(\cdot)$ denote the planar Lebesgue measure. Show that

$$m_2(E(p)) \leq 4\pi.$$

(In fact, $m_2(E(p)) \leq \pi$, which is exact for z^n; this is due to Pólya [28].)

E.6 Cartan's Lemma. *Let*

$$p(z) := \prod_{j=1}^{n}(z - z_j), \qquad z_j \in \mathbb{C}$$

and $\beta > 0$ be fixed. Then there exist at most n open disks, the sum of whose radii is at most 2β, so that if $z \in \mathbb{C}$ is outside the union of these open disks then

$$|p(z)| > \left(\frac{\beta}{e}\right)^n.$$

Prove Cartan's Lemma as follows:

a] Let $\alpha_1, \alpha_2, \dots, \alpha_\nu$ be fixed complex numbers. Let $\beta > 0$. Show that there exists a positive integer μ less than or equal to ν for which there exists an open disk with radius $\mu\beta/n$ containing α_j for exactly μ distinct values of $j = 1, 2, \dots, \nu$.

Hint: Suppose to the contrary that there is no such positive integer μ. Show that this would imply the existence of an open disk with radius $\nu\beta/n$ containing α_j for at least $\nu + 1$ distinct values of $j = 1, 2, \dots, \nu$, which is a contradiction. □

b] Show that there exist open disks $D_1, D_2, \dots, D_k$ and positive integers $m_1, m_2, \dots, m_k$ with the following properties:

(1) $\sum_{j=1}^{k} m_j = n, \quad m_1 \geq m_2 \geq \dots \geq m_k$;

(2) D_j has radius $\dfrac{m_j\beta}{n}$;

(3) $D_j \cap E_j$ contains exactly m_j zeros of p, where

$$E_j := \mathbb{C} \setminus (D_1 \cup D_2 \cup \ldots \cup D_{j-1});$$

(4) for every integer $m > m_j$, no open disk of radius $\frac{m\beta}{n}$ contains exactly m zeros of p in E_j.

Hint: Use part a]. □

c] Let $D_1, D_2, \dots, D_k$ be the open disks specified in part b]. For $j = 1, 2, \dots, k$, let D_j^* be the disk with the same center as D_j and with twice its radius. Show that for every

$$z \in E^* := \mathbb{C} \setminus (D_1^* \cup D_2^* \cup \dots \cup D_k^*)$$

there is a permutation $\widetilde{z}_1, \widetilde{z}_2, \dots, \widetilde{z}_n$ of the zeros $z_1, z_2, \dots, z_n$ such that

$$|z - \widetilde{z}_j| \geq \frac{j\beta}{n}, \qquad j = 1, 2, \dots, n.$$

Hint: Let $z \in E^*$ be fixed. Show, by induction on i, that for every $i = 0, 1, \dots, n-1$, there are at least $i+1$ zeros of p outside the open disk with center z and radius $\frac{(n-i)\beta}{n}$. Distinguish the cases

(1) $n - i > m_1$;

(2) $m_j \geq n - i > m_{j+1}, \qquad j = 1, 2, \dots, k-1$;

(3) $m_k \geq n - i > 0$. □

d] Finish the proof of Cartan's lemma.

E.7 The Length of the Boundary of $E(p)$. Let

$$p(z) := \prod_{j=1}^{n} (z - z_j)\,, \qquad z_j \in \mathbb{C}$$

and let

$$E := E(p) := \{z \in \mathbb{C} : |p(z)| \le 1\}\,.$$

Show that the boundary ∂E of E is of length at most $4e\pi n$. (In fact, with E.10 a], the estimate can be improved to $(5.2)\,\pi n$.)

Outline. Proceed as follows:

a] Let L be an arbitrary line in the complex plane. Show that the set $\partial E \cap L$ contains at most $2n$ distinct points.

Hint: By performing a translation and a rotation, if necessary, we may assume that $L = \mathbb{R}$. Now observe that there is a polynomial $P(x, y)$ of degree at most $2n$, in two real variables x and y, with complex coefficients, such that

$$\begin{aligned} \partial E &= \{z \in \mathbb{C} : |p(z)|^2 = 1\} \\ &= \{z \in \mathbb{C} : p(z)\overline{p(z)} = 1\} \\ &= \{z = x + iy : x, y \in \mathbb{R}\,, \;\; P(x, y) = 1\}\,. \end{aligned}$$

Hence

$$E \cap \mathbb{R} = \{x \in \mathbb{R} : P(x, 0) = 1\}$$

and since $P(x, 0) \in \mathcal{P}_{2n}^c$, the result follows. □

b] For $a \in \mathbb{C}$ and $r > 0$, let Q be the square

$$Q := \{z \in \mathbb{C} : |\text{Re}(z - a)| < r\,, \;\; |\text{Im}(z - a)| < r\}\,.$$

Show that $Q \cap \partial E$ is of length at most $8rn$.

Hint: Divide $Q \cap \partial E$ into subcurves $C_1, C_2, \ldots, C_m$ so that every vertical and horizontal line contains at most one point of each C_j. Let $l_x^{(j)}$ and $l_y^{(j)}$ denote the length of the interval obtained by projecting C_j to the x axes and y axes, respectively. Let

$$l_x := \sum_{j=1}^{m} l_x^{(j)} \qquad \text{and} \qquad l_y := \sum_{j=1}^{m} l_y^{(j)}\,.$$

Use part a] to show that $l_x \le 4nr$ and $l_y \le 4rn$. Hence, if l denotes the length of $Q \cap \partial E_n$, then

$$l \le l_x + l_y \le 8rn\,.$$

□

c] Show that the boundary ∂E of E is of length at most $16en$.

Hint: Combine Cartan's lemma (see E.7) and part b]. □

The rest of the exercise is about improving $16en$ to $4\pi en$.

d] Let B be the open disk with center $a \in \mathbb{C}$ and radius $r > 0$. Show that $B \cap \partial E$ is of length at most $2\pi rn$.

Outline. Let Q be the square

$$Q := \{z \in \mathbb{C} : |\mathrm{Re}(z)| < r\,, \;\; |\mathrm{Im}(z)| < r\}\,.$$

For $\alpha \in [0, 2\pi)$ let Q_α be the square

$$Q_\alpha := \{a + e^{i\alpha}(z - a) : z \in Q\}\,.$$

Let $l(\alpha)$ denote the length of $Q_\alpha \cap \partial E$. Let $l_x(\alpha)$ and $l_y(\alpha)$ denote the total length of the intervals obtained by projecting $Q_\alpha \cap \partial E$ to the lines

$$\left\{z = re^{i\alpha} : r \in \mathbb{R}\right\} \qquad \text{and} \qquad \left\{z = re^{i(\alpha+\pi/2)} : r \in \mathbb{R}\right\},$$

respectively (counting multiplicities). The precise definition of $l_x(\alpha)$ and $l_y(\alpha)$ can be formulated in the same way as in the hint to part b], which is left to the reader. Use part a] to show that

$$l_x(\alpha) + l_y(\alpha) \le 8rn\,, \qquad \alpha \in [0, 2\pi)\,.$$

Hence

$$\begin{aligned}\frac{4}{\pi}\, l(0) &= \frac{1}{2\pi}\int_0^{2\pi} l(0)(|\sin\alpha| + |\cos\alpha|)\, d\alpha \\ &= \frac{1}{2\pi}\int_0^{2\pi} (l_x(\alpha) + l_y(\alpha))\, d\alpha \le 8rn\,,\end{aligned}$$

and $l(0) \le 2\pi rn$ follows. □

e] Prove the initial statement of the exercise.

Hint: Combine Cartan's lemma (see E.6) and part d]. □

E.8 On the Length of Another Lemniscate. Suppose $p \in \mathcal{P}_n^c$. Show that the length of the lemniscate

$$F = F(p) := \left\{z \in \mathbb{C} : \left|\frac{p'(z)}{p(z)}\right| = n\right\}$$

is at most $16n(1 + \log n)$ (actually at most $4\pi n(1 + \log n)$).

Hint: The arguments are very similar to those given in the outline to E.7. First show that if L is an arbitrary line in the complex plane, then $F \cap L$ contains at most $2n$ distinct points. Next prove that if D_r is an open disk of radius r in the complex plane, then $D_r \cap F$ is of length at most $8rn$ (actually at most $2\pi rn$). Now use E.6 c] with $\beta = 1 + \log n$. □

The proof of the following exercise requires some familiarity with harmonic functions.

E.9 On the Diameter of $E(p)$. The *diameter* $\operatorname{diam}(A)$ of a nonempty set $A \subset \mathbb{C}$ is defined by

$$\operatorname{diam}(A) := \sup\{|z_1 - z_2| : z_1, z_2 \in A\}.$$

Let $p \in \mathcal{P}_n^c$ be an arbitrary monic polynomial of degree n, that is,

$$p(z) := \prod_{j=1}^{n} (z - z_j), \qquad z_j \in \mathbb{C}.$$

Then the diameter of the set

$$E(p) := \{z \in \mathbb{C} : |p(z)| \le 1\}$$

is at least 2.

Proceed as follows:

a] Let

$$\widehat{\mathbb{C}} := \mathbb{C} \cup \{\infty\} \qquad \text{and} \qquad \Delta := \{z \in \widehat{\mathbb{C}} : |z| > 1\}.$$

Let $p \in \mathcal{P}_n^c$ be monic. Let $E := \operatorname{co}(E(p))$, that is, the convex hull of $E(p)$. Use the Riemann mapping theorem (see Ahlfors [53]) to show that there exists a function g of the form

$$g(z) = bz + \sum_{j=0}^{\infty} b_j z^{-j}, \qquad z \in \Delta, \quad b, b_j \in \mathbb{C}$$

such that g is analytic and one-to-one on Δ, and

$$g(\Delta) = \widehat{\mathbb{C}} \setminus E.$$

Hint: Note that $\widehat{\mathbb{C}} \setminus E$ is simply connected. □

b] Let b be the same as in part a]. Show that $|b| \ge 1$.

Outline. Because of the definition of E, for every $\varepsilon > 0$ there exists a $\delta > 0$ such that

$$-\varepsilon \le \log|z^{-n}p(g(z))|, \qquad |z| = 1 + \delta.$$

Since $G(z) := \log|z^{-n}p(g(z))|$ is harmonic on Δ, we have

$$G(\infty) = \frac{1}{2\pi(1+\delta)} \int_0^{2\pi} G((1+\delta)e^{i\theta})\, d\theta \ge \frac{-\varepsilon}{2\pi(1+\delta)}$$

for every $\varepsilon > 0$, so $G(\infty) \ge 0$. On the other hand, since $p \in \mathcal{P}_n^c$ is a monic polynomial of degree n,

$$G(\infty) = \log |b|^n = n \log |b| ,$$

from which $|b| \geq 1$ follows. □

c] Show that $\operatorname{diam}(E) \geq 2$.

Outline. Assume to the contrary that $\operatorname{diam}(E) < 2$. Then there exists a $\delta > 0$ such that

$$|z^{-1}(g(z) - g(-z))| < 2, \qquad |z| = 1 + \delta .$$

Since

$$F(z) := z^{-1}(g(z) - g(-z)) = 2b + 2b_1 z^{-2} + 2b_3 z^{-4} + \cdots$$

is analytic on Δ, the maximum principle (E.1 d] of Section 1.2) yields

$$2|b| = |F(\infty)| \leq \max_{|z|=1+\delta} |F(z)| = \max_{|z|=1+\delta} |z^{-1}(g(z) - g(-z))| < 2,$$

that is, $|b| < 1$, which contradicts part b]. □

d] Note that $\operatorname{diam}(A) = \operatorname{diam}(\operatorname{co}(A))$ for every nonempty $A \subset \mathbb{C}$, in particular, $\operatorname{diam}(E(p)) = \operatorname{diam}(E)$.

The more general result that $\operatorname{diam}(A) \geq 2 \operatorname{cap}(A)$ for every nonempty $A \subset \mathbb{C}$ is observed in Pommerenke [75].

E.10 More on $E(p)$. Suppose p is a monic polynomial with complex coefficients. As before, let

$$E := E(p) := \{z \in \mathbb{C} : |p(z)| \leq 1\} .$$

a] It follows from E.6 (Cartan's lemma) that the set $E(p)$ can be covered by disks the sum of whose radii is at most $2e$. It is conjectured in Erdős, Herzog, and Piranian [58] that the correct value in this problem is 2. (The current best constant is less than 2.6.)

b] If $E(p)$ is connected, then it is contained in a disk with radius 2 centered at $\frac{1}{n}\sum_{k=1}^{n} z_k$, where $z_1, z_2, \ldots, z_n$ are the zeros of p.

Proof. This is conjectured in Erdős, Herzog, and Piranian [58] and proved in Pommerenke [59b]. □

c] If $E(p)$ is connected, then its circumference is at least 2π.

Proof. This is also conjectured in Erdős, Herzog, and Piranian [58], and proved in Pommerenke [59b]. □

A1

Algorithms and Computational Concerns

Overview

Appendix 1 presents some of the basic algorithms for computing with polynomials and rational functions and discusses some of the complexity issues. Included is a discussion of root finding methods. It requires very little background and can essentially be read independently.

Algorithms and Computational Concerns

Polynomials lend themselves to computation perhaps more than any other object of analysis. Algorithms that involve special functions, differential equations, series, and the like usually must reduce at some point to a finite polynomial or rational approximation or truncation. This often allows analytic problems to be reduced to algebraic ones. This appendix will present, as a series of exercises, some of the principal algorithmic concerns. The reader is encouraged to experiment with the algorithms. With current technology this is most comfortably done in any of the large symbolic manipulation packages available. Code, actual or schematic, is not presented. Indeed, methods rather than algorithms are presented. Current "practical" best methods date quickly in this rapidly evolving area. It is also the authors' belief that today's theoretical curiosities may be vital for tomorrow's algorithms as larger instances are calculated on faster machines. Historically we have already seen this happen repeatedly with algorithms such as the fast Fourier transform algorithm.

One of the most interesting lessons to be learned from the last few decades of revitalized interest in computational mathematics is that many of the most familiar mathematical algorithms, such as how to multiply large numbers, were very poorly understood, and indeed, still are incompletely analyzed. Very many of the familiar processes of mathematics, such as multiplication of large numbers or computation of determinants, can be computed far more expeditiously than allowed by the usual "school" algorithms. See, for example, Aho, Hopcroft, and Ullman [74]; Bini and Pan [92]; Borodin and Munro [75]; Borwein and Borwein [87]; Brent [74]; Knuth [81]; Pan [92]; Smale [85]; and Wilf [86] for the complexity side of the following exercises.

E.1 Complexity and Recursion. We are concerned with measuring the size of an algorithm given an input of a certain length. Unfortunately, there are many different ways of measuring this. (One could, for example, use the length of the tapes of some well-defined instantiation of a Turing machine.) We will settle for less. The measure of input size will usually be chosen to be a natural one, so for polynomials of degree n, the measure will often be n. The complexity measure then depends a bit on the problem. For example, it might count the number of additions of coefficients (we do not distinguish subtraction from addition) and multiplications of coefficients required to evaluate the polynomial at a point. (So, for example, by Horner's rule $O(n)$ operations suffice.) Care is already required to count naturally. Note that we have not specified the size of the coefficients (this may or may not be reasonable) and so we could cheat on addition of coefficients by doing two additions as one addition of twice the length. (Since $a+b$ and $c+d$ can be decoded from $(10^m a + c) + (10^m b + d)$, where m is larger than the number of digits in any of a, b, c, or d.) It is more reasonable in this context to fix a precision (or to think of working to infinite precision or over some polynomial ring). Our cases are fairly simple, and the measures should be clear in context.

We adopt the following notations:

$$f(n) = O(g(n)) \quad \text{means} \quad \limsup_{n\to\infty} \frac{f(n)}{g(n)} < \infty$$

and

$$f(n) = \Omega(g(n)) \quad \text{means} \quad \limsup_{n\to\infty} \frac{g(n)}{f(n)} < \infty .$$

Parts of these exercises are reprinted from Borwein and Borwein [87], with permission from Wiley.

So the first measure gives an upper bound, while the second gives a lower bound.

Many algorithms are analyzed recursively. For example, addition of two polynomials of degree at most $2n$ reduces to two additions of polynomials of degree at most n, plus perhaps an "overhead" for reassembling the pieces. In other words, for the complexity of addition, we have

$$A(2n) \le 2A(n) + c\,,$$

from which one can deduce that

$$A(n) = O(n)\,.$$

This general recursive strategy of breaking a problem in half is often called "divide and conquer."

We introduce the following functions. Here n is the maximum degree of the polynomials p and q. Additions, multiplications, and so on are performed in the underlying field of coefficients (in our case $\mathbb{C}$ or $\mathbb{R}$) and are all performed to some predetermined fixed precision (possibly infinite).

$$\begin{aligned}
A(n) &:= \text{the maximum number of } \pm\,,\ \times\,,\ \div\,,\ \text{ to compute } p \pm q\,;\\
M(n) &:= \text{the maximum number of } \pm\,,\ \times\,,\ \div\,,\ \text{ to compute } pq\,;\\
e(n) &:= \text{the maximum number of } \pm\,,\ \times\,,\ \div\,, \text{to evaluate}\\
&\qquad p(\alpha)\ \text{ for an arbitrary fixed } \alpha \in \mathbb{C}\,;\\
E(n) &:= \text{the maximum number of } \pm\,,\ \times\,,\ \div\,, \text{to evaluate}\\
&\qquad p(\alpha_1), \dots, p(\alpha_n)\ \text{ for arbitrary fixed } \alpha_1, \dots, \alpha_n \in \mathbb{C}\,.
\end{aligned}$$

These are the complexity functions for polynomial addition, polynomial multiplication, polynomial evaluation at a single point, and polynomial evaluation at n points, respectively. The input for the computation is the coefficients (and the evaluation points for $e(n)$ and $E(n)$). So the input may be considered to be in $\mathbb{C}^{n+1}$ (or more generally an $(n+1)$-dimensional vector space over an infinite field). In the first two cases the output is the sequence of coefficients. In the last two cases, respectively, the output is the evaluation and the sequence of evaluations.

a] Show that usual algorithms give

$$\begin{aligned}
A(n) &\le 2n + 2 = O(n)\,,\\
M(n) &= O(n^2)\,,\\
e(n) &= O(n) \qquad \text{(Horner's rule)}\,,\\
E(n) &= O(n^2)\,.
\end{aligned}$$

E.2 and E.3 of this appendix provide better upper bounds for the last three functions above.

b] Show that

$$\begin{aligned} A(n) &\geq n+1\,, \\ M(n) &\geq n+1\,, \\ e(n) &\geq n+1\,, \\ E(n) &\geq n+1\,. \end{aligned}$$

Hint: In all cases this is a uniqueness argument. At least one operation must be performed for each coefficient, otherwise the algorithm will not distinguish various different sequences of input. □

c] Some Recursive Bounds. Let $a, b > 0$ and $c, d > 1$. Suppose that f is monotone on $(0, \infty)$.

If $f(n) \leq af(n/c) + bn$, then

$$\begin{aligned} f(n) &= O(n) && \text{if } a < c\,, \\ f(n) &= O(n \log n) && \text{if } a = c\,, \\ f(n) &= O(n^{\log_c a}) && \text{if } a > c\,. \end{aligned}$$

If $f(n) \leq df(n/d) + bn(\log n)^{c-1}$, then

$$f(n) = O(n(\log n)^c)\,.$$

Hint: Analyze the equality case. Then establish the general principle that the equality solution is the maximal solution. □

E.2 The (Finite) Fast Fourier Transform (FFT). This is undoubtedly one of the most widely used algorithms. It has, in its various forms, tremendous practical and theoretical applications.

Let w be a primitive $(n+1)$th root of unity in either $\mathbb{C}$ or a finite field $\mathbb{F}_m$, that is, $w^{n+1} = 1$ and $w^k \neq 1$ for $k = 1, 2, \ldots, n$. In the complex case we may take $w := e^{2\pi i/(n+1)}$. Consider the following two problems.

Interpolation Problem. Given $n+1$ numbers, $\alpha_0, \alpha_1, \ldots, \alpha_n$, find the coefficients of the unique polynomial

$$p(z) := a_0 + a_1 z + \cdots + a_n z^n$$

of degree n that satisfies

$$p(w^k) = \alpha_k\,, \qquad k = 0, 1, \ldots, n\,.$$

Evaluation Problem. Given the coefficients of a polynomial p_n of degree at most n, calculate the $n+1$ values

$$p(w^k)\,, \qquad k = 0, 1, \ldots, n\,.$$

These are the two directions of the *(finite) Fourier transform*. The classical approaches to either part of the Fourier transform problem have complexity at least cn^2. This is the complexity, for example, of evaluating p_n at $n+1$ points using Horner's rule. Both directions can, however, be solved with complexity $O(n \log n)$.

a] If $n+1 = 2^m$, then both the interpolation and the evaluation problem have complexity $O(n \log n)$. (Here we are counting the number of additions and multiplications in the underlying coefficient field, which for most of our purposes is $\mathbb{C}$.)

Outline. We treat the evaluation first. Suppose

$$p(x) := a_0 + a_1 x + \cdots + a_n x^n .$$

Let

$$q(x^2) := a_0 + a_2 x^2 + a_4 x^4 + \cdots + a_{n-1} x^{n-1}$$

and

$$x\, r(x^2) := x(a_1 + a_3 x^2 + \cdots + a_n x^{n-1}) .$$

Then, with $y := x^2$,

$$p(x) = x\, r(y) + q(y) ,$$

where r and q are both polynomials of degree at most $2^{m-1} - 1$. The observation that makes the proof work is that for w an $(n+1)$th root of unity,

$$(w^k)^2 = \left(w^{(n+1)/2 + k}\right)^2 .$$

Hence, evaluating $p(x)$ at the $n+1$ roots of unity reduces to evaluating r and q each at the $\frac{1}{2}(n+1)$ points $(w^2)^1, (w^2)^2, \ldots, (w^2)^{(n+1)/2}$ and amalgamating the results. Observe that w^2 is a primitive (2^{m-1})th root of unity, so we can iterate this process. Let $F(2^m)$ be the number of additions and multiplications required to evaluate a polynomial of degree at most $2^m - 1$ at the 2^m points w^k, $k = 1, 2, \ldots, 2^m$, where w is a primitive (2^m)th root of unity. Then

$$F(1) = 1 \qquad \text{and} \qquad F(2^m) = 2F(2^{m-1}) + 2 \cdot 2^m , \qquad m = 1, 2, \ldots .$$

The second term comes from the single addition and multiplication required to calculate each $p(w^k)$ from $r(w^{2k})$ and $q(w^{2k})$. This recursion solves as

$$F(2^m) = 2^{m+1} m ,$$

and the bound for the evaluation problem is established.

The interpolation problem is equivalent to evaluation. This can be seen as follows. Let w be a primitive $(n+1)$th root of unity, and let

$$W := \begin{pmatrix} 1 & 1 & 1 & \dots & 1 \\ 1 & w & w^2 & \dots & w^n \\ 1 & w^2 & w^4 & \dots & w^{2n} \\ \vdots & \vdots & \vdots & \ddots & \vdots \\ 1 & w^n & w^{2n} & \dots & w^{n^2} \end{pmatrix}.$$

Then

$$W^{-1} = \frac{1}{n+1} \begin{pmatrix} 1 & 1 & 1 & \dots & 1 \\ 1 & w^{-1} & w^{-2} & \dots & w^{-n} \\ 1 & w^{-2} & w^{-4} & \dots & w^{-2n} \\ \vdots & \vdots & \vdots & \ddots & \vdots \\ 1 & w^{-n} & w^{-2n} & \dots & w^{-n^2} \end{pmatrix}$$

and w^{-1} is also a primitive $(n+1)$th root of unity. The interpolation problem can be formulated as follows. Find $(a_0, \dots, a_n)$ so that

$$W(a_0, a_1, \dots, a_n) = (\alpha_0, \alpha_1, \dots, \alpha_n).$$

However, this can be solved by

$$W^{-1}(\alpha_0, \alpha_1, \dots, \alpha_n) = (a_0, a_1, \dots, a_n),$$

which is exactly the evaluation problem. □

See Borodin and Munro [75] and Borwein and Borwein [87]. Versions of FFT exist in a plethora of shapes and sizes. We have just exposed the tip of the iceberg.

As an application we construct a fast polynomial multiplication.

b] Fast Polynomial Multiplication. Suppose the polynomials p, q of degree at most $n-1$ are given. Compute the coefficients of the product pq as follows:

b1] Use an FFT to evaluate p and q at the primitive $(2n)$th roots of unity $w^1, w^2, \dots, w^{2n}$.

b2] Form

$$p(w^k)q(w^k), \qquad k = 1, 2, \dots, 2n.$$

b3] Find the coefficients of the product pq by solving the interpolation problem once again by using the FFT. Show that this algorithm requires

$$O(n \log n)$$

additions, multiplications, and divisions (of complex numbers) and therefore $M(n) = O(n \log n)$.

This is the best-known upper bound on the serial complexity of polynomial multiplication. The only known lower bound is the trivial one $O(n)$. In parallel (on a PRAM) polynomial multiplication can be done in $O(\log n)$ time on $O(N)$ processors; see Pan [92]. The same bounds apply for the FFT in a].

E.3 Other Elementary Operations.

a] Fast Polynomial Division. For polynomials p of degree n and q of degree $m \le n$, it is possible to find polynomials u and r with $\deg(r) < \deg(q)$ such that

$$p(x) = u(x)q(x) + r(x)$$

in $O(n \log n)$ additions and multiplications (of complex numbers).

Outline. Simplify by observing that it suffices to calculate u since r may then be computed by E.2 b]. If we replace x by $1/x$ then

$$\frac{p(x^{-1})}{q(x^{-1})} = u(x^{-1}) + \frac{r(x^{-1})}{q(x^{-1})}$$

and so, for some $h \ge 1$,

$$\frac{p^*(x)}{q^*(x)} = u^*(x) + x^{n-m+h}\frac{r^*(x)}{q^*(x)}, \qquad \text{where} \qquad v^*(x) := x^{\deg(v)}v(x^{-1}).$$

To calculate u^* (and hence to calculate u) it suffices to calculate the first $n-m$ $(= \deg(u))$ Taylor coefficients of $1/q^*$. This can be done by Newton's method (see the next exercise) as follows: Suppose $\deg s_i = j-1$ and

$$\frac{1}{q^*(x)} - s_i(x) = O(x^j).$$

Establish that

$$\frac{1}{q^*(x)} - \left[2s_i(x) - s_i^2(x)q^*(x)\right] = \frac{1}{q^*(x)}\left[1 - s_i(x)q^*(x)\right]^2 = O(x^{2j}).$$

Note that we may assume $q^*(0) \neq 0$. Now the computation of

$$s_{i+1} := 2s_i - s_i^2 q^*$$

can be performed by using an FFT-based polynomial multiplication and it needs only be performed by using the first $2j-1$ coefficients of q^* and s_i. By starting with an appropriate first estimate of s_0 (say, $s_0(x) := 1/q^*(0)$), and proceeding inductively as above (doubling the number of coefficients used at each stage) we can show that the required number of terms of the expansion can be calculated in $O(n \log n)$ additions and multiplications (of complex numbers).

b] Fast Reversion of Power Series. Let

$$f(x) = \sum_{k=0}^{\infty} a_k x^k, \qquad a_0 \neq 0$$

be a formal power series with given coefficients. Show that the first n coefficients of the formal Taylor expansion of $1/f$ can be computed in $O(n \log n)$ additions, multiplications, and divisions (of complex numbers).

c] **Fast Polynomial Expansion.** Given $\alpha_1, \alpha_2, \ldots, \alpha_n$ show that the coefficients of $\prod_{i=1}^{n}(x - \alpha_i)$ can all be calculated in $O(n(\log n)^2)$ additions, multiplications, and divisions (of complex numbers).

Hint: Proceed recursively by dividing the problem into two parts of roughly half the size. Recombine the pieces using part b] of the previous exercise. □

d] **Fast Polynomial Expansion at Arbitrary Points.** Given a polynomial p of degree at most n, and $n+1$ distinct points $x_0, x_1, \ldots, x_n$, show that $p(x_0), p(x_1), \ldots, p(x_n)$ can all be evaluated in $O(n(\log n)^2)$ multiplications and additions.

Hint: Let

$$q_1(x) := \prod_{i=0}^{\lfloor n/2 \rfloor - 1} (x - x_i)$$

and let r_1 be the remainder on dividing p by q_1. Note that $r_1(x_i) = p(x_i)$ for each $i < n/2$. Similarly, use

$$q_2(x) := \prod_{\lfloor n/2 \rfloor}^{n} (x - x_i).$$

Thus two divisions reduce the problem to two problems of half the size. □

e] Extend d] to rational functions.

f] **Evaluation of x^n.** The S-and-X binary method for calculating x^n is the following algorithm. Suppose n has binary representation $\delta_0 \delta_1 \delta_2 \ldots \delta_k$ with $\delta_0 = 1$. Given symbols S and X, define

$$S_i := \begin{cases} SX & \text{if } \delta_i = 1 \\ S & \text{if } \delta_i = 0 \end{cases}$$

and construct the rule

$$S_1 S_2 \cdots S_k \,.$$

Now let S be the operation of squaring, and let X be the operation of multiplying by x. Let $S_1 S_2 \cdots S_k$ operate from left to right beginning with x. For example, for $n = 27$,

$$\delta_0 \delta_1 \delta_2 \delta_3 \delta_4 = 11011$$

and

$$S_1 S_2 S_3 S_4 = (SX)(S)(SX)(SX)\,.$$

The sequence of calculations of x^{27} is then

$$x \to x^2 \to x^3 \to x^6 \to x^{12} \to x^{13} \to x^{26} \to x^{27}.$$

f1] Prove that the above method computes x^n and observe that it only requires storing x, n, and one partial product.

f2] Show that the number of multiplications in the S-and-X method is less than $2\lfloor \log_2 n \rfloor$.

f3] Show that the S-and-X method is optimal for computation of x^{2m} (considering only multiplications).

f4] Show that the S-and-X method is not optimal for computing x^{15}.

An extended discussion of this interesting and old problem is presented in Knuth [81].

Much further material on the complexity of polynomial operations and complexity generally may be found in Aho, Hopcroft, and Ulman [74]; Borodin and Munro [75]; Pan [92]; and Wilf [86].

E.4 Newton's Method. One of the very useful algorithms, both in theory and practice, for zero finding is Newton's method.

a] Suppose f is analytic in a (complex) neighborhood of z_0, and suppose $f(z_0) = 0$ and $f'(z_0) \neq 0$. Show that the iteration

$$x_{n+1} := x_n - \frac{f(x_n)}{f'(x_n)}$$

converges locally uniformly quadratically, that is, with a constant c independent of n,

$$|x_{n+1} - z_0| < c\,|x_n - z_0|^2$$

for initial values x_0 in some neighborhood of z_0. As before, we call this *locally quadratic convergence.*

Hint: Note that

$$f(x_n) = f(z_0) + (x_n - z_0)f'(z_0) + O((x_n - z_0)^2),$$

which implies

$$x_{n+1} - z_0 = (x_n - z_0)\left[\frac{f'(x_n) - f'(z_0)}{f'(x_n)}\right] + O((x_n - z_0)^2).$$

□

b] Show that the iteration

$$x_{n+1} := x_n + \frac{f(x_n)}{f'(x_n)}$$

converges locally quadratically to the simple poles of a meromorphic function f. Note that this is Newton's method with the sign changed.

c] The iteration

$$x_{n+1} := x_n - \frac{f(x_n)f'(x_n)}{f'(x_n)^2 - f(x_n)f''(x_n)}$$

converges locally quadratically to a zero of an analytic f independent of its multiplicity; see Henrici [74].

d] Let $g := f^{-1}$. The iteration

$$x_{n+1} := x_n + (n+1)\,\frac{g^{(n)}(x_n)}{g^{(n+1)}(x_n)}$$

converges locally uniformly to a zero of an analytic function f with $(n+2)$th order. Newton's method is $n := 0$; Halley's method is $n := 1$ (see Householder [70]).

Newton's method and its variants work tremendously well provided that a good starting value can be found. This is a problem. On a real interval a bisection method can be used initially to localize the zeros. In the plane, life is more complicated as is seen in the next exercise. Another drawback to Newton's method is the need to compute the derivative. Of course, this is not a problem for polynomials, but in a general setting it is usually replaced by an approximation such as

$$\frac{f(x_n) - f(x_{n-1})}{x_n - x_{n-1}}$$

(which yields the so-called secant method).

e] Consider Newton's method for computing $\sqrt{x}$ starting with $x_0 := 1$. This gives

$$x_{n+1} := \frac{1}{2}\left(x_n + \frac{x}{x_n}\right).$$

Show that $r_n(x) := x_{n+1}$ is a rational function in x with numerator of degree 2^n and denominator of degree $2^n - 1$. Show that $r_n(x) - \sqrt{x}$ has a zero of order 2^{n+1} at 1. So r_n is in fact the $(2^n, 2^n - 1)$ Padé approximant to $\sqrt{x}$. (This implies, though not obviously, that r_n has all real negative roots and poles.)

An attractive feature of Newton's method is that it is "self-correcting." So, for example, to compute a root to a large precision, one can start at a small precision and double the precision at each iteration. This is a substantial savings both practically and theoretically. The same feature applies to Newton's method solutions over formal power series, as in E.3 a]. This allows for doubling the number of terms used at each stage. Much additional material on Newton's method is available in Borwein and Borwein [87], Henrici [74], Householder [70], and Traub [82].

E.5 Newton's Method in Many Variables.

a] Let $f : \mathbb{C}^n \to \mathbb{C}^n$, and suppose f has Jacobian

$$J := \left| \frac{\partial f_i}{\partial z_j} \right| ,$$

where $f := (f_1, f_2, \dots, f_n)$ with $f_i : \mathbb{C}^n \to \mathbb{C}$. Let

$$\mathbf{x} := (x_1, x_2, \dots, x_n) \in \mathbb{C}^n .$$

The function $J(\mathbf{x})$ is the Jacobian evaluated at $\mathbf{x}$. Then Newton's method becomes

$$\mathbf{x}_{k+1} = \mathbf{x}_k - \mathbf{s}_k$$

where $\mathbf{s}_k$ solves $J(\mathbf{x}_k)\mathbf{s}_k = f(\mathbf{x}_k)$. This iteration converges locally uniformly quadratically to a zero $\mathbf{z_0}$ of f, that is, with a constant c independent of n,

$$|\mathbf{x}_{n+1} - \mathbf{z}_0| < c\,|\mathbf{x}_n - \mathbf{z}_0|^2$$

for initial values $\mathbf{x_0}$ in some neighborhood of $\mathbf{z_0}$, provided in a neighborhood of $\mathbf{z}_0$, f is continuously differentiable, J^{-1} exists and is bounded in norm, and J satisfies a Lipschitz condition. We call this *locally quadratic convergence.* For a polynomial f, we require only that J^{-1} exists in a neighborhood of the zero $\mathbf{z}_0$. (For examples and detail, see Dennis and Schnabel [83].)

b] **Finding All Zeros of a Polynomial.** Let

$$p(z) := a_0 + a_1 z + \cdots + a_n z^n , \qquad a_n := 1 .$$

Let $f_i(x_1, \dots, x_n)$ be the ith coefficient of

$$g(x) := \prod_{i=1}^{n} (x - x_i) - \sum_{i=0}^{n} a_i x^i$$

and let

$$f(\mathbf{x}) := (f_1, f_2, \dots, f_n) .$$

Then the iteration of part a] applied to f converges locally quadratically to $\mathbf{z}_0 := (z_1, z_2, \ldots, z_n)$, where $z_1, z_2, \ldots, z_n$ are the zeros of p, provided the zeros of p are distinct.

c] **Another Approach to Finding All Zeros.** Let $x_j(k)$ denote the kth approximation to the jth root of p, where p is a polynomial of degree n with n distinct roots. Let

$$x_j(k+1) = x_j(k) - \frac{p(x_j(k))}{\prod_{i \neq j}^{k} (x_j(k) - x_i(x))}, \qquad j = 1, 2, \ldots, n.$$

Show that for sufficiently good choice of $(x_0(1), \ldots, x_0(k))$, the above iteration converges locally quadratically to a sequence of the n distinct zeros of p.

Hint: This is really just the single variable version of Newton's method for each root, where the derivative is approximated by the derivative of the kth estimate. □

d] Observe that the iteration of part c] fails to converge for $p(x) := x^n - 1$, $n > 2$, if the starting values are all taken to be real.

In practice, the iteration of part c] works rather well for reasonably chosen starting values. One can use the techniques of the next exercise to localize the zeros first. With care, an algorithm can be given that computes a zero of a polynomial with an error $< 2^{-b}$ in $O(n \log b \log n)$ time and all zeros in $O(n^2 \log b \log n)$ time; see Pan [92]. On a parallel machine (PRAM) an algorithm requiring $O(\log^3(nb))$ time and $O(nb)^{O(1)}$ processors can be given for computing all zeros.

e] Modify the iteration of E.4 c] as given in the previous exercise to get a method that computes all roots even in the presence of repeated roots.

For further discussion, see Aberth [73], Durand [60], Kerner [66], and Werner [82].

E.6 Localizing Zeros.

a] **Cauchy Indices.** Let r be a real rational function with a real pole α. The *Cauchy index* of r at α is

$$\begin{cases} 1 & \text{if } \lim_{x\to\alpha-} r(x) = -\infty \text{ and } \lim_{x\to\alpha+} r(x) = \infty \\ -1 & \text{if } \lim_{x\to\alpha-} r(x) = \infty \text{ and } \lim_{x\to\alpha+} r(x) = -\infty \\ 0 & \text{otherwise}. \end{cases}$$

The Cauchy index of r on an interval $[a, b]$ is the sum of the Cauchy indices of the poles of r in (a, b). (We demand that neither a nor b be poles of r.) We denote this by $I_a^b(r)$.

b] **The Euclidean Algorithm.** Let p_0 and p_1 be nonzero polynomials. Define polynomials $p_0, p_1, \ldots, p_m$ and $q_1, q_2, \ldots, q_m$ (by the usual division algorithm) so that

$$\begin{aligned}
p_0(z) &= p_1(z)q_1(z) - p_2(z)\,, &\qquad \deg(p_2) < \deg(p_1)\,;\\
p_1(z) &= p_2(z)q_2(z) - p_3(z)\,, &\qquad \deg(p_3) < \deg(p_2)\,;\\
&\vdots\\
p_{m-1}(z) &= p_m(z)q_m(z) - 0\,;
\end{aligned}$$

the algorithm stops the first time that the remainder is zero. This is called the *Euclidean algorithm.* Show that p_m is the greatest common factor of p_0 and p_1.

c] Let $p_0, p_1, \dots, p_m$ be the polynomials generated by the Euclidean algorithm in part b]. Let $P_i := p_i/p_m,\ \ i = 0, 1, \dots, m$. Suppose $p_0(a)p_0(b) \neq 0$. Show that if $P_k(\gamma) = 0$ for some k and $a \leq \gamma \leq \beta$, then

$$P_1(\gamma) \neq 0 \quad \text{if} \quad k = 0$$

and

$$P_{k-1}(\gamma)P_{k+1}(\gamma) < 0 \quad \text{if} \quad 1 \leq k \leq m-1\,.$$

d] Show that for real polynomials p_0 and p_1 with $p_0(a)p_0(b) \neq 0$,

$$I_a^b(p_1/p_0) = v(a) - v(b)\,, \qquad a < b\,,$$

where $v(\alpha)$ is the number of sign changes in the sequence

$$(p_0(\alpha), p_1(\alpha), \dots, p_m(\alpha))$$

and where the polynomials p_i are generated by the Euclidean algorithm as in part b]. (As before, by a sign change we mean that $p_i(\alpha)p_{i+k}(\alpha) < 0$ and $p_{i+1}(\alpha) = p_{i+2}(\alpha) = \cdots = p_{i+k-1}(\alpha) = 0$.)

Hint: Without loss of generality, we may assume that $p_m = 1$ (why?), so $P_k = p_k$ for each k. First, note that $v(x)$ may change magnitude only if $p_i(x) = 0$ for some i. By continuity of the polynomials p_i, and by part c], $v(x)$ is constant on any subinterval of $[a, b]$ that does not contain a zero of p_0.

We are now reduced to considering the behavior of $v(x)$ at the zeros of p_0. Consider the various possibilities for the behavior of p_1/p_0 at the zeros of p_0 by considering the four possible changes of signs of p_0 and p_1 at the zeros of p_0 and the effect this has on the Cauchy index and the increase and decrease of $v(x)$. (Note that $v(x)$ decreases by 1 if the Cauchy index is 1 while $v(x)$ increases by 1 if the Cauchy index is -1.) □

e] **Zeros on an Interval.** Suppose p is a real polynomial and $p(a)p(b) \neq 0$. Then $I_a^b(p'/p)$ equals the number of distinct zeros of p in $[a, b]$. This also equals $v(a) - v(b)$, where $v(x)$ is as in d] with $p_0 := p$ and $p_1 := p'$.

Hint: Use

$$\frac{p'(z)}{p(z)} = \sum \frac{m_k}{z - \zeta_k}$$

and note that each distinct pole in (a, b) contributes $+1$ to the Cauchy index. □

f] Show that the number of zeros of a real polynomial on the real line equals

$$\lim_{a \to \infty} (v(-a) - v(a)),$$

where $v(x)$ is computed as in part e]. (That is, $v(x)$ is the number of sign changes in the sequence $(p_0, p_1, \ldots, p_m)$ generated by the Euclidean algorithm with $p_0 := p$ and $p_1 := p'$.)

g] The Number of Zeros in $H := \{z \in \mathbb{C} : \mathrm{Im}(z) > 0\}$. Suppose that the monic polynomial $p \in \mathcal{P}_n^c$ has (exactly) k real zeros counting multiplicities. Write

$$p(x) = r(x) + i\, s(x), \qquad r, s \in \mathcal{P}_n.$$

Then the number of zeros of p in H is

$$\frac{1}{2}\left[n - k - I_{-\infty}^{\infty}(s/r)\right],$$

and this can be computed as in part d] by using

$$I_{-\infty}^{\infty}(s/r) = \lim_{\alpha \to \infty} I_{-\alpha}^{\alpha}(s/r).$$

Hint: First consider the case where $k = 0$. Consider p on a counterclockwise semicircular contour with base $[-\alpha, \alpha]$ and radius α. Consider the argument of p as the contour is traversed. On the half-circle, for large α, the argument increases by $n\pi$ asymptotically; while on the axis the change is $-\pi I_{-\alpha}^{\alpha}(s/r)$, from which the result follows. □

h] Budan-Fourier Theorem. *Let $p \in \mathcal{P}_n$. Let $V(x)$ be the number of sign changes in the sequence*

$$\left(p(x), p'(x), \ldots, p^{(n)}(x)\right).$$

Then the number of zeros of p in the interval $[a, b]$, counting multiplicities, is $V(a) - V(b) - 2m$ for some nonnegative integer m.

We have followed Henrici [74] in this discussion.

E.7 Zeros in a Disk.

a] The transform $\dfrac{z+i}{1+iz}$ maps the unit circle to the real axis and maps the open unit disk D to the upper half-plane $\{z \in \mathbb{C} : \mathrm{Im}(z) > 0\}$. So the algorithms of the previous exercise apply after transformation.

b] Zero Counting by Winding Number. Let $\omega_m := e^{2\pi i/m}$, where m is a positive integer. If $p \in \mathcal{P}_n^c$ then

$$z_p := \frac{1}{2\pi i}\int_{\partial D}\frac{p'(z)}{p(z)}\,dz \sim \sum_{k=1}^{m}\omega_m^k\frac{p'(\omega_m^k)}{p(\omega_m^k)}$$

counts the number of zeros of p in the open unit disk D (assuming no zeros on the boundary ∂D). More precisely, we have

$$z_p = \lim_{m\to\infty}\sum_{k=1}^{m}\omega_m^k\frac{p'(\omega_m^k)}{p(\omega_m^k)}\,.$$

(Note that this lends itself to rapid evaluation by FFT methods.)

c] Show that

$$\frac{1}{2\pi i}\int_{\partial D}\frac{1}{z-\alpha} = \sum_{k=1}^{m}\frac{\omega_m^k}{\omega_m^k-\alpha} + \epsilon_m(\alpha)\,,$$

where

$$\epsilon_m(\alpha) \le \frac{|\alpha|^{m-1}}{1-|\alpha|} \qquad \text{if} \quad |\alpha| < 1$$

and

$$\epsilon_m(\alpha) \le \frac{|\alpha|^{-m-1}}{1-|\alpha|^{-1}} \qquad \text{if} \quad |\alpha| > 1\,.$$

Hint: Write

$$\frac{1}{z-\alpha} = -\frac{1}{\alpha}\sum_{k=0}^{\infty}\frac{z^k}{\alpha^k}$$

and use the fact that

$$0 = \int_{\partial D} p(z)\,dz = \sum_{k=1}^{m}\omega_m^k p(\omega_m^k)$$

for every $p \in \mathcal{P}_{m-1}^c$. □

d] If $p \in \mathcal{P}_n^c$ has no zeros in the annulus $\zeta \le z \le 1/\zeta$ and if m in part b] is greater than

$$\left\lfloor \frac{\log\frac{1-\zeta}{n}}{\log\zeta}\right\rfloor + 1\,,$$

then the error in estimating z_p by the sum is less than 1. So this provides an algorithm.

E.8 Computing General Chebyshev Polynomials. Given a Chebyshev system $\mathcal{M} := (f_0, \ldots, f_n)$ of C^1 functions on, say, $[0,1]$ how does one compute the associated Chebyshev polynomial T_n? That is, how does one find the unique equioscillating form $T := \sum_{i=0}^n a_i f_i$ of Section 3.3?

a] The Remez Algorithm

Step 1. Choose

$$\mathbf{x}_0 := (0 =: x_0^{(0)} < x_1^{(0)} < \cdots < x_n^{(0)} := 1)$$

and find $P_0 \in \operatorname{span} \mathcal{M}$ such that

$$P_0(x_i^{(0)}) = (-1)^i \,, \qquad i = 0, 1, \ldots, n \,.$$

Step 2. Inductively set $\mathbf{x}_m := (0 =: x_0^{(m)} < x_1^{(m)} < \cdots < x_n^{(m)} := 1)$, where

$$P'_{m-1}(x_i^{(m)}) = 0 \,, \qquad i = 1, 2, \ldots, n-1 \,.$$

(That is, find the extrema of P_{m-1}.)

Step 3. Find $P_m \in \operatorname{span} \mathcal{M}$ with

$$P_m(x_i^{(m)}) = (-1)^i \,, \quad i = 0, 1, \ldots, n \,.$$

Then, provided the initial estimate $\mathbf{x}_0$ is sufficiently good, $P_m \to T_n$ quadratically (see Veidinger [60]).

This is reasonably easy to code. It involves solving an interpolation problem in Steps 1 and 3. The zero finding at Step 2 can be done quite easily since one can find very good starting values for Newton's method, namely $(x_i^{(m-1)})_{i=0}^n$.

b] This algorithm modifies to solve the best approximation problem

$$\min_{p \in \operatorname{span} \mathcal{M}} \|\omega p - f\|_{[a,b]}$$

for $\omega, f \in C[a,b]$, where ω is positive on $[a,b]$. One uses the Remez agorithm to find an equioscillating form

$$P(x) = f(x) - \omega(x) \sum_{i=0}^n c_i f_i(x)$$

at $n+2$ points. To do this, one solves the system

$$f(x_k) - \omega(x_k) \sum_{i=0}^n c_i f_i(x_k) + (-1)^k \lambda \,, \qquad k = 0, 1, \ldots, n+1$$

for both the c_i and λ. (This works reasonably well, provided that at each stage $\|P\|_{[a,b]}$ occurs at one of the x_k. If not, an extra point must be inserted where the maximum norm occurs and one of the original points must be dropped. This is effected in such a way as to maintain the alternations in sign of the error.) For details see Cody, Fraser, and Hart [68] and Veidinger [60].

A2

Orthogonality and Irrationality

Overview

This appendix is an application of orthogonalization of particular Müntz systems to the proof of the irrationality of $\zeta(3)$ and some other familiar numbers. It reproduces Apéry's remarkable proof of the irrationality of $\zeta(3)$ in the context of orthogonal systems.

Orthogonality and Irrationality

Apéry's wonderful proof of the irrationality of $\zeta(3)$ amounts to showing that

$$0 < |d_n^3 a_n \zeta(3) - b_n| \to 0\,,$$

where b_n is an integer,

$$a_n := \sum_{k=0}^{n} \binom{n+k}{k}^2 \binom{n}{k}^2\,,$$

and

$$d_n := \operatorname{lcm}\{1, 2, \ldots, n\}\,.$$

Here lcm denotes the least common multiple; see van der Poorten [79] and Beukers [79]. In [81] Beukers recast the proof using Padé approximations.

Many, maybe most, irrationality proofs may be based on approximation by Padé approximants and related orthogonal polynomials; see, for example, Borwein [91a], [92] or Chudnovsky and Chudnovsky [84]. It is the intention of this appendix to try to put the proof of the irrationality of $\zeta(3)$ into the framework of orthogonality. From the general orthogonalization of the system

$$(x^{\lambda_0}, x^{\lambda_1}, x^{\lambda_2}, \dots)$$

on $[0,1]$, where the numbers λ_j are nonnegative and distinct, specializing to the case when

$$\lambda_{2j} = j \qquad \text{and} \qquad \lambda_{2j+1} = j + \epsilon\,, \qquad j = 0, 1, \dots$$

where ϵ decreases to 0, is a very natural thing to do. This is how one should interpret orthogonalizing the system $(x^0, x^0, x^1, x^1, \dots)$. This leads to orthogonal functions that generalize the Legendre polynomials and are of the form

$$p_n(x) \log x + q_n(x)$$

with polynomials $p_n, q_n \in \mathcal{P}_n$ of degree n. Legendre polynomials are closely tied to irrationality questions concerning log (see Borwein and Borwein [87], Chapter 11), and higher-order analogs prove to be the basis of dealing with the irrationality of the trilog $\left(\sum_{n=1}^{\infty} x^n/n^3\right)$ for some values of x. We think that the proof of the irrationality of $\zeta(3)$ flows quite naturally from this point of view. Although in the end (Lemma A.2.3) we get back to Beukers' integral approach to the irrationality of $\zeta(3)$ (as indeed we must). What follows, up to one application of the prime number theorem, is a self-contained proof of the irrationality of $\zeta(3)$.

The orthogonalization in question is the content of the first theorem.

Theorem A.2.1. *Let*

$$G_n(x) := \frac{1}{2\pi i} \int_\Gamma \frac{\prod_{k=0}^{n-1} (t+k+1)^2}{\prod_{k=0}^{n} (t-k)^2} (t+n+1) x^t \, dt\,,$$

where Γ is any simple contour containing $t = 0, 1, \dots, n$. Then

$$G_n(x) = p_n(x) \log x + q_n(x)\,,$$

where

$$p_n(x) = \sum_{k=0}^{n} \binom{n}{k}^2 \binom{n+k}{k}^2 (n+k+1)\, x^k\,.$$

Furthermore, we have the orthogonality relations

$$\int_0^1 G_n(x) x^k \, dx = 0\,, \qquad k = 0, 1, \dots, n\,;$$

$$\int_0^1 G_n(x) (\log x) x^k \, dx = 0\,, \qquad k = 0, 1, \dots, n-1\,;$$

and

$$\int_0^1 G_n^2(x)\,dx = \frac{1}{2n+1}\,.$$

Proof. As in Section 3.4, the representation of G_n is just the evaluation of the integral at the poles, $t = 0, 1, \ldots, n$, by the residue theorem. The proof of the orthogonality conditions is a straightforward exercise on evaluating

$$\int_0^1 x^k (\log x)^j G_n(x)\,dx$$

by interchanging the order of integration as in the proof of Theorem 3.4.3. □

We need to modify these forms marginally to give a zero at 1, as in the next result.

Theorem A.2.2. *Let G_n and Γ be defined as in Theorem A.2.1. Then*

$$\begin{aligned} F_n(x) &:= \frac{1}{x^{n+1}} \int_0^x y^n G_n(y)\,dy \\ &= \frac{1}{2\pi i} \int_\Gamma \frac{\prod_{k=0}^{n-1}(t+k+1)^2}{\prod_{k=0}^{n}(t-k)^2} x^t\,dt = A_n(x)\log x + B_n(x)\,, \end{aligned}$$

where

$$A_n(x) := \sum_{k=0}^{n} \binom{n+k}{k}^2 \binom{n}{k}^2 x^k$$

and

$$B_n(x) := \sum_{k=0}^{n} c_k x^k$$

with

$$c_k := \binom{n+k}{k}^2 \binom{n}{k}^2 \left\{ \sum_{i=0}^{n-1} \frac{2}{k+i+1} - \sum_{\substack{i=0 \\ i\neq k}}^{n} \frac{2}{k-i} \right\}.$$

Furthermore,

$$\int_0^1 F_n(x)F_m(x)\,dx = \frac{2\delta_{n,m}}{(2n+1)^3}\,, \qquad \delta_{n,m} := \begin{cases} 0 & \text{if } n \neq m \\ 1 & \text{if } n = m\,; \end{cases}$$

$$\int_0^1 F_n(x)x^k(\log x)^j\,dx = 0\,, \qquad k = 0, 1, \ldots, n-1\,, \quad j = 0, 1\,;$$

and

$$F_n(1) = 0\,, \qquad F_n'(1) = 1\,, \qquad F_n''(1) = 2n^2 + 2n - 1\,.$$

Proof. This follows much as in Theorem A.2.1. The fact that $F_n(1) = 0$ is just the statement that G_n is orthogonal to x^n. □

The fact that $F_n(1) = 0$ is critical in what follows. It immediately gives that $B_n(x)$ has a zero at 1, so $B_n(x)/(1-x)$ is a polynomial, as in part c] of the following corollary. (This is the only part of the corollary we need, but the corollary is of some interest in its own right.)

Corollary A.2.3. *Let F_n be defined as in Theorem A.2.2. Then*

a] *F_n has $2n+1$ zeros on $(0,1]$.*

b] *$A_n(x)$ has all real negative zeros.*

c] *$B_n(x)/(1-x)$ is a polynomial with all real negative zeros that interlace the zeros of $A_n(x)$.*

Proof. The orthogonality conditions give $2n$ zeros of F_n on $(0,1)$ in a standard fashion, and there is one zero at 1. The real negative zeros of $A_n(x)$ and $B_n(x)$ and their interlacing follow from the fact that

$$\log x + \frac{B_n(x)}{A_n(x)}$$

has $2n+1$ zeros on $(0,1]$, and known results on interpolating Stieltjes transforms by rational functions (see, for example, Baker and Graves Morris [81] or Borwein [83]). □

One can, from the integral representation, deduce the next corollary, which is also not actually needed in the proof of the irrationality of $\zeta(3)$.

Corollary A.2.4. *Both F_n and A_n satisfy*

$$x^2(y_n'' - y_{n-1}'') - 2nx(y_n' + y_{n-1}') + x(y_n' - y_{n-1}') + n^2(y_n - y_{n-1}) = 0\,.$$

From Theorem A.2.2 we obtain (see E.2 b]) the following:

Corollary A.2.5. *Let $d_n := \operatorname{lcm}\{1, 2, \ldots, n\}$. Then the polynomial $d_n B_n(x)$ has integer coefficients.*

We get an approximation to $\zeta(3)$ by integration over the unit square.

Theorem A.2.6.

$$-\frac{1}{2}\int_0^1\int_0^1 \frac{F_n(xy)}{1-xy}\,dx\,dy = A_n(1)\zeta(3) + R_n\,,$$

where $2d_n^3 R_n$ is an integer.

Proof. Recall that

$$-\frac{1}{2}\int_0^1\int_0^1 \frac{\log(xy)}{1-xy} = \zeta(3)$$

since

$$\int_0^1\int_0^1 (xy)^n \log(xy)\, dx\, dy = \frac{-2}{(n+1)^3}\,.$$

We have

$$\frac{F_n(xy)}{1-xy} = \frac{A_n(xy)-A_n(1)}{1-xy}\log(xy) + \frac{A_n(1)}{1-xy}\log(xy) + \frac{B_n(xy)}{1-xy}\,.$$

The first term of the last expression is a polynomial multiple of $\log(xy)$, and the last term is a polynomial. Both have degree $n-1$ in xy. Here we use part c] of Corollary A.2.3 in an essential way. Integrating the above equation with respect to x on $[0,1]$ and with respect to y on $[0,1]$, we get the identity of the theorem. The fact that $d_n^3 R_n$ is an integer can be seen as follows. One d_n arises from each of the two integrations of a polynomial of degree $n-1$ with integer coefficients and one d_n comes from Corollary A.2.5. □

Theorem A.2.7. *The number $\zeta(3)$ is irrational.*

Proof. It now suffices to show that there is an $\epsilon > 0$ for which

$$0 < \left|\int_0^1\int_0^1 \frac{F_n(xy)}{1-xy}\, dx\, dy\right| = O\left(\frac{1}{e^{(3+\epsilon)n}}\right) = o\left(\frac{1}{d_n^3}\right)$$

since by the prime number theorem, $\operatorname{lcm}\{1,2,\ldots,n\} = O(e^{(1+\epsilon)n})$ for every $\epsilon > 0$; see Borwein and Borwein [87, p. 377]. (We use the notation $b_n = o(a_n)$ if $b_n = \epsilon_n a_n$ with $\lim_{n\to\infty}\epsilon_n = 0$.) This can be proven in a number of ways, we chose to connect this proof via Padé approximants to the integral estimate due to Beukers. This is the content of the following results and, in particular, Lemma A.2.10. From Lemma A.2.10, the above estimates are easy since the integrand in the right-hand side in Lemma A.2.10 satisfies

$$0 < \frac{xyv(1-x)(1-y)(1-v)}{1-(1-xy)v} \le (\sqrt{2}-1)^4$$

on the open unit cube $0 < x, y, v < 1$. □

The following lemma gives standard representations for the Padé approximants to log and can be checked by expanding the integrals; see E.3.

Lemma A.2.8 (Padé Approximants). *We have*

$$\begin{aligned}\frac{(n!)^2}{2\pi i}\int_\Gamma \frac{x^t\,dt}{\prod_{k=0}^n (t-k)^2} &= (x-1)^{2n+1}\int_0^1 \frac{v^n(1-v)^n\,dv}{(1-(1-x)v)^{n+1}}\\ &= p_n(x)\log x + q_n(x) = O((x-1)^{2n+1}),\end{aligned}$$

where Γ is a simple contour containing $t = 0, 1, \ldots, n$, and p_n and q_n are polynomials of degree n. (So p_n/q_n is the (n,n) Padé approximant to $\log x$ at 1.)

Lemma A.2.9 (Rodrigues-Type Formula). *With Γ as in Theorem A.2.1,*

$$\begin{aligned}F_n(xy) &= \frac{d^n}{dy^n}\frac{d^n}{dx^n}\frac{x^n y^n}{2\pi i}\int_\Gamma \frac{(xy)^t\,dt}{\prod_{k=0}^n (t-k)^2}\\ &= \frac{1}{(n!)^2}\frac{d^n}{dy^n}\frac{d^n}{dx^n}\int_0^1 \frac{(xy)^n v^n (1-v)^n (xy-1)^{2n+1}}{(1-(1-xy)v)^{n+1}}\,dv.\end{aligned}$$

Proof. This follows from differentiating the two representations derived from Lemma A.2.8 coupled with Theorem A.2.2. □

Lemma A.2.10. *We have*

$$\begin{aligned}&\int_0^1\int_0^1 \frac{F_n(xy)}{1-xy}\,dx\,dy\\ &\qquad = -\int_0^1\int_0^1\int_0^1 \frac{[xyv(1-x)(1-y)(1-v)]^n\,dx\,dy\,dv}{(1-(1-xy)v)^{n+1}}.\end{aligned}$$

Proof. For k, n nonnegative integers

$$\begin{aligned}&\int_0^1\int_0^1 (xy)^{n+k}(1-x)^n(1-y)^n\,dx\,dy\\ &\qquad = -\frac{1}{(n!)^2}\int_0^1\int_0^1 \frac{1}{1-xy}\frac{d^n}{dy^n}\frac{d^n}{dx^n}(xy)^{n+k}(xy-1)^{2n+1}\,dx\,dy.\end{aligned}$$

(Both sides equal

$$\left(\frac{n!(n+k)!}{(2n+k+1)!}\right)^2,$$

though this is not completely transparent; see E.4.) So

$$\begin{aligned}&\frac{1}{(n!)^2}\int_0^1\int_0^1 \frac{1}{1-xy}\frac{d^n}{dy^n}\frac{d^n}{dx^n}x^n y^n (xy-1)^{2n+1+k}\,dx\,dy\\ &\qquad = -\int_0^1\int_0^1 [xy(1-x)(1-y)]^n\,(xy-1)^k\,dx\,dy.\end{aligned}$$

Hence, on expanding $(1-(1-xy)v)^{-n-1}$ in the following integrands,

$$\frac{1}{(n!)^2}\int_0^1\int_0^1 \frac{1}{1-xy}\frac{d^n}{dy^n}\frac{d^n}{dx^n}\frac{x^n y^n (xy-1)^{2n+1}}{(1-(1-xy)v)^{n+1}}\,dx\,dy$$
$$= -\int_0^1\int_0^1 \frac{[xy(1-x)(1-y)]^n}{(1-(1-xy)v)^{n+1}}\,dx\,dy\,,$$

and with Lemma A.2.9 we are done. □

Comments, Exercises, and Examples.

The approach of this appendix follows Borwein, Dykshoorn, Erdélyi, and Zhang [to appear]. Beukers' [79] very elegant recasting of Apéry's proofs of the irrationality of $\zeta(2)$ and $\zeta(3)$ is also presented in Borwein and Borwein [87]. E.5 recasts the irrationality of $\zeta(2)=\frac{1}{6}\pi^2$ into a form similar to the proof of the irrationality of $\zeta(3)$. E.6 treats the irrationality of log 2. Mahler [31] casts transcendence results for exp and log in terms of general systems of Padé approximants.

E.1 Proof of Corollaries A.2.4 and A.2.5.

a] Prove Corollary A.2.4 from the explicit representations of Theorem A.2.2.

b] Prove Corollary A.2.5 from the formula for c_k in Theorem A.2.2.

Hint: Observe that if p is a prime and $n < p^\alpha \le n+k$ for some integer α, then p divides $\binom{n+k}{k}$. This is fairly straightforward from Euler's formula for the largest power of a prime dividing a factorial. □

E.2 Formulas for $\zeta(n)$.

a] Show that

$$\int_0^1\cdots\int_0^1\int_0^1 \frac{dx_1\,dx_2\,\cdots\,dx_n}{1-x_1x_2\cdots x_n} = \zeta(n) := \sum_{k=1}^{\infty}\frac{1}{k^n}\,.$$

b] Show that

$$\int_0^1 \frac{\log x}{1-x}\,dx = -\frac{\pi^2}{6}\,.$$

c] Find a closed formula for

$$\int_0^1\cdots\int_0^1\int_0^1 \frac{\log(x_1x_2\cdots x_n)}{1-x_1x_2\cdots x_n}\,dx_1\,dx_2\,\cdots\,dx_n$$

in terms of $\zeta(n+1)$.

E.3 Proof of Lemma A.2.8. Give a proof of Lemma A.2.8.

E.4 The Identity in the Proof of Lemma A.2.10. Prove the identity

$$\begin{aligned}
&\int_0^1 \int_0^1 (xy)^{n+k}(1-x)^n(1-y)^n \, dx\, dy \\
&\qquad = -\frac{1}{(n!)^2} \int_0^1 \int_0^1 \frac{1}{1-xy} \frac{d^n}{dy^n} \frac{d^n}{dx^n} (xy)^{n+k}(xy-1)^{2n+1} \, dx\, dy \\
&\qquad = \left(\frac{n!(n+k)!}{(2n+k+1)!} \right)^2 ,
\end{aligned}$$

where k and n are nonnegative integers. Follow parts a] to c].

Let k and n be nonnegative integers.

a] Show that

$$A_{n,k} := \int_0^1 \int_0^1 (xy)^{n+k}(1-x)^n(1-y)^n \, dx\, dy = \left(\frac{n!(n+k)!}{(2n+k+1)!} \right)^2 .$$

b] Let

$$B_{n,k} := -\frac{1}{(n!)^2} \int_0^1 \int_0^1 \frac{1}{(1-xy)} \frac{d^n}{dy^n} \frac{d^n}{dx^n} x^{n+k}(xy-1)^{2n+1} \, dx\, dy .$$

Show that

$$\begin{aligned}
n^2 B_{n,k} =& [(n+k)^2 + (2n+1)(2n+2k+1) + (2n+1)(2n)] B_{n-1,k+2} \\
& - [2(n+k)^2 + (2n+1)(2n+2k+1)] B_{n-1,k+1} \\
& + (n+k)^2 B_{n-1,k} .
\end{aligned}$$

c] Show that $A_{0,k} = B_{0,k}$. Show that the values

$$A_{n,k} = \left(\frac{n!(n+k)!}{(2n+k+1)!} \right)^2$$

satisfy the recurrence relation established by part b] for the values $B_{n,k}$.

E.5 The Irrationality of π^2. Consider

$$F_n(x) := \frac{n!}{(2\pi i)} \int_\Gamma \frac{\prod_{k=1}^n (t+k)}{\prod_{k=0}^n (t-k)^2} x^t \, dt ,$$

where Γ is any simple contour containing the poles at $0, 1, \ldots, n$.

a] Show that

$$\int_0^1 \frac{F_n(x)}{1-x}\,dx = \sum_{k=0}^{n} \binom{n+k}{k}\binom{n}{k}^2 \zeta(2) + R_n\,,$$

where $d_n^2 R_n$ is an integer ($d_n := \operatorname{lcm}\{1,2,\dots,n\}$ as before).
Hint: Show that

$$F_n(x) = A_n(x)\log x + B_n(x)\,,$$

where

$$A_n(x) := \sum_{k=0}^{n} \binom{n+k}{k}^2 \binom{n}{k}^2 x^k$$

and

$$B_n(x) := \sum_{k=0}^{n} c_k x^k$$

with

$$c_k := \binom{n+k}{k}\binom{n}{k}^2 \left\{ \sum_{i=1}^{n} \frac{1}{k+i} - \sum_{\substack{i=0\\ i\neq k}}^{n} \frac{2}{k-i} \right\}.$$

Write

$$\frac{F_n(x)}{1-x} = \frac{A_n(x)-A_n(1)}{1-x}\log x + \frac{A_n(1)}{1-x} + \frac{B_n(x)}{1-x}\,.$$

Show that $F_n(1)=0$, hence $B_n(1)=0$. Recall that

$$-\int_0^1 \frac{\log x}{1-x\,dx} = \zeta(2)\,.$$

Now the conclusions follow, as in the proof of Theorem A.2.6. □

b] Show that

$$\int_0^1 \frac{F_n(x)}{1-x}\,dx = \int_0^1\int_0^1 \frac{x^n y^n (1-x)^n (1-y)^n}{(1-xy)^{n+1}}\,dx\,dy\,.$$

Hint: Use Lemma A.2.8. □

c] Show that there exists a constant c independent of n such that

$$0 < \left| \int_0^1 \frac{F_n(x)}{1-x}\,dx \right| \leq c\left(\frac{\sqrt{5}-1}{2}\right)^{5n}$$

and deduce that $\zeta(2) = \frac{1}{6}\pi^2$ irrational. Hence π is irrational.

E.6 The Irrationality of $\log 2$. Let

$$p_n(x) := \frac{1}{n!}\frac{d^n}{dx^n}[x^n(1-x)^n] = \sum_{k=0}^{n}\binom{n+k}{k}\binom{n}{k}(-1)^k x^k$$

be the nth Legendre polynomial on $[0,1]$.

a] Show that

$$\int_0^1 \frac{p_n(x)}{1+x}\,dx = \int_0^1 \frac{p_n(x)-p_n(-1)}{1+x}\,dx + \int_0^1 \frac{p_n(-1)}{1+x}\,dx$$
$$= p_n(-1)\log 2 + R_n\,,$$

where $d_n R_n$ is an integer ($d_n := \operatorname{lcm}\{1,2,\dots,n\}$ as before).

b] Show that

$$\left|\int_0^1 \frac{p_n(x)}{1+x}\,dx\right| = \int_0^1 \frac{x^n(1-x)^n}{(1+x)^{n+1}}\,dx\,.$$

c] Use parts a] and b] to show that $\log 2$ is irrational.

A3
An Interpolation Theorem

Overview

Appendix 3 presents an interpolation theorem for linear functions that is used in Section 7.1. From this Haar's characterization of Chebyshev spaces follows, as do alternate proofs of many of the basic inequalities.

An Interpolation Theorem

The main result of this section, Theorem A.3.3, is an interpolation theorem that plays an important role in Section 7.1. Further applications are given in the exercises.

Throughout this appendix we use the following notation. Let Q be a compact Hausdorff space. Let $C(Q)$ be the space of all real- or complex-valued continuous functions defined on Q. Let P be a (usually finite-dimensional) linear subspace of $C(Q)$ over $\mathbb{R}$ if $C(Q)$ is real or over $\mathbb{C}$ if $C(Q)$ is complex. The function $f \in C(Q)$ is said to be *orthogonal* to P, written as $f \perp P$, if

$$\|f\|_Q \leq \|f + p\|_Q \qquad \text{for all } p \in P\,.$$

This is an L_∞ analog of the more usual L_2 notion of orthogonality, as in Section 2.2. The following two lemmas give necessary and sufficient conditions for the relation $f \perp P$.

Lemma A.3.1. *Let $0 \neq f \in C(Q)$. The function f is orthogonal to P if and only if there exists no $p \in \mathcal{P}$ such that*

$$\mathrm{Re}\big(\overline{f(x)}p(x)\big) > 0 \qquad (A.3.1)$$

holds on

$$E := E(f) := \{x \in Q : |f(x)| = \|f\|_Q\}\,. \qquad (A.3.2)$$

Proof. Assuming there exists $p \in \mathcal{P}$ satisfying (A.3.1) on E defined by (A.3.2), we show that

$$\|f - \epsilon p\|_Q < \|f\|_Q$$

for some $\epsilon > 0$. Since the set E defined by (A.3.2) is compact, $\mathrm{Re}\big(\overline{f(x)}p(x)\big)$ attains its positive minimum, say, $2\delta > 0$, on E, and there exists an open set G containing E such that

$$\mathrm{Re}\big(\overline{f(x)}p(x)\big) > \delta > 0\,, \qquad x \in G\,.$$

Since $G^c := Q \setminus G$ is also compact, there exists an $\alpha > 0$ such that

$$|f(x)| < (1-\alpha)\|f\|_Q\,, \qquad x \in G^c\,.$$

Thus, with a sufficiently small $\epsilon > 0$,

$$|f(x) - \epsilon p(x)|^2 \leq \|f\|_Q^2 - 2\epsilon\delta + \epsilon^2\|p\|_Q^2 < \|f\|_Q^2\,, \qquad x \in G\,,$$

while

$$|f(x) - \epsilon p(x)| \leq (1-\alpha)\|f\|_Q + \epsilon\|p\|_Q < \|f\|_Q\,, \qquad x \in G^c\,.$$

Therefore

$$\|f - \epsilon p\|_Q < \|f\|_Q$$

if $\epsilon > 0$ is small enough.

Conversely, if f is not orthogonal to P, then there exists $p \in P$ such that $\|f - p\|_Q^2 < \|f\|_Q^2$, so

$$2\,\mathrm{Re}\big(\overline{f(x)}p(x)\big) > \|p\|_Q^2 \geq 0\,, \qquad x \in E\,,$$

and the proof is finished. □

Lemma A.3.2. *Let $0 \neq f \in C(Q)$ and let P be an n-dimensional linear subspace of $C(Q)$ over $\mathbb{R}$ if $C(Q)$ is real or over $\mathbb{C}$ if $C(Q)$ is complex. Then the function f is orthogonal to P if and only if there exist points $x_1, x_2, \dots, x_r$ in $E(f)$ defined by (A.3.2) and positive real numbers $c_1, c_2, \dots, c_r$, where $1 \leq r \leq n+1$ when $C(Q)$ is real, and $1 \leq r \leq 2n+1$ when $C(Q)$ is complex, such that*

$$\sum_{i=1}^{r} c_i \overline{f(x_i)} p(x_i) = 0\,, \qquad p \in P\,. \tag{A.3.3}$$

Proof. Suppose (A.3.3) holds with some positive real $c_1, c_2, \dots, c_r$ satisfying $\sum_{i=1}^{r} c_i = 1$. As $|f(x_i)| = \|f\|_Q$, we have

$$\begin{aligned} \|f\|_Q^2 &= \sum_{i=1}^{r} c_i \overline{f(x_i)} f(x_i) = \sum_{i=1}^{r} c_i \overline{f(x_i)} (f(x_i) - p(x_i)) \\ &\leq \|f\|_Q \sum_{i=1}^{r} c_i \max_{1 \leq j \leq r} |f(x_j) - p(x_j)| \leq \|f\|_Q \|f - p\|_Q \end{aligned}$$

for every $p \in P$, so $f \perp P$. (Note that $r \leq n+1$ or $r \leq 2n+1$, respectively, was not needed for this part of the proof, so the sufficiency of (A.3.3) is valid with no hypothesis about r.)

Conversely, suppose $f \perp P$. Let $\{\varphi_1, \varphi_2, \dots, \varphi_n\}$ be a basis for P over $\mathbb{R}$ (or $\mathbb{C}$, respectively), and consider the map $T : Q \to \mathbb{R}^n$ (or $\mathbb{C}^n$, respectively) defined by

$$T(x) = \overline{f(x)} (\varphi_1(x), \varphi_2(x), \dots, \varphi_n(x))\,.$$

Observe that the origin is in the convex hull of

$$T(E) := \{T(x) : x \in E\}\,,$$

otherwise by the principle of separating hyperplanes (a corollary of the Hahn-Banach theorem; see Rudin [73]), there would exist complex numbers $a_1, a_2, \dots, a_n$ such that

$$\operatorname{Re}\left(\sum_{i=1}^{n} a_i \overline{f(x)} \varphi_i(x) \right) > 0\,, \qquad x \in E\,.$$

Hence, with $p := \sum_{i=1}^{n} a_i \varphi_i \in P$,

$$\operatorname{Re}(\overline{f(x)} p(x)) > 0\,, \qquad x \in E$$

and f is not orthogonal to P by Lemma A.3.1, which contradicts our assumption.

Now it follows from Caratheodory's lemma (see E.1) that there exist points $x_1, x_2, \ldots, x_r$ in E and positive real numbers $c_1, c_2, \ldots, c_r$, where $1 \le r \le n+1$ when $C(Q)$ is real and $1 \le r \le 2n+1$ when $C(Q)$ is complex, such that

$$\sum_{i=1}^{r} c_i \overline{f(x_i)} \varphi_k(x_i) = 0, \qquad k = 1, 2, \ldots, n.$$

Therefore

$$\sum_{i=1}^{r} c_i \overline{f(x_i)} p(x_i) = 0, \qquad p \in P,$$

and the lemma is proved. □

Theorem A.3.3 (Interpolation of Linear Functionals). *Let $C(Q)$ be the set of real- (complex-) valued continuous functions on the compact Hausdorff space Q. Let P be an n-dimensional linear subspace of $C(Q)$ over $\mathbb{R}$ $(\mathbb{C})$. Let $L \neq 0$ be a real- (complex-) valued linear functional on P. Then there exist points $x_1, x_2, \ldots, x_r$ in Q, and nonzero real (complex) numbers $a_1, a_2, \ldots, a_r$, where $1 \le r \le n$ in the real case and $1 \le r \le 2n-1$ in the complex case, such that*

$$(A.3.4) \qquad L(p) = \sum_{i=1}^{r} a_i p(x_i), \qquad p \in P$$

and

$$(A.3.5) \qquad \|L\| = \sum_{i=1}^{r} |a_i|,$$

where

$$\|L\| := \sup\{|L(p)| : p \in P, \ \|p\|_Q \le 1\}.$$

Proof. Because of the finite dimensionality of P, there exists an element $p^* \in P$ (called an *extremal element* for L) such that $\|p^*\|_Q = 1$ and $L(P^*) = \|L\|$. Let P_0 denote the null-space of L, so

$$P_0 := \{p \in P : L(p) = 0\}.$$

Now p^* is orthogonal to P_0 because if

$$\|p^* + p_0\|_Q < \|p^*\|_Q = 1$$

for some $p_0 \in P$, then $g := p^* + p_0$ satisfies $\|g\|_Q < 1$ and $L(g) = \|L\|$, which is impossible. Note that the dimension of P_0 over $\mathbb{R}$ is $n-1$ in the real case and $2n-2$ in the complex case. So by Lemma A.3.2 there exist points $x_1, x_2, \ldots, x_r$ in

$$E := \{x \in Q : |p^*(x)| = 1\}$$

and positive real numbers $c_1, c_2, \dots, c_r$, where $1 \le r \le n$ in the real case and $1 \le r \le 2n-1$ in the complex case, such that

$$\sum_{i=1}^{r} c_i \overline{p^*(x_i)} p_0(x_i) = 0, \qquad p_0 \in P_0.$$

Since $L(p)p^* - L(p^*)p \in P_0$ for all $p \in P$, we have

$$L(p) \sum_{i=1}^{r} c_i |p^*(x_i)|^2 = L(p^*) \sum_{i=1}^{r} c_i \overline{p^*(x_i)} p(x_i), \qquad p \in P.$$

Since $L(p^*) = \|L\|$ and $|p^*(x_i)| = 1$ for each i, we obtain (A.3.4) by taking

$$a_i = \frac{c_i \overline{p^*(x_i)}}{\sum_{j=1}^{r} c_j} \|L\|.$$

Using the fact that $|p^*(x_i)| = 1$ for each i in the above formula for a_i, we get (A.3.5). □

Comments, Exercises, and Examples.

We have followed Shapiro [71], which gives a long discussion of questions related to the best uniform approximation of a function $f \in C(Q)$ from a (usually finite-dimensional) linear subspace $P \subset C(Q)$. Some of these, together with other applications, are discussed in the exercises.

E.1 Caratheodory's Lemma. *If $A \subset \mathbb{R}^n$, then every point from the convex hull $co(A)$ of A can be written as a convex linear combination of at most $n+1$ points of A.*

Proof. Let $x \in A$. After a translation if necessary, we may assume that $x = 0$. Suppose

$$(A.3.6) \qquad 0 = \sum_{i=1}^{r} \alpha_i x_i, \qquad x_i \in A, \quad \alpha_i > 0, \quad r > n+1.$$

Since $r > n+1$, the elements $x_2, x_3, \dots, x_r$ are linearly dependent, so there exist real numbers β_i, $i = 2, 3, \dots, r$, not all zero, such that

$$\sum_{i=2}^{r} \beta_i x_i = 0.$$

Let $\beta_1 := 0$. For all $\lambda \in \mathbb{R}$, we have

$$0 = \sum_{i=1}^{r} \alpha_i x_i + \lambda \sum_{i=1}^{r} \beta_i x_i = \sum_{i=1}^{r} (\alpha_i + \lambda \beta_i) x_i \,.$$

When $\lambda = 0$, each term in the last sum is positive. We now define $c := \min |\alpha_i/\beta_i|$, where the minimum is taken for all indices i for which $\beta_i \neq 0$. If the index j is chosen so that $|\alpha_j/\beta_j| = c$, and if $\lambda := -\alpha_j/\beta_j$, then at least one of the numbers $\alpha_i + \lambda\beta_i$ is zero, and all are nonnegative. Also $\alpha_1 + \lambda\beta_1 = \alpha_1 > 0$. We have thus obtained a representation of the same form as (A.3.6), but with s terms, where $1 \leq s \leq r-1$. If $s > n+1$, then the process can be repeated, and after a finite number of steps we obtain the desired representation. □

E.2 Reformation in Terms of Integrals. Lemma A.3.2 can be reformulated as follows. Under the assumptions of Lemma A.3.2, $f \in C(Q)$ is orthogonal to P if and only if there exists a nonzero nonnegative Borel measure ρ on Q whose support consists of r points of $E(f)$ defined by (A.3.2), where $1 \leq r \leq n+1$ in the real case and $1 \leq r \leq 2n+1$ in the complex case, such that $\overline{f(x)}\, d\rho(x)$ annihilates P, that is,

$$\int_Q \overline{f(x)} p(x)\, d\rho(x) = 0\,, \qquad p \in P\,. \tag{A.3.7}$$

This reformation is not only a notational convenience, but it is essential in generalizations where P is no longer finite-dimensional. Moreover, (A.3.7) with any nonzero nonnegative Borel measure (not necessarily discrete) is sufficient for $f \perp P$. This is often useful, even when P is finite-dimensional; see Shapiro [71].

E.3 Haar's Characterization of Chebyshev Spaces.

a] Let $f_0, f_1, \ldots, f_n$ be real- or complex-valued continuous functions defined on a (not necessarily compact) Hausdorff space Q. Show that $P := \operatorname{span}\{f_0, \ldots, f_n\}$, where the span is taken over $\mathbb{R}$ (or $\mathbb{C}$), is a Chebyshev space if and only if there exists no real (or complex) measure on Q annihilating P whose support consists of less than $n+1$ points.

Hint: Use Proposition 3.1.2. □

b] Let P be an n-dimensional linear subspace of $C(Q)$, the space of real- (or complex-) valued continuous functions defined on a compact Hausdorff space Q containing at least n points. The space P is a Chebyshev space if and only if for each $f \in C(Q)$, there is a unique best uniform approximation to f from P.

Proof. First suppose P is a Chebyshev space of dimension n, and p_1 and p_2 are best uniform approximations to some $f \in C(Q)$ from P. Then $p_3 := \frac{1}{2}(p_1 + p_2)$ is also a best uniform approximation to f from P. As

$f - p_3 \perp P$, Lemma A.3.2 yields that $|f(x) - p_3(x)|$ attains its maximum on Q at r points, $x_1, x_2, \ldots, x_r$, that support an annihilating measure for P, where $r \geq n+1$ by part a]. Note that

$$p_1(x_j) - f(x_j) = p_2(x_j) - f(x_j)\,, \qquad j = 1, 2, \ldots, n+1\,,$$

and hence, since P is a Chebyshev space, $p_1 - p_2 = 0$.

Conversely, if P is not a Chebyshev space, then there exist n distinct points $x_1, x_2, \ldots, x_n$ in Q such that the system of homogeneous linear equations

$$\begin{pmatrix} g_1(x_1) & \ldots & g_1(x_n) \\ \vdots & \ddots & \vdots \\ g_n(x_1) & \ldots & g_n(x_n) \end{pmatrix} \begin{pmatrix} a_1 \\ \vdots \\ a_n \end{pmatrix},$$

where $\{g_1, \ldots, g_n\}$ is a basis for P, has a nontrivial solution. Then also the homogeneous system formed with the transposed matrix has a nontrivial solution, so there exist constants b_i, not all zero, so that

$$\sum_{i=1}^{n} b_i g_i(x_j) = 0\,, \qquad j = 1, 2, \ldots, n\,.$$

Thus, with $g := \sum_{i=1}^{n} b_i g_i$, we have

$$g(x_j) = 0\,, \qquad j = 1, 2, \ldots, n\,.$$

We may assume, without loss, that $\|g\|_Q = 1$. Some of the constants $a_1, a_2, \ldots, a_n$ may be zero; however, the set Γ of indices j for which $a_j \neq 0$ is not empty. By Tietze's theorem there exists an $f \in C(Q)$ such that $\|f\|_Q = 1$ and

$$f(x_j) = \frac{\overline{a}_j}{|a_j|}\,, \qquad j \in \Gamma\,.$$

Setting $h(x) := f(x)(1 - |g(x)|)$, we have

$$h(x_j) = \frac{\overline{a}_j}{|a_j|}\,, \qquad j \in \Gamma\,.$$

We claim that $\|h - p\|_Q \geq 1$ for every $p \in P$. Indeed, if $\|h - p\|_Q < 1$ for some $p \in P$, then

$$|f(x_j) - p(x_j)|^2 = |f(x_j)|^2 - 2\,\mathrm{Re}\big(p(x_j)\overline{f(x_j)}\big) + |p(x_j)|^2 < 1$$

for every $j \in \Gamma$; hence

$$\mathrm{Re}(a_j p(x_j)) > 0\,, \qquad j \in \Gamma\,.$$

Since $a_j = 0$ if $j \notin \Gamma$, we have $\mathrm{Re}(a_j p(x_j)) = 0$ for each $j \notin \Gamma$. Thus

$$\mathrm{Re}\left(\sum_{j=1}^{n} a_j p(x_j)\right) > 0 .$$

However, if $p := \sum_{i=1}^{n} c_i g_i$, then

$$\sum_{j=1}^{n} a_j p(x_j) = \sum_{j=1}^{n} a_j \sum_{i=1}^{n} c_i g_i(x_j) = \sum_{i=1}^{n} b_i \sum_{j=1}^{n} a_i g_i(x_j) = 0 ,$$

which contradicts the previous inequality and shows that $\|h - p\|_Q \geq 1$ for every $p \in P$.

Finally, for all $\lambda \in [0, 1]$, λg is a best uniform approximation to h from P because

$$\begin{aligned} |h(x) - \lambda g(x)| &\leq |f(x)|(1 - |g(x)|) + \lambda |g(x)| \\ &\leq 1 + (\lambda - 1)|g(x)| \leq 1 \end{aligned}$$

for all $x \in Q$, so the best uniform approximation to $h \in C(Q)$ from P is not unique. □

E.4 Unicity of the Extremal Function. Assume the notation of Theorem A.3.3. Show that if $P \subset C(Q)$ is an n-dimensional real Chebyshev space and $r = n$, then the extremal element $p^* \in P$ satisfying $\|p^*\|_Q = 1$ and $L(p^*) = \|L\|$ is unique.

The interesting relations of Theorem A.3.3 to the Riesz representation theorem, the Krein-Milman theorem, and the Hahn-Banach theorem are discussed in Shapiro [71].

E.5 Applications of the Interpolation Theorem. As before, let

$$D := \{z \in \mathbb{C} : |z| < 1\} \qquad \text{and} \qquad K := \mathbb{R} \pmod{2\pi} .$$

Prove the following statements. Each of them may be proven by characterizing the extremal polynomial for the given inequality with the help of Theorem A.3.3. A detailed hint is given only to part a].

a] **Bernstein's Inequality.**

$$|t'(\theta)| \leq n\|t\|_K , \qquad t \in \mathcal{T}_n , \quad \theta \in \mathbb{R} .$$

Hint: Let $\theta_0 \in \mathbb{R}$ be fixed, and study the linear functional $L(t) := t'(\theta_0)$, $t \in \mathcal{T}_n$. Observe that an extremal p in Lemma A.3.3 must satisfy $|p(x_i)| = 1$ for each $i = 1, 2, \ldots, r$ and r must equal $2n$. Note that $r \leq 2n$ holds by

Theorem A.3.3, while the argument for $r \geq 2n$ is similar to the corresponding step in the proof of Theorem 7.1.7. Finally, show that the extremal element t satisfying $L(t) = \|L\|$ is of the form

$$t(\theta) = \cos(n\theta - \alpha)$$

for some $\alpha \in K$. □

b] Markov's Inequality.

$$|p'(1)| \leq n^2 \|p\|_{[-1,1]}, \qquad p \in \mathcal{P}_n.$$

c] Chebyshev's Inequality.

$$|p(x)| \leq |T_n(x)| \, \|p\|_{[-1,1]}, \qquad p \in \mathcal{P}_n, \quad x \in \mathbb{R} \setminus [-1,1],$$

where T_n is the Chebyshev polynomial of degree n defined by (2.1.1).

d] Bernstein's Inequality.

$$|p(z)| \leq |z|^n \|p\|_D, \qquad p \in \mathcal{P}_n^c, \quad z \in \mathbb{C} \setminus \overline{D}.$$

e] Bernstein's Inequality.

$$|p'(z)| \leq n|z|^{n-1} \|p\|_D, \qquad p \in \mathcal{P}_n^c, \quad z \in \mathbb{C} \setminus \overline{D}.$$

Hint: Use Theorem 1.3.1 (Lucas' theorem). □

f] Riesz' Identity. There are real numbers a_i with $\sum_{i=1}^{2n} |a_i| = n$ such that

$$t'(\theta) = \sum_{i=1}^{2n} a_i t(\theta + \theta_i), \qquad t \in \mathcal{T}_n, \quad \theta \in \mathbb{R},$$

where

$$\theta_i := \tfrac{2i-1}{2n} \pi, \qquad i = 1, 2, \ldots, 2n.$$

(This is, apart from the explicit determination of the number a_i, an identity discovered by M. Riesz [14].)

g] Show that in part f],

$$a_i = (-1)^{i+1} \frac{1}{4n \sin^2(\theta_i/2)}, \qquad i = 1, 2, \ldots, 2n.$$

h] Bernstein's Inequality in L_p.

$$\int_0^{2\pi} |t'(\theta)|^p \, d\theta \leq n^p \int_0^{2\pi} |t(\theta)|^p \, d\theta, \qquad t \in \mathcal{T}_n, \quad p \geq 1.$$

Hint: Use part f] and Jensen's inequality (see E.20 of Appendix 4). □

Arestov [81] shows that the inequality of part h] is valid for all $p > 0$. Golitschek and Lorentz [89] gives a simpler proof of this.

i] Find all extremal polynomials in parts a] to e] and h].

E.6 An Inequality of Szegő. If $p \in \mathcal{P}_n^c$ and $z_1, z_2, \ldots, z_{2n}$ are any equally spaced points on the unit circle ∂D, then

$$\|p'\|_D \leq n \max_{1 \leq k \leq 2n} |p(z_k)|.$$

Proof. See Frappier, Rahman, and Ruscheweyh [85]. □

A4

Inequalities for Generalized Polynomials in L_p

Overview

Many inequalities for generalized polynomials are given in this appendix. Of particular interest are the extensions of virtually all the basic inequalities to L_p spaces. The principal tool is a generalized version of Remez's inequality.

Inequalities for Generalized Polynomials in L_p

Generalized (nonnegative) polynomials are defined by (A.4.1) and (A.4.3). The basic inequalities of Chapter 5 are extended to these functions by replacing the degree with the generalized degree. The crucial observation is that Remez's inequality extends naturally to this setting. This Remez inequality then plays a central role in the extensions of the other inequalities. These generalizations allow for a simple general treatment of L_p inequalities, which is one main feature of this appendix.

The function

$$f(z) = |\omega| \prod_{j=1}^{m} |z - z_j|^{r_j} \tag{A.4.1}$$

with $0 < r_j \in \mathbb{R}$, $z_j \in \mathbb{C}$, and $0 \neq \omega \in \mathbb{C}$ is called a generalized nonnegative (algebraic) polynomial of (generalized) degree

$$N := \sum_{j=1}^{m} r_j\,. \tag{A.4.2}$$

The set of all generalized nonnegative algebraic polynomials of degree at most N is denoted by GAP_N.

The function

$$f(z) = |\omega| \prod_{j=1}^{m} |\sin((z - z_j)/2)|^{r_j} \tag{A.4.3}$$

with $0 < r_j \in \mathbb{R}$, $z_j \in \mathbb{C}$, and $0 \neq \omega \in \mathbb{C}$ is called a generalized nonnegative trigonometric polynomial of degree

$$N := \frac{1}{2}\sum_{j=1}^{m} r_j\,. \tag{A.4.4}$$

The set of all generalized nonnegative trigonometric polynomials of degree at most N is denoted by GTP_N. Throughout this section we will study generalized nonnegative polynomials restricted to the real line. If the exponents r_j in (A.4.1) or (A.4.3) are even integers, then f is a nonnegative algebraic or trigonometric polynomial, respectively. Note that the classes GAP_N and GTP_N are not linear spaces. Note also that if $f \in \mathrm{GAP}_N$ or $f \in \mathrm{GTP}_N$ is of the form (A.4.1) or (A.4.3), respectively, with all $r_j \geq 1$, then the one-sided derivatives of f exist at every $x \in \mathbb{R}$ with the same modulus, hence $|f'(x)|$ is well-defined for every $x \in \mathbb{R}$. We use the notation $|f'(x)|$ for $f \in \mathrm{GAP}_N$ or $f \in \mathrm{GTP}_N$ and $x \in \mathbb{R}$ throughout this section with this understanding. If $f \in \mathrm{GAP}_N$ is of the form (A.4.1) or $f \in \mathrm{GTP}_N$ is of the form (A.4.3), where the zeros $z_j \in \mathbb{C}$, $j = 1, 2, \ldots, m$, are distinct, then r_j is called the multiplicity of z_j in f. Our intention in this section is to extend most of the classical inequalities of Section 5.1 to generalized nonnegative polynomials. In addition, we prove Nikolskii-type inequalities for GAP_N and GTP_N.

Theorem A.4.1 (Remez-Type Inequality for GAP_N**).** *The inequality*

$$\|f\|_{[-1,1]} \leq \left(\frac{\sqrt{2} + \sqrt{s}}{\sqrt{2} - \sqrt{s}}\right)^{N}$$

holds for every $f \in \mathrm{GAP}_N$ *and* $s \in (0, 2)$ *satisfying*

$$m(\{x \in [-1,1] : f(x) \leq 1\}) \geq 2 - s\,.$$

E.5 shows that this inequality is sharp. Note that if $0 < s \leq 1$, then

$$\left(\frac{\sqrt{2} + \sqrt{s}}{\sqrt{2} - \sqrt{s}}\right)^{N} \leq \exp(5N\sqrt{s}).$$

Throughout this section, as before, $K := \mathbb{R} \pmod{2\pi}$.

Theorem A.4.2 (Remez-Type Inequality for GTP_N). *The inequality*

$$\|f\|_K \le \exp\left(N\left(s + \tfrac{7}{4}s^2\right)\right) \le \exp(4Ns)$$

holds for every $f \in \text{GTP}_N$ and $s \in (0, \pi/2]$ satisfying

$$m(\{x \in [-\pi, \pi) : f(x) \le 1\}) \ge 2\pi - s\,. \tag{A.4.5}$$

The inequality

$$\|f\|_K \le \left(\frac{\sqrt{2} + \sqrt{\sigma}}{\sqrt{2} - \sqrt{\sigma}}\right)^N, \qquad \sigma = 1 - \cos(s/2)$$

holds for every even $f \in \text{GTP}_N$ and $s \in (0, 2\pi)$ satisfying (A.4.5).

We do not discuss what happens when $s \in (\pi/2, 2\pi)$ in the general case because the case when $s \in (0, \pi/2]$ is satisfactory for our needs.

Proof of Theorem A.4.1. First assume that $f \in \text{GAP}_N$ is of the form (A.4.1) with rational exponents $r_j = q_j/q$, where $q_j, q \in \mathbb{N}$. Let $k \in \mathbb{N}$ be an integer. Then (restricted to $\mathbb{R}$) $p := f^{2kq} \in \mathcal{P}_{2kqN}$ and

$$m\left(\{x \in [-1, 1] : |p(x)| \le 1\}\right) \ge 2 - s\,.$$

Hence Theorem 5.1.1 yields

$$\|f\|_{[-1,1]} = \|p\|_{[-1,1]}^{1/(2kq)} \le \left(T_{2kqN}\left(\frac{2+s}{2-s}\right)\right)^{1/(2kq)}.$$

Since by E.4,

$$\lim_{k \to \infty} \left(T_{2kqN}\left(\frac{2+s}{2-s}\right)\right)^{1/(2kq)} = \left(\frac{\sqrt{2} + \sqrt{s}}{\sqrt{2} - \sqrt{s}}\right)^N, \tag{A.4.6}$$

the theorem is proved. The case when the exponents $r_j > 0$ are arbitrary real numbers can be easily reduced to the already proved rational case by a straightforward density argument. □

Theorem A.4.2 follows from Theorem 5.1.2 in exactly the same way that Theorem A.4.1 follows from Theorem 5.1.1; see E.6.

Theorem A.4.3 (Nikolskii-Type Inequality for GTP_N). *Let χ be a nonnegative nondecreasing function defined on $[0, \infty)$ such that $\chi(x)/x$ is nonincreasing on $[0, \infty)$. Then there is an absolute constant $c_1 > 0$ such that*

$$\|\chi(f)\|_{L_p(K)} \le (c_1(1 + qN))^{1/q - 1/p} \|\chi(f)\|_{L_q(K)}$$

for every $f \in \text{GTP}_N$ and $0 < q < p \le \infty$. If $\chi(x) = x$, then $c_1 \le e(4\pi)^{-1}$.

Theorem A.4.4 (Nikolskii-Type Inequality for GAP_N**).** *Let χ be a nonnegative nondecreasing function defined on $[0,\infty)$ such that $\chi(x)/x$ is nonincreasing on $[0,\infty)$. Then there is an absolute constant $c_2 > 0$ such that*

$$\|\chi(f)\|_{L_p[-1,1]} \le (c_2(2+qN))^{2/q-2/p}\|\chi(f)\|_{L_q[-1,1]}$$

for every $f \in \text{GAP}_N$ and $0 < q < p \le \infty$. If $\chi(x) = x$, then $c_2 \le e^2(2\pi)^{-1}$.

In the proof of the second part of Theorem A.4.4 we will need the following Schur-type inequality, which is interesting in its own right.

Theorem A.4.5 (Schur-Type Inequality for GAP_N**).** *The inequality*

$$\|f\|^q_{[-1,1]} \le e(1+qN)\left\|\sqrt{1-x^2}f^q(x)\right\|_{[-1,1]}$$

holds for every $f \in \text{GAP}_N$ and $q > 0$.

According to Theorem 5.1.9 (Schur's inequality), if $N \in \mathbb{N}$, $f \in \mathcal{P}_N$, and $q \in \mathbb{N}$, then the constant e in the above inequality can be replaced by 1.

It is sufficient to prove Theorems A.4.3 and A.4.4 when $p = \infty$, and then a simple argument gives the required results for arbitrary exponents $0 < q < p < \infty$. To see this, say, in the trigonometric case, assume that there is a constant C_N such that

$$\|\chi(f)\|_K \le C_N^{1/q}\|\chi(f)\|_{L_q(K)}$$

for every $f \in \text{GTP}_N$ and $0 < q < \infty$. Then

$$\begin{aligned}\|\chi(f)\|^p_{L_p(K)} &= \|(\chi(f))^{p-q+q}\|_{L_1(K)}\\ &\le \|\chi(f)\|_K^{p-q}\|\chi(f)\|^q_{L_q(K)}\\ &\le C_N^{p/q-1}\|\chi(f)\|^{p-q}_{L_q(K)}\|\chi(f)\|^q_{L_q(K)},\end{aligned}$$

and therefore

$$\|\chi(f)\|_{L_p(K)} \le C_N^{1/q-1/p}\|\chi(f)\|_{L_q(K)}$$

for every $f \in \text{GTP}_N$ and $0 < q < p \le \infty$.

Proof of Theorem A.4.3 (when $p = \infty$). Since χ is nonnegative and nondecreasing and $(\chi(x)/x)^q$ is nonincreasing on $[0,\infty)$, we have

$$(\chi(f(\theta)))^q \ge \exp(-qNs)\|\chi(f)\|^q_K$$

whenever

$$f(\theta) \ge \exp(-Ns)\|f\|_K\,.$$

Hence, by E.7 b], we can deduce that

$$m(\{\theta \in [-\pi, \pi) : (\chi(f(\theta)))^q \geq \exp(-qNs)\|\chi(f)\|_K^q\}) \geq \frac{s}{4}$$

for every $f \in \mathrm{GTP}_N$ and $s \in (0, 2\pi)$. Choosing $s := (1+qN)^{-1}$, we get

$$m\left(\{\theta \in [-\pi, \pi) : (\chi(f(\theta)))^q \geq e^{-1}\|\chi(f)\|_K^q\}\right) \geq (4(1+qN))^{-1}.$$

Hence, integrating only on the subset I of K where

$$(\chi(f(\theta)))^q \geq e^{-1}\|\chi(f)\|_K^q,$$

we conclude that

$$\begin{aligned}\|\chi(f)\|_K^q &\leq 4e(1+qN)\int_I (\chi(f(\theta)))^q \, d\theta \\ &\leq 4e(1+qN)\|\chi(f)\|_{L_q(K)}^q,\end{aligned}$$

and the first part of the theorem is proved.

Now we turn to the second statement. Let

$$D := \{z \in \mathbb{C} : |z| < 1\} \qquad \text{and} \qquad \partial D := \{z \in \mathbb{C} : |z| = 1\}.$$

If h is analytic in the open unit disk D and continuous on the closed unit disk $\overline{D}$, then by Cauchy's integral formula we have

$$(1 - |rz|^2)h(rz) = \frac{1}{2\pi i}\int_{\partial D} h(u)\frac{1 - r\overline{z}u}{u - rz}\, du$$

whenever $z \in \overline{D}$ and $r \in [0, 1)$. Note that $u \in \partial D$ and $z \in \partial D$ imply $|1 - r\overline{z}u| = |u - rz|$ for all $r \in [0, 1)$. Hence, if $P \in \mathcal{P}_n^c$ and $0 < q < \infty$, then

$$(1 - r^2)|P^*(rz)|^q \leq \frac{1}{2\pi}\int_{\partial D} |P^*(u)|^q \, |du|$$

whenever $z \in \partial D$ and $r \in [0, 1]$, where P^* is obtained from the factorization of P by replacing each factor $(z - \alpha)$ of P with $|\alpha| < 1$ by $(1 - \overline{\alpha}z)$. Since

$$\tfrac{1}{2}(1+r)|z - \sigma| \leq |rz - \sigma|, \qquad |\sigma| > 1, \quad z \in \partial D, \quad r \in [0, 1],$$

we have

$$(1 - r^2)\left(\tfrac{1}{2}(1+r)\right)^{q\deg(P)}|P(z)|^q \leq \frac{1}{2\pi}\int_{\partial D} |P^*(u)|^q \, |du|$$

whenever $z \in \partial D$ and $r \in [0, 1]$. Maximizing the left-hand side for $r \in [0, 1]$ and using the fact that $|P^*(z)| = |P(z)|$ for $z \in \partial D$, we conclude that

$$|P(z)|^q \leq \frac{(2 + q\deg(P))e}{8\pi} \int_{-\pi}^{\pi} |P(e^{i\theta})|^q \, d\theta \,, \qquad z \in \partial D \,.$$

Hence, by E.8,

$$\|R\|_K^q \leq \frac{(1+qn)e}{4\pi} \int_{-\pi}^{\pi} |R(\theta)|^q \, d\theta$$

for every $R \in \mathcal{T}_n$. If $f \in \mathrm{GTP}_N$ is of the form (A.4.3) with rational exponents $r_j = \alpha_j/\alpha$, where $\alpha_j, \alpha \in \mathbb{N}$, then on applying the above inequality to $R := f^{2\alpha} \in \mathcal{T}_{2\alpha N}$ with q replaced by $q/(2\alpha)$, we conclude that

$$\|f\|_K^q \leq \frac{1+qN}{4\pi} \int_{-\pi}^{\pi} f(\theta)^q \, d\theta$$

and the second statement of the theorem is proved. The case when the exponents $r_j > 0$, $j = 1, 2, \ldots, m$, are arbitrary real numbers can be reduced to the already proved rational case by a straightforward density argument. □

Proof of Theorem A.4.5. Let $P \in \mathcal{P}_n$ and

$$M := \left\| \sqrt{1-x^2}\, |P(x)|^q \right\|_{[-1,1]} .$$

By E.8 c], there exists an $R \in \mathcal{P}_{2n}$ such that $|R(e^{i\theta})| = |P(\cos\theta)|$, $\theta \in \mathbb{R}$. We define $R^* \in \mathcal{P}_{2n}^c$ from the factorization of R by replacing all the factors $(z - \alpha)$ of R with $|\alpha| < 1$ by $(1 - \overline{\alpha} z)$. Note that $|1 - e^{2i\theta}| = 2|\sin\theta|$ and $|R(e^{i\theta})| = |R^*(e^{i\theta})|$ for all $\theta \in \mathbb{R}$. Hence the maximum principle yields that

$$\begin{aligned} |1-(rz)^2||R^*(rz)|^q &\leq \max_{z \in \partial D} |1 - z^2||R^*(z)|^q \\ &= \max_{\theta \in \mathbb{R}} 2|\sin\theta||P(\cos\theta)|^q = 2M \,. \end{aligned}$$

By E.9 we have

$$|R^*(z)| \leq \left(\frac{2}{1+r}\right)^{2n} |R^*(rz)| \,, \qquad z \in \partial D \,, \;\; r \in [0,1] \,.$$

Hence

$$|R^*(z)|^q \leq \frac{2^{2nq}}{(1+r)^{2nq}} \frac{1}{1-r^2} 2M \,, \qquad z \in \partial D \,, \;\; r \in [0,1] \,,$$

where the minimum on $[0,1]$ of the right-hand side is taken at

$$r := \frac{qn}{1+qn} \,.$$

Estimating the right-hand side at this value of r, we get

$$\|P\|_{[-1,1]}^q = \max_{z\in\partial D} |R^*(z)|^q \le e(1+qn)\left\|\sqrt{1-x^2}\,|P(x)|^q\right\|_{[-1,1]}.$$

If $f \in \mathrm{GAP}_N$ is of the form (A.4.1) with rational exponents $r_j = \alpha_j/\alpha$, $\alpha_j, \alpha \in \mathbb{N}$, then applying the above inequality to $P = f^{2\alpha} \in \mathcal{P}_{2\alpha N}$ with q replaced by $q/(2\alpha)$, we get

$$\|f\|_{[-1,1]}^q \le e(1+qN)\left\|\sqrt{1-x^2}\,f^q(x)\right\|_{[-1,1]},$$

and the theorem is proved. The case when the exponents $r_j > 0$ are arbitrary real numbers can be easily reduced to the already proved rational case once again by a standard approximation. □

Proof of Theorem A.4.4 (when $p = \infty$). Since χ is nonnegative and nondecreasing and $(\chi(x)/x)^q$ is nonincreasing on $[0,\infty)$, we have

$$(\chi(f(x)))^q \ge \exp(-qN\sqrt{s})\|\chi(f)\|_{[-1,1]}^q$$

whenever

$$f(x) \ge \exp(-N\sqrt{s})\|f\|_{[-1,1]}.$$

So by E.7 a] we can deduce that

$$m\left(\left\{x \in [-1,1] : (\chi(f(x)))^q \ge \exp(-qN\sqrt{s})\|\chi(f)\|_{[-1,1]}^q\right\}\right) \ge \frac{s}{8}$$

for all $s \in (0,2)$. Choosing $s := (1+qN)^{-2}$, we obtain

$$m\left(\{x \in [-1,1] : (\chi(f(x)))^q \ge e^{-1}\|\chi(f)\|_{[-1,1]}\}\right) \ge \frac{1}{8(1+qN)^2}.$$

Integrating on the subset I of $[-1,1]$ where

$$(\chi(f(x)))^q \ge e^{-1}\|\chi(f)\|_{[-1,1]},$$

we conclude that

$$\begin{aligned}\|\chi(f)\|_{[-1,1]}^q &\le 8e(1+qN)^2\int_I (\chi(f(x)))^q\,dx \\ &\le 8e(1+qN)^2\|\chi(f)\|_{L_q[-1,1]}^q.\end{aligned}$$

Thus the first part of the theorem is proved.

To show that the given constant works in the case that $\chi(x) = x$ we use another method. Let $h \in \mathrm{GAP}_M$. Then by E.10, $g(\theta) = h(\cos\theta) \in \mathrm{GTP}_M$. On using the substitution $x = \cos\theta$, from Theorem A.4.3 we get

$$\|h\|_{[-1,1]} \leq e(2\pi)^{-1}(1+qM) \int_{-1}^{1} |h(x)|^q (1-x^2)^{-1/2}\, dx\,.$$

If $f \in \mathrm{GAP}_N$, then $h(x) = f(x)(1-x^2)^{1/(2q)} \in \mathrm{GAP}_M$ with $M = N + q^{-1}$, so an application of the above inequality yields

$$(A.4.7) \qquad \left\|\sqrt{1-x^2} f^q(x)\right\|_{[-1,1]} \leq e(2\pi)^{-1}(2+qN) \int_{-1}^{1} f^q(x)\, dx\,.$$

(Note that the weaker assumption $h \in \mathrm{GAP}_M$ instead of $f \in \mathrm{GAP}_N$ already implies (A.4.7).)

Now a combination of Theorem A.4.5 and inequality (A.4.7) gives that if $\chi(x) = x$, then the inequality of Theorem A.4.4 holds with $c_2 := e^2(2\pi)^{-1}$. □

Now we prove extensions (up to multiplicative absolute constants) of Markov's and Bernstein's inequalities for generalized nonnegative polynomials.

Theorem A.4.6 (Bernstein-Type Inequality for GTP_N). *There exists an absolute constant $c_3 > 0$ such that*

$$\|f'\|_K \leq c_3 N \|f\|_K$$

for every $f \in \mathrm{GTP}_N$ of the form (A.4.3) with each $r_j \geq 1$.

Theorem A.4.7 (Bernstein-Type Inequality for GAP_N). *The inequality*

$$|f'(x)| \leq \frac{c_3 N}{\sqrt{1-x^2}} \|f\|_{[-1,1]}\,, \qquad x \in (-1,1)\,,$$

holds for every $f \in \mathrm{GAP}_N$ of the form (A.4.1) with each $r_j \geq 1$, where c_3 is as in Theorem A.4.6.

Theorem A.4.8 (Markov-Type Inequality for GAP_N). *There exists an absolute constant $c_4 > 0$ such that*

$$\|f'\|_{[-1,1]} \leq c_4 N^2 \|f\|_{[-1,1]}$$

for every $f \in \mathrm{GAP}_N$ of the form (A.4.1) with each $r_j \geq 1$.

To prove Theorem A.4.6 we need the following lemma.

Lemma A.4.9. *Suppose $g \in \mathrm{GTP}_N$ is of the form (A.4.3) with each $z_j \in \mathbb{R}$, and suppose at least one of any two adjacent (in K) zeros has multiplicity at least 1. Then there exists an absolute constant $c_5 > 0$ such that for every such g there is an interval $I \subset K$ of length at least $c_5 N^{-1}$ for which*

$$\min_{\theta \in I} g(\theta) \geq e^{-1} \|g\|_K\,.$$

Proof. Take a $g \in \mathrm{GTP}_N$ satisfying the hypothesis of the lemma. Because of the periodicity of g we may assume that

$$g(\pi) = \|g\|_K \,. \tag{A.4.8}$$

Define

$$Q_{n,\omega}(\theta) := T_{2n}\left(\frac{\sin(\theta/2)}{\sin(\omega/2)}\right) \tag{A.4.9}$$

with

$$n := \lfloor N \rfloor \qquad \text{and} \qquad \omega := \pi - (3N)^{-1}\,, \tag{A.4.10}$$

where T_{2n} is the Chebyshev polynomial of degree $2n$ defined by (2.1.1). By E.11 there exists an absolute constant $c_6 > 1$ such that $Q_{n,\omega}(\pi) \geq c_6$. Introduce the set

$$A := \left\{\theta \in [\pi - (3N)^{-1}, \pi + (3N)^{-1}] : g(\theta) \geq e^{-1}g(\pi)\right\}\,.$$

We study $h := g|Q_{n,\omega}| \in \mathrm{GTP}_{2N}$. The inequality $Q_{n,\omega}(\pi) \geq c_6$ and assumption (A.4.8) yield

$$h(\theta) \leq g(\theta) \leq c_6^{-1}Q_{n,\omega}(\pi)g(\pi) = c_6^{-1}\|h\|_K$$

for all $\theta \in [-\omega, \omega] = [-\pi + (3N)^{-1},\ \pi - (3N)^{-1}]$. Further, the definition of the set A, the fact that $\|Q_{n,\omega}\|_K = Q_{n,\omega}(\pi)$, and (A.4.8) imply that

$$h(\theta) \leq e^{-1}g(\pi)Q_{n,\omega}(\pi) = e^{-1}\|h\|_K$$

for all $\theta \in [\pi - (3N)^{-1}, \pi + (3N)^{-1}] \setminus A$. From the last two inequalities we conclude that

$$h(\theta) \leq c_7^{-1}\|h\|_{[-1,1]} \quad \text{for all } \theta \in [-\pi, \pi] \setminus A\,,$$

where $c_7 := \min\{c_6, e\} > 1$ is an absolute constant. Therefore, by E.7 b]

$$m(A) \geq c_8 N^{-1} \quad \text{with } c_8 := \tfrac{1}{17}\log c_7 > 0\,.$$

Since $g \in \mathrm{GTP}_N$ is of the form (A.4.3) with each $z_j \in \mathbb{R}$, and at least one of any two adjacent zeros of g has multiplicity at least 1, E.12 and assumption (8.1.8) imply that g cannot have two or more distinct zeros in $[\pi - (3N)^{-1},\ \pi + (3N)^{-1}]$. Hence A is the union of at most two intervals. Therefore there exists an interval $I \subset A$ such that $m(I) \geq c_8(2N)^{-1}$, and the lemma is proved. □

Proof of Theorem A.4.6. Let $f \in \text{GTP}_N$ be of the form (A.4.3) with each $r_j \geq 1$. Without loss of generality we may assume that $|f'(\pi)| = \|f'\|_K$, and it is sufficient to prove only that

$$|f'(\pi)| \leq c_3 N \|f\|_K \,.$$

By E.13 we may assume that each z_j is real in (A.4.3). Hence, by E.2, $g := |f'|$ satisfies the assumption of Lemma A.4.9. Denote the endpoints of the interval I coming from Lemma A.4.9 by $a < b$. We can now deduce that

$$\begin{aligned}\|f'\|_K = |f'(\pi)| &\leq \frac{e}{b-a}\int_a^b |f'(\theta)|\, d\theta \\ &\leq \frac{eN}{c_5}\int_a^b |f'(\theta)|\, d\theta = \frac{eN}{c_5}|f(b)-f(a)| \leq c_3 N \|f\|_K\end{aligned}$$

with $c_3 := ec_5^{-1}$, and the proof is finished. □

Proof of Theorem A.4.7. The theorem follows from Theorem A.4.6 by using the substitution $x = \cos\theta$ and E.10 b]. □

Proof of Theorem A.4.8. Let $\alpha := 1-(1+N)^{-2}$. Using Theorem A.4.7 and then E.14, we obtain

$$\begin{aligned}\|f'\|_{[-\alpha,\alpha]} &\leq c_3 N(N+1)\|f\|_{[-1,1]} \\ &\leq c_3 N(N+1) c_9 \|f\|_{[-\alpha,\alpha]} \leq c_4 N^2 \|f\|_{[-\alpha,\alpha]}\,,\end{aligned}$$

and then the theorem follows by a linear transformation. □

Now we establish Remez-, Bernstein-, and Markov-type inequalities for generalized nonnegative polynomials in L_p. In the proofs we use the inequalities proved in this appendix so far, and the methods illustrate how one can combine the "basic" inequalities in the proofs of various other inequalities for generalized nonnegative polynomials. First we state the main results.

Theorem A.4.10 (L_p Remez-Type Inequality for GAP_N). *Let χ be a nonnegative nondecreasing function defined on $[0,\infty)$ such that $\chi(x)/x$ is nonincreasing on $[0,\infty)$. There exists an absolute constant $c \leq 5\sqrt{2}$ such that*

$$\int_{-1}^1 (\chi(f(x)))^p\, dx \leq \left(1+\exp(cpN\sqrt{s})\right)\int_A (\chi(f(x)))^p\, dx$$

for every $f \in \text{GAP}_N$, $A \subset [-1,1]$ with $m([-1,1]\setminus A) \leq s \leq 1/2$, and for every $p \in (0,\infty)$.

Theorem A.4.11 (L_p Remez-Type Inequality for GTP_N). *Let χ be a nonnegative nondecreasing function defined on $[0, \infty)$ such that $\chi(x)/x$ is nonincreasing on $[0, \infty)$. There exists an absolute constant $c \le 8$ such that*

$$\int_{-\pi}^{\pi} (\chi(f(\theta)))^p \, d\theta \le (1 + \exp(cpNs)) \int_A (\chi(f(\theta)))^p \, d\theta$$

for every $f \in \mathrm{GTP}_N$, $A \subset [-\pi, \pi]$ with $m([-\pi, \pi] \setminus A) \le s \le \pi/2$, and for every $p \in (0, \infty)$.

Theorem A.4.12 (L_q Bernstein-Type Inequality for GTP_N). *Let χ be a nonnegative, nondecreasing, convex function defined on $[0, \infty)$. There exists an absolute constant c such that*

$$\int_{-\pi}^{\pi} \chi(N^{-q}|f'(\theta)|^q) \, d\theta \le \int_{-\pi}^{\pi} \chi(cf(\theta)^q) \, d\theta$$

for every $f \in \mathrm{GTP}_N$ of the form (A.4.3) with each $r_j \ge 1$, and for every $q \in (0, 1]$.

Corollary A.4.13 (L_p Bernstein-Type Inequality for GTP_N). *The inequality*

$$\int_{-\pi}^{\pi} |f'(\theta)|^p \, d\theta \le c^{p+1} N^p \int_{-\pi}^{\pi} |f(\theta)|^p \, d\theta$$

holds for every $f \in \mathrm{GTP}_N$ of the form (A.4.3) with each $r_j \ge 1$, and for every $p \in (0, \infty)$, where c is as in Theorem A.4.12.

Theorem A.4.14 (L_p Markov-Type Inequality for GAP_N). *There exists an absolute constant c such that*

$$\int_{-1}^{1} |f'(x)|^p \, dx \le c^{p+1} N^{2p} \int_{-1}^{1} |f(x)|^p \, dx$$

for every $f \in \mathrm{GAP}_N$ of the form (A.4.1) with each $r_j \ge 1$, and for every $p \in (0, \infty)$.

Theorems A.4.1 and A.4.2 can be easily obtained from their L_∞ analogs, Theorems A.4.1 and A.4.2, respectively; see E.15 and E.16, where hints are given.

Proof of Theorem A.4.12. For $n := \lfloor N \rfloor$ let

$$D_n(\theta) := \left| \sum_{j=-n}^{n} e^{ij\theta} \right|$$

be the modulus of the nth Dirichlet kernel. Choose $q \in (0, 1]$, and set $m := 2q^{-1} \ge 2$. Let $g \in \mathrm{GTP}_N$ be of the form (A.4.3) with each $r_j \ge 1$. On applying the Nikolskii-type inequality of Theorem A.4.3 to

$$G := gD_n^m \in \mathrm{GTP}_{N+2nq^{-1}}\,,$$

we obtain

$$(A.4.11)\qquad \begin{aligned}\|gD_n^m\|_K^q &\le c_1\left(1+q\left(N+2nq^{-1}\right)\right)\|gD_n^m\|_{L_q(K)}^q \\ &\le c_1(1+3N)\int_{-\pi}^{\pi}(g(\theta)D_n^m(\theta))^q\,d\theta \\ &= c_1(1+3N)\int_{-\pi}^{\pi}g^q(\theta)D_n^2(\theta)\,d\theta\,.\end{aligned}$$

If $g \in \mathrm{GTP}_N$ is of the form (A.4.3) with each $r_j \ge 1$, then $m \ge 2$ implies that $G \in \mathrm{GTP}_N$ is of the form (A.4.3) with each $r_j \ge 1$ as well. If we apply the Bernstein-type inequality of Theorem A.4.6 to G and use (A.4.11), we can deduce that

$$\begin{aligned}&\left|g'(\theta)D_n^m(\theta)+mD_n^{m-1}(\theta)D_n'(\theta)g(\theta)\right|_K^q \\ &\qquad\le\left(c_3\left(N+2nq^{-1}\right)\right)^q\|gD_n^m\|_K^q \\ &\qquad\le c_3^qN^q\left(1+2q^{-1}\right)^q c_1(1+3N)\int_{-\pi}^{\pi}g^q(\theta)D_n^2(\theta)\,d\theta\end{aligned}$$

for every $\theta \in K$ (we take one-sided derivatives everywhere). By putting $\theta = 0$, and noticing that

$$D_n'(0)=0 \quad\text{and } D_n(0)^m=(2n+1)^{2/q}\ge N^{2/q}\,,$$

we get

$$(A.4.12)\qquad |g'(0)|^q \le cN^q\int_{-\pi}^{\pi}g^q(\theta)(2\pi)^{-1}(2n+1)^{-1}D_n^2(\theta)\,d\theta$$

with an absolute constant c. Now let $f \in \mathrm{GTP}_n$ be of the form (A.4.3) with each $r_j \ge 1$. Let $\tau \in K$ be fixed. On applying (A.4.12) to $g(\theta) := f(\theta+\tau)$, we conclude that

$$|f'(\tau)|^q \le cN^q\int_{-\pi}^{\pi}f^q(\theta)(2\pi)^{-1}(2n+1)^{-1}D_n^2(\theta-\tau)\,d\theta\,.$$

Hence

$$(A.4.13)\qquad N^{-q}|f'(\tau)|^q \le \int_{-\pi}^{\pi}cf^q(\theta)(2\pi)^{-1}(2n+1)^{-1}D_n^2(\theta-\tau)\,d\theta\,.$$

Since

$$(A.4.14)\qquad \int_{-\pi}^{\pi}(2\pi)^{-1}(2n+1)^{-1}D_n^2(\theta-\tau)\,d\theta = 1\,,$$

Jensen's inequality (see E.7) and (A.4.3) imply that

$$\chi(N^{-q}|f'(\tau)|^q) \leq \int_{-\pi}^{\pi} \chi(cf^q(\theta))(2\pi)^{-1}(2n+1)^{-1}D_n^2(\theta-\tau)\,d\theta\,.$$

If we integrate both sides with respect to τ, Fubini's theorem and (A.4.14) (on interchanging the role of θ and τ) yield the inequality of the theorem. □

Proof of Corollary A.4.13. If $0 < p \leq 1$, then Theorem A.4.12 yields the corollary with $q = p$ and $\chi(x) = x$. If $1 \leq p < \infty$, then the corollary follows from Theorem A.4.12 again with $q = 1$ and $\chi(x) = x^p$. □

Proof of Theorem A.4.14. We distinguish two cases.

Case 1: $p \geq 1$. Let $f \in \text{GAP}_N$ be of the form (A.4.1) with each $r_j \geq 1$. Then by E.10 b], $g(\theta) := f(\cos\theta) \in \text{GTP}_N$ is of the form (A.4.3) with each $r_j \geq 1$. With the substitution $x = \cos\theta$, Corollary A.4.13 and Theorem A.4.11 imply that

$$\text{(A.4.15)}\qquad \begin{aligned}&\int_{-1}^{1} |f'(x)|^p(1-x^2)^{(p-1)/2}\,dx\\ &\qquad\leq c_1^p N^p \int_{-1}^{1} f^p(x)(1-x^2)^{-1/2}\,dx\\ &\qquad\leq c_1^p N^p \exp(c_2 pNN^{-1}) \int_{-\delta}^{\delta} f^p(x)(1-x^2)^{-1/2}\,dx\,,\end{aligned}$$

where $\delta := \max\{1-N^{-2}, \cos(\pi/16)\}$ and c_1 and c_2 are appropriate absolute constants. Since $p-1 \geq 0$, it follows from (A.4.15) that

$$\text{(A.4.16)}\qquad \begin{aligned}&\int_{-\delta}^{\delta} |f'(x)|^p\,dx\\ &\qquad\leq (1-\delta^2)^{(1-p)/2} \int_{-\delta}^{\delta} |f'(x)|^p(1-x^2)^{(p-1)/2}\,dx\\ &\qquad\leq (1-\delta^2)^{(1-p)/2} c_1^p N^p \exp(c_2 p)(1-\delta^2)^{-1/2} \int_{-\delta}^{\delta} f^p(x)\,dx\\ &\qquad\leq c_3^p N^{p-1} N^p N \int_{-\delta}^{\delta} f^p(x)\,dx\\ &\qquad= c_3^p N^{2p} \int_{-\delta}^{\delta} f^p(x)\,dx\,,\end{aligned}$$

where c_3 is also an absolute constant. Since (A.4.16) is valid for every $f \in \text{GAP}_N$ of the form (A.4.1) with each $r_j \geq 1$, the theorem follows by a linear shift from $[-\delta, \delta]$ to $[-1, 1]$.

Case 2: $0 < p \le 1$. Let $f \in \mathrm{GAP}_N$ be of the form (A.4.1) with each $r_j \ge 1$. Using the inequality $|a+b|^p \le |a|^p + |b|^p$ for $p \in (0,1]$, we can deduce that

$$\text{(A.4.17)} \qquad \begin{aligned} &\int_A \left(|f'(\cos\theta)||\sin\theta|^{1/p+1}\right)^p d\theta \\ &\qquad \le \int_A \left|\left(f(\cos\theta)|\sin\theta|^{1/p}\right)'\right|^p d\theta \\ &\qquad\quad + \int_A \left(f(\cos\theta)p^{-1}|\sin\theta|^{1/p-1}|\cos\theta|\right)^p d\theta \end{aligned}$$

for every measurable subset A of $[-\pi, \pi)$. Applying Theorem A.4.12 (with $\chi(x) = x$) to

$$g(\theta) := f(\cos\theta)|\sin\theta|^{1/p} \in \mathrm{GTP}_{N+1/p}\,,$$

then using (A.4.17) with $A := [-\delta, \delta]$, $\delta := 1 - (N+1)^{-2}$, we conclude, by the substitution $x = \cos\theta$, that

$$\text{(A.4.18)} \quad \begin{aligned} &\int_{-\delta}^{\delta} |f'(x)|^p (1-x^2)^{p/2}\, dx \\ &\qquad \le c_1(N+1/p)^p \int_{-1}^{1} f^p(x)\, dx + p^{-p} \int_{-\delta}^{\delta} f^p(x)(1-x^2)^{-p/2}\, dx\,, \end{aligned}$$

where c_1 is an absolute constant. Note that Theorem A.4.10, $0 < p \le 1$, and the choice of δ imply that

$$\text{(A.4.19)} \qquad \int_{-1}^{1} f^p(x)\, dx \le c_2 \int_{-\delta}^{\delta} f^p(x)\, dx$$

with an absolute constant c_2. A combination of (A.4.18) and (A.4.19) yields

$$\text{(A.4.20)} \quad \begin{aligned} &\int_{-\delta}^{\delta} |f'(x)|^p\, dx \\ &\le (1-\delta^2)^{-p/2} \int_{-\delta}^{\delta} |f'(x)|^p (1-x^2)^{p/2}\, dx \\ &\le (1-\delta^2)^{-p/2}\left(c_1c_2(N+1/p)^p + p^{-p}(1-\delta^2)^{-p/2}\right) \int_{-\delta}^{\delta} f^p(x)\, dx \\ &\le 2^{p/2}(N+1)^p\left(c_1c_2(N+1/p)^p + p^{-p}2^{p/2}(N+1)^p\right) \int_{-\delta}^{\delta} f^p(x)\, dx \\ &\le c_3 N^{2p} \int_{-\delta}^{\delta} f^p(x)\, dx\,, \end{aligned}$$

where c_3 is an absolute constant. Since (A.4.20) is valid for every $f \in \mathrm{GAP}_N$ of the form (A.4.1) with each $r_j \ge 1$, the theorem follows by a linear shift from $[-\delta, \delta]$ to $[-1, 1]$. □

Comments, Exercises, and Examples.

Most of the results of this section can be found in Erdélyi [91a] and [92a]; Erdélyi, Máté, and Nevai [92]; and Erdélyi, Li, and Saff [94]; however, the proofs are somewhat simplified here. For polynomials $f \in \mathcal{P}_n$ and for arbitrary $q \in (0, \infty)$, Theorem A.4.5 was also proved by Kemperman and Lorentz [79]. An early version of Markov's inequality in L_p for ordinary polynomials is proven in Hille, Szegő, and Tamarkin [37]. Weighted Markov- and Bernstein-type analogs of Theorems A.4.6 to A.4.8 are obtained in Erdélyi [92b]. Applications of the inequalities of this section are given in Erdélyi, Magnus, and Nevai [94] and in Erdélyi and Nevai [92], where bounds are established for orthonormal polynomials and related functions associated with (generalized) Jacobi weight functions. Further applications in the theory of orthogonal polynomials may be found in Freud [71] and Erdélyi [91d].

L_p extensions of Theorem 5.1.4 (Bernstein's inequality) and Theorem 5.1.8 (Markov's inequality) have been studied by a number of authors. The sharp L_p version of Bernstein's inequality for trigonometric polynomials was first established by Zygmund [77] for $p \geq 1$. Using an interpolation formula of M. Riesz, he proved that

$$(A.4.21) \qquad \int_{-\pi}^{\pi} |t'(\theta)|^p \, d\theta \leq n^p \int_{-\pi}^{\pi} |t(\theta)|^p \, d\theta$$

for every $t \in \mathcal{T}_n$ (see E.5 h] of Appendix 4). For $0 < p < 1$, first Klein [51] and later Osval'd [76] proved (A.4.21) with a multiplicative constant $c(p)$. Nevai [79a] showed that $c(p) \leq 8p^{-1}$. Subsequently, Máté and Nevai [80] showed the validity of (A.4.21) with a multiplicative absolute constant, and then Arestov [81] proved (A.4.21) (with the best possible constant 1) for every $0 < p < 1$. Golitschek and Lorentz [89] gave a very elegant proof of Arestov's theorem.

The L_p analog of Markov's inequality states that

$$(A.4.22) \qquad \int_{-1}^{1} |Q'(x)|^p \, dx \leq c^{p+1} n^{2p} \int_{-1}^{1} |Q(x)|^p \, dx$$

for every $Q \in \mathcal{P}_n$ and $0 < p < \infty$, where c is an absolute constant. This can be obtained from Arestov's theorem similarly to the way that Theorem A.4.14 is proven from Corollary A.4.13. To find the best possible constant in (A.4.22) is still an open problem even for $p = 2$ or $p = 1$.

The magnitude of

$$(A.4.23) \qquad \frac{\|f'w\|_{[-1,1]}}{\|fw\|_{[-1,1]}},$$

$$(A.4.24) \qquad \frac{|f'(y)w(y)|}{\|fw\|_{[-1,1]}}, \qquad -1 \leq y \leq 1,$$

and their corresponding L_p analogs for $f \in \mathcal{P}_n$ and generalized Jacobi weight functions

$$(A.4.25) \qquad w(z) = \prod_{j=1}^{k} |z - z_j|^{r_j}, \qquad z_j \in \mathbb{C}, \quad r_j \in (-1, \infty)$$

have been examined by several people. See, for example, Ditzian and Totik [88], Khalilova [74], Konjagin [78], Lubinsky and Nevai [87], Nevai [79a] and [79b], and Protapov [60], but a multiplicative constant depending on the weight function appears in these estimates. The magnitude of (A.4.23), (A.4.24), and their L_p analogs are examined in Erdélyi [92b] and [93], when both f and w are generalized nonnegative polynomials. In these inequalities only the degree of f, the degree of w, and a multiplicative absolute constant appear. The most general results are the following:

Theorem A.4.15. *There exists an absolute constant c such that*

$$\int_{-\pi}^{\pi} |f'(\theta)|^p w(\theta)\, d\theta \leq c^{p+1}(N+M)^p (Mp^{-1}+1)^p \int_{-\pi}^{\pi} |f(\theta)|^p w(\theta)\, d\theta$$

and

$$\|f'w\|_{[-\pi,\pi]} \leq c(N+M)(M+1)\|fw\|_{[-\pi,\pi]}$$

for every $f \in \mathrm{GTP}_N$ of the form (A.4.3) with each $r_j \geq 1$, for every $w \in \mathrm{GTP}_M$, and for every $p \in (0,\infty)$.

Theorem A.4.16. *There exists an absolute constant c such that*

$$\int_{-1}^{1} |f'(x)|^p w(x)\, dx \leq c^{p+1}(N+M)^{2p}(Mp^{-1}+1)^{2p} \int_{-1}^{1} |f(x)|^p w(x)\, dx$$

and

$$\|f'w\|_{[-1,1]} \leq c(N+M)^2 \|fw\|_{[-1,1]}$$

for every $f \in \mathrm{GAP}_N$ of the form (A.4.1) with each $r_j \geq 1$, for every $w \in \mathrm{GAP}_M$, and for every $p \in (0,\infty)$.

E.1 Another Representation of Generalized Nonnegative Polynomials.

a] Show that if $f = \prod_{j=1}^{k} |Q_j|^{r_j}$ with each $Q_j \in \mathcal{P}_{n_j}$ and $r_j > 0$, then $f \in \mathrm{GAP}_N$ with $N \leq \sum_{j=1}^{k} r_j n_j$. Similarly, if $f = \prod_{j=1}^{k} |Q_j|^{r_j}$ with each $Q_j \in \mathcal{T}_{n_j}$ and $r_j > 0$, then $f \in \mathrm{GTP}_N$ with $N \leq \sum_{j=1}^{k} r_j n_j$.

b] Show that if $f \in \mathrm{GAP}_N$ is of the form (A.4.1), then $f = \prod_{j=1}^{m} |Q_j|^{r_j/2}$ with each $Q_j \in \mathcal{P}_2$ and $0 \leq Q_j$ on $\mathbb{R}$. Similarly, if $f \in \mathrm{GTP}_N$ is of the form (A.4.3), then $f = \prod_{j=1}^{m} |Q_j|^{r_j/2}$ with each $Q_j \in \mathcal{T}_1$ and $0 \leq Q_j$ on $\mathbb{R}$.

E.2 The Derivative of an $f \in \text{GTP}_N$ with Real Zeros. Show that if $f \in \text{GTP}_N$ is of the form (A.4.3) with each $r_j \geq 1$ and $z_j \in \mathbb{R}$, then $|f'| \in \text{GTP}_N$ is of the form (A.4.3) with each $r_j > 0$ and $z_j \in \mathbb{R}$, and at least one of any two adjacent (in K) zeros of $|f'|$ has multiplicity exactly 1.

E.3 Generalized Nonnegative Polynomials with Rational Exponents. Show that if $f \in \text{GAP}_N$ is of the form (A.4.1) with rational exponents $r_j = q_j/q$, where $q_j, q \in \mathbb{N}$, then $f^{2q} \in \mathcal{P}_{2qN}$, while if $f \in \text{GTP}_N$ is of the form (A.4.3) with rational exponents r_j of the above form then $f^{2q} \in \mathcal{T}_{2qN}$.

E.4 Proof of (A.4.6). Prove (A.4.6) from the explicit formula (2.1.1) for the Chebyshev polynomial T_n.

E.5 Sharpness of the Remez-Type Inequality for GAP_N**.** Let

$$f_n(x) := \left|T_n\left(\frac{2x+s}{2-s}\right)\right|^{N/n} \in \text{GAP}_N\,, \qquad n = 1, 2, \ldots .$$

Show that

$$m(\{x \in [-1,1] : f_n(x) \leq 1\}) = m([-1, 1-s]) = 2 - s$$

and

$$\lim_{n\to\infty} f_n(1) = \left(\frac{\sqrt{2}+\sqrt{s}}{\sqrt{2}-\sqrt{s}}\right)^N .$$

The upper bound in Theorem A.4.1 is actually not achieved by an element of GAP_N; see Erdélyi, Li, and Saff [94].

E.6 Proof of Theorem A.4.2.

Hint: First assume that $f \in \text{GTP}_N$ is of the form (A.4.3) with rational exponents $r_j = q_j/q$, where $q_j, q \in \mathbb{N}$. Then $p := f^{2q} \in \mathcal{T}_{2qN}$, and the desired inequality can be obtained from Theorem 5.1.2 as in the proof of Theorem A.4.1. □

E.7 Corollaries of Theorems A.4.1 and A.4.2.

a] *The inequality*

$$m\left(\left\{x \in [-1,1] : f(x) \geq \exp(-N\sqrt{s})\|f\|_{[-1,1]}\right\}\right) \geq \frac{s}{8}$$

holds for every $f \in \text{GAP}_N$ *and* $0 < s < 2$. *In particular,*

$$m\left(\left\{x \in [-1,1] : f(x) \geq e^{-1}\|f\|_{[-1,1]}\right\}\right) \geq (8N^2+4)^{-1}$$

holds for every $f \in \text{GAP}_N$.

b] *The inequality*

$$m\left(\{\theta \in [-\pi, \pi) : f(\theta) \geq \exp(-Ns)\|f\|_K\}\right) \geq \frac{s}{4}$$

holds for every $f \in \mathrm{GTP}_N$ *and* $0 < s < 2\pi$. *In particular,*

$$m\left(\left\{\theta \in [-\pi, \pi) : f(\theta) \geq e^{-1}\|f\|_K\right\}\right) \geq (4(N+1))^{-1}$$

holds for every $f \in \mathrm{GTP}_N$.

E.8 Nonnegative Trigonometric Polynomials. Part a] restates E.3 c] of Section 2.4. Parts b] and c] discuss simple related observations.

a] Let $t \in \mathcal{T}_n$ be nonnegative on $\mathbb{R}$. Show that there is a $p \in \mathcal{P}_n^c$ such that $t(\theta) = |p(e^{i\theta})|^2$ for every $\theta \in \mathbb{R}$.

b] Let $p \in \mathcal{P}_n^c$ and $t(\theta) := |p(e^{i\theta})|^2$ for every $\theta \in \mathbb{R}$. Then $t \in \mathcal{T}_n$ and t is nonnegative on $\mathbb{R}$.

c] Show that if $t \in \mathcal{T}_n^c$, then there is a $p \in \mathcal{P}_{2n}^c$ such that $|t(\theta)| = |p(e^{i\theta})|$ for every $\theta \in \mathbb{R}$.

E.9 A Crucial Inequality in the Proof of Theorem A.4.5. Show that

$$|P(z)| \leq \left(\frac{2}{1+r}\right)^n |P(rz)|\,, \qquad z \in \partial D\,,\ r \in [0,1]$$

for every $P \in \mathcal{P}_n^c$ having all its zeros outside the open unit disk D.

Hint: Let $P(z) = c\prod_{j=1}^m (z - z_j)$, where $c \in \mathbb{C}$, $z_j \in \mathbb{C} \setminus D$, and $m \leq n$. Show that

$$|z - z_j| \leq \frac{2}{1+r}|rz - z_j|\,, \qquad j = 1, 2, \dots, m\,, \quad r \in [0,1]\,.$$

□

E.10 $f \in \mathrm{GAP}_N$ **Implies** $f(\cos\theta) \in \mathrm{GTP}_N$.

a] Show that if $f \in \mathrm{GAP}_N$, then $g(\theta) = f(\cos\theta) \in \mathrm{GTP}_N$.

b] Show that if $f \in \mathrm{GAP}_N$ is of the form (8.1.1) with each $r_j \geq 1$, then $g(\theta) = f(\cos\theta) \in \mathrm{GTP}_N$ is of the form (A.4.3) with each $r_j \geq 1$.

E.11 A Property of $Q_{n,\omega}$. Let $Q_{n,\omega}$ be defined by (A.4.9) and (A.4.10). Show that there is an absolute constant $c_6 > 1$ such that $Q_{n,\omega}(\pi) \geq c_6$.

Hint: Use the explicit formula (2.1.1) for the Chebyshev polynomial T_n. □

E.12 A Property of the Zeros of a $g \in \text{GTP}_N$. Assume that $g \in \text{GTP}_N$ is of the form (A.4.3) and let

$$M := \sum_{\substack{j=1 \\ z_j \in [\pi-\rho, \pi+\rho]}}^{m} r_j \,.$$

Show that $M \leq 3N\rho\|g\|_{[-1,1]}^{-1}$.

Hint: First assume that each r_j is rational with common denominator $q \in \mathbb{N}$, and apply E.12 of Section 5.1 to $p := g^{2q} \in \mathcal{T}_{2qN}$. □

E.13 Extremal Functions for the Bernstein-Type Inequality.

a] Let $r_j \geq 1$, $j = 1, 2, \ldots, m$, be fixed real numbers. Show that there exists an $\widetilde{f} \in \text{GTP}_N$ of the form

$$\text{(A.4.26)} \qquad \widetilde{f}(z) = \prod_{j=1}^{m} |\sin((z - \widetilde{\zeta}_j)/2)|^{r_j}\,, \qquad \widetilde{\zeta}_j \in \mathbb{C}$$

such that

$$\frac{|\widetilde{f}'(\pi)|}{\|\widetilde{f}\|_K} = \sup_f \frac{|f'(\pi)|}{\|f\|_K}\,,$$

where the supremum in the right-hand side is taken for all $f \in \text{GTP}_N$ of the form

$$\text{(A.4.27)} \qquad f(z) = |\omega| \prod_{j=1}^{m} |\sin((z - z_j)/2)|^{r_j}\,, \qquad z_j \in \mathbb{C}\,, \quad 0 \neq \omega \in \mathbb{C}\,.$$

Hint: Write each f of the form (A.4.27) for which the supremum is taken as

$$f(z) = |\omega_0| \prod_{j=1}^{m} |\omega_j \sin((z - z_j)/2) \sin((z - \overline{z}_j)/2)|^{r_j/2}\,,$$

where the numbers $\omega_j > 0$ are defined by

$$\|\omega_j \sin((z - z_j)/2) \sin((z - \overline{z}_j)/2)\|_K = 1\,, \qquad j = 1, 2, \ldots, m\,,$$

and then use a compactness argument for each factor separately. □

b] Let $\widetilde{f}$ be as in part a]. Show that each zero of $\widetilde{f}$ is real, that is, $\widetilde{\zeta}_j \in \mathbb{R}$ for each j in (A.4.26).

Hint: Assume that there is an index $1 \leq j \leq m$ such that $\zeta_j \in \mathbb{C} \setminus \mathbb{R}$. Then

$$\widetilde{f}_\epsilon(z) := \widetilde{f}(z) \left(1 - \frac{\epsilon \sin^2((z - \pi)/2)}{\sin((z - \zeta_j)/2) \sin((z - \overline{\zeta}_j)/2)}\right)^{r_j} \in \text{GTP}_N$$

with sufficiently small $\epsilon > 0$ contradicts the maximality of $\widetilde{f}$. □

E.14 A Corollary of the Remez-Type Inequality for GAP_N. *Suppose that* $\alpha := 1 - (1+N)^{-2}$. *Show that there is an absolute constant* $c_9 > 0$ *such that*

$$\|f\|_{[-1,1]} \le c_9 \|f\|_{[-\alpha,\alpha]}$$

for every $f \in \mathrm{GAP}_N$.

Hint: This is a corollary of Theorem A.4.1. □

E.15 Proof of Theorem A.4.10.

Outline. For $f \in \mathrm{GAP}_N$, let

$$I(f) := \left\{ x \in [-1,1] : (\chi(f(x)))^p \ge \exp(-5pN\sqrt{2s}) \|\chi(f)\|^p_{[-1,1]} \right\}.$$

From Theorem 5.1.1, $0 < s \le 1/2$, and the assumptions prescribed for χ it follows that $m(I(f)) \ge 2s$. Let $I := A \cap I(f)$. Since $m([-1,1] \setminus A) \le s$, $m(I) \ge s$. Therefore

$$\begin{aligned}
\int_{[-1,1]\setminus A} (\chi(f(x)))^p dx &\le \int_{[-1,1]\setminus A} \|\chi(f)\|^p_{[-1,1]}\, dx \\
&\le \exp\big(5pN\sqrt{2s}\big) \int_I (\chi(f(x)))^p\, dx \\
&\le \exp\big(5pN\sqrt{2s}\big) \int_A (\chi(f(x)))^p\, dx\,.
\end{aligned}$$

□

E.16 Proof of Theorem A.4.11.

Hint: For $f \in \mathrm{GTP}_N$, let

$$I(f) := \left\{ \theta \in [-\pi,\pi] : (\chi(f(\theta)))^p \ge \exp(-8pNs) \|\chi(f)\|^p_K \right\}.$$

From Theorem 5.2.2, $0 < s \le \pi/4$, and the assumptions for χ, it follows that $m(I(f)) \ge 2s$. Now finish the proof as in the hint for E.1. □

E.17 Sharpness of Theorem A.4.10.

a] Show that there exists a sequence of polynomials $Q_n \in \mathcal{P}_n$ and an absolute constant $c > 0$ such that

$$\int_{-1}^{1} |Q_n(x)|^p\, dx \ge s\big(1 + \exp\big(cpn\sqrt{s}\big)\big) \int_{-1}^{1-s} |Q_n(x)|^p\, dx$$

for every $n \in \mathbb{N}$, $s \in (0,1]$, and $p \in (0,\infty)$.

Hint: Study the Chebyshev polynomials T_n transformed linearly from $[-1,1]$ to $[-1,1-s]$. □

b] Show that there exist a sequence of trigonometric polynomials $t_n \in \mathcal{T}_n$ and an absolute constant $c > 0$ such that

$$\int_{-\pi}^{\pi} |t_n(\theta)|^p \, d\theta \ge s(1 + \exp(cpns)) \int_{-\pi+s/2}^{\pi-s/2} |t_n(\theta)|^p \, d\theta$$

for every $p \in \mathbb{N}$, $s \in (0, \pi]$, and $p \in (0, \infty)$.

Hint: Study $Q_{n,\omega}$ defined by (A.4.9) with $\omega := \pi - s/2$. □

E.18 Sharpness of Corollary A.4.13 and Theorem A.4.14.

a] Find a sequence of trigonometric polynomials $t_n \in \mathcal{T}_n$ that shows the sharpness of Corollary A.4.13 up to the constant $c > 0$ for every $p \in (0, \infty)$ simultaneously.

Hint: Take $t_n(\theta) := \cos n\theta$. □

b] For every $p \in (0, \infty)$, find a sequence of polynomials $Q_{n,p} \in \mathcal{P}_n$ which shows the sharpness of Theorem A.4.14 up to the constant $c > 0$.

Outline. Let $L_k \in \mathcal{P}_k$ be the kth orthonormal Legendre polynomial on $[-1, 1]$ (see E.5 of Section 2.3), and let

$$G_m(x) := c \sum_{k=0}^{m} L_k'(1) L_k(x) \,,$$

where c is chosen so that

$$\int_{-1}^{1} G_m^2(x) \, dx = 1 \,. \tag{A.4.28}$$

Show that there exist absolute constants $c_1 > 0$ and $c_2 > 0$ (independent of m) such that

$$|G_m(1)| \ge c_1 m \qquad \text{and} \qquad |G_m'(1)| \ge c_2 m^3 \,. \tag{A.4.29}$$

For a fixed $n \in \mathbb{N}$, let $u := \lfloor 2/p \rfloor + 1$ and $m := \lfloor n/u \rfloor$. If $m \ge 1$, then let $Q_{n,p} := G_m^u \in \mathcal{P}_n$, otherwise let $Q_{n,p}(x) := x \in \mathcal{P}_n$. If $m = 0$, then the calculation is trivial, so let $m \ge 1$. Using (A.4.29) and the Nikolskii-type inequality of Theorem A.4.4, show that there exists an absolute constant $c_3 > 0$ such that

$$\int_{-1}^{1} |Q_{n,p}'(x)|^p \, dx \ge c_3^{p+1} m^{(u+2)p} (1 + pn)^{-2} \,. \tag{A.4.30}$$

Use the inequality

$$\int_{-1}^{1} |Q_{n,p}(x)|^p \, dx = \int_{-1}^{1} |G_m(x)|^{up} \, dx \le \|G_m\|_{[-1,1]}^{up-2} \int_{-1}^{1} G_m^2(x) \, dx \,,$$

the Nikolskii-type inequality of Theorem A.4.4, and (A.4.28) to show that there is an absolute constant $c_4 > 0$ such that

$$\int_{-1}^{1} |Q_{n,p}(x)|^p \, dx \le c_4^{p+1} m^{up-2} . \tag{A.4.31}$$

Finally, combine (A.4.30) and (A.4.31). Note that if $p > 2$, then $m = \lfloor n \rfloor$, while if $p \le 2$, then $\frac{1}{4}p(n-1) \le m \le 2pn$. Note also that p^p is between two positive bounds for $p \in (0, 2]$. □

E.19 Sharpness of Theorems A.4.3 and A.4.4.

a] Let $q \in (0, \infty)$ be fixed. Show that there exists a sequence of polynomials $Q_{n,q} \in \mathcal{P}_n$ and an absolute constant $c > 0$ such that

$$\|Q_{n,q}\|_{[-1,1]} \ge c^{1+1/q}(1+qn)^{2/q}\|Q_{n,q}\|_{L_q[-1,1]}$$

for every $n \in \mathbb{N}$.

Hint: Study $Q_{n,p}$ with $p = q$ in the hint to the previous exercise. □

b] Let $Q_{n,q} \in \mathcal{P}_n$ be the same as in part a]. Show that there exist absolute constants $c_1 > 0$ and $c_2 > 0$ such that

$$\|Q_{n,q}\|_{L_p[-1,1]} \ge c_1^{1+1/q} c_2^{1/p}(1+qn)^{2/q}(1+pn)^{-2/p}\|Q_{n,q}\|_{L_p[-1,1]}$$

for every $n \in \mathbb{N}$ and $0 < q < p \le \infty$.

Hint: Combine part a] and the Nikolskii-type inequality of Theorem A.4.4. □

c] Let $q \in (0, \infty)$ be fixed. Show that there exists a sequence of trigonometric polynomials $t_{n,q} \in \mathcal{T}_n$ and an absolute constant $c > 0$ such that

$$\|t_{n,q}\|_K \ge c^{1+1/q}(1+qn)^{1/q}\|t_{n,q}\|_{L_q(K)}$$

for every $n \in \mathbb{N}$.

Hint: Let $t_{n,q}(\theta) := Q_{n,q}(\cos\theta)$, where $Q_{n,q}$ are the same as in part a]. Use part a] and the Schur-type inequality of Theorem A.4.5. □

d] Let $t_{n,q} \in \mathcal{T}_n$ be the same as in part c]. Show that there exist absolute constants $c_1 > 0$ and $c_2 > 0$ such that

$$\|t_{n,q}\|_{L_p(K)} \ge c_1^{1+1/q} c_2^{1/p}(1+qn)^{1/q}(1+pn)^{1/p}\|t_{n,q}\|_{L_q(K)}$$

for every $n \in \mathbb{N}$ and $0 < q < p \le \infty$.

Hint: Combine part a] and the Nikolskii-type inequality of Theorem A.4.3. □

E.20 Jensen's Inequality. Let χ be a real-valued convex function defined on $[0, \infty)$, and let f and w be nonnegative Riemann integrable functions on the interval $[a, b]$, where $\int_a^b w = 1$. Show that

$$\chi\left(\int_a^b f(x)w(x)\,dx\right) \le \int_a^b \chi(f(x))w(x)\,dx\,.$$

Hint: First assume that w is a step function. Use the fact that the convexity of χ implies its continuity, and hence the functions $\chi(f)w$ and fw are Riemann integrable. □

E.21 A Pointwise Remez-Type Inequality for GAP_N**.**

a] Show that there exists an absolute constant $c > 0$ such that

$$|p(y)| \le \exp\left(cn \min\left\{\frac{s}{\sqrt{1-y^2}}, \sqrt{s}\right\}\right), \qquad y \in [-1, 1]$$

for every $p \in \mathcal{P}_n$ and $s \in (0, 1]$ satisfying

$$m(\{x \in [-1, 1] : |p(x)| \le 1\}) \ge 2 - s\,,$$

that is, with the notation of Theorem 5.1.1, for every $p \in \mathcal{P}_n(s)$ with $s \in (0, 1]$.

Proof. Assume, without loss of generality, that $y \in [0, 1]$. Let

$$a := y + \tfrac{1}{2}(1 - y), \qquad \alpha := \arccos a\,,$$

$$\beta := 2\arccos y - \arccos a\,, \qquad \omega := \pi - \tfrac{1}{2}(\beta - \alpha)\,,$$

$$Q_{n,\omega}(\theta) := T_{2n}\left(\frac{\sin(\theta/2)}{\sin(\omega/2)}\right),$$

where T_{2n} is the Chebyshev polynomial of degree $2n$ defined by (2.1.1), and let

$$Q_{n,\alpha,\beta}(\theta) := Q_{n,\omega}\left(\tfrac{1}{2}(\theta - (2\pi - (\alpha + \beta)))\right).$$

Associated with $p \in \mathcal{P}_n(s)$, we introduce

$$g(\theta) := p(\cos\theta)Q_{n,\alpha,\beta}(\theta) \in \mathcal{T}_{2n}\,.$$

Obviously

$$|Q_{n,\alpha,\beta}(\theta)| \le 1\,, \qquad \theta \in [0, 2\pi) \setminus (\alpha, \beta)$$

and

$$\|Q_{n,\alpha,\beta}\|_{[0,2\pi]} = Q_{n,\alpha,\beta}\left(\tfrac{1}{2}(\alpha + \beta)\right) = Q_{n,\omega}(\pi)\,.$$

The definition of ω implies that there exist absolute constants $c_1 > 0$ and $c_2 > 0$ such that

$$c_1\sqrt{1-y^2} \le \pi - \omega \le c_2\sqrt{1-y^2}\,.$$

This, together with E.3 c], yields that there are absolute constants $c_3 > 0$ and $c_4 > 0$ such that

$$Q_{n,\omega}(\pi) \ge \exp\big(c_3 n\sqrt{1-y^2}\big) \ge \exp\big(5n\sqrt{s}\big)$$

whenever $y \in [0, 1 - c_4 s]$. It follows now from Theorem 5.1.1 that

$$|g(\theta)| \le \exp\big(5n\sqrt{s}\big) \le Q_{n,\omega}(\pi)$$

for every $\theta \in [0, 2\pi) \setminus (\alpha, \beta)$ and $y \in [0, 1 - c_4 s]$. Furthermore

$$|g(\theta)| \le Q_{n,\omega}(\pi)$$

for every $\theta \in (\alpha, \beta)$ for which $|p(\cos\theta)| \le 1$. Note that

$$|\cos\theta| \le 1 - \tfrac{1}{2}(1-y)\,, \qquad \theta \in (\alpha, \beta)$$

and hence $p \in \mathcal{P}_n(s)$ with $s \in (0, 1]$ implies that there exists an absolute constant $c_5 > 0$ such that

$$m(\{\theta \in (\alpha, \beta) : |p(\cos\theta)| \ge 1\}) \le \frac{c_5 s}{\sqrt{1-y^2}}\,.$$

Therefore

$$f := (Q_{n,\omega}(\pi))^{-1} g \in \mathcal{T}_{2n}$$

satisfies

$$m(\{\theta \in [0, 2\pi) : |f(\theta)| \le 1\}) \ge 2\pi - \widetilde{s}$$

with

$$\widetilde{s} := \frac{c_5 s}{\sqrt{1-y^2}}\,, \qquad y \in [0, 1 - c_4 s]\,.$$

Applying Theorem 5.1.2 with f and $\widetilde{s}$, we conclude that

$$\begin{aligned} |p(y)| &= \left|p\left(\cos\left(\tfrac{1}{2}(\alpha+\beta)\right)\right)\right| = (Q_{n,\omega}(\pi))^{-1} g\left(\tfrac{1}{2}(\alpha+\beta)\right) \\ &= f\left(\tfrac{1}{2}(\alpha+\beta)\right) \le \exp(4n\widetilde{s}) \le \exp\left(\frac{c_5 s}{\sqrt{1-y^2}}\right) \end{aligned}$$

whenever $\widetilde{s} \in (0, \pi/2]$ and $y \in [0, 1 - c_4 s]$. If $s \in (0, 1]$, but $\widetilde{s} > \pi/2$ or $y \in [1 - c_4 s, 1]$, then Theorem 5.1.1 yields the required inequality. □

b] Show that the inequality of part a] is sharp up to the absolute constant $c > 0$.

Hint: Assume, without loss, that $y \in [0, 1]$. Show that there exists an absolute constant $c_1 > 0$ such that the polynomials

$$W_{n,y,s}(x) := T_n \left(\frac{2x}{2-x} + \frac{s}{2-s} \right) \in \mathcal{P}_n(s)\,,$$

where T_n is the Chebyshev polynomial of degree n defined by (2.1.1), satisfy

$$|W_{n,y,s}(y)| \geq \exp(c_1 n \sqrt{s})$$

for every $n \in \mathbb{N}$, $y \in [1 - s/2, 1]$, and $s \in (0, 1]$.

If $y \in [0, 1 - s/2]$, then let

$$a := y + \frac{s}{4}\,, \qquad \alpha := \arccos a\,,$$

$$\beta := 2 \arccos y - \arccos a\,, \qquad \omega := \pi - \frac{\beta - \alpha}{2}\,,$$

$$Q_{n,\alpha,\beta}(\theta) := T_{2n} \left(\frac{\sin((\theta + \pi - (\alpha + \beta)/2)/2)}{\sin(\omega/2)} \right),$$
$$R_{n,\alpha,\beta}(\theta) := \tfrac{1}{2}(Q_{n,\alpha,\beta}(\theta) + Q_{n,\alpha,\beta}(-\theta))\,,$$

and we define $W_{n,y,s}$ for every $n \in \mathbb{N}$, $y \in [0, 1 - s/2)$, and $s \in (0, 1]$ by

$$R_{n,\alpha,\beta}(\theta) = W_{n,y,s}(\cos\theta)\,, \qquad W_{n,y,s} \in \mathcal{P}_n\,.$$

Show that $W_{n,y,s} \in \mathcal{P}_n(s)$ and that there exists an absolute constant $c > 0$ such that

$$|W_{n,y,s}(y)| \geq \exp\left(cn \frac{s}{\sqrt{1 - y^2}} \right)$$

for every $n \in \mathbb{N}$, $y \in [0, 1 - s/2)$, and $s \in (0, 1]$. □

c] Extend the validity of part a] to the class GAP_N; that is, prove that there exists an absolute constant $c > 0$ such that

$$|f(y)| \leq \exp\left(cN \min\left\{ \frac{s}{\sqrt{1 - y^2}}, \sqrt{s} \right\} \right), \qquad y \in [-1, 1]$$

for every $f \in \mathrm{GAP}_N$ and $s \in (0, 1]$ satisfying

$$m(\{x \in [-1, 1] : f(x) \leq 1\}) \geq 2 - s\,.$$

A5

Inequalities for Polynomials with Constraints

Overview

This appendix deals with inequalities for constrained polynomials. Typically the constraints are on the location of the zeros, though various coefficient constraints are also considered.

Inequalities for Polynomials with Constraints

For integers $0 \leq k \leq n$, let

$$\mathcal{P}_{n,k} := \{p \in \mathcal{P}_n : p \text{ has at most } k \text{ zeros in } D\}$$

where, as before, $D := \{z \in \mathbb{C} : |z| < 1\}$. For $a < b$, let

$$\mathcal{B}_n(a,b) := \left\{p \in \mathcal{P}_n : p(x) = \pm \sum_{j=0}^{n} \alpha_j (b-x)^j (x-a)^{n-j}\,, \quad \alpha_j \geq 0\right\}.$$

For integers $0 \leq k \leq n$, let

$$\widetilde{\mathcal{P}}_{n,k}(a,b) := \{p = hq : h \in \mathcal{B}_{n-k}(a,b)\,,\ \ q \in \mathcal{P}_k\}\,.$$

Two useful relations, given in E.1, are

$$\mathcal{P}_{n,0} \subset \mathcal{B}_n(-1,1) \qquad \text{and} \qquad \mathcal{P}_{n,k} \subset \widetilde{\mathcal{P}}_{n,k}(-1,1)\,.$$

Theorem A.5.1 (Markov-Type Inequality for $\widetilde{\mathcal{P}}_{n,k}$). *The inequality*

$$\|p^{(m)}\|_{[-1,1]} \leq (18n(k+2m+1))^m \|p\|_{[-1,1]}$$

holds for every $p \in \widetilde{\mathcal{P}}_{n,k}(-1,1)$. (When $m=1$, the constant 18 can be replaced by 9.)

Proof. First we study the case $m=1$. Applying Theorem 6.1.1 (Newman's inequality) with

$$(\lambda_0, \lambda_1, \ldots, \lambda_k) = (n-k, n-k+1, \ldots, n)$$

and using a linear shift from $[0,1]$ to $[-1,1]$, we obtain that

$$|p'(1)| \leq \tfrac{9}{2} n(k+1) \|p\|_{[-1,1]} \tag{A.5.1}$$

for every $p \in \widetilde{\mathcal{P}}_{n,k}(-1,1)$ of the form

$$p(x) = (x+1)^{n-k} q(x)\,, \qquad q \in \mathcal{P}_k\,. \tag{A.5.2}$$

Now let $p \in \widetilde{\mathcal{P}}_{n,k}(-1,1)$ be of the form $p = hq$ with $q \in \mathcal{P}_k$ and

$$h(x) = \sum_{j=0}^{n-k} \alpha_j (1-x)^j (x+1)^{n-k-j} \qquad \text{with each } \alpha_j \geq 0\,.$$

Without loss of generality, we may assume that $n-k \geq 1$, otherwise Theorem 5.1.8 (Markov's inequality) gives the theorem. Using (A.5.1) and the fact that each $\alpha_j \geq 0$, we get

$$\begin{aligned}
(A.5.3)\quad & |p'(1)| \\
&= \left| \sum_{j=0}^{n-k} \left(\alpha_j (1-x)^j (x+1)^{n-k-j} q(x)\right)'(1) \right| \\
&= \left| \sum_{j=0}^{1} (\alpha_j (1-x)^j (x+1)^{n-k-j} q(x))'(1) \right| \\
&\leq \tfrac{9}{2} n(k+2) \left\| (x+1)^{n-k-1} (\alpha_0 (x+1) + \alpha_1 (1-x)) q(x) \right\|_{[-1,1]} \\
&\leq \tfrac{9}{2} n(k+2) \left\| \sum_{j=0}^{n-k} \alpha_j (1-x)^j (x+1)^{n-k-j} q(x) \right\|_{[-1,1]} \\
&\leq \tfrac{9}{2} n(k+2) \|p\|_{[-1,1]}\,.
\end{aligned}$$

Now let $y \in [-1,1]$ be arbitrary. To estimate $|p'(y)|$ we distinguish two cases.

If $y \in [0,1]$, then by a linear shift from $[-1,1]$ to $[-1,y]$, we obtain from (A.5.3) that

$$(A.5.4) \qquad |p'(y)| \leq \frac{9}{y+1} n(k+2)\|p\|_{[-1,y]} \leq 9n(k+2)\|p\|_{[-1,1]}$$

for every $p \in \widetilde{\mathcal{P}}_{n,k}(-1,y)$. It follows from E.1 d] that

$$\widetilde{\mathcal{P}}_{n,k}(-1,1) \subset \widetilde{\mathcal{P}}_{n,k}(-1,y)\,.$$

So (A.5.4) holds for every $p \in \widetilde{\mathcal{P}}_{n,k}(-1,1)$.

If $y \in [-1,0]$, then by a linear shift from $[-1,1]$ to $[y,1]$, we obtain from (A.5.3) that

$$(A.5.5) \qquad |p'(y)| \leq \frac{9}{1-y} n(k+2)\|p\|_{[y,1]} \leq 9n(k+2)\|p\|_{[-1,1]}$$

for every $p \in \widetilde{\mathcal{P}}_{n,k}(y,1)$. By E.1 d] again,

$$\widetilde{\mathcal{P}}_{n,k}(-1,1) \subset \widetilde{\mathcal{P}}_{n,k}(y,1)\,.$$

So (A.5.5) holds for every $p \in \widetilde{\mathcal{P}}_{n,k}(-1,1)$, which finishes the case when $m = 1$.

Now we turn to the case when $m \geq 2$. Note that an induction on m does not work directly for an arbitrary $p \in \widetilde{\mathcal{P}}_{n,k}(-1,1)$. However, it follows by induction on m that

$$(A.5.6) \qquad \|p^{(m)}\|_{[-1,1]} \leq (9n(k+m+1))^m \|p\|_{[-1,1]}$$

for every $p \in \widetilde{\mathcal{P}}_{n,k}(-1,1)$ of the form (A.5.2). Now let $p \in \widetilde{\mathcal{P}}_{n,k}(-1,1)$ be of the form $p = hq$, where $q \in \mathcal{P}_k$ and

$$h(x) = \sum_{j=0}^{n-k} \alpha_j (1-x)^j (x+1)^{n-j} \quad \text{with each } \alpha_j \geq 0.$$

For notational convenience let $s := \min\{n-k, m\}$. Using (A.5.6) and the fact that each $\alpha_j \geq 0$, we get

$$\begin{aligned}
(A.5.7) \quad & |p^{(m)}(1)| \\
&= \left| \sum_{j=0}^{n-k} (\alpha_j (1-x)^j (1+x)^{n-k-j} q(x))^{(m)}(1) \right| \\
&= \left| \sum_{j=0}^{s} (\alpha_j (1-x)^j (1+x)^{n-k-j} q(x))^{(m)}(1) \right| \\
&\leq (9n(k+s+m+1))^m \left\| \sum_{j=0}^{s} \alpha_j (1-x)^j (x+1)^{n-k-j} q(x) \right\|_{[-1,1]} \\
&\leq (9n(k+2m+1))^m \left\| \sum_{j=0}^{n-k} \alpha_j (1-x)^j (x+1)^{n-k-j} q(x) \right\|_{[-1,1]} \\
&= (9n(k+2m+1))^m \|p\|_{[-1,1]}\,.
\end{aligned}$$

From (A.5.7), in a similar fashion to the case when $m = 1$, we conclude that

$$\|p^{(m)}\|_{[-1,1]} \le (18n(k+2m+1))^m \|p\|_{[-1,1]}$$

for every $p \in \widetilde{\mathcal{P}}_{n,k}(-1,1)$, which finishes the proof. □

Theorem A.5.1 is essentially sharp as is shown in E.2 and E.4 f]. However, a better upper bound can be given for $|p^{(m)}(y)|$ when $y \in (-1,1)$ is away from the endpoints -1 and 1.

Theorem A.5.2 (Markov-Bernstein Type Inequality for $\mathcal{P}_{n,k}$). *There exists a constant $c(m)$ depending only on m such that*

$$|p^{(m)}(y)| \le c(m) \left(\min\left\{ n(k+1)\,, \frac{\sqrt{n(k+1)}}{\sqrt{1-y^2}} \right\} \right)^m \|p\|_{[-1,1]}$$

for every $p \in \mathcal{P}_{n,k}$ and $y \in [-1,1]$.

Theorem A.5.2 has been proved in Borwein and Erdélyi [94]. Its proof is long, and we do not reproduce it here. However, the proof of a less sharp version, where $\sqrt{1-y^2}$ is replaced by $1-y^2$, is outlined in E.4 and E.5. The factor $\sqrt{n(k+1)}$ in Theorem A.5.2 is essentially sharp in the case that $m=1$; see E.7.

Theorem A.5.3 (Markov-Bernstein Type Inequality for $\mathcal{B}_n(-1,1)$). *There exists a constant $c(m)$ depending only on m such that*

$$|p^{(m)}(y)| \le c(m) \left(\min\left\{ n\,, \frac{\sqrt{n}}{\sqrt{1-y^2}} \right\} \right)^m \|p\|_{[-1,1]}$$

for every $p \in \mathcal{B}_n(-1,1)$ (hence for every $p \in \mathcal{P}_{n,0}$) and $y \in [-1,1]$.

Note that the uniform (Markov-type) upper bound of the above theorem is a special case ($k=0$) of Theorem A.5.1. Our proof of Theorem A.5.3 offers another way to prove the case of Theorem A.5.1 when $k=0$. First we need a lemma.

Lemma A.5.4. *For $n \in \mathbb{N}$ and $y \in \mathbb{R}$, let*

$$\Delta_{n,y} := \frac{1}{4}\left(\frac{\sqrt{(1-y^2)^+}}{\sqrt{n}} + \frac{1}{n} \right),$$

where $x^+ := \max\{x,0\}$. Then

$$|p(y+i\gamma\Delta_{n,y})| \le \sqrt{2}e\left|p\left(y \pm \tfrac{1}{2n}\right)\right|$$

for every $p \in \mathcal{B}_n(-1,1)$, $y \in \left[-1-\frac{1}{8n}, 1+\frac{1}{8n}\right]$, and $\gamma \in [0,1]$, where i is the imaginary unit. The $+$ sign is taken if $y \in \left[-1-\frac{1}{8n}, 0\right)$, and the $-$ sign is taken if $y \in \left[0, 1+\frac{1}{8n}\right]$.

Proof. It is sufficient to prove the lemma for the polynomials

$$p_{n,j}(x) := (1-x)^j(x+1)^{n-j}\,, \qquad n \in \mathbb{N}\,, \quad j = 0, 1, \ldots, n\,.$$

The general case when

$$p = \sum_{j=0}^{n} \alpha_j p_{n,j} \qquad \text{with each } \alpha_j \geq 0 \text{ or each } \alpha_j \leq 0$$

can then be obtained by taking linear combinations. Without loss of generality, we may assume that $y \in \left[0, 1 + \frac{1}{8n}\right]$. Then

$$\Delta_{n,y}^2 \leq \frac{1}{8}\left(\frac{(1-y^2)^+}{n} + \frac{1}{n^2}\right) \leq \frac{(1-y)^+}{4n} + \frac{1}{8n^2}$$

from which it follows that

$$\begin{aligned}
|p_{n,j}(y + i\gamma\Delta_{n,y})| &\leq |p_{n,j}(y + i\Delta_{n,y})| \\
&= \left((1-y)^2 + \Delta_{n,y}^2\right)^{j/2}\left((1+y)^2 + \Delta_{n,y}^2\right)^{(n-j)/2} \\
&\leq \left((1-y) + \tfrac{1}{2n}\right)^j\left((1+y) + \tfrac{1}{2n}\right)^{n-j} \\
&= \left(1 - \left(y - \tfrac{1}{2n}\right)\right)^j\left(1 + \left(y - \tfrac{1}{2n}\right)\right)^{n-j}\left(\frac{1 + y + \frac{1}{2n}}{1 + y - \frac{1}{2n}}\right)^{n-j} \\
&\leq \left(\frac{2n+1}{2n-1}\right)^n p_{n,j}\left(y - \tfrac{1}{2n}\right) \leq \sqrt{2}e p_{n,j}\left(y - \tfrac{1}{2n}\right)\,,
\end{aligned}$$

and the lemma is proved. □

Proof of Theorem A.5.3. For $n \in \mathbb{N}$ and $y \in [-1, 1]$, let $B_{n,y}$ denote the circle of the complex plane with center y and radius $\frac{1}{4}\Delta_{n,y}$. Using E.4, Lemma A.5.4, and the maximum principle, we obtain

$$|p(z)| \leq \sqrt{2}e\, \|p\|_{[-1,1]}$$

for every $p \in \mathcal{B}_n(-1, 1)$, $z \in B_{n,y}$, $n \in \mathbb{N}$, and $y \in [-1, 1]$. Hence, by Cauchy's integral formula,

$$\begin{aligned}
|p^{(m)}(y)| &= \left|\frac{m!}{2\pi i}\int_{B_{n,y}} \frac{p(\xi)}{(\xi - y)^{m+1}}\, d\xi\right| \\
&\leq \frac{m!}{2\pi}\int_{B_{n,y}} \left|\frac{p(\xi)}{(\xi - y)^{m+1}}\right| |d\xi| \\
&\leq \frac{m!}{2\pi}\, 2\pi\, \tfrac{1}{4}\Delta_{n,y}\left(\tfrac{1}{4}\Delta_{n,y}\right)^{-(m+1)}\sqrt{2}e\|p\|_{[-1,1]} \\
&\leq c(m)\left(\min\left\{n\,, \frac{\sqrt{n}}{\sqrt{1-y^2}}\right\}\right)^m \|p\|_{[-1,1]}
\end{aligned}$$

for every $p \in \mathcal{B}_n(-1, 1)$ and $y \in [-1, 1]$, which proves the theorem. □

For $r \in (0,1]$, let

$$D_r^{\pm} := \{z \in \mathbb{C} : |z \pm (1-r)| < r\}.$$

For $n \in \mathbb{N}$, $k = 0, 1, \dots, n$, and $r \in (0,1]$, let $\mathcal{P}_{n,k,r}^{\pm}$ be the set of all $p \in \mathcal{P}_n$ that have at most k zeros (counting multiplicities) in $D_r^{\pm}$, and let $\mathcal{P}_{n,k,r} := \mathcal{P}_{n,k,r}^{+} \cap \mathcal{P}_{n,k,r}^{-}$. The following result is proved in Erdélyi [89a]. The proof in a special case is outlined in E.9.

Theorem A.5.5 (Markov-Type Inequality for $\mathcal{P}_{n,k,r}$). *There exists a constant $c(m)$ depending only on m such that*

$$\|p^{(m)}\|_{[-1,1]} \le c(m) \left(\frac{n(k+1)^2}{\sqrt{r}} \right)^m \|p\|_{[-1,1]}$$

for every $p \in \mathcal{P}_{n,k,r}$, $m \in \mathbb{N}$, and $r \in (0,1]$.

We state, without proof, the L_q analogs of Theorem A.5.2 for $m = 1$, established in Borwein and Erdélyi [to appear 2].

Theorem A.5.6 (L_q Markov-Type Inequality for $\mathcal{P}_{n,k}$). *There exists an absolute constant c such that*

$$\int_{-1}^{1} |p'(x)|^q \, dx \le c^{q+1} (n(k+1))^q \int_{-1}^{1} |p(x)|^q \, dx$$

for every $p \in \mathcal{P}_{n,k}$ and $q \in (0, \infty)$.

Theorem A.5.7 (L_q Bernstein-Type Inequality for $\mathcal{P}_{n,k}$). *There exists an absolute constant c such that*

$$\int_{-\pi}^{\pi} |p'(\cos t) \sin t|^q \, dt \le c^{q+1} (n(k+1))^{q/2} \int_{-\pi}^{\pi} |p(\cos t)|^q \, dt$$

for every $p \in \mathcal{P}_{n,k}$ and $q \in (0, \infty)$.

Both of the above inequalities are sharp up to the absolute constant $c > 0$.

A Markov-type inequality for polynomials having at most k zeros in the disk

$$D_r := \{z \in \mathbb{C} : \ |z| < r\}$$

is given by the following theorem; see Erdélyi [90a].

Theorem A.5.8 (Markov-Type Inequality for Polynomials with At Most k Zeros in D_r). *Let $k \in \mathbb{N}$ and $r \in (0,1]$. Then there exist constants $c_1(k) > 0$ and $c_2(k) > 0$ depending only on k so that*

$$c_1(k)(n + (1-r)n^2) \le \sup_p \frac{\|p'\|_{[-1,1]}}{\|p\|_{[-1,1]}} \le c_2(k)(n + (1-r)n^2),$$

where the supremum is taken for all $p \in \mathcal{P}_n$ that have at most k zeros in D_r.

Comments, Exercises, and Examples.

Erdős [40] proved that

$$\|p'\|_{[-1,1]} \leq \frac{n}{2}\left(\frac{n}{n-1}\right)^{n-1} \|p\|_{[-1,1]} < \frac{e}{2}n \, \|p\|_{[-1,1]}$$

for every $p \in \mathcal{P}_{n,0}$ $n \geq 2$, having only real zeros. In this result, the assumption that p has only real zeros can be dropped. In E.11 we outline the proof of the above inequality for every $p \in \mathcal{P}_{n,0}$, $n \geq 2$. The polynomials

$$p_n(x) := (x+1)^{n-1}(1-x)\,, \qquad n = 1, 2, \dots\,,$$

show that Erdős's result is the best possible for $\mathcal{P}_{n,0}$. Erdős [40] also showed that there exists an absolute constant c such that

$$|p'(y)| \leq \frac{c\sqrt{n}}{(1-y^2)^2} \|p\|_{[-1,1]}$$

for every $p \in \mathcal{P}_{n,0}$ having only real zeros. Markov- and Bernstein-type inequalities for $\mathcal{B}_n(-1,1)$ were first established by Lorentz [63], who proved Theorem A.5.3. Lorentz's approach is outlined for $m = 1$ in E.8. The proof presented in the text follows Erdélyi [91c]. Up to the constant $c(m) > 0$ Theorem A.5.3 is sharp; see E.12. Scheick [72] found the best possible asymptotic constant in Lorentz's Markov-type inequality for $m = 1$ and $m = 2$. He proved the inequalities

$$\|p'\|_{[-1,1]} \leq \frac{e}{2}n \, \|p\|_{[-1,1]} \quad \text{and} \quad \|p''\|_{[-1,1]} \leq \frac{e}{2}n(n-1)\|p\|_{[-1,1]}$$

for every $p \in \mathcal{B}_n(-1,1)$ (and hence for every $p \in \mathcal{P}_{n,0}$). Note that with $p_n(x) := (x+1)^{n-1}(1-x)$,

$$\frac{\|p_n'\|_{[-1,1]}}{n\|p_n\|_{[-1,1]}} \to \frac{e}{2} \quad \text{and} \quad \frac{\|p_n''\|_{[-1,1]}}{n(n-1)\|p\|_{[-1,1]}} \to \frac{e}{2} \quad \text{as } n \to \infty\,.$$

A slightly weaker version of Theorem A.5.1 was conjectured by Szabados [81], who gave polynomials $p_{n.k} \in \mathcal{P}_{n,k}$ with only real zeros so that

$$|p_{n,k}'(1)| \geq \tfrac{1}{3}n(k+1)\|p_{n,k}\|_{[-1,1]}$$

for all integers $0 \leq k \leq n$. After some results of Szabados and Varma [80] and Máté [81], Szabados's conjecture has been settled in Borwein [85], where it is shown that

$$\|p'\|_{[-1,1]} \leq 9n(k+1)\|p\|_{[-1,1]}$$

for every $p \in \mathcal{P}_{n,k}$ having $n-k$ zeros in $\mathbb{R} \setminus (-1,1)$. The crucial part of this proof is outlined in E.4 d] and e]. The above inequality is extended to all $p \in \widetilde{\mathcal{P}}_{n,k}$ (and hence to all $p \in \mathcal{P}_{n,k}$) and to higher derivatives in Erdélyi [91b], whose approach is followed in our proof of Theorem A.5.1.

After partial results of Erdélyi and Szabados [88, 89b] and Erdélyi [91b], the "right" Markov-Bernstein-type analogs of Theorem A.5.2 for the classes $\mathcal{P}_{n,k}$ has been proved in Borwein and Erdélyi [94]. Note that $\mathcal{P}_{n,n} = \mathcal{P}_n$, and hence, up to the constant $c(m)$, Theorem A.5.2 contains the classical inequalities of Markov and Bernstein, and of course the case $k=0$ gives back Lorentz's inequalities for the classes $\mathcal{P}_{n,0} \subset \mathcal{B}_n(-1,1)$. The "right" Markov- and Bernstein-type inequalities of Theorems A.5.8 and A.5.9 for all classes $\mathcal{P}_{n,k}$ in L_q, $0 < q < \infty$, are established in Borwein and Erdélyi [to appear 2].

The Markov-type inequality for the classes $\mathcal{P}_{n,k,r}$ given by Theorem A.5.5 is proved in Erdélyi [89a]. A weaker version, when $k=0$ and the factor $r^{-1/2}$ is replaced by a constant $c(r)$ depending on r, is established in Rahman and Labelle [68]. When $k=0$, Theorem A.5.5 is sharp up to the constant $c(m)$ depending only on m; see E.10.

E.1 Relation Between Classes of Constrained Polynomials.

a] Show that $\mathcal{P}_{n,0}(-1,1) \subset \mathcal{B}(-1,1)$. (This is an observation of G. G. Lorentz.)

Hint: Use the identities

$$(x-\alpha)(x-\overline{\alpha}) = \tfrac{1}{4}|1+\alpha|^2(1-x)^2 + \tfrac{1}{2}(|\alpha|^2-1)(1-x^2) + \tfrac{1}{4}|1-\alpha|^2(1+x)^2$$

and

$$x-\alpha = \tfrac{1}{2}(1-\alpha)(x+1) - \tfrac{1}{2}(\alpha+1)(1-x)\,.$$

□

b] Show that $\mathcal{P}_{n,k} \subset \widetilde{\mathcal{P}}_{n,k}(-1,1)$.

c] Show that $\mathcal{B}_n(a,b) \subset \mathcal{B}_n(c,d)$ whenever $[c,d] \subset [a,b]$.

Hint: First show that

$$x-a = \alpha_1(x-c) + \alpha_2(d-x) \quad \text{and} \quad b-x = \alpha_3(x-c) + \alpha_4(d-x)$$

with some nonnegative coefficients. □

d] Show that $\widetilde{\mathcal{P}}_{n,k}(a,b) \subset \widetilde{\mathcal{P}}_{n,k}(c,d)$ whenever $[c,d] \subset [a,b]$.

E.2 Sharpness of Theorem A.5.1. Show that there exist polynomials $p_{n,k} \in \mathcal{P}_{n,k}$ of the form

$$p_{n,k}(x) = (x+1)^{n-k} q_{n,k}(x)\,, \qquad q_{n,k} \in \mathcal{P}_k$$

such that

$$|p'_{n,k}(1)| \geq \tfrac{1}{6} n(k+1) \|p_{n,k}\|_{[-1,1]}$$

for every $n \in \mathbb{N}$, $k = 0, 1, \ldots, n$.

Hint: Use the lower bound in Theorem 6.1.1 (Newman's inequality) with

$$(\lambda_0, \lambda_1, \dots, \lambda_k) = (n-k, n-k+1, \dots, n\} .$$

□

E.3 A Technical Detail in the Proof of Lemma A.5.4. For $n \in \mathbb{N}$ and $y \in [-1,1]$, let

$$F_n := \left\{z = a + ib : \ a \in \left[-1 - \tfrac{1}{8n}, 1 + \tfrac{1}{8n}\right] , \ b \in (-\Delta_{n,a}, \Delta_{n,a})\right\} ,$$

and

$$B_{n,y} := \left\{z \in \mathbb{C} : |z - y| = \tfrac{1}{4}\Delta_{n,y}\right\} .$$

Show that $B_{n,y} \subset F_n$ for every $y \in [-1,1]$.

E.4 Bernstein-Type Inequality for $\mathcal{P}_{n,k}$. Prove that there exists an absolute constant c such that

$$|p'(y)| \le \frac{c\sqrt{n(k+1)}}{1-y^2} \|p\|_{[-1,1]}$$

for every $p \in \mathcal{P}_{n,k}$, and $y \in (-1,1)$. Proceed as follows:

a] Show that for every $n \in \mathbb{N}$ and $k = 0, 1, \dots, n$, there exists a polynomial $Q \in \mathcal{P}_{n,k}$ such that

$$\frac{|Q'(0)|}{\|Q\|_{[-1,1]}} = \sup_{p \in \mathcal{P}_{n,k}} \frac{|p'(0)|}{\|p\|_{[-1,1]}} .$$

Hint: Use a compactness argument. Use Rouché's theorem to show that the uniform limit of a sequence of polynomials from $\mathcal{P}_{n,k}$ on $[-1,1]$ is also in $\mathcal{P}_{n,k}$. □

b] Show that Q has only real zeros, and at most $k+1$ of them (counting multiplicities) are different from ± 1.

Hint: Use a variational method. For example, if $z_0 \in \mathbb{C} \setminus \mathbb{R}$ is a zero of Q, then the polynomial

$$Q_\epsilon(x) := Q(x)\left(1 - \frac{\epsilon x^2}{(x - z_0)(x - \overline{z}_0)}\right)$$

with sufficiently small $\epsilon > 0$ is in $\mathcal{P}_{n,k}$ and contradicts the maximality of Q. □

c] Let $\delta := (36n(k+3))^{-1}$. Show that

$$\|p\|_{[-\delta,1]} \le 2\,\|p\|_{[0,1]}$$

for every polynomial $\mathcal{P}_{n,k}$ having all its zeros in $[0,\infty)$ with at most k of them (counting multiplicities) in $(0,1)$.

Hint: Use the Mean Value Theorem and the result of Theorem A.5.1 transformed linearly from $[-1,1]$ to $[0,1]$. □

d] Let $z_0 := i\gamma(36n(k+3))^{-1/2}$, where i is the imaginary unit and $\gamma \in [0,1]$. Show that

$$|p(z_0)| \leq \sqrt{2}\,\|p\|_{[-1,1]}$$

for every polynomial $p \in \mathcal{P}_n$ having only real zeros with at most k of them (counting multiplicities) in $(-1,1)$.

Outline. Let $p(x) = \prod_{j=1}^{s}(x-u_j)$ with some $s \leq n$. Applying part c] to

$$q(x) := \prod_{j=1}^{s}(x-u_j^2)\,,$$

we obtain

$$\begin{aligned}|q(-(36n(k+3))^{-1})| &\leq 2\|q\|_{[0,1]} = 2\|q(x^2)\|_{[0,1]} = 2\|p(x)p(-x)\|_{[-1,1]}\\ &\leq 2\|p\|^2_{[-1,1]}\,.\end{aligned}$$

Observe that

$$\begin{aligned}|p(z_0)|^2 \leq |p(i(36n(k+3))^{-1/2})|^2 &= \prod_{j=1}^{s}(u_j^2 + (36n(k+3))^{-1})\\ &= |q(-(36n(k+3))^{-1})|\,,\end{aligned}$$

which, together with the previous inequality, yields the desired result. □

e] Let $Q \in \mathcal{P}_{n,k}$ be the extremal polynomial of part a], and let

$$F_{n,k} := \left\{z = a + ib:\; |a| \leq 1,\; |b| \leq (36n(k+3))^{-1/2}(1-|a|)\right\}.$$

Show that

$$|Q(z)| \leq \sqrt{2}\,\|Q\|_{[-1,1]}$$

for every $z \in F_{n,k}$.

Hint: Use parts b] and d] and a linear shift from the interval $[-1,1]$ to $[2\,\mathrm{Re}(z)-1, 1]$ if $\mathrm{Re}(z) \geq 0$, or to $[-1, 2\,\mathrm{Re}(z)+1]$ if $\mathrm{Re}(z) < 0$. □

f] Show that there exists an absolute constant $c > 0$ such that

$$|p'(0)| \le c\sqrt{n(k+1)}\,\|p\|_{[-1,1]}$$

for every $p \in \mathcal{P}_{n,k}$.

Hint: By part a] it is sufficient to prove the inequality when $p = Q$, in which case use part e] and Cauchy's integral formula. □

g] Prove the main result stated in the beginning of the exercise.

Hint: In order to estimate $|p'(y)|$ when, for example, $y \in [0, 1]$ use a linear shift from $[-1, 1]$ to $[2y - 1, 1]$ and apply part f]. □

E.5 Bernstein-Type Inequality for $\mathcal{P}_{n,k}$ for Higher Derivatives. Prove that there exists a constant $c(m)$ depending only on m so that

$$|p^{(m)}(y)| \le c(m)\left(\frac{\sqrt{n(k+1)}}{1-y^2}\right)^m \|p\|_{[-1,1]}$$

for every $p \in \mathcal{P}_{n,k}$ and $y \in (-1, 1)$.

Hint: First show that for every $n, m \in \mathbb{N}$, $k = 0, 1, \ldots, n$, and $\delta \in (0, 1]$, there exists a polynomial $Q_\delta \in \mathcal{P}_{n,k}$ such that

$$\frac{|Q_\delta^{(m)}(0)|}{\|Q_\delta\|_{[-1,1]\setminus[-\delta,\delta]}} = \sup_{p\in\mathcal{P}_{n,k}} \frac{|p^{(m)}(0)|}{\|p\|_{[-1,1]\setminus[-\delta,\delta]}}.$$

Show that Q_δ has at most $k+m$ zeros different from ± 1. Show that there is a polynomial $Q \in \mathcal{P}_{n,k}$ having at most $k+m$ zeros (by counting multiplicities) different from ± 1 such that

$$\frac{|Q^{(m)}(0)|}{\|Q\|_{[-1,1]}} = \sup_{p\in\mathcal{P}_{n,k}} \frac{|p^{(m)}(0)|}{\|p\|_{[-1,1]}}.$$

If $y = 0$, then use E.5 and induction on m to prove the inequality of the exercise for all $p \in \mathcal{P}_{n,k}$ having at most $k+m$ zeros different from ± 1. For an arbitrary $y \in (-1, 1)$ use a linear shift as is suggested in the hint to E.5 g]. □

E.6 Sharpness of Theorem A.5.2. Show that there exist polynomials $p_{n,k} \in \mathcal{P}_{n,k}$ and an absolute constant $c > 0$ such that

$$|p'_{n,k}(0)| \ge c\sqrt{n(k+1)}\|p_{n,k}\|_{[-1,1]}$$

for every $n \in \mathbb{N}$ and $k = 0, 1, \ldots, n$.

Hint: If $k = 0$, then let $m := \left\lfloor \frac{1}{3}(n-1) \right\rfloor$ and

$$p_{n,k}(x) := (x-1)^m (x+1)^{m+1+\lfloor \sqrt{m} \rfloor}\,.$$

If $1 \leq k \leq \frac{1}{3}n$, then let $m := \left\lfloor \frac{1}{3}n \right\rfloor$, $s := \left\lfloor \frac{1}{3}(k-1) \right\rfloor$, and

$$p_{n,k}(x) := (x^2-1)^m \, T_{2s+1}\left(\sqrt{\frac{m}{2s+1}}\, x \right)\,.$$

If $\frac{1}{3}n < k \leq n$, then let $m := \left\lfloor \frac{1}{3}n \right\rfloor$ and $p_{n,k} := p_{n,m}$. □

E.7 Some Inequalities of Lorentz. (See Lorentz [63].)

a] Show that

$$|p'(x)| \leq \left(\frac{n}{n-1} \right)^n n \left| p\left(x \pm \tfrac{1}{n}\right) \right| \leq 4n \left| p\left(x \pm \tfrac{1}{n}\right) \right|$$

for every $p \in \mathcal{B}_n(-1,1)$ (hence for every $p \in \mathcal{P}_{n,0}$), $n \geq 2$, and $x \in [-1,1]$, where in $x \pm \frac{1}{n}$ the $+$ sign is taken if $x \in [-1,0)$, while the $-$ sign is taken if $x \in [0,1]$.

Hint: Observe that it is sufficient to prove the inequality only for

$$p_{n,j}(x) := (1-x)^j (x+1)^{n-j}\,, \qquad n \in \mathbb{N}\,, \quad j = 0, 1, \ldots, n\,.$$

□

b] Let

$$\delta_n(x) := \sqrt{\frac{1-x^2}{n}}\,, \qquad n \in \mathbb{N}\,, \quad x \in [-1,1]\,.$$

Show that there exists an absolute constant $c > 0$ such that

$$|p'(x)| \leq (\delta_n(x))^{-1} \max\left\{ |p(x)|,\ \left| p\left(x \pm \tfrac{1}{2}\delta_n(x)\right) \right| \right\}$$

for every $p \in \mathcal{B}_n(-1,1)$ (hence for every $p \in \mathcal{P}_{n,0}$) and $x \in \left[-1 + \frac{c}{n}, 1 - \frac{c}{n}\right]$.

Hint: Note that it is sufficient to prove the inequality only for

$$p_{n,j}(x) := (1-x)^j (x+1)^{n-j}\,, \qquad n \in \mathbb{N}\,, \quad j = 0, 1, \ldots, n\,.$$

□

c] Show that

$$\int_{-1}^{1} |p'(x)|^q \, dx \leq 2 \cdot 4^q n^q \int_{-1}^{1} |p(x)|^q \, dx$$

for every $p \in \mathcal{B}_n(-1,1)$ (hence for every $p \in \mathcal{P}_{n,0}$) and $q \in (0, \infty)$.

Hint: Use part a]. □

d] Show that there is an absolute constant $c > 0$ such that

$$\int_{-1}^{1} |p'(x)\sqrt{1-x^2}|^q \, dx \le 5c^q n^{q/2} \int_{-1}^{1} |p(x)|^q \, dx$$

for every $p \in \mathcal{B}_n(-1,1)$ (hence for every $p \in \mathcal{P}_{n,0}$) and $q \in (0,\infty)$.

Hint: Use parts b] and c]. □

Analogs of parts a] and b] for higher derivatives are established by Lorentz [63]. From these, analogs of parts c] and d] for higher derivatives can be proven.

E.8 Theorem A.5.5 in a Special Case. Show that there is an absolute constant c_1 such that

$$|p'(1)| \le \frac{c_1 n}{\sqrt{r}} \|p\|_{[-1,1]}$$

for every $p \in \mathcal{P}_n$ having all its zeros in $(-\infty, 1-2r]$, $r \in (0,1]$.

Proof. First assume that

$$(A.5.8) \qquad |p(1)| = \|p\|_{[-1,1]} \, .$$

Without loss of generality, we may assume that $p \in \mathcal{P}_n \setminus \mathcal{P}_{n-1}$. Denote the zeros of p by $(-\infty <) x_1 \le x_2 \le \cdots \le x_n (\le 1-2r)$. Let

$$I_\nu := (1 - 2(\nu+1)^4 r, 1 - 2\nu^4 r] \, , \qquad \nu = 1, 2, \dots \, .$$

Using E.12 of Section 5.1, we obtain

$$\begin{aligned}\frac{|p'(1)|}{|p(1)|} &= \sum_{j=1}^{n} \frac{1}{1-x_j} = \sum_{\nu=1}^{\infty} \sum_{x_j \in I_\nu} \frac{1}{1-x_j} \\ &\le \sum_{\upsilon=1}^{\infty} \frac{e}{\sqrt{2}} n \sqrt{2(\nu+1)^4 r} \, \frac{1}{2\nu^4 r} \\ &\le \frac{e}{2} \sum_{\upsilon=1}^{\infty} \frac{(\nu+1)^2}{\nu^4} \, \frac{n}{\sqrt{r}} \le \frac{c_1 n}{\sqrt{r}}\end{aligned}$$

with an absolute constant c_1.

Now we can drop assumption (A.5.8) as follows: Since p has all its zeros in $(-\infty, 1-2r]$, $|p|$ and $|p'|$ are increasing on $[1-2r, \infty)$. Pick the unique $y \in [1,\infty)$ satisfying $|p(y)| = \|p\|_{[-1,1]} = \|p\|_{[-1,y]}$. Using a linear transformation, from the already proved case we easily obtain

$$\begin{aligned}|p'(1)| \le |p'(y)| &\le c_1 \frac{2n}{y+1} \left(\frac{2r+y-1}{y+1} \right)^{-1/2} \|p\|_{[-1,y]} \\ &\le \frac{c_1 n}{\sqrt{r}} \|p\|_{[-1,1]} \, .\end{aligned}$$

□

E.9 Sharpness of Theorem A.5.5. For $n \in \mathbb{N}$ and $r \in (0, \infty)$, let $S_{n,r}$ be the family of all $p \in \mathcal{P}_n$ that have no zeros in the strip

$$\{z \in \mathbb{C} : |\mathrm{Im}(z)| < r\}\,.$$

a] Show that there exist polynomials $p_{n,m,r} \in S_{n,r}$ and a constant $c(m) > 0$ depending only on m such that

$$|p_{n,m,r}^{(m)}(1)| \geq c(m) \left(\min \left\{ \frac{n}{\sqrt{r}} \,,\, n^2 \right\} \right)^m \|p_{n,m,r}\|_{[-1,1]}$$

for every $n \in \mathbb{N}$, $r \in (0, 1]$, and $m = 1, 2, \ldots, n$.

Outline. If $0 < r \leq \frac{\pi}{4m}$, then with

$$x_j := \left(1 - \tfrac{4}{\pi} mr\right) \cos\left(\tfrac{2j-1}{2n}\pi\right), \qquad j = 1, 2, \ldots, n$$

and

$$z_j := x_j + ir\,, \qquad j = 1, 2, \ldots, n$$

let

$$p_{n,m,r}(x) := \prod_{j=1}^{n} (x - z_j)(x - \overline{z}_j) \in S_{n,r}\,.$$

By E.1 of Section 5.2, $|p_{n,m,r}(1)| = \|p_{n,m,r}\|_{[-1,1]}$. Prove that

$$\begin{aligned}
\frac{|p_{n,m,r}^{(m)}(1)|}{\|p_{n,m,r}\|_{[-1,1]}} &= \frac{|p_{n,m,r}^{(m)}(1)|}{|p_{n,m,r}(1)|} \\
&\geq 2 \left(1 + \frac{\pi}{4m}\right)^{-m} \frac{1}{\sqrt{2}} \frac{|q_{n,m,r}^{(m)}(1)|}{|q_{n,m,r}(1)|} \\
&\geq \frac{\sqrt{2}}{e} \left(\sum_{j=m}^{n} \frac{1}{1 - x_j} \right)^m \\
&\geq c(m) \left(\min \left\{ \frac{n}{\sqrt{r}} \,,\, n^2 \right\} \right)^m,
\end{aligned}$$

where $q_{n,m,r}(x) := \prod_{j=1}^{n} (x - x_j)$ and $c(m) > 0$ depends only on m. If $\frac{\pi}{4m} < r \leq 1$, then let $p_{n,m,r} := p_{n,m,\widetilde{r}}$, where $\widetilde{r} := \frac{\pi}{4m}$. □

b] Conclude from Theorem A.5.5 and part a] that there exist two constants $c_1(m) > 0$ and $c_2(m) > 0$ depending only on m such that

$$\begin{aligned}
c_1(m) \left(\min \left\{ \frac{n}{\sqrt{r}} \,,\, n^2 \right\} \right)^m &\leq \sup \frac{\|p^{(m)}\|_{[-1,1]}}{\|p\|_{[-1,1]}} \\
&\leq c_2(m) \left(\min \left\{ \frac{n}{\sqrt{r}} \,,\, n^2 \right\} \right)^m,
\end{aligned}$$

where the supremum is taken either for all $p \in S_{n,r}$ or for all $p \in \mathcal{P}_{n,0,r}$.

E.10 An Inequality of Erdős. (See Erdős [40].) Prove that

$$\|p'\|_{[-1,1]} \le \frac{n}{2}\left(\frac{n}{n-1}\right)^{n-1}\|p\|_{[-1,1]}$$

for every $p \in \mathcal{P}_{n,0}$, $n \ge 2$. This extends a result of Erdős [40] where the above inequality is proven under the additional assumption that p has only real zeros.

a] Suppose $p \in \mathcal{P}_{n,0}$, all the zeros of p are real, $p(1) = p(-1) = 0$, and $\|p\|_{[-1,1]} = 1$. Show that

$$p(x) \le \left(\frac{n}{n-1}\right)^{n-1}\frac{1+x}{1+x_0}$$

for every $x \in [-1,1]$, where x_0 is the only point in $(-1,1)$ with $p'(x_0) = 0$.

Proof. Without loss of generality, we may assume that $\deg(p) = n$ and $-1 < x < x_0$. Let $d := x_0 - x$. Let $x_1 := -1, x_2, \dots, x_n$ denote the zeros of p. Then

$$p(x) = \frac{p(x)}{p(x_0)} = \frac{1+x}{1+x_0}\prod_{j=2}^{n}\left(1 - \frac{d}{x_0 - x_j}\right).$$

Since the geometric mean of $n-1$ nonnegative numbers is not greater than their arithmetic mean, we have

$$\begin{aligned}\prod_{j=2}^{n}\left(1-\frac{d}{x_0-x_j}\right) &\le \left(\frac{1}{n-1}\left(n-1-\sum_{j=2}^{n}\frac{d}{x_0-x_j}\right)\right)^{n-1}\\ &\le \left(\frac{1}{n-1}\left(n-1+\frac{d}{x_0+1}\right)\right)^{n-1} \le \left(\frac{n}{n-1}\right)^{n-1}.\end{aligned}$$

□

b] Under the assumptions of part a] show that

$$\|p'\|_{[-1,1]} \le \frac{n}{2}\left(\frac{n}{n-1}\right)^{n-1}\|p\|_{[-1,1]}\,.$$

Proof. Note that

$$\sum_{x_j \ge 1}\frac{1}{x_j - x_0} = \sum_{x_j \le -1}\frac{1}{x_0 - x_j} \le \min\left\{\frac{k}{1-x_0}, \frac{n-k}{x_0+1}\right\} \le \frac{n}{2},$$

where k denotes the number of zeros of p in $[1,\infty)$. Hence, by part a],

$$
\begin{aligned}
p'(x) &= p(x) \sum_{j=1}^{n} \frac{1}{x - x_j} \\
&\leq \left(\frac{n}{n-1} \right)^{n-1} \frac{1+x}{1+x_0} \sum_{x_j \leq -1} \frac{1}{x - x_j} \\
&\leq \left(\frac{n}{n-1} \right)^{n} \frac{1+x}{1+x_0} \frac{1+x_0}{1+x} \sum_{x_j \leq -1} \frac{1}{x_0 - x_j} \\
&\leq \frac{n}{2} \left(\frac{n}{n-1} \right)^{n-1}
\end{aligned}
$$

for every $x \in [-1, 1]$. Similarly

$$p'(x) \geq -\frac{n}{2} \left(\frac{n}{n-1} \right)^{n-1}$$

for every $x \in [-1, 1]$. □

c] Suppose $p \in \mathcal{P}_{n,0}$ has only real zeros, p' does not vanish in $[-1, 1]$, $p(-1) = 0$, and $p(1) = 1$. Show that

$$\|p'\|_{[-1,1]} \leq \frac{n}{2} \|p\|_{[-1,1]} \,.$$

Hint: Use the relation $\mathcal{P}_{n,0} \subset \mathcal{B}_n(-1, 1)$ to show that

$$p(x) \leq \left(\frac{x+1}{2} \right)^n, \qquad x \in [-1, 1] \,.$$

Denote the zeros of p by $x_1 = -1, x_2, x_3, \ldots, x_n$. Then

$$0 \leq p'(x) = p(x) \sum_{j=1}^{n} \frac{1}{x - x_j} \leq \left(\frac{x+1}{2} \right)^n \frac{n}{x+1} \leq \frac{n}{2} = \frac{n}{2} \|p\|_{[-1,1]}$$

for every $x \in [-1, 1]$. □

d] Prove Erdős's inequality for every $p \in \mathcal{P}_{n,0}$ having only real zeros.

Hint: Reduce the general case to either part b] or part c] by a linear transformation. □

e] Prove Erdős's inequality for every $p \in \mathcal{P}_{n,0}$.

Hint: Show that for every $n \in \mathbb{N}$ and $y \in [-1, 1]$, there exists a polynomial $Q \in \mathcal{P}_{n,0}$ such that

$$\frac{|Q'(y)|}{\|Q\|_{[-1,1]}} = \sup_{p \in \mathcal{P}_{n,0}} \frac{|p'(y)|}{\|p\|_{[-1,1]}} \,.$$

Show by a variational method that if $y \in (-1, 1)$, then Q has only real zeros. Now part d] finishes the proof. □

f] Show that

$$\|p'\|_{[-1,1]} = \frac{n}{2}\left(\frac{n}{n-1}\right)^{n-1} \|p\|_{[-1,1]}$$

holds for a $p \in \mathcal{P}_{n,0}$ having only real zeros if and only if either

$$p(x) = c(x+1)^{n-1}(1-x) \qquad \text{or} \qquad p(x) = c(1-x)^{n-1}(x+1)$$

with some $0 \neq c \in \mathbb{R}$.

E.11 Sharpness of Theorem A.5.3. For every $n \in \mathbb{N}$, $m = 1, 2, \dots, n$, and $y \in [-1,1]$, there are polynomials $p_{n,m,y} \in \mathcal{B}_n(-1,1)$ having zeros only in $\mathbb{R} \setminus (-1,1)$ such that

$$|p^{(m)}_{n,m,y}(y)| \geq c(m)\left(\min\left\{n\,, \frac{\sqrt{n}}{\sqrt{1-y^2}}\right\}\right)^m \|p_{n,m,y}\|_{[-1,1]}$$

with a constant $c(m) > 0$ depending only on m.

Outline. If

$$y \in [-1,1] \setminus \left[-1 + \tfrac{2m}{n}\,, 1 - \tfrac{2m}{n}\right],$$

then let

$$p_{n,m,y}(x) := (x+1)^n\,.$$

In what follows, assume that

$$y \in \left[-1 + \tfrac{2m}{n}\,, 1 - \tfrac{2m}{n}\right].$$

Let $q_{n,j}(x) := (1-x)^j(x+1)^{n-j}$, $n \in \mathbb{N}$, $j = 0, 1, \dots, n$. Show that

$$\frac{q^{(m)}_{n,j}(x)}{q_{n,j}(x)} = \frac{Q_{n,j,m}(x)}{(x^2-1)^m}\,, \qquad m \leq j \leq n-m\,,$$

where $Q_{n,j,m}$ is a polynomial of degree m with only real zeros and with leading coefficient $\frac{n!}{(n-m)!}$. Let

$$\Delta_{n,y} := \max\left\{\frac{1}{n}\,, \frac{\sqrt{1-y^2}}{\sqrt{n}}\right\}, \qquad n \in \mathbb{N}\,, \quad y \in [-1,1]\,.$$

Use the Mean Value Theorem and Theorem A.5.3 to show that there exists an absolute constant $c \in (0,1)$ such that

$$q_{n,j}(x) \geq \tfrac{1}{2}\|q_{n,j}\|_{[-1,1]}\,, \qquad m \leq j \leq n-m$$

for every

$$x \in I_y := [y - c\Delta_{n,y}, y + c\Delta_{n,y}] \cap [-1,1]\,, \qquad y = 1 - \tfrac{2j}{n}\,.$$

Let $m \le j \le n - m$ and $y = 1 - \frac{2j}{n}$ be fixed. Choose a point $\xi \in I_y$ such that

$$|\xi - \alpha_i| \ge \frac{c\Delta_{n,y}}{2(m+1)}, \qquad i = 1, 2, \dots, m,$$

where the numbers α_i are the zeros of $Q_{n,j,m}$. Now show that there exists a constant $c_1(m) > 0$ depending only on m such that

$$(A.5.9) \qquad |q_{n,j}^{(m)}(\xi)| \ge c_1(m) \left(\min\left\{ n, \frac{\sqrt{n}}{\sqrt{1-y^2}} \right\} \right)^m \|q_{n,j}\|_{[-1,1]} .$$

Next show that if

$$y \in \left[-1 + \tfrac{2m}{n}, 1 - \tfrac{2m}{n}\right],$$

then there exists a point

$$\xi \in \left[y - \tfrac{1}{2}(1 - |y|), y + \tfrac{1}{2}(1 - |y|)\right]$$

and a value of j, $m \le j \le n - m$, such that (A.5.9) holds. Polynomials $p_{n,m,y}$ with the desired properties can now be easily defined by using linear transformations. □

E.12 An Inequality of Turán. (See Turán [39].) Show that

$$\frac{\|p'\|_{[-1,1]}}{\|p\|_{[-1,1]}} > \frac{1}{6}\sqrt{n}$$

for every $p \in \mathcal{P}_n \setminus \mathcal{P}_{n-1}$ having all its zeros in $[-1, 1]$.

Outline. Assume that $p \in \mathcal{P}_n$ has all its zeros in $[-1, 1]$, and $\|p\|_{[-1,1]} = 1$. Choose an $a \in [-1, 1]$ such that $|p(a)| = 1$. Without loss of generality, assume that $p(a) = 1$.

a] Show that if $a = \pm 1$, then $|p'(1)| \ge \frac{1}{2}n > \frac{1}{6}\sqrt{n}$.

If $a \in (-1, 1)$, then $p'(a) = 0$. Without loss of generality, assume that $a \in [-1, 0]$. Let

$$I := \left[a, a + 2n^{-1/2}\right] \subset [-1, 1].$$

If $n \le 3$, then the result follows by the Mean Value Theorem; let $n \ge 4$.

b] Use the Mean Value Theorem to show that if $|p'(x)| \le \frac{1}{6}\sqrt{n}$ on I, then $p(x) \ge \frac{2}{3}$ on I.

c] Show that if $|p''(x)| > \frac{1}{12}n$ on I, then $|p'(a + 2n^{-1/2})| > \frac{1}{6}\sqrt{n}$.

d] The proof of Turán's inequality can now be finished as follows. Suppose $p(x) > \frac{2}{3}$ on I, and there exists a $\xi \in I$ such that $p''(\xi) \le \frac{1}{12}n$. Note that

$$p'(x)^2 - p(x)p''(x) = p(x)^2 \sum_{k=1}^{n} \frac{1}{(x - x_k)^2},$$

where $x_1, x_2, \dots, x_n$ denote the zeros of p. Since each x_k lies in $[-1, 1]$, the inequality $\sum_{k=1}^n (x - x_k)^{-2} \geq \frac{1}{4}n$ holds for every $x \in I$. Since $p(x) \geq \frac{2}{3}$ on I, this implies that

$$p'(x)^2 - p(x)p''(x) \geq \frac{n}{9}, \qquad x \in I$$

and hence

$$p'(\xi)^2 \geq \frac{n}{9} - |p(\xi)||p''(\xi)| > \frac{n}{9} - \frac{n}{12} = \frac{n}{36}.$$

□

An extension of Turán's inequality to L_p norms is given by Zhou [92b].

E.13 An Inequality of Erdős and Turán. Let $p \in \mathcal{P}_n$ be of the form

$$p(x) = \pm \prod_{j=1}^n (x - x_j), \qquad -1 \leq x_1 \leq x_2 \leq \cdots \leq x_n \leq 1.$$

Suppose p is convex on $[x_{k-1}, x_k]$ for some index k. Then

$$x_k - x_{k-1} \leq \frac{16}{\sqrt{n}}.$$

Proceed as follows: Let a be the only point in $[x_{k-1}, x_k]$ for which $p'(a) = 0$.

a] Show that there exist $\xi_1, \xi_2 \in \mathbb{R}$ such that

$$a - 2n^{-1/2} \leq \xi_1 < a < \xi_2 \leq a + 2n^{-1/2},$$

$$|p'(\xi_1)| \geq \tfrac{1}{6}\sqrt{n}\,|p(a)|, \qquad \text{and} \qquad |p'(\xi_2)| \geq \tfrac{1}{6}\sqrt{n}\,|p(a)|.$$

Hint: Modify the outline of the proof of E.12. □

b] Show that

$$\xi_1 - x_{k-1} \leq \frac{6}{\sqrt{n}} \qquad \text{and} \qquad x_k - \xi_2 \leq \frac{6}{\sqrt{n}}.$$

Outline. To prove, say, the first inequality, we may assume that $x_{k-1} < \xi_1$, otherwise the inequality is trivial. Using the convexity of p on $[x_{k-1}, x_k]$, we get

$$|p'(x)| \geq |p'(\xi_1)|, \qquad x \in [x_{k-1}, \xi_1].$$

Hence

$$\begin{aligned} |p(a)| \geq |p(\xi_1)| = |p(\xi_1) - p(x_{k-1})| &= \left| \int_{x_{k-1}}^{\xi_1} p'(x)\,dx \right| = \int_{x_{k-1}}^{\xi_1} |p'(x)|\,dx \\ &\geq (\xi_1 - x_{k-1})|p'(\xi_1)| \geq (\xi_1 - x_{k-1})\tfrac{1}{6}\sqrt{n}\,|p(a)|, \end{aligned}$$

and the result follows. □

c] Conclude that

$$a - x_{k-1} \le (a - \xi_1) + (\xi_1 - x_{k-1}) \le \frac{2}{\sqrt{n}} + \frac{6}{\sqrt{n}} = \frac{8}{\sqrt{n}},$$

and similarly

$$x_k - a \le (\xi_2 - a) + (x_k - \xi_2) \le \frac{2}{\sqrt{n}} + \frac{6}{\sqrt{n}} = \frac{8}{\sqrt{n}}.$$

□

d] Erőd [39] establishes the sharp inequalities

$$x_k - x_{k-1} \le \frac{2}{\sqrt{2n-3}} \qquad \text{if } n \text{ is even,}$$

$$x_k - x_{k-1} \le \frac{2}{\sqrt{2n-3}} \frac{\sqrt{n^2-2n}}{n-1} \qquad \text{if } n \ge 3 \text{ is odd.}$$

E.14 Schur-Type Inequality for $\mathcal{B}_n(-1,1)$. Let α be an arbitrary positive real number. Show that

$$\sup_{0 \ne p \in \mathcal{B}_n(-1,1)} \frac{\|p\|_{[-1,1]}}{\|p(x)(1-x^2)^\alpha\|_{[-1,1]}} = \frac{(n+2\alpha)^{n+2\alpha}}{(4\alpha)^\alpha (n+\alpha)^{n+\alpha}} < \left(\frac{e}{4\alpha}(n+2\alpha)\right)^\alpha.$$

The supremum is attained if and only if $p(x) = c(1 \pm x)^n$, $0 \ne c \in \mathbb{R}$.

Hint: Let $x_1 := \frac{n}{n+2\alpha}$. If $|y| \le x_1$, then

$$\begin{aligned}\frac{|p(y)|}{\|p(x)(1-x^2)^\alpha\|_{[-1,1]}} &\le \frac{1}{(1-y^2)^\alpha} \le \frac{1}{(1-x_1^2)^\alpha} \\ &= \frac{(n+2\alpha)^{2\alpha}}{(4\alpha)^\alpha (n+\alpha)^\alpha} < \frac{(n+2\alpha)^{n+2\alpha}}{(4\alpha)^\alpha (n+\alpha)^{n+\alpha}}.\end{aligned}$$

If $x_1 < |y| \le 1$, say $x_1 < y \le 1$, then

$$\begin{aligned}(1-y)^j (y+1)^{n-j} &\le \|(1-x)^j (x+1)^{n-j}\|_{[-1,1]} = \frac{2^n j^j (n-j)^{n-j}}{n^n} \\ &= \frac{j^j (n-j)^{n-j} (n+2\alpha)^n}{\alpha^j n^n (n+\alpha)^{n-j}} (1-x_1)^j (x_1+1)^{n-j} \\ &\le \left(\frac{j(n+\alpha)}{\alpha(n-j)}\right)^j \left(\frac{n+2\alpha}{n+\alpha}\right)^n (1-x_1)^j (x_1+1)^{n-j} \\ &\le \left(\frac{n+2\alpha}{n+\alpha}\right)^n (1-x_1)^j (x_1+1)^{n-j}\end{aligned}$$

whenever $0 \le j \le \frac{\alpha n}{n+2\alpha}$.

On the other hand, since the function $(1-x)^j(x+1)^{n-j}$ is monotone decreasing in $\left[1-\frac{2j}{n},1\right]$, the inequality

$$(1-y)^j(y+1)^{n-j} < (1-x_1)^j(x_1+1)^{n-j}$$

follows whenever $\frac{\alpha n}{n+2\alpha} < j \leq n$. Finally, use the representation

$$p(x) = \sum_{j=0}^{n} a_j(1-x)^j(x+1)^{n-j}$$

with each $a_j \geq 0$ or each $a_j \leq 0$ to show that

$$|p(y)| \leq \frac{(n+2\alpha)^{n+2\alpha}}{(4\alpha)^\alpha(n+\alpha)^{n+\alpha}} \left|p(x_1)(1-x_1^2)^\alpha\right|$$

for every $p \in \mathcal{B}_n(-1,1)$. □

E.15 Schur-Type Inequality for $\mathcal{T}_n(-\omega,\omega)$. For $n \in \mathbb{N}$ and $\omega \in (0,\pi]$, let

$$\mathcal{T}_n(-\omega,\omega) := \{t \in \mathcal{T}_n : d_\omega(t) \leq n\},$$

where the Lorentz degree $d_\omega(t)$ is defined in E.5 of Section 2.4. Let α be an arbitrary positive real number. Show that

$$\sup_{0 \neq t \in \mathcal{T}_n(-\omega,\omega)} \frac{\|t\|_{[-\omega,\omega]}}{\left\|t(\theta)\left(\frac{1}{2}(\cos\theta - \cos\omega)\right)^\alpha\right\|_{[-\omega,\omega]}}$$

$$\sim \begin{cases} \left(n(\pi-2\omega)+\sqrt{\omega n}\right)^\alpha, & 0 < \omega \leq \pi/2 \\ (2\omega-\pi+n^{-1/2})^{-\alpha}, & \pi/2 \leq \omega \leq \pi, \end{cases}$$

and the supremum is attained if and only if

$$t(\theta) = c\sin^{2n}\frac{\omega \pm \theta}{2}, \qquad 0 \neq c \in \mathbb{R}.$$

Here the $\sim$ symbol means that the ratio of the two sides is between two positive constants depending only on α (and independent of $n \in \mathbb{N}$ and $\omega \in (0,\pi]$).

E.16 Extensions and Variations of Lax's Inequality. Theorem 7.1.11 contains, as a limiting case, an inequality of Lax [44] conjectured by Erdős; see part a]. Various extensions of this inequality are given by Ankeny and Rivlin [55], Govil [73], Malik [69], and others. Parts b] to e] discuss some of these. As before, let

$$D := \{z \in \mathbb{C} : |z| < 1\} \qquad \text{and} \qquad \partial D := \{z \in \mathbb{C} : |z| = 1\}.$$

a] Lax's Inequality. The inequality

$$\|p'\|_D \le \frac{n}{2}\,\|p\|_D$$

holds for all $p \in \mathcal{P}_n^c$ that have no zeros in the open unit disk.

Proof. This follows from Theorem 7.1.11 as a limiting case. □

b] Malik's Extension. Associated with

$$p(z) = c\prod_{j=1}^{n}(z - z_j)\,, \qquad z_j \in \mathbb{C}\,, \quad 0 \ne c \in \mathbb{C}\,,$$

let

$$p^*(z) := \overline{c}\prod_{j=1}^{n}(1 - z\overline{z}_j) = z^n\overline{p(1/\overline{z})}\,.$$

Then

$$\max_{z\in\partial D}\left(|p'(z)| + |p^{*\prime}(z)|\right) = n$$

for every $0 \ne p \in \mathcal{P}_n^c$.

Proof. See Malik [69]. □

c] An Observation of Kroó. Suppose $p \in \mathcal{P}_n^c$ satisfies that if $p(z) = 0$ for some $z \in D$, then $p(1/\overline{z}) = 0$ (there is no restriction for the zeros of p outside D). Then

$$\|p'\|_D \le \frac{n}{2}\,\|p\|_D\,.$$

Hint: Show that if $p \in \mathcal{P}_n^c$ satisfies the assumption of the lemma, then $|p'(z)| \le |p^{*\prime}(z)|$ for every $z \in \partial D$. Use part b] to finish the proof. □

d] An Inequality of Ankeny and Rivlin. Let $r \ge 1$. The inequality

$$\max_{|z|=r}|p(z)| \le \frac{r^n+1}{2}\max_{|z|=1}|p(z)|$$

holds for all $p \in \mathcal{P}_n^c$ that have no zeros in the open unit disk.

Proof. See Ankeny and Rivlin [55]. □

e] An Inequality of Govil. Let $r \ge 1$. The inequality

$$\|p'\|_D \le \frac{n}{1+r}\,\|p\|_D$$

holds for all $p \in \mathcal{P}_n^c$ that have no zeros in the disk $\{z \in \mathbb{C} : |z| < r\}$.

Proof. See Govil [73]. □

f] Let

$$p(z) := c\prod_{k=1}^{n}(z - z_k)\,, \qquad 0 \neq c \in \mathbb{C}\,.$$

Show that

$$\|p'\|_D\| \geq \left(\sum_{k=1}^{n} \frac{1}{1+|z_k|}\right) \|p\|_D\,.$$

Proof. If $z \in \partial D$, then

$$\operatorname{Re}\left(\frac{zp'(z)}{p(z)}\right) = \sum_{k=1}^{n} \operatorname{Re}\left(\frac{z}{z - z_k}\right) \geq \sum_{k=1}^{n} \frac{1}{1+|z_k|}\,,$$

and the result follows. □

g] Let $r \in (0, 1]$. The inequality

$$\|p'\|_D \geq \frac{n}{1+r}\,\|p\|_D$$

holds for all $p \in \mathcal{P}_n^c$ that have all their zeros in the disk $\{z \in \mathbb{C} : |z| \leq r\}$.

Proof. Use part f]. □

h] Another Inequality of Govil. Let $r > 1$. The inequality

$$\|p'\|_D \geq \frac{n}{1+r^n}\,\|p\|_D$$

holds for all $p \in \mathcal{P}_n^c$ which have no zeros in the disk $\{z \in \mathbb{C} : |z| < r\}$.

Proof. See Govil [73]. □

E.17 Markov-Type Inequality for Nonnegative Polynomials. Show that

$$\|p'\|_{[-1,1]} \leq \frac{n^2}{2}\,\|p\|_{[-1,1]}$$

for every $p \in \mathcal{P}_n$ positive on $[-1, 1]$.

Proof. Suppose $p \in \mathcal{P}_n$ is positive on $[-1, 1]$ and $\|p\|_{[-1,1]} = 2$. Apply Theorem 5.1.8 (Markov's inequality) to $q := p - 1$. □

E.18 Markov's Inequality for Monotone Polynomials. It has been observed by Bernstein that Markov's inequality for monotone polynomials is not essentially better than for arbitrary polynomials. He proved that if n is odd, then

$$\sup_{0 \neq p} \frac{\|p'\|_{[-1,1]}}{\|p\|_{[-1,1]}} = \left(\frac{n+1}{2}\right)^2,$$

where the supremum is taken for all $p \in \mathcal{P}_n$ that are monotone on $[-1, 1]$.

For even n, the inequality

$$\sup_{0 \neq p} \frac{\|p'\|_{[-1,1]}}{\|p\|_{[-1,1]}} \leq \left(\frac{n+1}{2}\right)^2$$

still holds. Parts a] and b] of this exercise outline a proof.

a] Show that for every odd n, there is a $p \in \mathcal{P}_n$ monotone on $\mathbb{R}$ for which

$$\frac{|p'(1)|}{\|p\|_{[-1,1]}} = \left(\frac{n+1}{2}\right)^2.$$

Proof. Let $m := \frac{1}{2}(n-1)$. Use E.2 e] of Section 6.1 to show that there is a $q \in \mathcal{P}_m$ for which

$$q^2(1) = \left(\sum_{k=0}^{m}(1+2k)\right) \int_0^1 q^2(t)\,dt = (m+1)^2 \int_0^1 q^2(t)\,dt.$$

Now let

$$p(x) := \int_0^x q^2(t)\,dt - \frac{1}{2}\int_0^1 q^2(t)\,dt.$$

Obviously $p \in \mathcal{P}_n$, p is monotone on $\mathbb{R}$, and

$$\frac{|p'(1)|}{\|p\|_{[-1,1]}} = \frac{q^2(1)}{\frac{1}{2}\int_0^1 q^2(t)\,dt} = 2(m+1)^2 = 2\left(\frac{n+1}{2}\right)^2.$$

Now the proof can be finished by a linear transformation mapping $[-1,1]$ to $[0,1]$. □

b] Let n be odd. Show that

$$\sup_{0 \neq p} \frac{\|p'\|_{[-1,1]}}{\|p\|_{[-1,1]}} \leq \left(\frac{n+1}{2}\right)^2,$$

where the supremum is taken for all $p \in \mathcal{P}_n$ that are monotone on $[-1,1]$.

Hint: It is sufficient to prove that if $n := 2m$ is even, then

$$\|p\| \leq \frac{1}{2}\left(\frac{n+2}{2}\right)^2 \int_{-1}^1 p(t)\,dt$$

for every $p \in \mathcal{P}_n$ nonnegative on $[-1,1]$. Show that there is an extremal polynomial $\widetilde{p}$ for the above inequality for which $\|\widetilde{p}\|_{[-1,1]}$ is achieved at 1. Show, by a variational method, that this $\widetilde{p}$ must have at least $2m$ zeros (counting multiplicities) in $[-1,1)$. Since $\widetilde{p}$ is nonnegative, it is of the form $\widetilde{p} = q^2$ with a $q \in \mathcal{P}_m$. Now use E.2 e] of Section 6.1 to show that

$$q^2(1) \le \frac{1}{2}(m+1)^2 \int_{-1}^{1} q^2(t)\,dt = \frac{1}{2}\left(\frac{n+2}{2}\right)^2 \int_{-1}^{1} q^2(t)\,dt\,,$$

and the result follows. □

c] Let $r = p/q$, where $p, q \in \mathcal{P}_n$ and q is positive and monotone nondecreasing on an interval $[a,b]$. Show that

$$\|r'\|_{[a+\epsilon,b]} \le \frac{4n^2}{\epsilon}\,\|r\|_{[a,b]}$$

for every $\epsilon \in (0, b-a)$.

Proof. The proof follows Borwein [80]. Let $c \in [a+\epsilon, b]$ be a point where $|r'(c)| = \|r'\|_{[a+\epsilon,b]}$. Let $d \in [a,c]$ be a point $|p(d)| = \|p\|_{[a,c]}$. Then

$$r'(c) = \frac{p'(c)}{q(c)} - \frac{q'(c)}{q(c)}\,r(c)\,.$$

Theorem 5.1.8 (Markov's inequality) and the monotonicity of q imply that

$$\frac{|p'(c)|}{|q(c)|} \le \frac{2n^2\|p\|_{[a,c]}}{(c-a)|q(c)|} = \frac{2n^2|p(d)|}{(c-a)|q(c)|} \le \frac{2n^2}{c-a}\,\frac{|p(d)|}{|q(d)|} \le \frac{2n^2}{\epsilon}\,|r(c)|$$

and

$$\frac{|q'(c)|}{|q(c)|}\,|r(c)| \le \frac{2n^2\|q\|_{[a,c]}}{|q(c)|}|r(c)| \le \frac{2n^2}{\epsilon}\,|r(c)|\,.$$

Thus

$$\|r'\|_{[a+\epsilon,b]} = |r'(c)| \le \frac{4n^2}{\epsilon}\,\|r\|_{[a,b]}\,.$$

□

d] Sharpness of Part c]. Let n be odd. By part a], there exists a $p \in \mathcal{P}_n$ such that p is monotone increasing on $\mathbb{R}$, $p(a) = -1$, $p(a+\epsilon) = 1$, and

$$|p'(a+\epsilon)| = \frac{1}{2\epsilon}\left(\frac{n+1}{2}\right)^2.$$

Let $q := p+2$ and $r := 1/q$. Show that

$$\|r\|_{[a,\infty]} = 1 \qquad \text{and} \qquad |r'(a+\epsilon)| = \frac{1}{9}\cdot\frac{1}{2\epsilon}\left(\frac{n+1}{2}\right)^2.$$

Weighted polynomial inequalities and their applications are beyond the scope of this book. A thorough discussion requires serious potential theoretic background, and proofs are usually quite long. Some of the main results in this area include von Golitschek, Lorentz, and Makovoz [92]; He [91]; Ivanov and Totik [90]; Levin and Lubinsky [87a] and [87b]; Lubinsky [89], Lubinsky and Saff [88a], [88b], and [89]; Mhaskar and Saff [85]; Nevai and Totik [86] and [87]; Saff and Totik [to appear]; and Totik [94]. The following exercise treats a weighted Markov-type inequality with respect to the Laguerre weight on $[0, \infty)$. The method presented below works in more general cases; however, the technical details coming from the "right" polynomial approximation of the weight function is typically far more complicated.

E.19 A Weighted Markov-Type Inequality of Szegő. Part a] presents a simple weighted polynomial inequality of Szegő [64].

a] Show that

$$\|p'(x)e^{-x}\|_{[0,\infty)} \leq (8n+2)\|p(x)e^{-x}\|_{[0,\infty)}$$

for every $p \in \mathcal{P}_n$.

Proof. Let $p \in \mathcal{P}_n$. First prove the inequality

$$|p'(0)| \leq (8n+2)\|p(x)e^{-x}\|_{[0,\infty)}$$

as follows. Apply Theorem 5.1.8 (Markov's inequality) on $[0, n/2]$ (by a linear transformation) to

$$q(x) := p(x)\left(1 - \frac{x}{n}\right)^n$$

and note that

$$\left(1 - \frac{x}{n}\right)^n \leq e^{-x} \quad \text{and} \quad \left(1 - \frac{x}{n}\right)^{-1} \leq 2\,, \qquad x \in [0, n/2]\,.$$

The general case can be easily reduced to the case discussed above. If $y \in [0, \infty)$ is fixed, then let $q(x) := p(x - y) \in \mathcal{P}_n$. On applying the already proved inequality with q, we obtain

$$\begin{aligned}
|p'(y)e^{-y}| &= |q'(0)e^{-y}| \\
&\leq e^{-y}(8n+2) \max_{x\in[0,\infty)} |q(x)e^{-x}| \\
&\leq (8n+2) \max_{x\in[0,\infty)} |p(x+y)e^{-(x+y)}| \\
&\leq (8n+2) \max_{x\in[0,\infty)} |p(x)e^{-x}|\,,
\end{aligned}$$

which finishes the proof. □

b] There is an absolute constant $c > 0$ such that

$$\|p'(x)e^{-x}\|_{[0,\infty)} \le c\sqrt{n}\, \|p(x)e^{-x}\|_{[0,\infty)}$$

holds for every $p \in \mathcal{P}_n$ having no zeros in the disk with diameter $[0, 2]$.

Proof. See Erdélyi [89c], where this result is extended to various other general classes of weight functions and constrained polynomials. □

E.20 Inequalities for Generalized Polynomials with Constraints. For $s \in (0, 2)$, let

$$\widetilde{\mathcal{P}}_{n,k}(s) := \left\{ p \in \widetilde{\mathcal{P}}_{n,k} : m(\{x \in [-1,1] : |p(x)| \le 1\}) \ge 2 - s \right\}.$$

The polynomials $T_{n,k} \in \mathcal{P}_{n,k}$ are defined by

$$T_{n,k}(x) := T_n\{n-k, n-k+1, \dots, n; [0,1]\} \left(\tfrac{1}{2}(x+1)\right),$$

where $0 \le k \le n$ are integers. (Note that the notation introduced in Section 3.3 defines $T_{n,k}$ only on $[-1, 1]$, and, to be precise, $T_{n,k}$ denotes the polynomial defined above on $[-1, 1]$.)

a] Remez-Type Inequality for $\widetilde{\mathcal{P}}_{n,k}$. The inequality

$$\|p\|_{[-1,1]} \le T_{n,k}\left(\frac{2+s}{2-s}\right)$$

holds for every $p \in \widetilde{\mathcal{P}}_{n,k}(s)$ and $s \in (0, 2)$. Equality holds if and only if

$$p(x) \equiv \pm T_{n,k}\left(\frac{\pm 2x}{2-s} + \frac{s}{2-s}\right).$$

Proof. The proof is quite similar to that of Theorem 5.1.1 (Remez Inequality); see Borwein and Erdélyi [92]. □

b] A Numerical Version of Part a]. Show that

$$T_{n,k}\left(\frac{2+s}{2-s}\right) \le \exp\left(5\left(\sqrt{nks} + ns\right)\right)$$

for every $s \in (0, 1]$.

Proof. See Borwein and Erdélyi [92]. □

Let $\mathrm{GAP}_{N,K}$, $0 \leq K \leq N$, be the set of all $f \in \mathrm{GAP}_N$ of the form (A.4.1) for which

$$\sum_{\substack{j=1 \\ |z_j| \leq 1}}^{n} r_j \leq K\,.$$

Note that if $0 \leq K \leq N$ are integers and $p \in \mathcal{P}_{N,K}$, then $|p| \in \mathrm{GAP}_{N,K}$.

For $0 \leq K \leq N$ and $0 < s < 2$, let $\mathrm{GAP}_{N,K}(s)$ denote the collection of all $f \in GAP_{N,K}$ for which

$$m(\{x \in [-1,1] : f(x) \leq 1\}) \geq 2 - s\,.$$

The next part of the exercise extends a numerical version of part a] to the classes $\mathrm{GAP}_{N,K}(s)$.

c] **Remez-Type Inequality for** $\mathrm{GAP}_{N,K}$. There exists a constant $c_1 \leq 5$ such that

$$\|f\|_{[-1,1]} \leq \exp\left(c_1\left(\sqrt{NKs} + Ns\right)\right)$$

for every $f \in \mathrm{GAP}_{N,K}(s)$ and $s \in (0,1]$.

Hint: Let $q \in \mathbb{N}$ be the common denominator of the numbers r_j. Apply part a] with $p := f^{2q} \in \widetilde{\mathcal{P}}_{2qN,2qK}(s)$, and then use part b]. □

d] **Nikolskii-Type Inequality for** $\mathrm{GAP}_{N,K}$. Let χ be a nonnegative nondecreasing function defined on $[0,\infty)$ such that $\chi(x)/x$ is nonincreasing on $[0,\infty)$. Then there exists an absolute constant $c_2 \leq 25e^2$ such that

$$\|\chi(f)\|_{L_p[-1,1]} \leq (c_2 \max\{1\,,\, q^2NK\,,\, qN\})^{1/q-1/p}\|\chi(f)\|_{L_q[-1,1]}$$

for every $f \in \mathrm{GAP}_{N,K}$ and $0 < q < p \leq \infty$.

The case $K = N$ (when there is no restriction on the zeros) of part d] is the content of Theorem A.4.4. If $qK \geq 1$, then the Nikolskii factor in the unrestricted case ($K = N$) is like $(\sqrt{c_2}\,qN)^{2/q-2/p}$, while in our restricted cases it improves to $(\sqrt{c_2}\,q\sqrt{NK})^{2/q-2/p}$.

Proof. It is sufficient to prove the inequality when $p = \infty$, and then a similar argument, as in the proof of Theorems A.4.3 and A.4.4, gives the result for arbitrary $0 < q < p \leq \infty$. Thus, in the sequel, let $0 < q < p = \infty$.

Using the inequality of part c] with

$$s := \min\{1\,,\, (c_1^2q^2NK)^{-1}\,,\, (c_1qN)^{-1}\}$$

and recalling the conditions prescribed for χ, we conclude that

$$\begin{aligned}
m&\big(\{x \in [-1,1] : (\chi(f(x)))^q \geq e^{-2}\|\chi(f)\|^q_{[-1,1]}\}\big) \\
&\geq m\big(\{x \in [-1,1] : f(x) \geq e^{-2/q}\|f\|_{[-1,1]}\}\big) \\
&\geq m\big(\{x \in [-1,1] : f(x) \geq \exp(-c_1(\sqrt{NKs} + Ns))\|f\|_{[-1,1]}\}\big) \\
&\geq s
\end{aligned}$$

for every $f \in \text{GAP}_{N,K}$. Now integrating only on the subset E of $[-1,1]$ where

$$(\chi(f(x)))^q \geq e^{-2} \|\chi(f)\|^q_{[-1,1]},$$

we obtain

$$\begin{aligned}\|\chi(f)\|^q_{[-1,1]} &\leq \frac{e^2}{m(E)} \int_E (\chi(f(x)))^q \, dx \\ &\leq c_2 \max\{1\,,\, q^2NK\,,\, qN\} \|\chi(f)\|^q_{L_q[-1,1]}\end{aligned}$$

for every $f \in \text{GAP}_{N,K}$, where $c_2 := c_1^2 e^2$ (assuming $c_1 \geq 1$). Since $c_1 \leq 5$, we have $c_2 \leq 25e^2$. □

e] Remez-Type Inequality for $\mathcal{B}_n(-1,1)$. Show that

$$\|p\|_{[-1,1]} \leq \left(\frac{2}{2-s}\right)^n$$

for every $p \in \mathcal{B}_n(-1,1)$ satisfying

$$m(\{x \in [-1,1] : |p(x)| \leq 1\}) \geq 2 - s\,.$$

Equality holds if and only if p is of the form

$$p(x) = \pm\left(\frac{1 \pm x}{2-s}\right)^n \,.$$

Proof. See Erdélyi [90b]. □

f] Markov-Type Inequality for $\text{GAP}_{N,K}$. There exists an absolute constant $c > 0$ such that

$$\|f'\|_{[-1,1]} \leq cN(K+1)\|f\|_{[-1,1]}$$

for every $f \in \text{GAP}_{N,K}$ of the form (A.4.1) with each $r_j \geq 1$.

Recall that $|f'(x)|$ is well-defined for every $f \in \text{GAP}_{N,K}$ and $x \in \mathbb{R}$, as the modulus of the one-sided derivative of f at x. The condition that $r_j \geq 1$ for each j in (A.4.1) is needed to ensure that $|f'(z_j)| < \infty$ if $z_j \in \mathbb{R}$. The above result generalizes the corresponding polynomial inequality for the classes $\mathcal{P}_{n,k}$.

Proof. See Borwein and Erdélyi [92]. □

E.21 The Ilyeff-Sendov Conjecture. The following problem is known as the Ilyeff-Sendov conjecture. As before, let $D := \{z \in \mathbb{C} : |z| < 1\}$.

Suppose $P \in \mathcal{P}_n$ has all its zeros in the closed unit disk $\overline{D}$. Then each closed disk centered at a zero of P contains a zero of of P'.

Although the problem in this generality is still unanswered, several special cases have been settled. The case outlined in part a] was first proved in Rubinstein [68]. Vâjâitu and Zaharescu [93] contains stronger results; see also Miller [90]. Milovanović, Mitrinović, and Rassias [94] has a discussion about the recent status of the conjecture.

a] Suppose

$$P(z) = \prod_{k=1}^{n} (z - z_k), \qquad |z_k| \le 1, \quad |z_1| = 1.$$

Show that P' has at least one zero in the closed disk centered at z_1.

Proof. Without loss of generality, we may assume that $z_1 := 1$. Then P is of the form $P(z) = (z-1)Q(z)$, where

$$Q(z) := \prod_{k=2}^{n} (z - z_k), \qquad |z_k| \le 1.$$

Suppose the statement is false. Then $R(z) := P'(z+1)$ has no zero in the closed unit disk $\overline{D}$. Hence

$$\left|\frac{R'(0)}{R(0)}\right| < n - 1.$$

Observe that

$$R(0) = Q(1) \qquad \text{and} \qquad R'(0) = 2Q'(1).$$

Hence

$$\left|\frac{Q'(1)}{Q(1)}\right| = \frac{1}{2}\left|\frac{R'(0)}{R(0)}\right| < \frac{n-1}{2}.$$

However,

$$\operatorname{Re}\left(\frac{Q'(1)}{Q(1)}\right) = \operatorname{Re}\left(\sum_{k=2}^{n} \frac{1}{1 - z_k}\right) \ge \frac{n-1}{2}.$$

This contradiction finishes the proof. □

b] The polynomial $P(z) := z^n - 1$ shows the sharpness of part a].

It follows from Theorem A.5.3 that there exists an absolute constant $c > 0$ such that

$$\|p'\|_{[-1,1]} \leq cn \, \|p\|_{[-1,1]}$$

for every $p \in \mathcal{P}_n$ with no zeros in the open unit disk. The polynomials $p_n(x) := (x+1)^n$ show that up to the absolute constant $c > 0$ this inequality is the best possible. The next exercise shows that the "right" Markov factor on $[-1, 1]$ for polynomials of degree at most n *with complex coefficients* and with no zeros in the open unit disk is $cn(1 + \log n)$ rather than cn. This is an observation of Halász.

E.22 Markov Inequality for $\mathcal{P}_n^c$ with No Zeros in the Unit Disk. Show that there exists an absolute constant $c_1 > 0$ so that

$$\|p'\|_{[-1,1]} \leq c_1 n(1 + \log n)\|p\|_{[-1,1]}$$

for every $p \in \mathcal{P}_n^c$ with no zeros in the open unit disk.

Show also that there exist polynomials $p_n \in \mathcal{P}_n^c$ with no zeros in the open unit disk such that

$$\|p_n'\|_{[-1,1]} \geq c_2 n(1 + \log n)\|p_n\|_{[-1,1]} \,, \qquad n = 1, 2, \ldots$$

with an absolute constant $c_2 > 0$.

Proceed as follows:

a] Show that if $z \in \mathbb{C}$ is outside the open unit disk, then

$$\frac{|x - z|}{|x^{-1} - z|} \leq |x|$$

for every $x \in \mathbb{R} \setminus [-1, 1]$.

b] Suppose $p \in \mathcal{P}_n^c$ has no zeros in the open unit disk. Let $x \in \mathbb{R} \setminus [-1, 1]$. Show that

$$|p(x)| \leq |x|^n |p(x^{-1})| \,.$$

c] Prove the upper bound of the exercise.

Hint: Use part b] and E.11 a]. □

d] Let

$$z_k := e^{2ik\pi/(2n+1)} \,, \qquad k = 1, 2, \ldots, n$$

be the $(2n + 1)$th roots of unity in the open upper half-plane. Let

$$p_{2n+1}(z) := p_{2n+2}(z) := (z + 1) \prod_{k=1}^{n} (z - z_k)^2 \,.$$

Show that $|p_{2n+1}(x)| = |x^{2n+1} - 1|$ for every $x \in \mathbb{R}$. Note that this implies

$$|p_{2n+1}(-1)| = \|p_{2n+1}\|_{[-1,1]} = 2 \,.$$

Show also that

$$\left| \frac{p_{2n+1}'(-1)}{p_{2n+1}(-1)} \right| \geq cn(1 + \log n)$$

with an absolute constant $c > 0$.

Bibliography

Aberth, O., *Iteration methods for finding all zeros of a polynomial simultaneously*, Math. Comp. **27** (1973), 339–344.

Abramowitz, M., & I.A. Stegun, *Handbook of Mathematical Functions*, Dover, New York, NY, 1965.

Achiezer, N.I., *Theory of Approximation*, Frederick Ungar, New York, NY, 1956.

Ahlfors, L.V., *Complex Analysis*, McGraw-Hill, New York, NY, 1953.

Aho, A.V., J.E. Hopcroft, & J.D. Ullman, *The Design and Analysis of Computer Algorithms*, Addison-Wesley, Reading, MA, 1974.

Anderson, J.M., *Müntz-Szász type approximation and the angular growth of lacunary integral functions*, Trans. Amer. Math. Soc. **169** (1972), 237–248.

Anderson, N., E.B. Saff, & R.S. Varga, *On the Eneström-Kakeya theorem and its sharpness*, Linear Algebra and Appl. **28** (1979), 5–16.

Anderson, N., E.B. Saff, & R.S. Varga, *An extension of the Eneström-Kakeya theorem and its sharpness*, SIAM J. Math. Anal. **28** (1981), 10–22.

Ankeny, N.C., & T.J. Rivlin, *On a theorem of Bernstein*, Pacific J. Math. **5** (1955), 849–852.

Arestov, V.V., *On integral inequalities for trigonometric polynomials and their derivatives*, Izv. Akad. Nauk SSSR Ser. Mat. **45** (1981), 3–22.

Ash, R.B., *Complex Variables*, Academic Press, New York, NY, 1971.

Askey, R., & M. Ismail, *Recurrence Relations, Continued Fractions and Orthogonal Polynomials*, vol. 300, Mem. Amer. Math. Soc., Amer. Math. Soc., Providence, RI, 1984.

Aumann, G., *Satz über das Verhalten von Polynomen auf Kontinuen*, Sitz. Preuss. Akad., Wiss. Phys-Math. Kl. (1933), 924–931.

Aziz, A., & Q.M. Dawood, *Inequalities for a polynomial and its derivative*, J. Approx. Theory **54** (1988), 306–313.

Badalyan, G.V., *Generalization of Legendre polynomials and some of their applications*, Akad. Nauk Armyan. SSR Izv. Ser. Fiz.-Mat. Estest. Tekhn. Nauk **8** (1955), 1–28 (in Russian, Armanian summary).

Baishanski, B., *Approximation by polynomials of given length*, Illinois J. Math. **27** (1983), 449–458.

Baishanski, B., & R. Bojanic, *An estimate for the coefficients of polynomials of given length*, J. Approx. Theory **24** (1980), 181–188.

Bak, J., & D.J. Newman, *Rational combinations of $x^{\lambda_k}, \lambda_k \geq 0$ are always dense in $C[0,1]$*, J. Approx. Theory **23** (1978), 155–157.

Baker, G.A., & P. Graves Morris, *Padé Approximants, Part I, Basic Theory, Part II, Extensions and Applications*, Addison-Wesley, Reading, MA, 1981.

Barbeau, E.J., *Polynomials*, Springer-Verlag, New York, NY, 1989.

Barnard, R.W., W. Dayawansa, K. Pearce, & D. Weinberg, *Polynomials with nonnegative coefficients*, Proc. Amer. Math. Soc. **113** (1991), 77–85.

Beauzamy, B., E. Bombieri, P. Enflo, & H.L. Montgomery, *Products of polynomials in many variables*, J. Number Theory **36** (1990), 219–245.

Beauzamy, B., & P. Enflo, *Estimations de produits de polynômes*, J. Number Theory **21** (1985), 390–412.

Beauzamy, B., P. Enflo, & P. Wang, *Quantitative estimates for polynomials in one or several variables: From analysis and number theory to symbolic and massively parallel computation*, Math. Mag. **67** (1994), 243–257.

Bernstein, S.N., *Sur l'ordre de la meilleure approximation des functions continues par des polynômes de degré donné*, Memoires de l'Académie Royale de Belgique **4** (1912), 1–103.

Bernstein, S.N., *Sur la meilleure approximation de $|x|$ par des polynômes de degrés donnés*, Acta. Math. **37** (1913), 1–57.

Bernstein, S.N., *Sur la représentation des polynômes positifs*, Soobshch. Kharkow matem. ob-va, series 2 **14** (1915), 227–228.

Bernstein, S.N., *Sur une propriété des fonctions entières*, C. R. Acad. Sci. Paris **176** (1923), 1603–1605.

Bernstein, S.N., *Collected Works: Vol. I, Constr. Theory of Functions (1905–1930), English Translation*, Atomic Energy Commision, Springfield, VA, 1958.

Beukers, F., *A note on the irrationality of* $\zeta(2)$ *and* $\zeta(3)$, Bull. London Math. Soc. **11** (1979), 268–272.

Beukers, F., *The value of polylogarithms*, Colloq. Math. Soc. János Bolyai **34** (1981), 219–228.

Bini, D., & V. Pan, *Numerical and Algebraic Computations with Matrices and Polynomials*, Birkhäuser, Boston, MA, 1992.

Block, H.D., & H.P. Thielman, *Commutative polynomials*, Quart. J. Math. Oxford Ser. (2) **2** (1951), 241–243.

Bloom, T., *A spanning set for* $C(I^n)$, Trans. Amer. Math. Soc. **321** (1990), 741–759.

Boas, R.P., *Entire Functions*, Academic Press, New York, NY, 1954.

Bohman, H., *On approximation of continuous and of analytic functions*, Ark. Mat. **2** (1952), 43–56.

Bojanov, B.D., *An extension of the Markov inequality*, J. Approx. Theory **35** (1982a), 181–190.

Bojanov, B.D., *Proof of a conjecture of Erdős about the longest polynomial*, Proc. Amer. Math. Soc. **84** (1982b), 99–103.

Borodin A., & I. Munro, *The Computational Complexity of Algebraic and Numeric Problems*, American Elsevier, New York, NY, 1975.

Borosh, I., C.K. Chui, & P.W. Smith, *Best uniform approximation from a collection of subspaces*, Math. Z. **156** (1977), 13–18.

Borwein, J.M., & P.B. Borwein, *Pi and the AGM – A Study in Analytic Number Theory and Computational Complexity*, Wiley, New York, NY, 1987.

Borwein, P.B., *Inequalities and inverse theorems in restricted rational approximation theory*, Canad. J. Math. **32** (1980), 354–361.

Borwein, P.B., *The size of* $\{x : r_n'/r_n \geq 1\}$ *and lower bounds for* $\|e^{-x} - r_n\|$, J. Approx. Theory **36** (1982), 73–80.

Borwein, P.B., *Rational approximations to Stieltjes transforms*, Math. Scand. **53** (1983), 114–124.

Borwein, P.B., *Markov's inequality for polynomials with real zeros*, Proc. Amer. Math. Soc. **93** (1985), 43–48.

Borwein, P.B., *Zeros of Chebyshev polynomials in Markov systems*, J. Approx. Theory **63** (1990), 56–64.

Borwein, P.B., *On the irrationality of* $\sum 1/(q^n + r)$, J. Number Theory **37** (1991a), 253–259.

Borwein, P.B., *Variations on Müntz's theme*, Bull. Canad. Math. Soc. **34** (1991b), 305–310.

Borwein, P.B., *On the irrationality of certain series*, Math. Proc. Cambridge Philos. Soc. **112** (1992), 141–146.

Borwein, P.B., *Exact inequalities for the norms of factors of polynomials*, Can. J. Math. **46** (1994), 687–698.

Borwein, P.B., *The arc length of the lemniscate* $|p(z)| = 1$, Proc. Amer. Math. Soc. **123** (1995), 797–799.

Borwein, P.B., W. Chen, & K. Dilcher, *Zeros of iterated integrals of polynomials*, Can. J. Math. **11** (1995), 65–87.

Borwein, P.B., W. Dykshoorn, T. Erdélyi, & J. Zhang, *Orthogonality and irrationality* (to appear).

Borwein, P.B., & T. Erdélyi, *Notes on lacunary Müntz polynomials*, Israel J. Math. **76** (1991), 183–192.

Borwein, P.B., & T. Erdélyi, *Remez-, Nikolskii- and Markov-type inequalities for generalized nonnegative polynomials with restricted zeros*, Constr. Approx. **8** (1992), 343–362.

Borwein, P.B., & T. Erdélyi, *Lacunary Müntz systems*, Proc. Edinburgh Math. Soc. **36** (1993), 361–374.

Borwein, P.B., & T. Erdélyi, *Sharp Markov-Bernstein type inequalities for classes of polynomials with restricted zeros*, Constr. Approx. **10** (1994), 411–425.

Borwein, P.B., & T. Erdélyi, *Dense Markov spaces and unbounded Bernstein inequalities*, J. Approx. Theory **81** (1995a), 66-77.

Borwein, P.B., & T. Erdélyi, *Müntz spaces and Remez inequalities*, Bull. Amer. Math. Soc. **32** (1995b), 38–42.

Borwein, P.B., & T. Erdélyi, *Upper bounds for the derivative of exponential sums*, Proc. Amer. Math. Soc. **123** (1995c), 1481–1486.

Borwein, P.B., & T. Erdélyi, *Generalizations of Müntz's theorem via a Remez-type inequality for Müntz spaces* (to appear 1).

Borwein, P.B., & T. Erdélyi, *Markov and Bernstein type inequalities in* L_p *for classes of polynomials with constraints*, J. London Math. Soc. (to appear 2).

Borwein, P.B., & T. Erdélyi, *Newman's inequality for Müntz polynomials on positive intervals*, J. Approx. Theory (to appear 3).

Borwein, P.B., & T. Erdélyi, *Sharp extensions of Bernstein's inequality to rational spaces*, Mathematika (to appear 4).

Borwein, P.B., & T. Erdélyi, *The full Müntz theorem in* $C[0,1]$ *and* $L_1[0,1]$, J. London Math. Soc. (to appear 5).

Borwein P.B., & T. Erdélyi, *The L_p Version of Newman's Inequality for lacunary polynomials*, Proc. A.M.S. (to appear 6).

Borwein, P.B., T. Erdélyi, & J. Zhang, *Chebyshev polynomials and Markov-Bernstein type inequalities for rational spaces*, J. London Math. Soc. **50** (1994a), 501–519.

Borwein, P.B., T. Erdélyi, & J. Zhang, *Müntz systems and orthogonal Müntz-Legendre polynomials*, Trans. Amer. Math. Soc. **342** (1994b), 523–542.

Borwein P.B., E.A. Rakhmanov, & E.B. Saff, *Rational approximation with varying weights I*, Constr. Approx. (to appear).

Borwein, P.B., & E.B. Saff, *On the denseness of weighted incomplete approximations*, in A.A. Gonchar & E.B. Saff, Eds., Progress in Approximation Theory, Springer, New York, NY (1992), 419–429.

Borwein, P.B., & B. Shekhtman, *The density of rational functions: A counterexample to a conjecture of D.J. Newman*, Constr. Approx. **9** (1993), 105–110.

Boyd, D.W., *Two sharp inequalities for the norm of a factor of a polynomial*, Mathematika **39** (1992), 341–349.

Boyd, D.W., *Bounds for the height of a factor of a polynomial in terms of Bombieri's norms: I. the largest factor*, J. Symbolic Comput. **16** (1993a), 115–130.

Boyd, D.W., *Bounds for the height of a factor of a polynomial in terms of Bombieri's norms: II. The smallest factor*, J. Symbolic Comput. **16** (1993b), 131–145.

Boyd, D.W., *Large factors of small polynomials*, Contemp. Math. **166** (1994a).

Boyd, D.W., *Sharp inequalities for the product of polynomials*, Bull. London Math. Soc. **26** (1994b).

Boyer, C.B., *A History of Mathematics*, Wiley, New York, NY, 1968.

Braess, D., *Nonlinear Approximation Theory*, Springer-Verlag, Berlin, 1986.

Brent, R., *The parallel evaluation of general arithmetic expressions*, J. Assoc. Comput. Mach. **21** (1974), 201–206.

Browder, A., *On Bernstein's inequality and the norm of Hermitian operators*, Amer. Math. Monthly **78** (1971), 871–873.

Brown, G., & E. Hewitt, *A class of positive trigonometric sums*, Math. Ann. **268** (1984), 91–122.

de Bruijn, N.G., *Inequalities concerning polynomials in the complex domain*, Nederl. Akad. Wetensch. Proc. **50** (1947), 1265–1272; Indag. Math. **9** (1947), 591–598.

de Bruijn, N.G., & T.A. Springer, *On the zeros of composition-polynomials*, Nederl. Akad. Wetensch. Proc. **50** (1947), 895–903; Indag. Math. **9** (1947), 406–414.

Bultheel, A., P. González-Vera, H. Hendriksen, & O. Njåstad, *Orthogonal rational functions similar to Szegö polynomials*, in: Orthogonal Polynomials and their Applications (C. Brezinski, L. Gori, and A. Ronveaux, Eds.), Scientific Publishing Co., Singapore, 1991, pp. 195–204.

Burton, D.M., *The History of Mathematics*, Allyn and Bacon, Newton, MA, 1985.

Cauchy, A.L., *Exercises de Mathématique, Oeuvres (2)*, vol. 9, p. 122, 1829.

Cheney, E.W., *Introduction to Approximation Theory*, McGraw-Hill, New York, NY, 1966.

Chihara, T.S., *An Introduction to Orthogonal Polynomials*, Gordon and Breach, New York, NY, 1978.

Chudnovsky, D.V., & G.V. Chudnovsky, *Padé and rational approximation to systems of functions and their arithmetic applications*, in: Number Theory, New York, 1982 Seminar, Lecture Notes in Mathematics, vol. 1052, Springer-Verlag, Berlin, 1984, pp. 37–84.

Clark, A., *Elements of Abstract Algebra*, Wadsworth, Belmont, CA, 1971.

Clarkson, J.A., & P. Erdős, *Approximation by polynomials*, Duke Math. J. **10** (1943), 5–11.

Cody, W.J., W. Fraser, & J.F. Hart, *Rational Chebyshev approximation using linear equations*, Numer. Math. **12** (1968), 242–251.

van der Corput, J.G., & G. Schaake, *Ungleichungen für Polynome und trigonometrische Polynome*, Compositio Math. **2** (1935), 321–361.

Craven, T., G. Csordas, & W. Smith, *The zeros of derivatives of entire functions and the Pólya-Wiman Conjecture*, Ann. of Math. **125** (1987), 405–431.

Davis, P.J., *Interpolation and Approximation*, Dover, New York, NY, 1975.

Dennis, J.E., & R.B. Schnabel, *Numerical Methods for Unconstrained Optimization and Nonlinear Equations*, Prentice-Hall, Englewood Cliffs, NJ, 1983.

DeVore, R.A., & G.G. Lorentz, *Constructive Approximation*, Springer-Verlag, Berlin, 1993.

Dieudonné, M.J., *La Théorie Analytique des Polynômes d'une Variable*, Gauthier-Villars, Paris, 1938.

Dilcher, K., *Real Wronskian zeros of polynomials with nonreal zeros*, J. Math. Anal. Appl. **154** (1991), 164–183.

Ditzian, Z., *Multivariate Landau-Kolmogorov-type inequality*, Math. Proc. Cambridge Philos. Soc. **105** (1989), 335–350.

Ditzian, Z., & V. Totik, *Moduli of Smoothness*, Springer-Verlag, Berlin, 1988.

Dolženko, E.P., *Estimates of derivatives of rational functions*, Izv. Akad. Nauk SSSR Ser. Mat. **27** (1963), 9–28.

Duffin, R.J., & A.C. Schaeffer, *Some inequalities concerning functions of exponential type*, Bull. Amer. Math. Soc. **80** (1937), 1053–1070.

Duffin, R.J., & A.C. Scheaffer, *Some properties of functions of exponential type*, Bull. Amer. Math. Soc. **44** (1938), 236–240.

Duffin, R.J., & A.C. Scheaffer, *A refinement of an inequality of the brothers Markoff*, Trans. Amer. Math. Soc. **50** (1941), 517–528.

Durand, E., *Solutions Numériques des Équations Algébriques, Tome I: Equations du Type $F(x) = 0$: Racines d'un Polynôme*, Masson, Paris, 1960.

Eneström, G., *Härledning af en allmän formel för antalet pensionärer...*, Öfv. af. Kungl. Vetenskaps-Akademiens Förhandlingar **50** (1893), 405–415.

Erdélyi, A., et al., *Higher Transcendental Functions, Vols. I–III*, McGraw-Hill, New York, NY, 1953.

Erdélyi, T., *Markov type estimates for derivatives of polynomials of special type*, Acta Math. Hung. **51** (1988), 421–436.

Erdélyi, T., *Markov-type estimates for certain classes of constrained polynomials*, Constr. Approx. **5** (1989a), 347–356.

Erdélyi, T., *The Remez inequality on the size of polynomials*, in: Approximation Theory VI, Vol. 1 (C. K. Chui, L. L. Schumaker, & J. D. Ward, Eds.), Academic Press, Boston, MA, 1989b, pp. 243–246.

Erdélyi, T., *Weighted Markov-type estimates for the derivatives of constrained polynomials on* $[0, \infty)$, J. Approx. Theory **58** (1989c), 213–231.

Erdélyi, T., *A Markov-type inequality for the derivatives of constrained polynomials*, J. Approx. Theory **63** (1990a), 321–334.

Erdélyi, T., *A sharp Remez inequality on the size of constrained polynomials*, J. Approx. Theory **63** (1990b), 335–337.

Erdélyi, T., *Bernstein and Markov type inequalities for generalized non-negative polynomials*, Canad. J. Math. **43** (1991a), 495–505.

Erdélyi, T., *Bernstein-type inequalities for the derivative of constrained polynomials*, Proc. Amer. Math. Soc. **112** (1991b), 829–838.

Erdélyi, T., *Nikolskii-type inequalities for generalized polynomials and zeros of orthogonal polynomials*, J. Approx. Theory **67** (1991c), 80–92.

Erdélyi, T., *Estimates for the Lorentz degree of polynomials*, J. Approx. Theory **67** (1991d), 187–198.

Erdélyi, T., *Remez-type inequalities on the size of generalized polynomials*, J. London Math. Soc. **45** (1992a), 255–264.

Erdélyi, T., *Weighted Markov and Bernstein type inequalities for generalized non-negative polynomials*, J. Approx. Theory **68** (1992b), 283–305.

Erdélyi, T., *Remez-type inequalities and their applications*, J. Comput. Appl. Math. **47** (1993), 167-209.

Erdélyi, T., X. Li, & E.B. Saff, *Remez and Nikolskii type inequalities for logarithmic potentials*, SIAM J. Math. Anal. **25** (1994), 365–383.

Erdélyi, T., A. Magnus, & P. Nevai, *Generalized Jacobi weights, Christoffel functions and Jacobi polynomials*, SIAM J. Math. Anal. **25** (1994), 602–614.

Erdélyi T., A. Máté, & P. Nevai, *Inequalities for generalized non-negative polynomials*, Constr. Approx. **8** (1992), 241–255.

Erdélyi, T., & P. Nevai, *Generalized Jacobi weights, Christoffel functions and zeros of orthogonal polynomials*, J. Approx. Theory **69** (1992), 111–132.

Erdélyi, T., & J. Szabados, *On polynomials with positive coefficients*, J. Approx. Theory **54** (1988), 107–122.

Erdélyi, T., & J. Szabados, *Bernstein-type inequalities for a class of polynomials*, Acta Math. Hungar. **52** (1989a), 237–251.

Erdélyi, T., & J. Szabados, *On trigonometric polynomials with positive coefficients*, Studia Sci. Math. Hungar. **24** (1989b), 71–91.

Erdős, P., *An extremum-problem concerning trigonometric polynomials*, Acta. Sci. Math. Szeged **9** (1939), 113–115.

Erdős, P., *On extremal properties of the derivatives of polynomials*, Ann. of Math. **2** (1940), 310–313.

Erdős, P., *Extremal problems on polynomials*, in Approximation Theory II, Academic Press (1976), New York, NY, 347–355.

Erdős, P., F. Herzog, & G. Piranian, *Metric properties of polynomials*, J. Analyse Math. **6** (1958), 125–148.

Erdős, P., & P. Turán, *On the distribution of roots of polynomials*, Ann. of Math. **51** (1950), 105–119.

Erőd, J., *On the lower bound of the maximum of certain polynomials*, Mat. Fiz. Lap. **46** (1939), 58–83 (in Hungarian).

Favard, J., *Sur les polynômes de Tchebycheff*, C. R. Acad. Sci., Paris **200** (1935), 2052–2055.

Feinerman, R.P., & D.J. Newman, *Polynomial Approximation*, Williams and Wilkins, Baltimore, MD, 1976.

Filaseta, M., *Irreducibility criteria for polynomials with non-negative coefficients*, Can. J. Math. **40** (1988), 339–351.

Fischer, B., *Chebyshev polynomials for disjoint compact sets*, Constr. Approx. **8** (1992), 309–329.

Frappier, C., & Q.I. Rahman, *On an inequality of S. Bernstein*, Canad. J. Math. **34** (1982), 932–944.

Frappier, C., Q.I. Rahman, & St. Ruscheweyh, *New inequalities for polynomials*, Trans. Amer. Math. Soc. **288** (1985), 65–99.

Freud G., *Orthogonal Polynomials*, Pergamon Press, Oxford, 1971.

Fuchs, W.H.J., *On the closure of* $\{e^{-t}t^{a_\nu}\}$, Proc. Cambridge Phil. Soc. **42** (1946), 91–105.

Gasper, G., & M. Rahman, *Basic Hypergeometric Series*, Cambridge University Press, Cambridge, 1990.

Gauss, C.F., *Werke*, Göttingen, 12 vols., 1863-1929.

Gel'fond, A.O., *Transcendental and Algebraic Numbers*, Dover, New York, NY, 1960.

Gencev, T.G., *Inequalities for asymmetric entire functions of exponential type*, Dokl. Akad. Nauk SSSR **241** (1978), 1261–1264 (in Russian); Soviet Math. Dokl. **19** (1978), 981–985 (English translation).

Giroux, A., & Q.I. Rahman, *Inequalities for polynomials with a prescribed zero*, Trans. Amer. Math. Soc. **193** (1974), 67–98.

Glesser, P., *Nouvelle majoration de la norme des facteurs d'un polynôme*, C.R. Math. Rep. Acad. Sci. Canada **12** (1990), 224–228.

von Golitschek, M., *A short proof of Müntz's theorem*, J. Approx. Theory **39** (1983), 394–395.

von Golitschek, M., & G.G. Lorentz, *Bernstein inequalities in* $L_p, 0 < p \leq \infty$, Rocky Mountain J. Math. **19** (1989), 145–156.

von Golitschek, M., G.G. Lorentz, & Y. Makovoz, *Asymptotics of weighted polynomials*, in A.A. Gonchar, & E.B. Saff, Eds., Progress in Approximation Theory, Springer, New York, NY (1992), 431–451.

Govil, N.K., *On the derivative of a polynomial*, Proc. Amer. Math. Soc. **41** (1973), 543–546.

Grace, J.H., *The zeros of a polynomial*, Proc. Cambridge Philos. Soc. **11** (1902), 352–357.

Granville, A., *Bounding the coefficients of a divisor of a given polynomial*, Monatsh. Math. **109** (1990), 271–277.

Hardy, G.H., & J.E. Littlewood, *A further note on the converse of Abel's theorem*, Proc. London Math. Soc. **25** (1926), 219–236.

He, X., *Weighted Polynomial Approximation and Zeros of Faber Polynomials*, Ph.D. Dissertion, University of South Florida, Tampa, 1991.

He, X., & X. Li, *Uniform convergence of polynomials associated with varying Jocobi weights*, J. Math **21** (1991), 281–300.

Henrici, P., *Applied and Computational Complex Analysis, Vol. I*, Wiley, New York, NY, 1974.

Hille, E., G., Szegő, & J.D. Tamarkin, *On some generalizations of a theorem of A. Markov*, Duke Math. J. **3** (1937), 729–739.

Hölder, O, *Über einen Mittelwertsatz*, Göttinger Nachrichten (1889), 38–47.

Householder, A.S., *The Numerical Treatment of a Single Nonlinear Equation*, McGraw-Hill, New York, NY, 1970.

Ivanov, K.G., & V. Totik, *Fast decreasing polynomials*, Constr. Approx. **6** (1990), 1–20.

Ivanov, V.I., *Certain inequalities for trigonometric polynomials and their derivatives in various metrics*, Mat. Zametki **18** (1975), 489–498; Translation in Math. Notes **18** (1975), 880–885.

Jensen, J.L.W., *Recherches sur la théorie des équations*, Acta. Math. **36** (1913), 181–185.

Kakeya, S., *On the limits of the roots of an algebraic equation with positive coefficients*, Tôhoku Math. J. **2** (1912), 140–142.

Karlin, S., *Total Positivity*, Stanford University Press, Stanford, CA, 1968.

Karlin, S., & W.J. Studden, *Tchebycheff Systems with Applications in Analysis and Statistics*, Wiley, New York, NY, 1966.

Karp, R., & V. Ramachandran, *Parallel algorithms for shared memory machines*, in: Handbook of Theoretical Computer Science, Vol. A, Elsevier, Amsterdam, 1990, pp. 869–941.

Kemperman, J.H.B., & G.G. Lorentz, *Bounds for polynomials with applications*, Proc. Kon. Nederl. Akad. Wetensch. **A 82** (1979), 13–26.

Kerner, I.O., *Ein Gesamtschrittverfahren zur Berechnung der nullstellen von Polynomen*, Numer. Math. **8** (1966), 290–294.

Khalilova, B.A., *On some estimates for polynomials*, Izv. Akad. Nauk Azerbaĭdzhan. SSSR Ser. Fiz.-Tekhn. Mat. Nauk **2** (1974), 46–55 (in Russian).

Klein, G., *On the approximation of functions by polynomials* (1951), Ph.D. Dissertation, University of Chicago, IL.

Kneser, H., *Das Maximum des Produkts zweier Polynome*, Sitz. Preuss. Akad. Wiss., Phys. - Math. Kl. (1934), 429–431.

Knuth, D.E., *The Art of Computer Programming. Vol. 2: Seminumerical Algorithms*, Addison-Wesley, Reading, MA, 1981.

Kolmogorov, A.N., *On inequalities between the upper bounds of successive derivatives of an arbitrary function on an infinite interval*, Amer. Math. Soc. Transl. Series **1, 2** (1962), 233–243.

Konjagin, S.V., *Bounds on the derivatives of polynomials*, Dokl. Akad. Nauk SSSR **243** (1978), 1116–1118 (in Russian); English translation: Soviet Math. Dokl. **19** (1978), 1477-1480.

Korovkin, P., *On convergence of linear positive operators in the space of continuous functions*, Doklady Akad. Nauk SSSR **90** (1953), 961–964.

Kristiansen, G.K., *Proof of a polynomial conjecture*, Proc. Amer. Math. Soc. **44** (1974), 58–60.

Kristiansen, G.K., *Some inequalities for algebraic and trigonometric polynomials*, J. London Math. Soc. **20** (1979), 300–314.

Kristiansen, G.K., *Solution to Problem* $80 - 16$, Siam Review **24** (1982), 77–78.

Kroó, A., & F. Peherstorfer, *On the distribution of extremal points of general Chebyshev polynomials*, Trans. Amer. Math. Soc. **329** (1992), 117–130.

Kroó, A., & J.J. Swetits, *On density of interpolation points a Kadec-type theorem and Saff's principle of contamination in* L_p *approximation*, Constr. Approx. **8** (1992), 87–103.

Kroó, A., & J. Szabados, *Constructive properties of self-reciprocal polynomials*, Analysis **14** (1994a), 319–339.

Kroó, A., & J. Szabados, *Müntz-type problems for Bernstein polynomials*, J. Approx. Theory **78** (1994b), 446–457.

Lachance, M.A., *Bernstein and Markov inequalities for constrained polynomials*, in: Rational Approximation and Interpolation, Lecture Notes in Mathematics, vol. 1045, Springer-Verlag, 1984, pp. 125–135.

Lachance, M.A., E.B. Saff, & R.S. Varga, *Inequalities for polynomials with a prescribed zero*, Math. Z. **168** (1979), 105–116.

Lax, P.D., *Proof of a conjecture of P. Erdős on the derivative of a polynomial*, Bull. Amer. Math. Soc. **50** (1944), 509–513.

Lebesgue, H., *Sur l'approximation des fonctions*, Bull. Sci. Math. **22** (1898), 278–287.

Leviatan, D., *Improved estimates in Müntz-Jackson theorems*, in: Progress in Approximation Theory (P. Nevai & A. Pinkus, Eds.), Academic Press, 1991, pp. 575–582.

Levin, A.L., & D.S. Lubinsky, *Canonical Products and the weights* $\exp(-|x|^\alpha)$, $\alpha > 1$, *with applications*, J. Approx. Theory **49** (1987a), 149–169.

Levin, A.L., & D.S. Lubinsky, *Weights on the real line that admit good relative polynomial approximation, with applications*, J. Approx. Theory **49** (1987b), 170–195.

Loomis, L.H., *A note on the Hilbert transform*, Bull. Amer. Math. Soc. **50** (1946), 1082–1086.

Lorentz, G.G., *The degree of approximation by polynomials with positive coefficients*, Math. Ann. **151** (1963), 239–251.

Lorentz, G.G., *Approximation of Functions,* 2nd ed., Chelsea, New York, NY, 1986a.

Lorentz, G.G., *Bernstein Polynomials,* 2nd ed., Chelsea, New York, NY, 1986b.

Lorentz, G.G., *Notes on Approximation*, J. Approx. Theory **56** (1989), 360–365.

Lorentz, G.G., K. Jetter, & S.D. Riemenschneider, *Birkhoff Interpolation*, Addison-Wesley, Reading, MA, 1983.

Lorentz, G.G., M. von Golitschek, & Y. Makovoz, *Constructive Approximation: Advanced Problems* (to appear).

Lubinsky, D.S., *Strong Asymptotics for Extremal Errors and Polynomials Associated with Erdős Weights*, Pitman Research Notes in Mathematics Series, vol. 202, Longman House, Burnt Mill, Harlow, Essex, UK, 1989.

Lubinsky, D.S., & P. Nevai, *Markov-Bernstein inequalities revisited*, Approx. Theory Appl. **3** (1987), 98–119.

Lubinsky, D.S., & E.B. Saff, *Strong asymptotics for extremal polynomials associated with weights on* $\mathbb{R}$, Lecture Notes in Mathematics, vol. 1305, Springer-Verlag, Berlin, 1988a.

Lubinsky, D.S., & E.B. Saff, *Uniform and mean approximation by certain weighted polynomials, with applications*, Constr. Approx. **4** (1988b), 21–64.

Lubinsky, D.S., & E.B. Saff, *Szegő asymptotics for non-Szegő weights on* $[-1, 1]$, in: Approximation Theory VI, Vol. 2 (C.K. Chui, L.L. Schumaker, & J.D. Ward, Eds.), Academic Press, Boston, MA, 1989, pp. 409–412.

Lubinsky, D.S., & E.B. Saff, *Markov-Bernstein and Nikolskii inequalities, and Christoffel functions for exponential weights on* $[-1, 1]$, SIAM J. Math. Anal. **24** (1993), 528–556.

Lubinsky, D.S., & Z. Ziegler, *Coefficient bounds in the Lorentz representation of a polynomial*, Canad. Math. Bull. **32** (1990), 197–206.

Lucas, F., *Propriétés géométriques des fractions rationelles*, C. R. Acad. Sci., Paris **77** (1874), 631–633.

Luxemburg, W.A.J., & J. Korevaar, *Entire functions and Müntz-Szász type approximation*, Trans. Amer. Math. Soc. **157** (1971), 23–37.

Mahler, K., *Zur Approximation der Exponentialfunktion und des Logarithmus*, J. Reine Angew. Math. **166** (1931), 118–150.

Mahler, K., *An application of Jensen's formula to polynomials*, Mathematika **7** (1960), 98–100.

Mahler, K., *On some inequalities for polynomials in several variables*, J. London Math. Soc. **37** (1962), 341–344.

Mahler, K., *A remark on a paper of mine on polynomials*, Illinois J. Math. **8** (1964), 1–4.

Mairhuber, J., *On Haar's theorem concerning Chebyshev approximation problems having unique solutions*, Proc. Amer. Math. Soc. **7** (1956), 609–615.

Malik, M.A., *On the derivative of a polynomial*, J. London Math. Soc. **8** (1969), 57–60.

Marden, M., *Geometry of Polynomials, Math. Survey Monographs, No. 3*, Amer. Math. Soc., Providence, RI, 1966.

Markov, A.A., *Sur une question posée par D.I. Mendeleieff*, Izv. Acad. Nauk St. Petersburg **62** (1889), 1–24.

Markov, V.A., *Über Polynome die in einen gegebenen Intervalle möglichst wenig von Null abweichen*, Math. Ann. **77** (1916), 213–258; the original appeared in Russian in 1892.

Máté, A., *Inequalities for derivatives of polynomials with restricted zeros*, Proc. Amer. Math. Soc. **88** (1981), 221–224.

Máté, A., & P. Nevai, *Bernstein inequality in* L_p *for* $0 < p < 1$, *and* $(C, 1)$ *bounds of orthogonal polynomials*, Ann. of Math. **111** (1980), 145–154.

Máté, A., P. Nevai, & V. Totik, *Asymptotics for the ratio of leading coefficients of orthogonal polynomials on the unit circle*, Constr. Approx. **1** (1985), 63–69.

Máté, A., P. Nevai, & W. van Assche, *The support of measures associated with orthogonal polynomials and the spectra of the real self-adjoint operators*, Rocky Mountain J. Math. **21** (1991), 501–527.

McCarthy, P.C., J.E. Sayre, & B.L.R. Shawyer, *Generalized Legendre polynomials*, J. Math. Anal. Appl. **177** (1993), 530–537.

Mhaskar, M.N., & E.B. Saff, *Where does the sup norm of a weighted polynomial live?*, Constr. Approx. **1** (1985), 71–91.

Mignotte, M., *Some useful bounds*, in: Computer Algebra: Symbolic and Algebraic Computation (Burchberger, B., et al., Eds.) Springer, New York, NY, 1982, pp. 259–263.

Mignotte, M., *Mathematics for Computer Algebra*, Spinger-Verlag, New York, NY, 1992.

Miller, M., *Maximal polynomials and the Ilieff-Sendov conjecture*, Trans. Amer. Math. Soc. **321** (1990), 285–303.

Milovanović, G.V., D.S. Mitrinović, & Th.M. Rassias, *Topics in Polynomials: Extremal Problems, Inequalities, Zeros*, World Scientific, Singapore, 1994.

Mitrinović, D.S., J.E. Pecaric, & A.M. Fink, *Classical and New Inequalities in Analysis*, Kluwer, Dordrecht, 1993.

Müntz, C., *Über den Approximationssatz von Weierstrass*, H.A. Schwarz Festschrift, Berlin (1914).

Natanson, I.P., *Constructive Function Theory, Vol. 1*, Frederick Ungar, New York, NY, 1964.

Nevai, P., *Mean convergence of Lagrange interpolation, I*, J. Approx. Theory **18** (1976), 363–377.

Nevai, P., *Bernstein's inequality in* $L_p, 0 < p < 1$, J. Approx. Theory **27** (1979a), 239–243.

Nevai, P., *Orthogonal Polynomials*, vol. 213, Mem. Amer. Math. Soc., Amer. Math. Soc., Providence, RI, 1979b.

Nevai, P., *Géza Freud, Orthogonal polynomials and Christoffel functions. A case study*, J. Approx. Theory **48** (1986), 3–167.

Nevai, P., *Weakly convergent sequences of functions and orthogonal polynomials*, J. Approx. Theory **65** (1991), 332–340.

Nevai, P., & V. Totik, *Weighted polynomial inequalities*, Constr. Approx. **2** (1986), 113–127.

Nevai, P., & V. Totik, *Sharp Nikolskii inequalities with exponential weights*, Anal. Math. **13** (1987), 261–267.

Newman, D.J., *Derivative bounds for Müntz polynomials*, J. Approx. Theory **18** (1976), 360–362.

Newman, D.J., *Approximation with Rational Functions*, CBMS Regional Conf. Ser. in Math. **41**, Conf. Board Math. Sci., Washington, DC, 1978.

Newman, D.J., & T.J. Rivlin, *On polynomials with curved majorants*, Canad. J. Math. **34** (1982), 961–968.

Nürnberger, G., *Approximation by Spline Functions*, Springer-Verlag, Berlin, 1989.

Ogawa, S., & K. Kitahara, *An extension of Müntz's theorem in multivariables*, Bull. Austral. Math. Soc. **36** (1987), 375–387.

Olivier, P.F., & A.O. Watt, *Polynomials with a prescribed zero and the Bernstein's inequality*, Canad. J. Math. **45** (1993), 627–637.

Operstein, V., *Full Müntz theorem in* $L_p[0, 1]$, J. Approx. Theory (to appear).

Osval'd, P., *Some inequalities for trigonometric polynomials in* $L_p, 0 < p < 1$, Izv. Vyssh. Uchebn. Zaved. Mat. (7) **20** (1976), 65–75.

Pan, V., *Complexity of computations with matrices and polynomials*, SIAM Rev. **34** (1992), 225–262.

Peherstorfer, F., *On Tchebycheff polynomials on disjoint intervals*, in: Haar Memorial Conference, Vol. II (J. Szabados and K. Tandori, Eds.), North-Holland, Amsterdam, 1987, pp. 737–751.

Petrushev, P.P., & V.A. Popov, *Rational Approximation of Real Functions*, Cambridge University Press, Cambridge, 1987.

Pinkus, A., *On* L^1*-Approximation*, Cambridge University Press, Cambridge, 1989.

Pinkus, A., & Z. Ziegler, *Interlacing properties of the zeros of the error functions in best* L^p*-approximations*, J. Approx. Theory **27** (1979), 1–18.

Pólya, G., *Beitrag zur Verallgemeinerung des Verzerrungssatzes auf mehrfach Zusammenhängende Gebiete*, S.B. Preuss. Akad. Wiss., Berlin, K.L. Math. Phys. Tech. (1928), 228–232 and 280–282.

Pólya, G., & G. Szegő, *Problems and Theorems in Analysis, Vols. I and II,* 4th ed., Springer-Verlag, New York, NY, 1976.

Pommerenke, Ch., *On some problems by Erdős, Herzog and Piranian*, Michigan Math. J. **6** (1959a), 221–225.

Pommerenke, Ch., *On some metric properties of polynomials with real zeros*, Michigan Math. J. **6** (1959b), 377–380.

Pommerenke, Ch., *On the derivative of a polynomial*, Michigan Math. J. **6** (1959c), 373–375.

Pommerenke, Ch., *On some metric properties of complex polynomials*, Michigan Math. J. **8** (1961), 97–115.

Pommerenke, Ch., *On some metric properties of polynomials with real zeros, I and II,* Michigan Math. J. **6** (1959), 25–27; and **8** (1961), 49–54.

Pommerenke, Ch., *Univalent Functions*, Vandenhoeck & Ruprecht, Göttingen, 1975.

Protapov, M.K., *Some inequalities for polynomials and their derivatives*, Vestnik Moskov. Univ. Ser. I Mat. Mekh. **2** (1960), 10–19 (in Russian).

Rahman, Q.I., *On extremal properties of the derivatives of polynomials and rational functions*, J. London Math. Soc. **35** (1960), 334–336.

Rahman, Q.I., *On extremal properties of the derivatives of polynomials and rational functions*, Amer. J. Math. **113** (1991), 169–177.

Rahman, Q.I., & G. Labelle, *L'influence des zéros sur la borne dans le théoréme de Markoff pour les polynômes sur l'intervalle unité*, J. London Math. Soc. **43** (1968), 183–185.

Rahman, Q.I., & G. Schmeisser, *Some inequalities for polynomials with a prescribed zero*, Trans. Amer. Math. Soc. **216** (1976), 91–103.

Rahman, Q.I., & G. Schmeisser, *Les Inégalités de Markoff et de Bernstein*, Les Presses de L'Université de Montreal, 1983.

Rahman, Q.I., & G. Schmeisser, *L^p inequalities for entire functions of exponential type*, Trans. Amer. Math. Soc. **320** (1990), 91–103.

Rainville, E.D., *Special Functions*, Chelsea, New York, NY, 1971.

Rassias, G.M., J.M. Rassias, & Th.M. Rassias, *A counter-example to a conjecture of P. Erdős*, Proc. Japan Acad. Sci. **53** (1977), 119–121.

Remez, E.J., *Sur une propriété des polyômes de Tchebyscheff*, Comm. Inst. Sci. Kharkow **13** (1936), 93–95.

Reznick, B., *An inequality for products of polynomials*, Proc. Amer. Math. Soc. **117** (1993).

Riesz, M., *Eine trigonometrische Interpolations-formel und einige Ungleichungen für Polynome*, Jahresbericht des Deutschen Mathematiker-Vereinigung **23** (1914), 354–368.

Ritt, J.F., *Permutable rational functions*, Trans. Amer. Math. Soc. **25** (1923), 399–448.

Rivlin, T.J., *Chebyshev Polynomials,* 2nd ed., Wiley, New York, NY, 1990.

Rogers, L.J., *An extension of a certain theorem in inequalities*, Messenger of Math. **17** (1888), 145–150.

Rogosinski, W.W., *Extremum problems for polynomials and trigonometric polynomials*, J. London Math. Soc. **29** (1954), 259–275.

Royden, H.L., *Real Analysis,* 3rd ed., MacMillan, New York, NY, 1988.

Rubinstein, Z., *On a problem of Ilyeff*, Pacific J. Math. **26** (1968), 159–161.

Rudin, W., *Functional Analysis*, McGraw-Hill, New York, NY, 1973.

Rudin, W., *Real and Complex Analysis,* 3rd ed., McGraw-Hill, New York, NY, 1987.

Saff, E.B., & T. Sheil-Small, *Coefficient and integral mean estimates for algebraic and trigonometric polynomials with restricted zeros*, J. London Math. Soc. **9** (1974), 16–22.

Saff, E.B., & V. Totik, *Logarithmic Potentials with External Fields*, Springer-Verlag (to appear).

Saff, E.B., & R.S. Varga, *Uniform approximation by incomplete polynomials*, Internat. J. Math. Math. Sci. **1** (1978), 407–420.

Saff, E.B., & R.S. Varga, *On lacunary incomplete polynomials*, Math. Z. **177** (1981), 297–314.

Salem, R., *Algebraic Numbers and Fourier Analysis*, Heath, Boston, MA, 1963.

Schaeffer, A.C., & R.J. Duffin, *On some inequalities of S. Bernstein and V. Markoff for derivatives of polynomials*, Bull. Amer. Math. Soc. **44** (1938), 289–297.

Scheick, J.T., *Inequalities for derivatives of polynomials of special type*, J. Approx. Theory **6** (1972), 354–358.

Schmidt, E., *Zur Kompaktheit der Exponentialsummen*, J. Approx. Theory **3** (1970), 445–459.

Schönhage, A., *Asymptotically fast algorithms for the numerical multiplication and division of polynomials with complex coefficients*, in: EUROCAM (1982) Marseille, Springer Lecture Notes in Computer Science, vol. 144, 1982, pp. 3–15.

Schur, I., *Über das Maximum des absoluten Betrages eines Polynoms in einen gegebenen Intervall*, Math. Z. **4** (1919), 271–287.

Schwartz, L., *Etude des Sommes d'Exponentielles*, Hermann, Paris, 1959.

Shapiro, H.S., *Topics in Approximation Theory*, Springer-Verlag, Berlin, 1971.

Smale, S., *On the efficiency of the algorithms of analysis*, Bull. Amer. Math. Soc. **13** (1985), 87–121.

Smith, P.W., *An improvement theorem for Descartes systems*, Proc. Amer. Math. Soc. **70** (1978), 26–30.

Somorjai, G., *A Müntz-type problem for rational approximation*, Acta. Math. Hungar. **27** (1976), 197–199.

Stahl, H., & V. Totik, *General Orthogonal Polynomials*, Encyclopedia of Mathematics, vol. 43, Cambridge University Press, New York, NY, 1992.

Stewart, I., *Galois Theory*, Chapman and Hall, London, 1973.

Stieltjes, T.J., *Oevres Complétes*, vol. 2, Groningen, P. Noordhoff, 1914.

Stromberg, K.R., *Introduction to Classical Real Analysis*, Wadsworth, Belmont, CA, 1981.

Szabados, J., *Bernstein and Markov type estimates for the derivative of a polynomial with real zeros*, in: Functional Analysis and Approximation, Birkhäuser, Basel, 1981, pp. 177–188.

Szabados, J., & A.K. Varma, *Inequalities for derivatives of polynomials having real zeros*, in: Approximation Theory III (E. W. Cheney, Ed.), Academic Press, New York, NY, 1980, pp. 881–888.

Szabados, J., & P. Vértesi, *Interpolation of Functions*, World Scientific, Singapore, 1990.

Szász, O., *Über die Approximation stetiger Funktionen durch lineare Aggregate von Potenzen*, Math. Ann. **77** (1916), 482–496.

Szegő, G., *Bemerkungen zu einen Satz von J. H. Grace über die Wurzeln algebraischer Gleighungen*, Math. Z. **13** (1922), 28–55.

Szegő, G., *Über einen Satz von A. Markoff*, Math. Z. **23** (1925), 45–61.

Szegő, G., *Über einen Satz des Herrn Serge Bernstein*, Schriften der Königsberger Gelehrten Gesellschaft **22** (1928), 59–70.

Szegő, G., *On some problems of approximation*, Magyar Tud. Akad. Mat. Kut. Int. Közl. **2** (1964), 3–9.

Szegő, G., *Orthogonal Polynomials*, 4th ed., Amer. Math. Soc., Providence, RI, 1975.

Szegő, G., *Collected Papers, 3 Vols.*, Ed. R. Askey, Birkhäuser, Boston MA, 1982.

Taslakyan, A.K., *Some properties of Legendre quasi-polynomials with respect to a Müntz system*, Mathematics, Erevan University, Erevan (Russian, Armenian summary) **2** (1984), 179-189.

Todd, J., *A Legacy from E.I. Zolotarev (1847–1878)*, Math. Intelligencer **10** (1988), 50–53.

Totik, V., *Weighted Approximation with Varying Weight*, Lecture Notes in Mathematics, vol. 1569, Springer-Verlag, Berlin, 1994.

Traub, J.F., *Iterative Methods for the Solutions of Equations*, Chelsea, New York, NY, 1982.

Trent, T.T., *A Müntz-Szász theorem for C(D)*, Proc. Amer. Math. Soc. **83** (1981), 296–298.

Tsuji, M., *Potential Theory in Modern Function Theory*, Dover, New York, NY, 1959.

Turán, P., *Über die Ableitung von Polynomen*, Compositio Math. **7** (1939), 89–95.

Turán, P., *On an inequality of Čebyšev*, Ann. Univ. Sci. Budapest. Eötvös Sect. Math. **9** (1968), 15–16.

Turán, P., *On a New Method of Analysis and Its Applications*, Wiley, New York, NY, 1984.

Vâjâitu, V., & A. Zaharescu, *Ilyeff's conjecture on a corona*, Bull. London. Math. Soc. **25** (1993), 49–54.

van Assche, W., & I. Vanherwegen, *Quadrature formulas based on rational interpolation*, Math. Comp. **61** (1993), 765–783.

van der Poorten, A., *A proof that Euler missed... Apéry's proof of the irrationality of* $\zeta(3)$, Math. Intelligencer **1** (1979), 195–203.

van der Waerden, B.L., *Modern Algebra, Volume II*, Frederick Ungar, New York, NY, 1950.

Varga, R., *Scientific Computation on Mathematical Problems and Conjectures*, SIAM, Philadelphia, PA, 1990.

Varga, R.S., & Wu Wen-da, *On the rate of overconvergence of the generalized Eneström-Kakeya functional for polynomials*, J. Comput. Math. **3** (1985), 275–288.

Veidinger, L., *On the numerical determination of the best approximation in the Chebyshev sense*, Numer. Math. **2** (1960), 99–105.

Videnskii, V.S., *On estimates of the derivative of a polynomial*, Izv. Akad. Nauk SSSR Ser. Mat. **15** (1951), 401–420.

Videnskii, V.S., *Markov and Bernstein type inequalities for derivatives of trigonometric polynomials on an interval shorter than the period*, Dokl. Akad. Nauk SSSR **130** (1960), 13–16 (in Russian).

Walker, P., *Separation of the zeros of polynomials*, Amer. Math. Monthly **100** (1993), 272–273.

Walsh, J.L., *On the location of the roots of the derivative of a rational function*, Ann. of Math. **22** (1920), 128–144.

Walsh, J.L., *On the location of the roots of the Jacobian of two binary forms, and the derivative of a rational function*, Trans. Amer. Math. Soc. **22** (1921), 101–116.

Walsh, J.L., *The Location of Critical Points of Analytic and Harmonic Functions*, Amer. Math. Soc., New York, NY, 1950.

Walsh, J.L., *Interpolation and Approximation by Rational Functions in the Complex Domain,* 5th ed., Amer. Math. Soc., Providence, RI, 1969.

Weierstrass, K., *Mathematische Werke*, Berlin, 1915.

Werner, W., *On the Simultaneous Determination of Polynomial Roots*, Lecture Notes in Mathematics, vol. 953, Spinger-Verlag, Berlin, 1982, pp. 188–202.

Wilf, H.S., *Algorithms and Complexity*, Prentice Hall, Englewood Cliffs, NJ, 1986.

Zhou, S.P., *Rational Approximation and Müntz Approximation*, Ph. D. Thesis, Dalhousie University, 1992a.

Zhou, S.P., *Some remarks on Turán's inequality*, J. Approx. Theory **68** (1992b), 45–48.

Zielke, R., *Discontinuous Chebyshev Systems*, Lecture Notes in Mathematics, vol. 707, Spinger-Verlag, Berlin, 1979.

Zwick, D., *Characterization of WT-spaces whose derivatives form a WT-space*, J. Approx. Theory **38** (1983), 188–191.

Zygmund, A., *Trigonometric Series, Vols. I and II,* 2nd ed., Cambridge Univ. Press, Cambridge, 1977.

Notation

Definitions of the more commonly used spaces are given. The equation numbers here are the same as the equation numbers in the text.

Throughout the book, unless otherwise stated, spans should be assumed to be real. Likewise, in function spaces, unless otherwise stated, the functions should be assumed to be real valued.

The Basic Spaces.

$$\mathcal{P}_n^c := \left\{ p : p(z) = \sum_{k=0}^{n} a_k z^k, \quad a_k \in \mathbb{C} \right\}. \tag{1.1.1}$$

$$\mathcal{P}_n := \left\{ p : p(z) = \sum_{k=0}^{n} a_k z^k, \quad a_k \in \mathbb{R} \right\}. \tag{1.1.2}$$

$$\mathcal{R}_{m,n}^c := \left\{ \frac{p}{q} : p \in \mathcal{P}_m^c, q \in \mathcal{P}_n^c \right\}. \tag{1.1.3}$$

$$\mathcal{R}_{m,n} := \left\{ \frac{p}{q} : p \in \mathcal{P}_m, q \in \mathcal{P}_n \right\}. \tag{1.1.4}$$

(1.1.5)
$$\mathcal{T}_n^c := \left\{ t : t(\theta) := \sum_{k=-n}^{n} a_k e^{ik\theta}, \quad a_k \in \mathbb{C} \right\}$$
$$= \left\{ t : t(z) = a_0 + \sum_{k=1}^{n} (a_k \cos kz + b_k \sin kz)\,, \quad a_k, b_k \in \mathbb{C} \right\}.$$
$$\mathcal{T}_n := \left\{ t : t(z) = a_0 + \sum_{k=1}^{n} (a_k \cos kz + b_k \sin kz)\,, \quad a_k, b_k \in \mathbb{R} \right\}.$$

Müntz Spaces.

(3.4.2)
$$M_n(\Lambda) := \operatorname{span}\{x^{\lambda_0}, x^{\lambda_1}, \dots, x^{\lambda_n}\}\,.$$

(3.4.3)
$$M(\Lambda) := \bigcup_{n=0}^{\infty} M_n(\Lambda) = \operatorname{span}\{x^{\lambda_0}, x^{\lambda_1}, \dots\}\,.$$

Associated with a sequence $(\lambda_i)_{i=0}^{\infty}$ with $\mathrm{Re}(\lambda_i) > -1/2$ for each i, the nth Müntz-Legendre polynomial is:

(3.4.5)
$$L_n(x) := L_n\{\lambda_0, \dots, \lambda_n\}(x)$$
$$:= \frac{1}{2\pi i} \int_\Gamma \prod_{k=0}^{n-1} \frac{t + \overline{\lambda}_k + 1}{t - \lambda_k} \frac{x^t\, dt}{t - \lambda_n}\,.$$

For distinct λ_i,

(3.4.6)
$$L_n\{\lambda_0, \dots, \lambda_n\}(x) = \sum_{k=0}^{n} c_{k,n} x^{\lambda_k}\,, \qquad x \in (0, \infty)$$

with
$$c_{k,n} := \frac{\prod_{j=0}^{n-1} (\lambda_k + \overline{\lambda}_j + 1)}{\prod_{j=0, j\neq k}^{n} (\lambda_k - \lambda_j)}\,.$$

Also,

(3.4.8)
$$L_n^* := (1 + \lambda_n + \overline{\lambda}_n)^{1/2} L_n$$

is the nth orthonormal Müntz-Legendre polynomial.

Rational Spaces.

$K := \mathbb{R} \pmod{2\pi}, \quad D := \{z \in \mathbb{C} : |z| < 1\}, \quad \partial D := \{z \in \mathbb{C} : |z| = 1\}.$

$$\text{(3.5.3)} \qquad \mathcal{P}_n(a_1, a_2, \ldots, a_n) := \left\{ \frac{p(x)}{\prod_{k=1}^n |x - a_k|} : p \in \mathcal{P}_n \right\}.$$

$$\text{(3.5.4)} \qquad \mathcal{T}_n(a_1, a_2, \ldots, a_n) := \left\{ \frac{t(\theta)}{\prod_{k=1}^n |\cos\theta - a_k|} : t \in \mathcal{T}_n \right\}.$$

Also

$$\mathcal{T}_n(a_1, a_2, \ldots, a_{2n}; K) := \mathcal{T}_n(a_1, a_2, \ldots, a_{2n}) := \left\{ \frac{t(\theta)}{\prod_{k=1}^{2n} |\sin((\theta - a_k)/2)|} : t \in \mathcal{T}_n \right\}$$

and

$$\mathcal{T}_n^c(a_1, a_2, \ldots, a_{2n}; K) := \left\{ \frac{t(\theta)}{\prod_{k=1}^{2n} \sin((\theta - a_k)/2)} : t \in \mathcal{T}_n^c \right\}$$

on K with $a_1, a_2, \ldots, a_{2n} \in \mathbb{C} \setminus \mathbb{R}$;

$$\mathcal{P}_n(a_1, a_2, \ldots, a_n; [-1, 1]) := \left\{ \frac{p(x)}{\prod_{k=1}^n |x - a_k|} : p \in \mathcal{P}_n \right\}$$

and

$$\mathcal{P}_n^c(a_1, a_2, \ldots, a_n; [-1, 1]) := \left\{ \frac{p(x)}{\prod_{k=1}^n (x - a_k)} : p \in \mathcal{P}_n^c \right\}$$

on $[-1, 1]$ with $a_1, a_2, \ldots, a_n \in \mathbb{C} \setminus [-1, 1]$;

$$\mathcal{P}_n^c(a_1, a_2, \ldots, a_n; \partial D) := \left\{ \frac{p(z)}{\prod_{k=1}^n (z - a_k)} : p \in \mathcal{P}_n^c \right\}$$

on ∂D with $a_1, a_2, \ldots, a_n \in \mathbb{C} \setminus \partial D$; and

$$\mathcal{P}_n(a_1, a_2, \ldots, a_n; \mathbb{R}) := \left\{ \frac{p(x)}{\prod_{k=1}^n |x - a_k|} : p \in \mathcal{P}_n \right\}$$

and

$$\mathcal{P}_n^c(a_1, a_2, \ldots, a_n; \mathbb{R}) := \left\{ \frac{p(z)}{\prod_{k=1}^n (z - a_k)} : p \in \mathcal{P}_n^c \right\}$$

on $\mathbb{R}$ with $a_1, a_2, \ldots, a_n \in \mathbb{C} \setminus \mathbb{R}$.

Generalized Polynomials.

The function

$$f(z) = |\omega| \prod_{j=1}^{m} |z - z_j|^{r_j} \tag{A.4.1}$$

with $0 < r_j \in \mathbb{R}$, $z_j \in \mathbb{C}$, and $0 \neq \omega \in \mathbb{C}$ is called a generalized nonnegative (algebraic) polynomial of (generalized) degree

$$N := \sum_{j=1}^{m} r_j \,. \tag{A.4.2}$$

The set of all generalized nonnegative algebraic polynomials of degree at most N is denoted by GAP_N.

The function

$$f(z) = |\omega| \prod_{j=1}^{m} |\sin((z - z_j)/2)|^{r_j} \tag{A.4.3}$$

with $0 < r_j \in \mathbb{R}$, $z_j \in \mathbb{C}$, and $0 \neq \omega \in \mathbb{C}$ is called a generalized nonnegative trigonometric polynomial of degree

$$N := \frac{1}{2} \sum_{j=1}^{m} r_j \,. \tag{A.4.4}$$

The set of all generalized nonnegative trigonometric polynomials of degree at most N is denoted by GTP_N.

Constrained Polynomials.

The following classes of polynomials with constraints appear in Appendix 5:

$$\mathcal{P}_{n,k} := \{p \in \mathcal{P}_n : p \text{ has at most } k \text{ zeros in } D\}\,, \qquad 0 \leq k \leq n\,,$$

$$\mathcal{B}_n(a,b) := \left\{p \in \mathcal{P}_n : p(x) = \pm \sum_{j=0}^{n} \alpha_j (b-x)^j (x-a)^{n-j}\,, \ \alpha_j \geq 0\right\}, \quad a < b\,,$$

$$\widetilde{\mathcal{P}}_{n,k}(a,b) := \{p = hq : h \in \mathcal{B}_{n-k}(a,b)\,, \ q \in \mathcal{P}_k\}\,, \qquad 0 \leq k \leq n\,.$$

Function Spaces.

The *uniform* or *supremum norm* of a complex-valued function f defined on a set A is defined by

$$\|f\|_A := \sup_{x \in A} |f(x)| .$$

The space of all real-valued continuous functions on a topological space A equipped with the uniform norm is denoted by $C(A)$. If $A := [a, b]$ is equipped with the usual metric topology, then the notation $C[a, b] := C(A)$ is used.

Let (X, μ) be a measure space (μ nonnegative) and $p \in (0, \infty]$. The space $L_p(\mu)$ is defined as the collection of equivalence classes of real-valued measurable functions for which $\|f\|_{L_p(\mu)} < \infty$, where

$$\|f\|_{L_p(\mu)} := \left(\int_X |f|^p \, d\mu \right)^{1/p}, \qquad p \in (0, \infty)$$

and

$$\|f\|_{L_\infty(\mu)} := \sup\{\alpha \in \mathbb{R} : \mu(\{x \in X : |f(x)| > \alpha\}) > 0\} < \infty .$$

The equivalence classes are defined by the equivalence relation $f \sim g$ if $f = g$ μ-almost everywhere on X. When we write $L_p[a, b]$ we always mean $L_p(\mu)$, where μ is the Lebesgue measure on $X = [a, b]$. The notations $L_p(a, b), L_p[a, b)$, and $L_p(a, b]$ are also used analogously to $L_p[a, b]$. Again, it is always our understanding that the space $L_p(\mu)$ is equipped with the $L_p(\mu)$ norm.

Sometimes $C(A)$ denotes the space of all *complex-valued* continuous functions defined on A equipped with the uniform norm. Similarly, $L_p(\mu)$ may denote the space all *complex-valued* continuous functions for which $\|f\|_{L_p(\mu)} < \infty$. This should always be clear from the context; many times, but not always, the reader is reminded if $C[a, b]$ or $L_p(\mu)$ is complex.

Index

Graduate Texts in Mathematics

continued from page ii

119 ROTMAN. An Introduction to Algebraic Topology.
120 ZIEMER. Weakly Differentiable Functions: Sobolev Spaces and Functions of Bounded Variation.
121 LANG. Cyclotomic Fields I and II. Combined 2nd ed.
122 REMMERT. Theory of Complex Functions. *Readings in Mathematics*
123 EBBINGHAUS/HERMES et al. Numbers. *Readings in Mathematics*
124 DUBROVIN/FOMENKO/NOVIKOV. Modern Geometry—Methods and Applications. Part III.
125 BERENSTEIN/GAY. Complex Variables: An Introduction.
126 BOREL. Linear Algebraic Groups.
127 MASSEY. A Basic Course in Algebraic Topology.
128 RAUCH. Partial Differential Equations.
129 FULTON/HARRIS. Representation Theory: A First Course. *Readings in Mathematics*
130 DODSON/POSTON. Tensor Geometry.
131 LAM. A First Course in Noncommutative Rings.
132 BEARDON. Iteration of Rational Functions.
133 HARRIS. Algebraic Geometry: A First Course.
134 ROMAN. Coding and Information Theory.
135 ROMAN. Advanced Linear Algebra.
136 ADKINS/WEINTRAUB. Algebra: An Approach via Module Theory.
137 AXLER/BOURDON/RAMEY. Harmonic Function Theory.
138 COHEN. A Course in Computational Algebraic Number Theory.
139 BREDON. Topology and Geometry.
140 AUBIN. Optima and Equilibria. An Introduction to Nonlinear Analysis.
141 BECKER/WEISPFENNING/KREDEL. Gröbner Bases. A Computational Approach to Commutative Algebra.
142 LANG. Real and Functional Analysis. 3rd ed.
143 DOOB. Measure Theory.
144 DENNIS/FARB. Noncommutative Algebra.
145 VICK. Homology Theory. An Introduction to Algebraic Topology. 2nd ed.
146 BRIDGES. Computability: A Mathematical Sketchbook.
147 ROSENBERG. Algebraic *K*-Theory and Its Applications.
148 ROTMAN. An Introduction to the Theory of Groups. 4th ed.
149 RATCLIFFE. Foundations of Hyperbolic Manifolds.
150 EISENBUD. Commutative Algebra with a View Toward Algebraic Geometry.
151 SILVERMAN. Advanced Topics in the Arithmetic of Elliptic Curves.
152 ZIEGLER. Lectures on Polytopes.
153 FULTON. Algebraic Topology: A First Course.
154 BROWN/PEARCY. An Introduction to Analysis.
155 KASSEL. Quantum Groups.
156 KECHRIS. Classical Descriptive Set Theory.
157 MALLIAVIN. Integration and Probability.
158 ROMAN. Field Theory.
159 CONWAY. Functions of One Complex Variable II.
160 LANG. Differential and Riemannian Manifolds.
161 BORWEIN/ERDÉLYI. Polynomials and Polynomial Inequalities.
162 ALPERIN/BELL. Groups and Representations.
163 DIXON/MORTIMER. Permutation Groups.